Algebraic Geometry and Topology

A SYMPOSIUM IN HONOR OF S. LEFSCHETZ

PRINCETON MATHEMATICAL SERIES

Editors: MARSTON MORSE and A. W. TUCKER

1. The Classical Groups, Their Invariants and Representatives. By HERMANN WEYL.
2. Topological Groups. By L. PONTRJAGIN. Translated by EMMA LEHMER.
3. An Introduction to Differential Geometry with Use of the Tensor Calculus. By LUTHER PFAHLER EISENHART.
4. Dimension Theory. By WITOLD HUREWICZ and HENRY WALLMAN.
5. The Analytical Foundations of Celestial Mechanics. By AUREL WINTNER.
6. The Laplace Transform. By DAVID VERNON WIDDER.
7. Integration. By EDWARD JAMES MCSHANE.
8. Theory of Lie Groups: I. By CLAUDE CHEVALLEY.
9. Mathematical Methods of Statistics. By HARALD CRAMÉR.
10. Several Complex Variables. By SALOMON BOCHNER and WILLIAM TED MARTIN.
11. Introduction to Topology. By SOLOMON LEFSCHETZ.
12. Algebraic Geometry and Topology. Edited by R. H. FOX, D. C. SPENCER, and A. W. TUCKER.
13. Algebraic Curves. By ROBERT J. WALKER.
14. The Topology of Fibre Bundles. By NORMAN STEENROD.
15. Foundations of Algebraic Topology. By SAMUEL EILENBERG and NORMAN STEENROD.
16. Functionals of Finite Riemann Surfaces. By MENAHEM SCHIFFER and DONALD C. SPENCER.
17. Introduction to Mathematical Logic, Vol. I. By ALONZO CHURCH.
18. Algebraic Geometry. By SOLOMON LEFSCHETZ.
19. Homological Algebra. By HENRI CARTAN and SAMUEL EILENBERG.
20. The Convolution Transform. By I. I. HIRSCHMAN and D. V. WIDDER.
21. Geometric Integration Theory. By HASSLER WHITNEY.

Algebraic Geometry and Topology

A Symposium in honor of S. Lefschetz

Edited by R. H. Fox, D. C. Spencer, A. W. Tucker
for the Department of Mathematics
Princeton University

PRINCETON, NEW JERSEY
PRINCETON UNIVERSITY PRESS
1957

Published, 1957, by Princeton University Press
London: Geoffrey Cumberlege, Oxford University Press

L.C. CARD: 57-8376

Composed by the University Press, Cambridge, England
Printed in the United States of America

Foreword

THIS volume contains papers in Algebraic Geometry and Topology contributed by mathematical colleagues of S. Lefschetz to celebrate his seventieth birthday (September 3, 1954). The goal has been to feature contemporary research that has developed from the vital basic work of Professor Lefschetz.

Eight of the papers, including the special surveys by W. V. D. Hodge and N. E. Steenrod, were presented at the Conference on Algebraic Geometry and Topology held in honor of Professor Lefschetz April 8–10, 1954, at Fine Hall, Princeton University.

The editing of the volume has been a joint enterprise of the members of the Princeton Department of Mathematics. In addition, the following kindly served as referees: E. G. Begle, M. P. Gaffney, Jr., V. K. A. M. Gugenheim, A. P. Mattuck, F. P. Peterson, H. Samelson, E. Snapper, and O. Zariski. The Departmental secretaries, Mrs. Agnes Henry and Mrs. Virginia Nonziato, and the *Annals of Mathematics* secretaries, Mrs. Ellen Weber and Mrs. Bettie Schrader, have ably looked after the many details of correspondence, manuscripts, and proofs. To all these, and to the Princeton University Press and its Director, Mr. H. S. Bailey, Jr., the Editors express their gratitude for the unstinted cooperation that has produced this volume.

A. W. TUCKER

Contents

PAGE

PART III: PAPERS IN TOPOLOGY

Part I

An appreciation of the work and influence of S. Lefschetz

Professor Lefschetz's Contributions to Algebraic Geometry: An Appreciation

W. V. D. Hodge

IF one attempts to give a systematic account of the work of a distinguished mathematician paper by paper, the final impression produced is apt to make this work appear as a museum piece, and this may be the opposite of what is really intended. If any such impression should be given by my remarks on Professor Lefschetz's work on algebraic geometry, it would be particularly unfortunate and misleading; for it is a fact that a number of the discoveries which he has made are of more vital interest to mathematicians at the present day than they have been since they aroused widespread excitement on their first appearance some thirty or thirty-five years ago. I therefore think it would be wiser for me not to follow the historical method, but to select a number of the outstanding contributions which Lefschetz has made to algebraic geometry and to discuss these in relation to current mathematical developments; and I venture to suggest that this procedure will have the approval of Lefschetz himself, since, though no one is more generous in recognizing the merits of others, to him mathematics is more important than the mathematician.

Without a word of warning, however, the method which I propose to follow may equally give a wrong impression, for in treating different aspects of Lefschetz's work separately I may give the idea that the ideas were developed independently, and this is far from being the case. Indeed, one of the most striking features of the whole range of Lefschetz's contributions to the topological and transcendental theories of algebraic varieties is the interplay between the various ideas, and it is clear that the next step in one line was inspired by some achievement in another direction. The essential fact is that Lefschetz clearly regarded the whole range which he covered as a single subject, and the keynote is its unity—a unity which he eventually carried into his work in pure topology, where his fixed-point theorems are clearly derived from ideas familiar in algebraic geometry. But having made this point, it will be convenient to take up a number of aspects of his work separately.

The topology of algebraic varieties

One of the great lessons to be learned from a study of Lefschetz's work on algebraic varieties is that before proceeding to the investigation of transcendental properties it is necessary first to acquire a thorough understanding of the topological properties of a variety. Not only does this greatly simplify the technical problems encountered in developing the transcendental theory, but it is an absolute necessity if one is to appreciate the true significance of the difficulties to be overcome. It is therefore natural to begin an appreciation of Lefschetz's work with a discussion of his investigations of the topological structure of varieties.

The importance of the topological structure of curves in the study of Abelian integrals of algebraic functions of one variable was made clear by Riemann, and when Picard came to study the integrals attached to an algebraic surface he naturally made use of all the topological methods available. His difficulty, of course, was the fact that at that time our knowledge of topology was extremely primitive, and he had to reinforce his weak topological weapons with more powerful analytic ones. One of the most impressive features of his celebrated treatise is the way in which, by such primitive and indirect means, he did succeed in obtaining a deep understanding of the topological nature of an algebraic surface, though naturally, owing to the use of transcendental methods, some of the finer aspects of the topology, such as torsion, were lost. Nevertheless, his final analysis of the topology of an algebraic surface provided Lefschetz with the scheme for a direct investigation of the topology, not only of surfaces but of varieties of any dimension. Lefschetz's achievement was to obtain all of Picard's topological results by direct methods, and then to use them to simplify the transcendental theory. I wish to emphasize the directness of Lefschetz's methods, for it does seem to me to be most satisfactory to know the topological structure of a variety thoroughly before embarking on transcendental considerations. This is not intended as a criticism of elegant methods which have been employed to obtain results similar to those of Lefschetz by using harmonic integrals or the theory of stacks (though it must be remembered these methods only obtain the homology groups with complex (or real) coefficients, whereas the direct method enables us to use integer coefficients), but merely reflects my view that without a direct method of investigating the topology of an algebraic variety something important would be missing.

The method used by Lefschetz is a direct generalization of the classical process for studying the topology of an ordinary Riemann surface by reducing it to an open 2-cell by means of a system of cuts. Lefschetz similarly introduced cuts into a non-singular variety V of complex dimension d and reduced it to an open $2d$-cell. Representing this $2d$-cell as the interior C_{2d} of a solid $2d$-sphere, the problem is then to determine the image of the boundary of C_{2d} in V, which, as a point set, coincides with the cuts. Once this is done, it is possible to read off all the necessary results.

An induction argument is used to reduce V to a cell, making use of the fact that on V there are systems of subvarieties of dimension $d-1$. A suitable subsystem is selected, for example, a pencil $|S_z|$ of prime sections which has a non-singular subvariety B of dimension $d-2$ as base and contains only a finite number of singular sections $S_1, \ldots, S_N$, where the singular point P_i of S_i is not in B. The varieties of $|S_z|$ can be represented by the points of the complex plane Σ, and the singular sections by the critical values $z_1, \ldots, z_N$ of z. It is quite simple to prove rigorously that a homeomorphism can be established between two non-singular sections $S_{z'}$ and $S_{z''}$ in which points in B are self-corresponding. What Lefschetz then does is to introduce on Σ a set of cuts $z_0 z_i$ $(i=1, \ldots, N)$ from a suitable point z_0 to the critical points z_i, and to try to establish a uniquely determined homeomorphism $T_{z'z''}$ between $C_{z'}$ and $C_{z''}$, where z' and z'' are in the cut plane Σ' with the properties: (a) points in B are unaltered in $T_{z'z''}$, and (b) $T_{z''z'''}T_{z'z''} = T_{z'z'''}$. By the hypothesis of induction, if z is any point in Σ', C_z can be reduced to a cell C'_z by introducing cuts in such a way that B lies on the boundary of the cell, and the homeomorphism $T_{zz'}$ enables us to reduce $C_{z'}$ to a cell. Thus we can reduce each C_z $(z \in \Sigma')$ to a cell homeomorphic to C'_z, and then V is reduced to a cell.

The next step required is to discuss the nature of the boundary of $C'_z \times \Sigma'$. This requires (a) a discussion of the behavior of C'_z as $z \to z_i$ $(i=1, \ldots, N)$; the essential point here is that we obtain a mapping of C_z *onto* C_{z_i}; (b) a discussion of the homeomorphism induced on $C_{z'}$, where z' is on $z_0 z_i$ and z describes a circuit round z_i beginning and ending at z' going from the right to the left of $z_0 z_i$; in fact, it is sufficient to consider the effect of this homeomorphism on the various homology groups of $C_{z'}$ and this Lefschetz examined in detail. By investigating these questions Lefschetz was able to extract all the information required about the topology of V.

The guiding principles of the whole of this investigation are clear; the difficulties lie in the details. It must be recalled that at the time this work was being done, while the study of algebraic topology was getting under way, the topological tools available were still primitive, and one would expect that nowadays, when we have so much more powerful tools at our disposal, it would be possible to use them to straighten out the difficulties which arise over details. It is indeed surprising that we have had to wait almost to the present moment for this to be done. In a forthcoming paper in the ANNALS OF MATHEMATICS, A. H. Wallace has used modern singular homology theory to rewrite that part of Lefschetz's work which deals with the homology groups of dimension less than d, and he has shown me a preliminary account of a subsequent paper in which he deals with the remaining homology groups. It is most striking to see how Wallace's methods in principle follow so closely the methods of Lefschetz and indeed provide a triumphant justification of them.

One of the main difficulties of the Lefschetz argument is to establish the unique homeomorphism $T_{z'z''}$ referred to above, for all points z', z'' of Σ'. It is quite easy to do this when z' is any point of Σ' different from the z_i and z'' is sufficiently near z'. This suggests that instead of using direct products we should use fibre bundles. Let K be any closed set in Σ not containing a critical point. Wallace shows that it is possible to construct a fibre bundle X_K over K, whose fibre is $C_{z'}$, and an embedded bundle X'_K over K equivalent to $B \times K$, such that, if V_K denotes the part of V covered by C_z $(z \in K)$, there is a continuous mapping ψ of X_K onto V_K which, when restricted to $X_K - X'_K$, is a homeomorphism on $V_K - B$, and projects the direct product X'_K on B. If K includes a critical point z_i, we can construct X_K, X'_K as before, but in this case ψ restricted to $X_K - X'_K$ ceases to be a homeomorphism at the part of $X_K - X'_K$ over z_i. The behavior of ψ at the critical points can be determined. Then taking $K = \Sigma$ it is possible to determine the singular homology properties of the pair (X_Σ, X'_Σ) and to use the properties of ψ to deduce the singular homology properties of V. With a number of obvious points of difference, Wallace's treatment bears a great resemblance to Lefschetz's original methods.

Thus Lefschetz's pioneer work comes into its own. In addition to its own intrinsic interest, it has a key position in the literature of algebraic varieties as the basis of a great deal of the transcendental theory of varieties. But it also has a unique historical interest, in being almost the first account of the topology of a construct of importance in general mathematics which is not trivial. And it settled a number of questions which now seem trivial, but which at one time caused a good deal of speculation. For instance, the fact the Betti numbers R_p of odd dimension are even, and that $R_p \geqq R_{p-2}$ $(2 \leqq p \leqq m)$, showed at once that not all orientable manifolds of even dimension are the carrier manifolds of algebraic varieties. Moreover, Lefschetz's work is the direct inspiration of all the researches which have taken place subsequently in the theory of complex manifolds. In fact, it is not too much to say that our greatest debt to Lefschetz lies in the fact that he showed us that a study of topology was an essential for all algebraic geometers.

Integrals of the second kind on an algebraic variety

One of the first applications of his work on the topology of algebraic varieties which Lefschetz made was to the theory of integrals of the second kind. Some of his work on this subject preceded the work on the topology of varieties, and it seems fairly clear that he was led to the topological work in order to make progress possible in the study of integrals. However that may be, it is certain that his most important contribution to our knowledge of integrals of the second kind depends essentially on his previous study of topology. In this he was, essentially, reversing the order followed by Picard, who used the theory of integrals to get to the topology.

It is as well to point out that there are several definitions of integrals of the second kind on an algebraic variety, and that they are not all equivalent. The notion derives, of course, from the notion of an integral of the second kind on a Riemann surface, where we have three possible definitions. It will be recalled that in discussing integrals the fields of integration must be in the open manifold obtained by removing the locus of singularities. The integrals under discussion are assumed to be integrals of forms which are meromorphic everywhere (rational forms).

(a) An integral is of the second kind if all its residues are zero. (A residue is an integral over a cycle which bounds on the Riemann surface);

(b) An integral is of the second kind if in the neighborhood of any point its integrand is equal to the derived of a local meromorphic 0-form (function);

(c) An integral is of the second kind if in the neighborhood of any point its integrand differs from the derived of a rational 0-form on the Riemann surface by a form holomorphic at the point.

The equivalence of these three definitions is almost trivial when $p=1$. But when we extend our investigation to exact p-fold integrals on a variety V of dimension d, they are no longer equivalent. With trivial alterations in wording, (a) and (b) give unambiguous definitions on V; but there are two ways in which (c) can be generalized:

(c_1) A p-fold integral is of the second kind if in the neighborhood of any point its integrand differs from the derived of a rational $(p-1)$-form on V by a locally holomorphic p-form;

(c_2) If $\Gamma_1, \ldots, \Gamma_k$ are the irreducible components of the locus of the singularities of the integral, $\int Q$ is an integral of the second kind if there exists a rational $(p-1)$-form P_i such that Γ_i is not a component of the locus of singularities of $Q - dP_i$.

In his investigations Picard used the definition (c_1), and Lefschetz uses (c_2). In the case $p=2$, $d=2$ both prove that their definitions are equivalent to (a), and hence, indirectly, we deduce the equivalence of (c_1) and (c_2). But even in this case, (b) is not equivalent to any of the others, and is a much weaker condition. For example, let $f(x,y,z)=0$ be the equation of a surface V of order n with only ordinary singularities and suppose that $x=0$ is a general section of this. Then it is easily seen that if $P(x,y,z)$ is an adjoint polynomial of order $n-3$, the integral

$$\int \frac{P(x,y,z)}{xf_z}\,dx\,dy$$

satisfies condition (b), but has residues, namely, the periods of the integral

$$\int \left(\frac{P(x,y,z)}{f_z}\right)_{x=0} dy$$

on the section $x=0$. More generally, it can be shown that *any* meromorphic integral of multiplicity q ($q \geqq 2$) satisfies condition (b) at any *non-singular* point of the singular locus of the integral. Neither Picard nor Lefschetz ever used (b); at the time at which they worked on this field it was more natural to think always in terms of globally defined forms on a variety, and it is impossible to say whether they were aware of the difficulties which attach to definition (b); but I feel it is necessary to point out the difference between (b) and the others, since a number of modern writers do use the term 'integral of the second kind' for the integral of a form which satisfies (b).

For general values of p and d, it has, as far as I am aware, never been proved that (c_1) and (c_2) are equivalent, though this may well be the case. But Lefschetz has shown that for $p=3$, $d=3$, an integral may have non-zero residues and yet be the integral of a derived form, contrary to what happens when $p=1, 2$. Hence in any discussion of integrals of the second kind it is necessary to make clear at the outset which definition is being used. Each definition can lead to an interesting theory, but one must not expect the same theory from different definitions.

If definition (c_1) or (c_2) is used, two p-fold integrals of the second kind are to be regarded as equivalent if their difference is equal to the derived of a rational $(p-1)$-form on V. The main problem is to determine the group of equivalence classes of p-fold integrals of the second kind. The first step, both with Picard and with Lefschetz, is to show that any integral of the second kind is equivalent to one having as its singular locus a fixed non-singular prime section C of V—usually taken as the section by the prime at infinity. The second stage is to show that any p-fold integral of the second kind having C as its complete singular locus which has zero periods on all the p-cycles of V in $V-C$ is equivalent to zero, and that there are integrals of the second kind having arbitrarily assigned periods on the p-cycles of V in $V-C$.

There is not much to be said here about the case $p=1$, which is really classical. Picard and Lefschetz both treated the case $p=2$, $d=2$ by somewhat similar but not identical methods, and Lefschetz extended the results to the case $p=2$, any d. He also gave a brief account of the case $p=3$, $d=3$, along lines which, in theory, should be applicable to any p, d, though there are a number of difficulties of a topological nature which still require detailed study. This work is extremely ingenious, and merits much more attention than it has ever had. But there is still a third stage in the process of computing the groups of p-fold integrals of the second kind which we have still to mention, and it is here that Lefschetz's topological approach produced an urgently needed fresh idea. We have seen that a base for the p-fold integrals of the second kind can be constructed by taking a set of r_p integrals having singularities only on C and having independent periods on the r_p independent cycles of V which lie in $V-C$. But these r_p integrals may not be independent. The case $p=2$, $d=2$ was fully investigated by Picard, and the results he obtained constitute one of his most famous contributions to the theory of surfaces. His solution related the non-zero combinations of his r_p integrals whose integrands are derived with the simple integrals of the third kind, but it is more convenient to explain his results in terms of the theory of algebraic equivalence of curves on a surface due to Severi. On the algebraic surface V there exists a set of curves $\Gamma_1, \ldots, \Gamma_\rho$ which are algebraically independent, and are such that any other curve on the surface is algebraically dependent on them. Without loss of generality, we may take $\Gamma_\rho = C$, and suppose that $\Gamma_1, \ldots, \Gamma_{\rho-1}$ are virtual curves of order zero. Then each Γ_i $(i \leqq \rho - 1)$ can be represented as a 2-cycle in $V - C$, and a base $\Delta_1, \ldots, \Delta_{r_2}$ for $V - C$ can be chosen so that $\Delta_i = \Gamma_i$ $(i \leqq \rho - 1)$, $(\Delta_i, \Delta_j) = 0$ $(i \leqq \rho - 1, j \geqq \rho)$. Picard's result may be stated as follows: an integral of the second kind with singularities only on C is equivalent to zero if and only if its periods on the cycles Δ_i $(i \geqq \rho)$ are zero.

The problem of extending this result to p-fold integrals on a variety of dimension d is not an easy one, and must have been extremely formidable around 1920, when the topological nature of an algebraic variety was not well understood. The difficulty is that Picard's result, as I have stated it, puts perhaps too much emphasis on the algebraic nature of the exceptional 2-cycles, and this obscures the essential properties of the cycles which make them exceptional. A further complication is due to the accident that the dimension of the period cycles is equal to the dimension of the cycles determined by the singular loci of an integral of the second kind in the case $p = 2$, $d = 2$. Lefschetz succeeded in seeing the essential nature of the exceptional cycles. What he has shown, in fact, is that there exists a maximal subset $\Delta_1, \ldots, \Delta_\sigma$ of p-cycles on an algebraic variety V such that, if $\bar{C}$ is any subvariety (possibly reducible) of dimension $d - 1$ on V, $\Delta_1, \ldots, \Delta_\sigma$ are homologous on V to cycles of $V - \bar{C}$, and an integral of the second kind is equivalent to zero if and only if it has zero periods on the cycles Δ_i $(i = 1, \ldots, \sigma)$.

Lefschetz's contribution to the theory of integrals of the second kind may therefore be summed up as follows: (1) by beginning with a clear understanding of the topology of an algebraic surface, he greatly simplified the work of Picard on the double integrals on a surface; (2) in extending the theory to double integrals on a variety of any dimension, and then to triple integrals, he made abundantly clear the pattern which the theory should follow in the case of p-fold integrals on a variety of d dimensions. That for many years nothing further was written on the subject was due to the fact that geometers felt they knew all that they wanted to know about integrals of the second kind, though some of the formal proofs were not available, owing to difficulties of a purely technical nature. Just as in the case of the study of the topology of an algebraic variety, I think that most geometers felt that there was little point in pursuing a subject the results of which were clear to them until a new technique appeared which would be capable of giving an elegant and comprehensive treatment of the whole subject.

Such a technique has, indeed, presented itself in the theory of stacks, and in the last few weeks a complete account of integrals of the second kind has been developed by this means. In developing this method, a careful study of Lefschetz's definitions and results has been necessary in order that the treatment should proceed on the right lines, and it is not too much to say that without Lefschetz's work as a guide it would have been much more difficult for the investigators to set their target so that an elegant and complete theory would result. In particular, it has turned out that it is the necessity of fitting the theory of stacks to the classical notions of algebraic integrals on a variety which has led to the rejection of definition (b) as a suitable definition of integrals of the second kind.

The definition of a p-form of the second kind which is adopted in the stack-theoretic approach can be briefly described. On the algebraic variety V of d dimensions, let Σ be any divisor, given locally by the equation $f = 0$. We denote by $\Omega^p(k\Sigma)$ the stack of germs of p-forms ω such that $f^k\omega$ is holomorphic, and we let $\Omega^p(*\Sigma)$ be the direct limit of these stacks as $k \to \infty$, $\Omega^p(*)$ the direct limit of the stacks $\Omega^p(*\Sigma)$ as Σ runs through the directed set of divisors. The essential results used in the theory are that $H^q(V, \Omega^p(*)) = 0$ $(q > 0)$, and that if Σ is ample, $H^q(V, \Omega^p(*\Sigma)) = 0$ $(q > 0)$. Let $\Phi^p(*)$ be the substack of $\Omega^p(*)$ formed by the germs of closed p-forms. Then any globally defined meromorphic closed p-form on V is an element of the cohomology group $H^0(V, \Phi^p(*))$, and the group of equivalent closed p-forms is isomorphic to $H^0(V, \Phi^p(*))/dH^0(V, \Omega^{p-1}(*))$.

We construct an exact sequence of stacks

$$0 \to d\Omega^{p-1}(*) \to \Phi^p(*) \to \mathscr{R}^p \to 0,$$

and the associated cohomology sequence

$$\dots \to H^q(V, d\Omega^{p-1}(*)) \to H^q(V, \Phi^p(*)) \to H^q(V, \mathscr{R}^p) \to H^{q+1}(V, d\Omega^{p-1}(*)) \to \dots.$$

If ψ is any closed meromorphic p-form on V, that is, an element of $H^0(V, \Phi^p(*))$, we call its image in $H^0(V, \mathscr{R}^p)$ in this sequence the 0-residue of ψ. Assume that the 0-residue is zero. Then ψ is the image of an element of $H^0(V, d\Omega^{p-1}(*))$ and hence the class of equivalent p-forms whose 0-residue is zero can be represented as

$$H^0(V, d\Omega^{p-1}(*))/dH^0(V, \Omega^{p-1}(*)),$$

and it can be shown that this quotient group is isomorphic to

$$H^1(V, \Phi^{p-1}(*)).$$

Let ψ_1 be the element of this group representing the class of equivalent forms to which ψ belongs. We consider the same cohomology sequence as above, but with p replaced by $p-1$, and define the 1-residue of ψ as the image of ψ_1 in $H^1(V, \mathscr{R}^{p-1})$. If this is zero, ψ_1 is the image of an element of $H^1(V, d\Omega^{p-1}(*))$, and this group is isomorphic to

$$H^2(V, \Phi^{p-2}(*)).$$

We can then repeat the argument and define a 2-residue of ψ_2, and if this is zero, a 3-residue, and so on, up to a $(p-1)$-residue. If it is possible to carry on the process so that we can define the $(p-1)$-residue of ψ and if this is zero, ψ is said to be of the second kind, and it is possible to show that this is equivalent to definition (a). Thus, according to this definition, classes of equivalent p-forms of the second kind are obtained as the images of the elements of $H^0(V, C)$ in

$$H^0(V, \Phi^p(*))/dH^0(V, \Omega^{p-1}(*))$$

in the sequence of homomorphisms

$$\begin{aligned} H^p(V,C) &\cong H^{p-1}(V, d\Omega^0(*)) \to H^{p-1}(V, \Phi^1(*)) \\ &\cong H^{p-2}(V, d\Omega^1(*)) \to H^{p-2}(V, \Phi^2(*)) \\ &\cong \ldots \to H^1(V, \Phi^{p-1}(*)) \\ &\cong H^0(V, d\Omega^{p-1}(*))/dH^0(V, \Omega^{p-1}(*)) \\ &\qquad \to H^0(V, \Phi^p(*))/dH^0(V, \Omega^{p-1}(*)). \end{aligned}$$

In this sequence of homomorphisms, the arrow always represents injection. The dimension of the image can be determined, and the result agrees with the results obtained by Lefschetz.

The subvarieties of an algebraic variety

One of Lefschetz's theorems which is of great current interest is that which gives necessary and sufficient conditions for an integral $(2d-2)$-dimensional homology class of a variety V of d complex dimension to contain the cycle of a divisor on V. These conditions are, of course, that the dual cohomology class can be represented by a 2-form of type $(1, 1)$. Lefschetz's proof concerns the case $d=2$; the generalization of this case to the case of a variety of d dimensions is an easy matter, and we need only consider the simple case. We recall that Lefschetz stated his results (referred to in what follows as his main result) in the equivalent form: *On an algebraic surface V a*

2-dimensional homology class contains the carrier cycle of a virtual algebraic curve if and only if all the algebraic double integrals of the first kind have zero periods with respect to it; and we shall consider the theorem in this form.

The necessity of the condition is trivial, though it is worth noting that in his earlier writings Lefschetz made rather heavy weather of it; later, as his understanding of the problem grew, he was able to reduce the proof to a couple of lines. There are indications that this part of the result was known before, though it is difficult to trace an explicit statement of it. But his great achievement was his proof of the sufficiency.

His proof is an extremely ingenious deduction from an investigation due to Poincaré. Poincaré considered a surface V, with ordinary singularities, whose equation we may take to be

$$f(x, y, z) = 0,$$

referred to axes in general position. A base for the integrals of the first kind attached to the curves $x = \text{const.}$ can be taken to be of the form

$$\int \frac{P_i(x, y, z)}{f_z} dy \quad (i = 1, \ldots, p),$$

where P_i is a polynomial in (x, y, z). Let (x, y_j, z_j) $(j = 1, \ldots, m)$ be m points in $x = \text{const.}$ which vary in some manner with x, and let A be any fixed base point of the pencil $x = \text{const.}$ We consider the Abelian sums

$$\text{(1)} \qquad \Sigma_{j=1}^{m} \int_A^{(x, y_j, z_j)} \frac{P_i(x, y, z)}{f_z} dy = \nu_i(x) \quad (i = 1, \ldots, p),$$

where the same path of integration in $x = \text{const.}$ is taken for each i. Poincaré considered the behavior of the functions $\nu_i(x)$ when the points (x, y_j, z_j) are the intersections of $x = \text{const.}$ with an algebraic curve Γ on V, or, more generally, when (x, y_j, z_j) $(j = 1, \ldots, m)$ for general x are the parts common to the section and Γ which do not lie at infinity (the base of the pencil $x = \text{const.}$).

Poincaré's determination of the form of the functions $\nu_i(x)$ has a close connection with the description of the 2-cycles on V given by Picard and Lefschetz. We again consider the critical values $x^1, \ldots, x^N$ of x for which $x = \text{const.}$ is tangent to V, and on the sphere Σ representing x we draw the cuts x^0x^j $(j = 1, \ldots, N)$. Picard and Lefschetz showed that, associated with each point of the cut x^0x^k, there is a uniquely determined 1-cycle δ^k varying continuously with x on the cut, and reducing to a point at $x = x^k$. Its locus is a 2-cell Δ_k, whose boundary is the position δ_0^k of δ^k as $x \to x^0$. Let $\omega_{ik}(x)$ be the period of $\int (P_i(x, y, z)\, dy)/f_z$ on δ^k, for x on $x^0 x^k$. Then Poincaré showed that if (x, y_j, z_j) $(i = 1, \ldots, m)$ describe an algebraic curve Γ as x varies, then

$$\nu_i(x) = \frac{1}{2\pi i} \Sigma_{k=1}^N \lambda_k \int_{x^0}^{x^k} \frac{\omega_{ik}(\xi)}{\xi - x^k} d\xi + \phi_i(x) \quad (j = 1, \ldots, p),$$

where $\lambda_1, \ldots, \lambda_N$ are integers, and $\phi_i(x)$ is zero if $P_i(x, y, z)$ is of degree not exceeding $n-3$, and a polynomial of degree m_i if $P_i(x, y, z)$ is of degree $n - 2 + m_i$. Conversely, he showed that, given integers $\lambda_1, \ldots, \lambda_N$ and any polynomials $\phi_i(x)$ of degree m_i, equations (1) determine a set of points (x, y_j, z_j) which describe an algebraic curve Γ if and only if the following conditions are satisfied:

(1) $$\Sigma_{k=1}^N \lambda_k \omega_{ik}(x) = 0;$$

(2) if for any point ξ of Σ there exist constants $a_1, \ldots, a_p$ such that

$$\Sigma_{i=1}^p a_i P_i(\xi, y, z) = 0,$$

then $$\Sigma_{i=1}^p a_i \nu_i(\xi) = 0.$$

Lefschetz derived his theorem by suitably interpreting these conditions. The first simply tells us that $\Sigma \lambda_k \delta_0^k \sim 0$ in $x = x^0$. The second simply states how the functions $\nu_i(x)$ are related at a point at which the integrals selected cease to be independent. The conditions can be simplified a good deal by a judicious choice of the integrals. Following Severi, Lefschetz devotes a considerable part of one chapter of his Borel tract to a discussion of the polynomials $\phi_i(x)$, and thereby obtains the Picard integrals of the first kind on V_2, but since this part of his work has been largely replaced by using the theory of harmonic integrals to discuss the Picard integrals, we shall shorten this account by making use of results on Picard integrals and adjoint systems obtained by other methods.

We then take the integrals

$$\int \frac{P_i(x, y, z)}{f_z} dy \quad (i = p - q + 1, \ldots, p),$$

where q is the irregularity of V $(2q = R_1)$, to be the integrals on $x = \text{const.}$ deduced from the Picard integrals. We have $m_i = 0$, $\omega_{ik}(x) = 0$ $(i \geqq p - q)$. In the remaining integrals $P_i(x, y, z)$ is an adjoint polynomial of degree $n - 3$ at most. If, for $x = \xi$ (ξ finite), we have

$$\Sigma_{i=1}^p a_i P_i(\xi, y, z) = 0, \quad \text{we have} \quad a_{p-q+1} = \ldots = a_p = 0,$$

and $$\Sigma_{i=1}^{p-q} a_i P_i(x, y, z) = (x - \xi) P(x, y, z),$$

and, by replacing one of the P_i by P, we can lower the degree of one of these polynomials. We can proceed in this way, and finally we reach a stage where no further reduction of degree is possible; this means that the polynomials $P_i(x, y, z)$ have been chosen to be linearly independent in $x = \text{const.}$ for all values of the constant except infinity. When we have so chosen the integrals

$$\int \frac{P_i(x, y, z)}{f_z} dy \quad (i = 1, \ldots, p),$$

Poincaré's second condition can be interpreted as

$$\Sigma_{k=1}^{N} \lambda_k \int_{x^0}^{x^k} \omega_k(x)\, dx = 0,$$

or

$$\Sigma_{k=1}^{N} \lambda_k \int_{\Delta_k} \frac{Q(x, y, z)}{f_z} dx\, dy = 0,$$

whenever $Q(x, y, z)$ is an adjoint polynomial of order $n-4$ (or less), and

$$\omega_k(x) = \int_{\delta^k} \frac{Q(x, y, z)}{f_z} dy.$$

The condition $\Sigma \lambda_k \delta_0^k \sim 0$ enables us to assert that there is a 2-chain M in $x = x^0$ such that $\Gamma_2 = \Sigma \lambda_k \Delta_k + M$ is a 2-cycle of V, and Lefschetz's form of Poincaré's second condition states that the double integrals of the first kind on V all have zero periods on Γ_2.

If the points (x, y_j, z_j) $(j = 1, \ldots, m)$ describe a curve Γ, then Lefschetz shows that $\Gamma \sim \Gamma_2 + \rho C$ (C being a plane section of V, and ρ an integer), and we obtain a new proof that the integrals of the first kind have zero periods on the algebraic cycles. Now consider any 2-cycle on V. Lefschetz showed that it is homologous to a cycle of the form $\Gamma_2 \sim \Sigma \lambda_k \Delta_k + M$. We then construct functions $\nu_j(x)$ with these λ_k, ϕ_i $(i > p - q)$ being arbitrary constants. By the Poincaré-Lefschetz theorem, there exists a curve Γ with these Abelian sums, and we have $\Gamma \sim \Gamma_2 + \rho C$. Hence Γ_2, being homologous to $\Gamma - \rho C$, is therefore homologous to the cycle of a virtual algebraic curve, and this establishes the theorem. The method has the directness which is typical of all Lefschetz's contributions to algebraic geometry.

Given a curve Γ on V, the set of corresponding integers λ_k is not uniquely determined, but the different sets correspond to the homologically different choices of the paths for A to (x, y_j, z_j) $(j = 1, \ldots, m)$.

In particular, if Γ and Γ' are two curves (of order m) which are homologous as cycles, we can choose the paths so that

$$\Sigma_{j=1}^{m}\int_{A}^{(x,y_j,z_j)}\frac{P_i(x,y,z)}{f_z}dy=\Sigma_{j=1}^{m}\int_{A}^{(x,y\,,z_j')}\frac{P_i(x,y,z)}{f_z}dy \quad (i=1,\ldots,p-q),$$

$$\Sigma_{j=1}^{m}\int_{A}^{(x,y_j,z_j)}\frac{P_i(x,y,z)}{f_z}dy=\Sigma_{j=1}^{m}\int_{A}^{(x,y_j',z_j')}\frac{P_i(x,y,z)}{f_z}dy+\rho_i$$
$$(i=p-q+1,\ldots,p),$$

where ρ_i is a constant. If $\rho_i=0$ for all i, the two curves cut linearly equivalent sets on a general plane section, hence they are linearly equivalent; if the ρ_i are not all zero, the curves are not linearly equivalent, but in fact belong to the same algebraic system, and hence we have the result that algebraic equivalence of curves and topological equivalence of the corresponding cycles are two aspects of the same relationship. This fundamental result of Lefschetz throws a great light on the geometric theory of the base for curves on a surface—a theory which was, of course, already known from the work of Severi.

It is scarcely necessary to emphasize the importance of this group of theorems to the algebraic geometer. Those of us who are old enough to remember our feelings on first discovering these results after years of study of the more geometrical methods of getting the results will recall the thrill with which we discovered that we had been presented with a new and powerful tool for attacking our problems, and how high our hopes were that we could use these new methods to attack new problems. Could these results be used to build up a satisfactory theory of equivalence for varieties of dimension r on a variety V of dimension d? As we have already remarked, the case $r=d-1$ presents no serious difficulties (though the method is more that of deduction from the case $d=2$ than a generalization of the argument), and the case $r=2$ any d can also be squeezed out of the Lefschetz argument, but it is unfortunately the fact that all other cases have defied all attempts to deal with them.

A direct generalization of Lefschetz's method would involve first the study of the d-cycles on a variety of dimension d. It is then an easy matter to set up the equation which corresponds to the equation

$$\Sigma_{j=1}^{m}\int_{A}^{(x,y_j,z_j)}\frac{P_i(x,y,z)}{f_z}dy=\frac{1}{2\pi i}\Sigma_{k=1}^{N}\lambda_k\int_{x^0}^{x^k}\frac{\omega_{ik}}{\xi-x^k}(\xi)\,d\xi+\phi_i(x),$$

but the difficulty is that the left-hand side is now a $(d-1)$-fold integral over a chain, and we have to determine a suitable boundary cycle for this chain which will satisfy the equation; and no generalization of the theorem of inversion of Abelian integrals on a curve has ever been found that is suitable for this problem. Many other devices have been tried for generalizing Lefschetz's arguments, all without success, and it has been clear for some time that the best hope of making progress was to find a new method of proving the main result established by Lefschetz.

In the last year such a method has been found, by Spencer and Kodaira. They begin by observing that any divisor D on an algebraic variety V of dimension d defines a fibre bundle with base D whose group is the multiplicative group C^* of non-zero complex numbers: If $\{U_\alpha\}$ is a suitable locally finite covering of V, and D is given by $f_\alpha = 0$ in U_α, $f_{\alpha\beta} = f_\alpha / f_\beta$, the transformation functions of the bundle are these $f_{\alpha\beta}$. Two divisors define analytically equivalent bundles if and only if they are linearly equivalent. The characteristic cocycle of such a bundle is an integral 2-cocycle, and it is shown that an integral 2-cocycle is the characteristic cocycle of a fibre bundle with group C^* if and only if its cohomology class contains a 2-form of type $(1, 1)$. This main result is that any such bundle is equivalent to one defined by a divisor, which is dual, in the classical sense, to the given 2-form. This new method of obtaining Lefschetz's main theorem (which applies without restriction on d) raises hopes of progress towards the generalization so eagerly sought. It has led to the formulation of a number of other problems, the solution of any one of which would yield the solution of the main problem, but so far none has been solved; but it is to be remembered that this method of proceeding has not yet had the thorough examination that the problem of generalizing Lefschetz's methods directly has had.

An important point to notice is that Lefschetz's method of proceeding links his main result with algebraic equivalence, while Spencer's and Kodaira's method links it with linear equivalence. If and when their methods yield the required generalization of the main theorem, many extremely interesting problems on equivalence, the answer to which is at present wide open, will be ripe for close examination.

The theory of correspondences

We have already seen the importance in the classical geometry of algebraic surfaces of Lefschetz's investigations on the curves on a surface. They have other equally important applications, and we now refer briefly to the application he made of his results to the theory of correspondences between curves. Severi had used the surface which is the product of two curves C and D with great effect to give a geometrical treatment of the theory of correspondences between C and D. Lefschetz used his theorems to give an extremely elegant treatment of the transcendental theory, obtaining all the results which Hurwitz had obtained.

On $C \times D$ any correspondence Λ between C and D is represented by a curve L, representing the pairs of points x, y of C, D which correspond in Λ. The 2-cycle defined by L is sufficient to determine the mapping of the homology group of C into the homology group of D, and that of D into that of C, and hence the usual equations of Hurwitz. Thus we have only to consider the algebraic cycles of $C \times D$. If x is a point of C, and $\gamma_1, \ldots, \gamma_{2p}$ is a minimal basis for the 1-cycles of C, and y is a point of D, $\delta_1, \ldots, \delta_{2q}$ a minimal basis for the 1-cycles of D, any 2-cycle of $C \times D$ is homologous to

$$\Gamma_2 \sim \alpha x \times D + \beta C \times y + \textstyle\sum_i \sum_j \mu_{ij} \gamma_i \times \delta_j,$$

where α, β, μ_{ij} are integers. A necessary and sufficient condition that there should be a curve on $C \times D$ homologous to this is that the double integrals of the first kind should have zero periods on this cycle. Let $\int du_i$ $(i = 1, \ldots, p)$ be a basis for the integrals of the first kind on C, and let

$$\int_{\gamma_j} du_i = \omega_{ij};$$

and similarly let $\int dv_i$ $(i = 1, \ldots, q)$ be a basis for the integrals of the first kind on D, and let

$$\int_{\delta_j} dv_i = \nu_{ij}.$$

Then $\int du_i \times dv_j$ $(i = 1, \ldots, p;\ j = 1, \ldots, q)$ is a basis for the double integrals of the first kind on $C \times D$, and the necessary and sufficient condition that Γ_2 is homologous to an algebraic cycle is

$$\textstyle\sum_i \sum_j \mu_{ij} \omega_{hi} \nu_{kj} = 0 \quad (h = 1, \ldots, p;\ k = 1, \ldots, q).$$

If there are k independent sets of relations of this type, there are exactly $k+2$ independent algebraic curves on $C \times D$, and hence exactly $k+2$ algebraically independent correspondences between C and D. This gives a new proof of the central result of Hurwitz's theory.

It is not necessary to enlarge here on the details of this theory, nor to discuss the application of the same arguments to the case in which D is a copy of C to obtain the Riemann relations between the periods of the integrals of the first kind on C. The brief sketch we have given should be sufficient to show that a most fruitful field of application of Lefschetz's result lies in the theory of correspondences. It is natural to apply similar methods to the theory of correspondences between irreducible algebraic varieties U and V of dimension r and s respectively. An algebraic correspondence between U and V, of dimension d, is represented by a variety L of dimension d on $U \times V$, and one easily sees that the problem is largely one of determining the conditions for a $2d$-cycle on $U \times V$ to be homologous to the cycle defined by an algebraic variety of dimension d, and the notion of equivalence of varieties of dimension d on $U \times V$ plays a fundamental role. In the case $d = r + s - 1$, as we have seen, we have the necessary knowledge and much progress has been made. In particular, D. B. Scott has been able to make very effective use of Lefschetz's methods, particularly in the case $r = s = 2$, but the general problem brings us back to the problem we discussed at the end of the last section, and gives an additional reason why we should use every endeavor to solve it.

Abelian varieties

No appreciation of Lefschetz's contributions to algebraic geometry would be complete without a reference to the extensive contributions he has made to the theory of Abelian varieties, though the present position of this branch of geometry is different from that of the other branches mentioned above. Although the theory of Abelian varieties in modern algebraic geometry is a matter of the greatest interest, I do not think it can be claimed that it derives to any extent from Lefschetz's work. The fact is that the study of Abelian varieties and Riemann matrices developed by Scorza and Rosati, and carried on by Lefschetz, has been essentially completed by Albert and Weyl, and, as happens too often now in modern mathematics, has then been left aside. In spite of this, however, Lefschetz's contributions call for some mention in the present setting.

Apart from the actual results achieved, Lefschetz's work on Abelian varieties is well worth attention from those who derive inspiration from the study of the way in which a master works; for I have formed the impression that his Bordin Prize memoir, which contains an account of nearly all his work on Abelian varieties, contains many indications of the manner in which his ideas on the topology of varieties, integrals of the second kind, and on the characterization of algebraic cycles, developed. The memoir is divided into two parts: the first is a forerunner of his Borel Tract, dealing with the properties of general varieties, and the second deals explicitly with Abelian varieties, but in fact it falls naturally into two sections, one dealing with the geometry of the varieties, and the other with the complex multiplication of Riemann matrices. This last is much the longer, and takes the subject a very long way, running on parallel lines to the work of Scorza. But it is not so typical of the work of Lefschetz as the part dealing with the geometry of Abelian varieties, and if I do not devote as much space to this part of the memoir this is not to be taken as implying in any way a lack of appreciation of its achievement. It did, indeed, set out a scheme for determining all possible complex multiplications and carried the application of the scheme a long way; but its very success carried its own condemnation, since in the hands of Albert it completely answered the question it was designed to solve and left little more to be done.

For me the fascination of the section of the Bordin Prize memoir dealing with the geometry of Abelian varieties lies in seeing Lefschetz's general ideas in action for the first time, and in seeing them verified in special cases. Of course, a mathematical theorem, rigorously proved, needs no such verification; but it is a common human instinct to see how a novel idea works in practice in a simple case, and Abelian varieties, where the topology is so simple, are ideal for applying Lefschetz's general theory. But there is more to it than this; by seeing how this theory works out on an Abelian variety, where the results can be obtained by using the theory of θ-functions which is available, one gets a deeper understanding of the general theory, and so a greater mastery of it. In the present case, one may even speculate on how far the study of Abelian varieties led to the theory for general varieties. Only Lefschetz can give a definite answer, but it seems to me to be quite possible, for instance, that he first observed that on an Abelian variety a 2-cycle could be represented as an algebraic curve if and only if the double integrals of the first kind had zero periods on it, and from this was led to his general theorem. Whether this is so or not, the significance of the general theorem when applied to Abelian varieties is most illuminating, and one cannot but wonder whether we could not, with profit, use Abelian varieties to test out a number of the conjectures which are now current in the theory of algebraic varieties.

Conclusion

In the foregoing sections my purpose has been mainly to discuss the relation to current problems of mathematics of the more important contributions which Lefschetz has made to algebraic geometry. I have, I hope, shown that this work bears heavily on a great deal of present-day research in algebraic geometry, and that its great quality is that it is as alive today as it ever was. But, lest it should seem that

I have merely selected a number of isolated contributions which happen to be of interest today in order to establish my thesis that his work counts a great deal in modern algebraic geometry, I should like to deal briefly in more general terms with Lefschetz's influence on algebraic geometry. The fact that so many of his contributions are concerned with present-day problems says a great deal, but I believe that his greatness lies even more in the permanent change in direction which his influence has produced in algebraic geometry.

To speculate on what might have been, had some historical event not taken place, is a singularly useless occupation, and any opinion on how algebraic geometry would have developed without Lefschetz's intervention can only be a personal one. I am, however, in a position to state one incontrovertible fact. The idea of generalizing the notion of an algebraic integral to give a theory of harmonic integrals on an algebraic variety arose out of a study of Chapter IV of Lefschetz's Borel Tract, and an attempt to carry the work of that chapter further, and but for the influence that Tract had on me I should never have thought of the idea. Whether others would have arrived at the same, or a similar, idea by other ways I cannot say, but it is surely not unreasonable to claim that Lefschetz's influence is responsible for that particular development, and led, through the fusion of the notion of harmonic integrals with Schwartz's theory of distributions, to de Rham's theory of currents, which in the hands of Kodaira, Spencer and others has done so much for algebraic geometry. I do not think that it can be claimed that the theory of stacks owes anything to Lefschetz, but, on the other hand, the number of applications which have been made of that theory, particularly to algebraic geometry, owes beyond doubt a very great deal to him.

In considering the development of algebraic geometry since 1924 (the date of the publication of the Borel Tract) one is conscious of three main streams: first, the classical theory which we owe to the Italian school, in which the ideas of system of equivalence and canonical systems have emerged; secondly, the abstract algebraic geometry which we owe to Zariski and Weil; and thirdly, the transcendental-topological theory, which has, to a certain extent, been merged into the theory of complex manifolds, in which, however, the algebraic varieties keep on distinguishing themselves as the only manifolds on which some of the operations of the theory can be performed. No single person can claim to be the sole founder of the theory of complex manifolds, but when one considers how many of the properties originally derived by Lefschetz for algebraic varieties now hold their place in the theory of complex manifolds, and how he has influenced decisively so many who have contributed to this theory, one must accord him an honored place among the founders of a great branch of mathematics; and this without even taking account of his influence, through his work in pure topology, on the topologists who have helped to build the theory. It seems clear to me that Lefschetz by his work on the topology and transcendental theory of algebraic varieties has been a major influence in turning the minds of geometers in new and fruitful directions, and in so doing he has achieved what it is given to few to do.

Bibliographical note

I have not given references to individual papers in this article. The plan I have adopted for my appreciation of Lefschetz's services to algebraic geometry involves a great deal of combining part of one paper with a part of another, and, at times, it has led me to reinterpret some of his results in a more modern idiom. Hence most quotations would have involved a multiplicity of references, and without a detailed guide to the relevant section of the papers quoted, the references would not be much help to the reader. A complete bibliography of Lefschetz's writings may be found on pp. 44–9 of this volume; the titles of these are usually sufficient to indicate to the reader which are relevant to the various sections of this article. Moreover, we are fortunate in the fact that Lefschetz has given us, in the form of books and tracts, a number of invaluable expositions of his work, and in these will be found copious references to his relevant papers, and to the writings of other mathematicians to whom he is indebted.

Of these, the Borel Tract of 1924 is the most famous. This sets out the essentials of his work on the topology of algebraic varieties, his theory of curves on a surface, the theory of integrals of the second kind, and on Abelian varieties (but not complex multiplication). The Mémorial des Sciences Mathématiques (1929) reviews, mainly without proofs, from a wider point of view, the whole range of algebraic geometry which is related to his own particular work, and ends with a suggestive list of unsolved problems, some of which have been solved in the meantime, but others are still a challenge. The chapters which he contributed to the National Research Council Report on Birational Transformations (1932) contains a good bibliographical account of the transcendental theory of varieties, and, in particular, the last chapter is an invaluable record of researches on Abelian varieties and Riemann matrices.

It is appropriate to refer here to another book on algebraic varieties which we owe to Lefschetz. In 1935–8 he gave at Princeton a series of lectures on algebraic geometry, particularly the birational theory of curves. In these he made consistent use of the theory of formal power series so that he could extend to any algebraically closed field of characteristic zero the classical treatment given to the theory of curves over the complex field. These lectures were issued at the time in mimeographed form, and in 1953 they were included in a book on algebraic geometry, which also contains an account of the basic principles of abstract algebraic geometry in space of n dimensions, and concludes with his latest reflections on the theory of integrals of the second kind on a surface. This book is of such recent date that there is no need to discuss its contents here, save only to say that it provides an astounding example of the way in which Lefschetz, in his seventieth year, retains his grasp on recent developments in a wide field.

CAMBRIDGE UNIVERSITY

The Work and Influence of Professor S. Lefschetz in Algebraic Topology

Norman E. Steenrod

1. Introduction

Of the many contributions of Lefschetz to algebraic topology, the two which are best known and most often remarked are the fixed-point formula and the duality theorem for manifolds with boundaries. Although these two results appear on the surface to be quite unrelated, this is not the case. They possess a common underlying theme, namely, intersections and products. It was in the development of intersection theory and the algebraic techniques connected with products that Lefschetz made contributions of the highest order.

This assertion is substantiated in part by a survey of subsequent developments in algebraic topology. The fixed-point and duality theorems are nearly terminal results. Much of their beauty lies in their completeness and finality. It is true that they have found numerous applications and have been extended in various directions. However, they have not inspired any large-scale new trends. In contrast, the theory of products has had an extensive development culminating in the cohomology ring and the reduced power operations associated with the ring structure. These appear as vital tools in all phases of algebraic topology beyond the most elementary.

The fixed-point problem seems to have dominated nearly all of Lefschetz's work in topology. Some dozen papers appearing during the period 1923–38 were concerned directly with the problem. It is probable that much of the inspiration and impetus for his investigations in topology was provided by his discovery in 1923 of the fixed-point formula and the proof of its validity for a self-mapping of a closed manifold. The problem he faced was the generalization of the result to a wider class of spaces than closed manifolds. This involved an extension of the meaning of the terms of the formula as well as an extension of the proof. Successive papers marked successive steps in the process from closed manifolds to relative manifolds, to general complexes, to the final form for locally connected spaces. In consequence he was a central participant in one of the major trends of the period 1925–35, namely, the extension of the methods of combinatorial analysis situs to general topological spaces.

It is indicative of the influence of Lefschetz that the present-day usage of the terms 'topology' and 'algebraic topology' is due to him. Before the appearance in 1930 of his first Colloquium Publication entitled *Topology*, the subject was known as analysis situs. When his second Colloquium Publication entitled *Algebraic Topology* appeared in 1942, the adjective 'combinatorial' fell into disuse.

2. The fixed-point formula

The results on fixed points which preceded the work of Lefschetz were the theorems of Brouwer for continuous mappings of the n-cell and n-sphere, and a theorem of Alexander on topological mappings of a 2-dimensional manifold.

The basic step toward a full-fledged result was Lefschetz's discovery in 1923 [25]† of a formula. Its description runs as follows. Let f be a continuous map of a topological space X into itself. For each dimension n, f induces an endomorphism f_n of the homology group $H_n(X)$ based on the rational numbers R as coefficient group. Now $H_n(X)$ is a vector space over R. If its rank is finite, there is assigned a numerical invariant of f_n called its *trace* and denoted by $\mathrm{Tr}(f_n)$. The trace is computed by choosing a base for H_n and taking the trace of the corresponding matrix representation of f_n. Then the Lefschetz number of f, denoted by $L(f)$, is given by

$$L(f) = \textstyle\sum_{n=0}^{\infty} (-1)^n \mathrm{Tr}(f_n). \tag{2.1}$$

It is clear that restrictions must be imposed if $L(f)$ is to be well defined. It suffices, for example, to require that X be the space of a finite complex. Then it can be shown that $\mathrm{Tr}(f_n)$ is an integer, and it is zero in dimensions exceeding that of the complex; hence $L(f)$ is defined and is an integer.

The conclusion of the fixed-point theorem reads: If $L(f) \neq 0$, then f has at least one fixed point (i.e. there is a point $x \in X$ such that $f(x) = x$).

† References are to the Bibliography of the Publications of S. Lefschetz, pp. 44–9, this volume.

The conclusion is valid whenever X is the space of a finite complex. This result was proved by Lefschetz in 1928.

His initial theorem in 1923 asserted the conclusion only when X is a compact orientable manifold (without boundary). In this case he was able to prove more by assigning a geometric significance to the numerical value of $L(f)$ for an arbitrary f in the following manner. For each f and each $\epsilon > 0$, there exists an ϵ-approximation g to f such that (i) the fixed points of g are finite in number, and (ii) for each fixed point, g maps some neighborhood topologically onto a neighborhood. Then, for each fixed point x_i $(i = 1, \ldots, k)$ of g, there is a local degree or index $a_i = \pm 1$ according as g preserves or reverses the orientation of the neighborhood. For such a g, he proved the basic relation

$$L(g) = \textstyle\sum_{i=1}^{k} a_i. \tag{2.2}$$

The method of proof will be discussed in the next section. If ϵ is sufficiently small, g is homotopic to f so that $f_n = g_n$ in each dimension n. Hence $L(f) = L(g)$. These facts justify calling $L(f)$ *the sum of the indices of the fixed points of f*. It is a count of the 'algebraic' number of fixed points.

3. Product spaces and intersections

As stated in the introduction, the techniques introduced by Lefschetz have had a major influence on the development of algebraic topology. His proof of the first fixed-point theorem (for manifolds, 1923) initiated the use of product spaces and intersections as basic tools.

A map f of a manifold M into itself determines its graph $G(f)$ in $M \times M$ consisting of the pairs (x, fx) for $x \in M$. The fixed points of f correspond to the points in which $G(f)$ intersects the diagonal D of $M \times M$. Assuming that M is oriented, we may regard $G(f)$ and D as n-dimensional cycles on the $2n$-dimensional manifold $M \times M$; then there is defined an intersection number (Kronecker index) of the two cycles. It can be computed by choosing a cellular decomposition of $M \times M$ in which the diagonal D appears as a subcomplex, and by deforming the cycle $G(f)$ into a cycle of the dual subdivision where the intersections of dual n-cells with the diagonal are regular and may be counted with due regard for signs.

If one observes that the deformation of f into a map g, as described in § 2, corresponds to the above deformation of $G(f)$ into a cycle which intersects the diagonal in general position, it becomes intuitively evident that the intersection number of $G(f)$ with D is precisely the sum of the indices of fixed points as defined in (2.2).

The final step is the computation of the intersection number in terms of the induced homomorphism f_*. To accomplish this, Lefschetz gave a formula for the homology class of $G(f)$ on $M \times M$ as a sum of products of cycles on the two factors as follows. In each dimension q, choose an independent base x_i^q $(i = 1, \ldots, R_q)$ for the homology group $H_q(M)$ with rational coefficients. Poincaré duality provides a dual base y_i^{n-q} $(i = 1, \ldots, R_q = R_{n-q})$ such that the intersection number $x_i^q \cdot y_j^{n-q} = 1$ or 0 according as $i = j$ or not. The cross-product $x_i^q \times f_* y_i^{n-q}$ is an n-dimensional homology class of $M \times M$. Summing over all i and q gives the homology class of the cycle $G(f)$

$$G(f) \sim \textstyle\sum_{q,i} x_i^q \times f_* y_i^{n-q}. \tag{3.1}$$

Taking f to be the identity, the same formula gives the diagonal cycle

$$D \sim \textstyle\sum_{q,i} x_i^q \times y_i^{n-q}.$$

Since D is symmetric under the interchange of the two factors, we have also

$$D \sim \textstyle\sum_{p,j} (-1)^{p(n-p)} y_j^{n-p} \times x_j^p. \tag{3.2}$$

The intersection number $G(f) \cdot D$ can now be computed from (3.1) and (3.2) using intersection relations in M. Since $x_j^q \cdot f_* y_i^{n-q}$ is the coefficient of y_j^{n-q} in the representation of $f_* y_i^{n-q}$ in terms of the y-base, the trace formula follows immediately.

4. Coincidences

The basic idea in the preceding construction was the observation that the fixed points of a map correspond to the intersections of the graph with the diagonal. Observing that the intersection number of two n-cycles in a $2n$-manifold is always defined, Lefschetz generalized the problem and the result as follows. Let M and N be two n-manifolds, and let f and g be two maps of M into N. A *coincidence* of f and g is a point x in M such that $fx = gx$. In the product manifold $M \times N$ both f and g have graphs $G(f)$ and $G(g)$. Clearly a coincidence corresponds to a common point of the two graphs. So if the intersection number $L(f, g)$ of the two n-cycles $G(f)$ and $G(g)$ is non-zero, the two maps will have a coincidence.

The reduction of the computation of $L(f, g)$ to a sum of traces is carried out much as before. The result, however, is not easily expressed in invariant form in the original language of homology. The invariant form needs also the language of cohomology. Let μ, ν

denote the duality isomorphisms between homology and cohomology in M and N with rational coefficients:

$$\mu_q\colon H_q(M) \approx H^{n-q}(M), \quad \nu_q\colon H_q(N) \approx H^{n-q}(N).$$

Let $\quad f_q\colon H_q(M) \to H_q(N), \quad g^{n-q}\colon H^{n-q}(N) \to H^{n-q}(M)$

be the homomorphisms of homology and cohomology induced by f and g. Then

$$\theta_q = \mu_q^{-1} g^{n-q} \nu_q f_q \tag{4.1}$$

is an endomorphism of $H_q(M)$ for each q. The main result becomes

$$L(f,g) = \Sigma_{q=0}^{n} (-1)^q \operatorname{Tr} \theta_q. \tag{4.2}$$

If $M = N$ and g is the identity, a coincidence of f, g is just a fixed point of f. Also in (4.1), $\mu = \nu$ and g^{n-q} is the identity so that $\theta_q = f_q$. Thus (4.2) reduces to the previous formula for $L(f)$.

5. The intersection ring

Consider now the case $\dim N < \dim M$. The graphs of f and g are still m-cycles, but $M \times N$ has dimension $m+n < 2m$. The geometric intersection of the two cycles is no longer a finite set of points, i.e. a 0-cycle with a Kronecker index. If the two cycles are in general position, the dimension of the intersection is $m-n$.

It is quite probable that this consideration was one of those which led Lefschetz to develop the theory of the intersection ring of a manifold [31, 32, 33]. Perhaps more important considerations were the parallelism between cycles on a manifold and subvarieties of an algebraic variety, and the applications real and potential, of homology theory to algebraic geometry.

In any case Lefschetz undertook the task of defining an intersection homology class corresponding to any two homology classes of an orientable manifold. If A and B are representative cycles of two homology classes on the *same* triangulation of M there is no obvious way of defining a combinatorial operation leading to an intersection cycle. Several triangulations are necessary.

The first solution was the introduction of the concept of *general position*. A *polyhedral* complex is a complex imbedded in a Euclidean n-space so that each q-cell σ is a convex subset of a q-dimensional linear subspace $L(\sigma)$. Two linear subspaces L_1, L_2 of dimensions p, q are in general position in n-space if their intersection $L_1 \cap L_2$ has the least possible dimension, namely, $p+q-n$. Two polyhedrons K_1, K_2 in n-space are in general position if, for each $\sigma \in K_1$ and each $\tau \in K_2$, $L(\sigma)$ and $L(\tau)$ are in general position. In this case a simple induction shows that if the closed cells $\bar{\sigma}$, $\bar{\tau}$ have a non-vacuous intersection, then $\sigma \cap \tau$ is an open convex set in $L(\sigma) \cap L(\tau)$ whose boundary is a polyhedron whose cells are the intersections of σ and its faces with τ and its faces (excluding $\sigma \cap \tau$).

An orientation of a vector space is an equivalence class of sets of basis vectors, two bases being equivalent if the matrix relating the two bases has a positive determinant. There are thus two orientations, and an orientation is specified by a set of basis vectors. A vector subspace is oriented in the same manner. A *linear* subspace, i.e. a coset of a vector subspace, is oriented by selecting an orientation of the parallel vector subspace. Let a fixed orientation of n-space be chosen. If L_1, L_2 are oriented linear subspaces, of dimensions p, q, a unique orientation of $L_1 \cap L_2$ is defined as follows. Select base vectors $x_1, \ldots, x_p$ for L_1 agreeing with the orientation of L_1 and so that $x_{n-q}, \ldots, x_p$ is a base for $L_1 \cap L_2$. Complete the latter base to a base $x_{n-q}, \ldots, x_n$ for L_2 agreeing with its orientation. Then $x_{n-q}, \ldots, x_p$ is or is not the orientation of $L_1 \cap L_2$ according as $x_1, \ldots, x_n$ is or is not the orientation of the n-space.

An orientation of a polyhedral cell σ is just an orientation of the linear space $L(\sigma)$. The concept of an oriented intersection extends naturally to an intersection $\sigma \cap \tau$. This defines an *algebraic* intersection for the generators of the chain groups of K_1 and K_2 with the added agreement that $\sigma \cap \tau = 0$ if the geometric intersection is empty. Then there is a unique bilinear extension to an intersection of chains: if c is a p-chain of K_1 and d is a q-chain of K_2, then $c \cap d$ is a $(p+q-n)$-chain of $K_1 \cap K_2$.

A most important step is to prove the boundary formula

$$\partial(c \cap d) = \partial c \cap d + (-1)^p c \cap \partial d. \tag{5.1}$$

It is sufficient of course to prove this for an intersection of oriented cells. This is not difficult if the incidence numbers occurring as coefficients in the boundary formula are defined as follows. Let ρ be a $(p-1)$-face of the p-cell σ. If $x_1, \ldots, x_{p-1}$ are vectors defining the orientation of ρ, adjoin a vector x_p in $L(\sigma)$ extending toward the side of $L(\rho)$ not containing σ (e.g. the outward normal). Define the incidence number of ρ and σ to be ± 1 according as $x_1, \ldots, x_p$ does or does not agree with the orientation of σ.

The importance of (5.1) is that it implies that the intersection of two cycles is a cycle, and also that the intersection of a cycle and a boundary is a boundary. These facts are vital if an intersection of homology classes is to be defined.

Recall that the original problem was to define intersections of cycles in a manifold. We have sketched above an intersection theory for two polyhedrons in general position in a linear space. Several more ideas are needed. Most manifolds do not admit a linear structure. However, a triangulation of a manifold defines a local 'piecewise' linear structure. Each cell has a linear structure and the linear structure of a face of a cell agrees with that of the cell. This proves to be sufficient for extending the concept of general position to triangulated manifolds.

The final step is to show that two homology classes of M can be represented by cycles in general position. This is done by constructing the *dual* complex K^* of a simplicial triangulation K of M. Let K' be the first barycentric subdivision of K. The vertices of K' are the barycenters of the simplexes of K; the *weight* of a barycenter is the dimension of the corresponding simplex. Then the *dual* of a q-simplex σ is the union σ^* of the simplexes of K' whose vertex of least weight is the barycenter of σ. Conditions on the triangulation K which insure that M is a manifold imply that σ^* is an $(n-q)$-cell; and the collection of dual cells forms a complex K^* in general position relative to K. Furthermore, $K \cap K^*$ has K' as a subdivision. Thus the intersections of cells of K and K^* can be defined directly as chains in K'.

The topological invariance of the homology groups implies that any two homology classes $\bar{b}, \bar{c}$ can be represented by cycles b on K and c on K^*, then the intersection cycle $b \cap c$ on K' represents the homology class $\bar{b} \cap \bar{c}$.

The foregoing sketch is given with some detail so that the boldness of the procedure is in evidence. One complicated construction is piled on top of another before even the basic definition can be given. Actually the sketch of the definition is incomplete; it must still be shown that the intersection class $\bar{b} \cap \bar{c}$ is independent of the choice of the initial triangulation K of M.

The basic properties of the intersection ring established by Lefschetz are the associative and distributive laws, the commutation law

$$\bar{b} \cap \bar{c} = (-1)^{pq} \bar{c} \cap \bar{b}, \tag{5.2}$$

where p, q are the dimensions of $\bar{b}, \bar{c}$; and the fact that the fundamental n-dimensional class of M is a unit of the ring. Although the results are simple the proofs were exceedingly complicated. The reader can easily convince himself of this by making a rough estimate of the number of triangulations involved in a proof of the associative law.

6. The relative homology groups

Although the fixed-point theorem for manifolds is an extremely beautiful result, Lefschetz must have been dissatisfied by the fact that it did not include the fixed-point theorem of Brouwer for an n-cell. A cell is not a manifold. However, it is a manifold with a boundary which is itself a manifold, i.e. it is a *relative* manifold. If the techniques used in proving the fixed-point formula for manifolds could be extended to relative manifolds, then the Brouwer theorem might be included. The techniques in question were products, intersections and duality.

It is likely that Lefschetz was impelled by these considerations into an investigation which resulted in two fundamental contributions: the introduction of the concept of relative homology groups, and the discovery of the relative form of the Poincaré duality theorem [34, 36, 44].

It is an idle occupation but highly fascinating to speculate as to what might have happened if Lefschetz had suspected in 1924, say, that his fixed-point theorem was valid for an arbitrary complex, and not just for manifolds and relative manifolds. Would he have headed straight for the goal of proving this general result, and thereby missed the relative duality theorem? When and in what connection would the basic concept of relative homology have been developed?

It is hardly necessary to describe in detail the definition of the relative homology groups. In modern language, if K is a chain complex, L a subcomplex, then the factor groups $C_q(K)/C_q(L)$ with the induced boundary operator give a chain complex K/L whose homology groups are the relative groups of K modulo L. In the original form, the relative cycles are chains of K whose boundaries are in L. Two relative cycles are homologous mod L if their difference plus a chain of L is a boundary.

7. Duality in a relative manifold

The foundation of duality is the construction of the dual complex K^* described in § 5. If the same construction is applied to a triangulated n-manifold K with boundary L all steps carry through as before for the cells of $K-L$. The dual of a q-cell σ in $K-L$ is a transverse $(n-q)$-cell σ^* contained in $K-L$ and incidence relations $\sigma < \tau$ in $K-L$ are inverted: $\tau^* < \sigma^*$. For a cell σ of L, the dual σ^* is not an open cell; it is only half of an open cell. Using the idea of relative homology, the cells of L are neglected, so that K^* is defined to be the complex of cells dual to the cells of $K-L$. Since $K-L$ is an open complex, K^* is a closed complex. As a geometric point set it is closed and contained in the open set $K-L$.

In modern language, the correspondence $\sigma \to \sigma^*$ defines isomorphisms of homology with cohomology: $H_q(K/L) \approx H^{n-q}(K^*)$. This is hardly satisfactory as a result, since K^* is a new geometric configuration, and as a point set it depends on the triangulation. However, it contains all of K except for a neighborhood of L, and this can be made arbitrarily small by using a sufficiently fine triangulation. Using the regularity conditions on cells of L, Lefschetz showed that any chain of K could be deformed into K^*. In modern language, K^* is a *deformation retract* of K. Therefore the inclusion map $K^* \subset K$ induces isomorphisms $H^{n-q}(K) \approx H^{n-q}(K^*)$. Combining the two isomorphisms gives the final result

$$H_q(K/L) \approx H^{n-q}(K).$$

Of course, the original result was not stated in this form since the language of cohomology had not been developed. In the original form the statement reads: the qth Betti number of $K \bmod L$ equals the $(n-q)^{\text{th}}$ Betti number of K, and the q-dimensional torsion numbers of $K \bmod L$ coincide with the $(n-q-1)$-dimensional torsion numbers of K.

If L is vacuous, it is clear that we obtain the Poincaré duality theorem, and therefore the relative theorem is a true generalization.

8. The Alexander duality theorem

One of the principal achievements of the Lefschetz duality theorem was that it revealed clearly the intimate relationship between the duality theorems of Poincaré and Alexander. Suppose K is an n-dimensional sphere and L is a closed subcomplex. As above we can construct the dual K^* of the open complex $K-L$, and we have

$$H_q(K/L) \approx H^{n-q}(K^*). \tag{8.1}$$

In this case L is not a regular boundary, and K^* is not a deformation retract of K. However, it is a deformation retract of the open set $K-L$. This is easily seen if one observes that each simplex of K' which is neither in L' nor in K^* is the join of a simplex σ of L' with a simplex τ contained in K^*. The complement of σ in the join is a set of half-open line segments each with an end in τ; hence there is a canonical deformation of this complement into τ. Therefore the inclusion map $K^* \subset K-L$ induces an isomorphism

$$H^{n-q}(K-L) \approx H^{n-q}(K^*). \tag{8.2}$$

Since $K-L$ is an open set, the type of cohomology group must be specified. It is of course the *singular* cohomology group.

Combining (8.1) and (8.2) gives the Lefschetz duality

$$H_q(K/L) \approx H^{n-q}(K-L) \tag{8.3}$$

based only on the weaker assumption that $K-L$ is an open manifold.

The final step, in current language, is to examine the homology sequence of (K, L):

$$\to H_q(K) \to H_q(K/L) \xrightarrow{\partial} H_{q-1}(L) \to H_{q-1}(K) \to .$$

Since K is an n-sphere, we have $H_q(K)=0$ if $q \neq 0$ or n. Since the sequence is exact,

$$\partial\colon\ H_q(K/L) \approx H_{q-1}(L) \quad (2 \leqq q < n).$$

Combining this with (8.3) gives the Alexander duality theorem

$$H_q(L) \approx H^{n-q-1}(K-L) \quad (1 \leqq q \leqq n-2). \tag{8.4}$$

The cases $q=0$ and $n-1$ are included by inserting the known values of $H_0(K)$ and $H_n(K)$, with the result

$$\tilde{H}_0(L) \approx H^{n-1}(K-L), \quad H_{n-1}(L) \approx \tilde{H}^0(K-L)$$

where the tilde means to take the *reduced* 0-dimensional group.

This, of course, is not the full story since it is assumed that L is a subcomplex of a triangulation of K. If it is only assumed that L is topologically imbedded in K, an additional argument is needed which runs as follows. Form the sequence of successive barycentric subdivisions of K. In the jth subdivision, let L_j be the smallest closed subcomplex containing L. Then $\{L_j\}$ is a decreasing sequence of closed complexes converging to L, and $\{K-L_j\}$ is an expanding sequence of open sets whose union is $K-L$. The groups $\{H_q(L_j)\}$ form an inverse sequence converging to $H_q(L)$ as limit. The groups $\{H^{n-q-1}(K-L_j)\}$ form an inverse sequence converging to $H^{n-q-1}(K-L)$ as limit. The duality relations $H_q(L_j) \approx H^{n-q-1}(K-L_j)$ give an isomorphism of the two inverse systems which in turn induces an isomorphism of the limit groups.

The reader should take heed that this account is much oversimplified by the use of concepts such as cohomology and inverse limits which had not been developed at the time Lefschetz was engaged in these investigations. In fact the homology group itself was barely recognized, for the language of groups and homomorphisms was not used generally in the subject. Instead one dealt with chains and cycles, systems of linear relations on cycles given by boundaries, and then passed directly to the Betti numbers and torsion coefficients. During the period 1925–35 there was a gradual shift of interest from the numerical invariants to the homology groups themselves. This shift was due in part to the influence of E. Noether, and developments in abstract algebra. It was also enforced by two directions of generalization: (1) from complexes to general spaces where the homology groups may not be characterized by numerical invariants, and (2) from integer coefficients to arbitrary coefficients where, again, numerical invariants are inadequate.

Because of these language difficulties, it is not possible to give a simple answer to a question such as: Did Lefschetz use inverse limits, and did he define the homology groups usually attributed to Čech? He did define Betti numbers for a compact set L imbedded in a sphere by the use of a decreasing sequence $\{L_j\}$ of polyhedra converging to L; and he did remark that these were topological invariants of L, and that the Alexander duality for Betti numbers holds in this general case.

9. The fixed-point theorem for a complex

As stated earlier, Lefschetz extended the validity of his fixed-point formula from closed manifolds to manifolds with a regular boundary. In 1928, Hopf showed that it held for an arbitrary n-dimensional complex if the transformation is restricted so that the fixed points are isolated, and are contained in n-cells. This breach in the wall of manifold-like assumptions probably convinced Lefschetz that the formula has very general validity. In any case he set about an investigation of the homology theory of finite-dimensional separable spaces. Just when and where he succeeded in extending to complexes the validity of the fixed-point formula is difficult to determine. In 1930 he over-optimistically claimed its validity for a compact metric space of finite dimension having finite Betti numbers [46, 48]. But certainly by 1933, he had devised a simple and elegant proof for a complex which runs as follows [57].

Let f be a map of a complex K into itself which has no fixed point. A simplicial triangulation of K is taken so fine that each closed star fails to intersect its image under f. Denote this triangulation by K. By the simplicial approximation theorem, there exists a subdivision K' of K and a simplicial map g: $K' \to K$ such that, for each x, gx lies on the closure of the simplex of K containing fx. Denote the subdivision chain-mapping $K \to K'$ by ϕ. Then $g\phi$ is a chain-mapping $K \to K$, and it induces the homomorphisms f_q: $H_q(K) \to H_q(K)$. Because of the fineness of the initial triangulation K and the nearness of g to f, each cell σ of K fails to intersect the chain $g\phi\sigma$. Therefore, in each dimension q, the trace of the chain-map $g\phi_q: C_q(K) \to C_q(K)$ is zero. Hence the alternating sum of these traces is zero. The final step is provided by a simple algebraic lemma which asserts in general that the alternating sum of the traces of an endomorphism of a chain complex equals the same sum for the induced endomorphisms of the homology groups. Thus, no fixed points implies $L(f) = 0$.

10. Pseudo-cycles

The proof sketched above bears little resemblance to the initial proof for manifolds. The product space $K \times K$ and intersections are not used. However, these ideas were used in an earlier proof [48] which is notable in that it marked the first appearance in the literature (1930) of a kind of *cocycle*. Lefschetz called it a *pseudo-cycle*.

The construction of pseudo-cycles is based on duality in a sphere. Let K be a complex of dimension n. Using rational coefficients, let δ be any homomorphism of $H_q(K)$ into the rational numbers. K can be imbedded topologically in a sphere of sufficiently high dimension ($2n+1$ is adequate). Consider any imbedding of K in a sphere S; and let $r = \dim S$. By duality, there exists an $(r-q)$-chain γ of S whose boundary does not meet K and such that the intersection number $\gamma \cdot z$ for any q-cycle z of K coincides with the value of δ on the homology class of z. γ is called a *pseudo-cycle* of K of dimension $n-q$ of K, and δ is called its class. Two pseudo-cycles γ, γ' for the same imbedding $K \subset S$ are homologous modulo $S - K$ if and only if they have the same class. If γ, γ' are pseudo-cycles of different imbeddings, they are called homologous if they have the same class. The pseudo-homology group of dimension $n-q$ is the set of homology classes of pseudo-cycles of dimension $n-q$. Clearly this is isomorphic to the vector space of linear maps of $H_q(K)$ into the rational numbers, i.e. to the q-dimensional cohomology group.

Lefschetz regarded the pseudo-cycle as a tool in the proof of the fixed-point theorem. He probably observed that two pseudo-cycles, for the same imbedding $K \subset S$, intersect in a pseudo-cycle; but there is no indication that he considered the question of the topological invariance of the resulting ring. This was proved first in a paper presented by Gordon at the Moscow conference in 1935. At the same conference Alexander and Kolmogoroff, independently, presented intrinsic definitions of the cohomology groups (i.e. without using imbeddings $K \subset S$). Alexander also presented a definition of a cup-product. It was not satisfactory since it did not correspond under duality to the intersection of pseudo-cycles. It deviated from this by a numerical factor. Subsequently Alexander, Čech and Whitney found independently the definition which did correspond.

11. Singular homology theory

As has already been remarked in § 8, Lefschetz did define cycles and homologies which give the Vietoris-Čech homology groups for a closed set in a sphere. He attempted to use these to extend the meaning and validity of his fixed-point formula to such spaces. It soon became clear however that the fixed-point theorem in this form is false for such general spaces. The best counter-example was given by Borsuk in 1932. He constructed a compact connected set X in Euclidean 3-space which was the intersection of a decreasing sequence of 3-cells. Its Čech homology is therefore the same as that of a 3-cell. Hence $L(f) = 1$ for any map f. Borsuk showed that X admits arbitrarily small deformations without fixed points.

The inadequacy of the Vietoris-Čech homology theory for the fixed-point theorem probably led Lefschetz to investigate other methods of defining homology groups. In any case, he gave the first formal definition of the singular homology theory [45]. Singular cycles had been used by Veblen in his proof of the topological invariance of the homology group. The term *singular* was applied by Veblen to emphasize the fact that his cells and chains were continuous images of polyhedral cells and chains, and were not themselves polyhedral.

Lefschetz defined a singular q-cell of a space X to consist of a geometric q-simplex σ and a map $f:\sigma\rightarrow X$. The cell was oriented by orienting σ, i.e. choosing an order of the vertices of σ unique up to even permutations. Two oriented q-cells $f:\sigma\rightarrow X$ and $g:\tau\rightarrow X$ are called equivalent if there exists a 1-1 barycentric map $h:\sigma\rightarrow\tau$ preserving orientation such that $gh = f$. He took the free group generated by the oriented singular q-cells and reduced by the relations which identified the negative of an oriented singular cell with the oppositely oriented singular cell. The resulting group was the singular chain group $C_q(X)$. The boundary operator $\partial : C_q(X)\rightarrow C_{q-1}(X)$ was defined by specifying its values on the generating cells in the obvious manner. The singular homology groups were those of the resulting chain complex.

With his definition, as outlined above, Lefschetz assumed as obvious that the chain groups $C_q(X)$ were free Abelian, and therefore that operators such as ∂ and the deformation operator $\mathscr{D}$ could be defined on $C_q(X)$ by merely specifying their values on the generators. Čech pointed out this oversight by giving an example of an oriented singular q-cell equivalent to its oppositely oriented cell (i.e. a 1-simplex which is folded about its mid-point and then mapped into X). As a consequence $C_q(X)$ contains elements of order 2 and is not free Abelian.

The oversight was repaired in a subsequent paper [53] in which Lefschetz called these cells and the group they generate *degenerate*. He showed that ∂ and the various other operations, when applied to degenerate chains, gave degenerate chains; and therefore one could reduce modulo the degenerate chains. Subsequently Tucker observed that a degenerate cycle was always a boundary so that it was not necessary to factor out degenerate chains.

In spite of this repair work, the neatness of the original definitions had been impaired. In consequence the singular groups were neglected until 1944 when Eilenberg re-established their popularity by using *ordered* singular simplexes instead of oriented. This modification gave free Abelian chain groups, and thereby restored simplicity to the fundamental constructions.

12. Locally connected spaces

The singular homology groups likewise proved inadequate for an extension of the fixed-point theorem. One can give a fairly simple example of a compact, connected 1-dimensional space with a fixed-point free map f such that $L(f) \neq 0$ where $L(f)$ is computed using singular homology. All such examples exhibit bad local structure, e.g. are not locally connected. If the fixed-point theorem cannot be extended to arbitrary compact metric spaces, perhaps it can still be extended to a useful class of spaces larger than triangulable spaces. Such thoughts may well have influenced Lefschetz; for he undertook an intensive investigation of local-connectedness.

It is not easy to communicate in a few words the spirit of the time, that is, the influential ideas, and the problems considered to be important. One outstanding problem was (and still is) the extension to higher dimensions of the topological characterizations of the 1-cell and 2-cell. Local connectedness in the sense of point-set topology had an important role in these characterizations. It was natural to reformulate this property in the language of algebraic topology so as to obtain an automatic generalization to higher dimensions. Such a generalization would surely be needed for a successful characterization of the n-cell.

Another important influence was the thesis of van Kampen in 1928. He showed that the combinatorial conditions for a manifold could be relaxed without losing the duality property. Instead of requiring that the transverse of a q-cell be an $(n-q)$-cell, he showed that it was sufficient for it to have the homology structure of a cell. This raised the problem of finding the largest (or, at least, a large) class of spaces for which the duality theorems could be proved. A number of definitions of *generalized manifold* were promulgated by Alexandroff, Čech, Lefschetz [55] and Wilder. In all cases, the more important conditions were local in nature.

In 1931, Borsuk intrrduced the concepts of the absolute retract (AR) and the absolute neighborhood retract (ANR). These spaces are generalizations of cells and complexes respectively. Borsuk extended to them numerous theorems valid for the latter. The definition of an ANR space is not exactly local in nature, however it was clear that such spaces are locally smooth in some sense.

It was in the presence of these cross-currents that Lefschetz made his contribution to the theory of local-connectedness [57]. A space X is *locally-connected in dimension* q, if for each point x of X, and each neighborhood V of x, there is a neighborhood U of x such that each continuous map of a q-sphere into U can be contracted to a point in V. This property is abbreviated by the symbols LC_q. A space is LC^n if it is LC_q for $q=0, 1, \ldots, n$; and it is LC^∞ if it is LC_q for all q. An LC^∞-space, for which the neighborhood U corresponding to V can be chosen independently of q is called an LC-space. Lefschetz showed [58] that the class of compact metric LC-spaces coincides with the ANR spaces of Borsuk.

Lefschetz defined an even broader concept of local-connectedness *in the sense of homology* [64], denoted by HLC_q, HLC^n, etc. He simply modified the above definition to read: 'any q-cycle in U bounds a chain in V'. He showed that the class of compact metric HLC spaces enjoys many of the properties of complexes. In particular the fixed-point theorem is valid for such a space [7].

13. The cohomology ring

As remarked earlier, the period 1936–9 saw the development of cohomology and cup products at the hands of Alexander, Čech and Whitney. It was not likely that Lefschetz, who had enjoyed a monopoly on products and intersections for ten years, would have nothing to say on the subject. In his 1942 Colloquium Volume [79], he gave a full and fresh treatment of the topic. It is only recently that the importance of his methods have been fully appreciated.

Alexander and Čech constructed products only for the simplicial type of complex. They derived the cup-product of cohomology classes from a cup-product of cochains. The basic law the cochain product satisfies is

$$\delta(u \cup v) = \delta u \cup v + (-1)^{\dim u} u \cup \delta v. \tag{13.1}$$

Its analogy with the formula for the boundary of a cross-product of two cells should not be missed.

There are many cochain cup-products which induce the same products of the cohomology classes. This is the basic difficulty, choices must be made, and then the result must be proved independent of the choice.

Alexander's fundamental choice was a simple ordering of the vertices of the simplicial complex K. If u is a p-cochain, v a q-cochain, and $A^0A^1 \ldots A^{p+q}$ is a $(p+q)$-simplex with vertices in the prescribed order, then he defined

$$(u \cup v)(A^0A^1 \ldots A^{p+q}) = u(A^0 \ldots A^p) \cdot v(A^p \ldots A^{p+q}). \tag{13.2}$$

The coboundary formula (13.1) is easily verified; hence there is an induced product of cohomology classes. This cochain product is associative, so the product of classes is associative. The really difficult argument was his proof that a rearrangement of the order and the corresponding new product of cochains gave the same product of classes. Topological invariance was not difficult since for any simplicial map $f: K \to K'$ one can choose orders in K, K' so that f is order preserving, hence $f(u \cup v) = fu \cup fv$.

Čech axiomatized a concept he called a *construction*. Any construction determined a unique cup-product of cochains satisfying (13.1). It was not necessarily associative. He then gave a proof that any two constructions determine the same product of classes. This argument was of the acyclic-carrier type, and was much easier than Alexander's explicit construction. Čech showed that any simple order of the vertices gave a construction which determined the cochain product (13.2).

Whitney considered cell complexes which were subject to the sole restriction that the closure of any cell is acyclic. He axiomatized the concept of a cochain cup-product; and he proved existence of such products and uniqueness of induced products of classes using the acyclic-carrier type of argument.

Whitney's method was the most general and the most conceptual. His cochain products were likewise not necessarily associative. He could prove associativity of the cohomology product only by passing to a simplicial subdivision, using invariance, and the Alexander formula (13.2) in the subdivision.

The three methods were alike in that they involved constructions entirely within the initial complex.

Lefschetz's procedure was still more general; but entirely in keeping with methods he had used earlier. If K, K' are two complexes, he defined an *external* cross-product of cochains u on K, v on K' by

$$(u \times v)\cdot(\sigma \times \tau) = (u\cdot\sigma)(v\cdot\tau), \tag{13.3}$$

where σ, τ are cells of K, K' respectively. The formula

$$\delta(u \times v) = \delta u \times v + (-1)^{\dim u} u \times \delta v$$

is an immediate consequence of the standard formula

$$\partial(\sigma \times \tau) = \partial\sigma \times \tau + (-1)^{\dim \sigma} \sigma \times \partial\tau.$$

Hence there is an induced cross-product of cohomology classes which is a pairing of the groups $H^p(K)$ and $H^q(K')$ to $H^{p+q}(K \times K')$.

Notice the uniqueness of the procedure. There is no choice involved in the formula (13.3). Topological invariance is proved readily, since chain maps $f: K \to K_1$, $g: K' \to K_1'$ determine a unique chain map $f \times g: K \times K' \to K_1 \times K_1'$ leading to the formula for the induced cochain maps

$$(f \times g)(u_1 \times v_1) = (fu_1) \times (gv_1).$$

Associativity of the external cross-product (for three complexes) is entirely obvious. The commutation law is easily derived. Let $T: K \times K' \to K' \times K$ be defined by $T(x, x') = (x', x)$. Then T maps cells onto cells; hence it induces a unique chain map. A simple induction based on $\partial T = T\partial$ yields

$$T(\sigma \times \tau) = (-1)^{pq} \tau \times \sigma,$$

where p, q are the dimensions of σ, τ. This gives immediately

$$T(u \times v) = (-1)^{pq} v \times u. \tag{13.4}$$

Thus the external cross-products are easily and uniquely defined, and all properties are readily derivable. Now comes the essential feature of Lefschetz's method: *the derivation of the cohomology cup-product from the cross-product.* Let K be a complex, let $d: K \to K \times K$ be the diagonal map, and let d^* be the induced homomorphism of the cohomology of $K \times K$ into that of K. If u, v are cohomology classes of K, define

$$u \cup v = d^*(u \times v).$$

The properties of cup-products follow quickly from those of cross-products using obvious properties of d. For example, since $Td = d$, the formula (13.4) implies the commutation law

$$u \cup v = (-1)^{pq} v \cup u.$$

In retrospect it is easy to see the source of the difficulties Alexander, Čech and Whitney had to overcome in their purely internal constructions. The map d is not a cellular map. Hence to compute d^*, it is necessary to choose a chain approximation $d_\#$ to the diagonal map. Any such $d_\#$ induces a cochain cup-product by the formula

$$(u \cup v) \cdot \sigma = (u \times v) \cdot d_\# \sigma.$$

Thus the multiplicity of cochain cup-products comes from the multiplicity of chain approximations to d. The various proofs that two cochain cup-products induce the same products of cohomology classes were nothing more than proofs that two chain approximations to d induce the same d^*. But this is a special case of a well-known basic theorem.

It should be remarked that Lefschetz handled the internal cap-product in a similar fashion.

14. Subsequent developments

The moral of the preceding discussion is that the external product is simple and geometrically intuitive; and the less obvious internal product can be obtained as an image of the external product under a homomorphism induced by a mapping. This lesson has been applied in several cases.

An example is the J. H. C. Whitehead product in homotopy groups. Let $f:(E^p, S^{p-1}) \to (X, x_0)$ be a map of a p-cell into X which sends its boundary into x_0 so that f represents an element α of the homotopy group $\pi_p(X, x_0)$. Similarly, let a map g represent an element β of $\pi_q(Y, y_0)$. Form the product mapping $f \times g$ of $E^p \times E^q$ into $X \times Y$. It maps the boundary of $E^p \times E^q$ into the union of two cross-sections (which are copies of X and Y):

$$X \vee Y = (X \times y_0) \cup (x_0 \times Y).$$

Hence $f \times g$ represents an element $\alpha \times \beta$ of the relative group

$$\pi_{p+q}(X \times Y, X \vee Y).$$

The properties of this external cross-product are easily established. Using the boundary homomorphism

$$\partial : \pi_{p+q}(X \times Y, X \vee Y) \to \pi_{p+q-1}(X \vee Y),$$

the *external* Whitehead product $[\alpha, \beta]$ is defined by

$$[\alpha, \beta] = \partial(\alpha \times \beta)$$

and $[\alpha, \beta]$ lies in the union of X and Y. In case $X = Y$, the *internal* Whitehead product is derived as the image of the external product under the map $X \vee X \to X$ which folds together the two copies of X. Most of the properties of the internal product become entirely obvious with this definition.

Another example is provided by the definition of addition in the Borsuk-Spanier cohomotopy groups. Here again product spaces are used liberally.

A final example is provided by the cup-i-products which this reporter defined in 1946. They are a sequence of bilinear cochain products $u \cup_i v$ defined for $i = 0, 1, \ldots$, such that $u \cup_0 v = u \cup v$. These cochain operations led to topologically invariant squaring operations which were used in the classification of mappings.

In my original construction of the cup-i-products, I followed the simplicial method of Alexander based on ordering the vertices. I was aware at the time of the advantages of the Lefschetz method; but it seemed to have no obvious extension to cover the new products. I experimented fruitlessly with possible external cross-i-products. Subsequent experiments of a different nature gradually revealed what I should have done. The external cross-product must be left untouched; it is the diagonal map d which must be generalized. This is done as follows.

If W is any complex, the projection $\lambda : W \times K \to K$ into the second factor can be composed with d to give a map $d\lambda : W \times K \to K \times K$. Let π be the group of order 2 generated by the interchange of the factors of $K \times K$. If W is chosen so that π operates also in W, then $d\lambda$ is an *equivariant* map. In particular, W can be taken to be π-free and acyclic, so that the homology groups of the collapsed complex W/π are the homology groups of the group π in the sense of Eilenberg and MacLane.

The above procedure, which is strictly in the Lefschetz spirit, revealed that the squaring operations on cohomology groups were nothing more than the non-zero elements of the homology groups of the group π of order 2. The generalization to n^{th} power operations then became obvious. Replace $K \times K$ by the product K^n of n factors K. Let $d : K \to K^n$ be the diagonal map; and let λ be as before. Then let π be the symmetric group of all permutations of the factors of K^n, and let W be a π-free acyclic complex. The final result is that any homology class of any symmetric group determines a topologically invariant operation on the cohomology groups of any complex.

15. Concluding remarks

Although Lefschetz's labors were crowned with a number of outstanding results, by far the greater part of his work was devoted to the development and polishing of the basic tools of algebraic topology. Although it is easy to trace the impact of some of his ideas in recent developments, the full story of his influence would require a review of the entire field. Over a period of twenty years, he and a few others nurtured algebraic topology from a small sprout to a husky and vital branch of mathematics.

PRINCETON UNIVERSITY

Bibliography of the Publications of S. Lefschetz to June 1955

BIBLIOGRAPHY†

1912

[1] *Two theorems on conics.* Ann. of Math. (2), 14: 47–50.

[2] *On the V_3^3 with five nodes of the second species in S_4.* Bull. Amer. Math. Soc., 18: 384–386.

[3] *Double curves of surfaces projected from space of four dimensions.* Bull. Amer. Math. Soc., 19: 70–74.

1913

[4] *On the existence of loci with given singularities.* Trans. Amer. Math. Soc., 14: 23–41. Doctoral dissertation, Clark University, 1911.

[5] *On some topological properties of plane curves and a theorem of Möbius.* Amer. J. Math., 35: 189–200.

1914

[6] *Geometry on ruled surfaces.* Amer. J. Math., 36: 392–394.

[7] *On cubic surfaces and their nodes.* Kansas Univ. Science Bull., 9: 69–78.

1915

[8] *The equation of Picard-Fuchs for an algebraic surface with arbitrary singularities.* Bull. Amer. Math. Soc., 21: 227–232.

[9] *Note on the n-dimensional cycles of an algebraic n-dimensional variety.* Rend. Circ. Mat. Palermo, 40: 38–43.

1916

[10] *The arithmetic genus of an algebraic manifold immersed in another.* Ann. of Math. (2), 17: 197–212.

[11] *Direct proof of De Moivre's formula.* Amer. Math. Monthly, 23: 366–368.

[12] *On the residues of double integrals belonging to an algebraic surface.* Quart. J. Pure and Appl. Math., 47: 333–343.

1917

[13] *Note on a problem in the theory of algebraic manifolds.* Kansas Univ. Science Bull., 10: 3–9.

[14] *Sur certains cycles à deux dimensions des surfaces algébriques.* Accad. dei Lincei. Rend. (5), 26, 1° sem.: 228–234.

† This bibliography was prepared by Miss Mary L. Carll, Princeton University, Departmental Librarian.

[15] *Sur les intégrales multiples des variétés algébriques.* Acad. des Sci. Paris, C.R., 164: 850–853.

[16] *Sur les intégrales doubles des variétés algébriques.* Annali di Mat. (3), 26: 227–260.

1919

[17] *Sur l'analyse situs des variétés algébriques.* Acad. des Sci. Paris, C.R., 168: 672–674.

[18] *Sur les variétés abéliennes.* Acad. des Sci. Paris, C.R., 168: 758–761.

[19] *On the real folds of Abelian varieties.* Proc. Nat. Acad. Sci. U.S.A., 5: 103–106.

[20] *Real hypersurfaces contained in Abelian varieties.* Proc. Nat. Acad. Sci. U.S.A., 5: 296–298.

1920

[21] *Algebraic surfaces, their cycles and integrals.* Ann. of Math. (2), 21: 225–228. *A correction.* Ibid., 23: 333.

1921

[22] *Quelques remarques sur la multiplication complexe.* p. 300–307 of Comptes Rendus du Congrès International des Mathématiciens, Strasbourg, Sept. 1920. Toulouse, É. Privat, 1921.

[23] *Sur le théorème d'existence des fonctions abéliennes.* Accad. dei Lincei. Rend. (5), 30, 1° sem.: 48–50.

[24] *On certain numerical invariants of algebraic varieties with application to Abelian varieties.* Trans. Amer. Math. Soc., 22: 327–482. Awarded the Bôcher Memorial Prize by the American Mathematical Society in 1924. A translation, with minor modifications, of the memoir awarded the Prix Bordin by the Académie des Sciences, Paris, for the year 1919; for announcement, see Acad. des Sci. Paris, C.R., 169: 1200–1202, and Bull. Sci. Math., 44: 5–7.

1923

[25] *Continuous transformations of manifolds.* Proc. Nat. Acad. Sci. U.S.A., 9: 90–93.

[26] *Progrès récents dans la théorie des fonctions abéliennes.* Bull. Sci. Math., 47: 120–128.

[27] *Sur les intégrales de seconde espèce des variétés algébriques.* Acad. des Sci. Paris, C.R., 176: 941–943.

[28] *Report on curves traced on algebraic surfaces.* Bull. Amer. Math. Soc., 29: 242–258.

1924

[29] L'Analysis Situs et la Géométrie Algébrique. Paris, Gauthier-Villars. vi, 154 pp. (Collection de Monographies publiée sous la Direction de M. Émile Borel.) Based in part on a series of lectures given in Rome in 1921 under the auspices of the Institute of International Education, and also on research conducted under the auspices of the American Association for the Advancement of Science. Nouveau tirage, 1950.

[30] *Sur les intégrales multiples des variétés algébriques.* J. Math. Pures Appl. (9), 3: 319–343.

1925

[31] *Intersections of complexes on manifolds.* Proc. Nat. Acad. Sci. U.S.A., 11: 287–289.

[32] *Continuous transformations of manifolds.* Proc. Nat. Acad. Sci. U.S.A., 11: 290–292.

1926

[33] *Intersections and transformations of complexes and manifolds.* Trans. Amer. Math. Soc., 28: 1–49.

[34] *Transformations of manifolds with a boundary.* Proc. Nat. Acad. Sci. U.S.A., 12: 737-739.

1927

[35] *Un théorème sur les fonctions abéliennes.* p. 186–190 of *In Memoriam N. I. Lobatschevskii.* Kazan', Glavnauka.

[36] *Manifolds with a boundary and their transformations.* Trans. Amer. Math. Soc., 29: 429–462, 848.

[37] *Correspondences between algebraic curves.* Ann. of Math. (2), 28: 342–354.

[38] *The residual set of a complex on a manifold and related questions.* Proc. Nat. Acad. Sci. U.S.A., 13: 614–622, 805–807.

[39] *On the functional independence of ratios of theta functions.* Proc. Nat. Acad. Sci. U.S.A., 13: 657–659.

1928

[40] *Transcendental theory; Singular correspondences between algebraic curves; Hyperelliptic surfaces and Abelian varieties.* Chap. 15–17, p. 310–395, vol. 1, of Selected Topics in Algebraic Geometry; Report of the Committee on Rational Transformations of the National Research Council. Washington (NRC Bulletin no. 63).

[41] *A theorem on correspondences on algebraic curves.* Amer. J, Math., 50: 159–166.

[42] *Closed point sets on a manifold.* Ann. of Math. (2), 29: 232–254.

1929

[43] Géométrie sur les Surfaces et les Variétés Algébriques. Paris, Gauthier-Villars. 66 pp. (Mémorial des Sciences Mathématiques, Fasc. 40.)

[44] *Duality relations in topology.* Proc. Nat. Acad. Sci. U.S.A., 15: 367–369.

1930

[45] Topology. New York, American Mathematical Society. ix, 410 pp. (Colloquium Publications, vol. 12.)

[46] *Les transformations continues des ensembles fermés et leurs points fixes.* Acad. des Sci. Paris. C.R., 190: 99–100.

[47] *On the duality theorems for the Betti numbers of topological manifolds* (with W. W. Flexner). Proc. Nat. Acad. Sci. U.S.A., 16: 530–533.

[48] *On transformations of closed sets.* Ann. of Math. (2), 31: 271–280.

1931

[49] *On compact spaces.* Ann. of Math. (2), 32: 521–538.

1932

[50] *On certain properties of separable spaces.* Proc. Nat. Acad. Sci. U.S.A., 18: 202–203.

[51] *On separable spaces.* Ann. of Math. (2), 33: 525–537.

[52] *Invariance absolue et invariance relative en géométrie algébrique.* Rec. Math. (Mat. Sbornik), 39, no. 3: 97–102.

1933

[53] *On singular chains and cycles.* Bull. Amer. Math. Soc., 39: 124–129.

[54] *On analytical complexes* (with J. H. C. Whitehead). Trans. Amer. Math. Soc., 35: 510–517.

[55] *On generalized manifolds.* Amer. J. Math., 55: 469–504.

1934

[56] Elementary One- and Two-Dimensional Topology (a course given by Prof. Lefschetz, Spring 1934; notes by H. Wallman). Princeton University. 95 pp., mimeographed.

[57] *On locally connected and related sets.* Ann. of Math. (2), 35: 118–129.

1935

[58] Topology (lectures 1934–35; notes by N. Steenrod and H. Wallman). Princeton University. 203 pp., mimeographed.

[59] *Algebraicheskaia͡ geometriia͡: metody, problemy, tendentsii.* pp. 337–349, vol. 1, of Trudy Vtorogo Vsesoiu͡znogo Matematicheskogo S''ezda, Leningrad, 24–30 June 1934. Leningrad-Moscow. An invited address at the Second All-Union Mathematical Congress.

[60] *Chain-deformations in topology.* Duke Math. J., 1: 1–18.

[61] *Application of chain-deformations to critical points and extremals.* Proc. Nat. Acad. Sci. U.S.A., 21: 220–222.

[62] *A theorem on extremals. I, II.* Proc. Nat. Acad. Sci. U.S.A., 21: 272–274, 362–364.

[63] *On critical sets.* Duke Math. J., 1: 392–412.

1936

[64] *On locally-connected and related sets* (second paper). Duke Math. J., 2: 435–442.

[65] *Locally connected sets and their applications.* Rec. Math. (Mat. Sbornik) n.s., 1: 715–717. A contribution to the First International Topological Conference, September 1935.

[66] *Sur les transformations des complexes en sphères.* Fund. Math., 27: 94–115.

[67] *Matematicheskaia͡ deia͡tel'nost' v Prinstone.* Uspekhi Mat. Nauk vyp. 1, pp. 271–273.

1937

[68] Lectures on Algebraic Geometry 1936–37; notes by M. Richardson and E. D. Tagg. Princeton University. 69 pp., planographed.

[69] *Algebraicheskaia͡ geometriia͡.* Uspekhi Mat. Nauk vyp. 3, pp. 63–77.

[70] *The role of algebra in topology.* Bull. Amer. Math. Soc., 43: 345–359. Address of retiring president of the American Mathematical Society.

[71] *On the fixed point formula.* Ann. of Math. (2), 38: 819–822.

1938

[72] Lectures on Algebraic Geometry (Part II) 1937–1938. Princeton University Press. 73 pp., planographed.

[73] *On chains of topological spaces.* Ann. of Math. (2), 39: 383–396.

[74] *On locally connected sets and retracts.* Proc. Nat. Acad. Sci. U.S.A., 24: 392–393.

[75] *Sur les transformations des complexes en sphères (note complémentaire).* Fund. Math., 31: 4–14.

[76] *Singular and continuous complexes, chains and cycles.* Rec. Math. (Mat. Sbornik) n.s., 3: 271–285.

1939

[77] *On the mapping of abstract spaces on polytopes.* Proc. Nat. Acad. Sci. U.S.A., 25: 49–50.

1941

[78] *Abstract complexes.* pp. 1–28 of Lectures in Topology; the University of Michigan Conference of 1940. Ann Arbor, University of Michigan Press; London, Oxford University Press.

1942

[79] Algebraic Topology. New York, American Mathematical Society. vi, 389 pp. (Colloquium Publications, vol. 27.)

[80] Topics in Topology. Princeton University Press. 137 pp. (Annals of Mathematics Studies, no. 10). A second printing, 1951. London, Oxford University Press.

[81] *Émile Picard (1856–1941): Obituary.* Amer. Phil. Soc. Yearbook 1942, pp. 363–365.

1943

[82] Introduction to Non-linear Mechanics, by N. Kryloff and N. Bogoliuboff; a free translation by S. Lefschetz of excerpts from two Russian monographs. Princeton University Press. 105 pp. (Annals of Mathematics Studies, no. 11). London, Oxford University Press.

[83] *Existence of periodic solutions for certain differential equations.* Proc. Nat. Acad. Sci. U.S.A., 29: 29–32.

1946

[84] Lectures on Differential Equations. Princeton University Press. viii, 210 pp. (Annals of Mathematics Studies, no. 14.) London, Oxford University Press.

1949

[85] Introduction to Topology. Princeton University Press. viii, 218 pp. (Princeton Mathematical Series, no. 11.) London, Oxford University Press. A work originating from a short course delivered in 1944 before the Institute of Mathematics of the National University of Mexico.

[86] Theory of Oscillations, by A. A. Andronow and C. E. Chaikin; English language edition edited under the direction of S. Lefschetz. Princeton University Press. ix, 358 pp.

[87] *Scientific research in the U.S.S.R.: Mathematics.* Amer. Acad. Polit. and Soc. Sci. Annals, 263: 139–140.

1950

[88] Contributions to the Theory of Nonlinear Oscillations, edited by S. Lefschetz. Princeton University Press. ix, 350 pp. (Annals of Mathematics Studies, no. 20.) London, Oxford University Press.

[89] *The structure of mathematics.* American Scientist, 38: 105–111.

1951

[90] *Numerical calculations in nonlinear mechanics.* pp. 10–12 of Problems for the Numerical Analysis of the Future. Washington, Govt. Printing Office. (National Bureau of Standards, Applied Math. Series, no. 15.)

1952

[91] Contributions to the Theory of Nonlinear Oscillations, vol. 2, edited by S. Lefschetz. Princeton University Press. 116 pp. (Annals of Mathematics Studies, no. 29.) London, Oxford University Press.

[92] *Notes on differential equations.* pp. 61–73 of Contributions to the Theory of Nonlinear Oscillations, vol. 2.

1953

[93] Algebraic Geometry. Princeton University Press. ix, 233 pp. (Princeton Mathematical Series, no. 18.)

[94] *Algunos trabajos recientes sobre ecuaciones diferenciales.* pp. 122–123, vol. 1 of Memoria de Congreso Cientifico Mexicano, U.N.A.M., Mexico.

[95] *Las grades corrientes en las matemáticas del siglo XX.* pp. 206–211, vol. 1 of Memoria de Congreso Cientifico Mexicano, U.N.A.M., Mexico.

1954

[96] *Russian contributions to differential equations.* pp. 68–74 of Proceedings of the Symposium on Nonlinear Circuit Analysis, New York, 1953. New York, Polytechnic Institute of Brooklyn.

[97] *Complete families of periodic solutions of differential equations.* Comment. Math. Helv. 28: 341–345.

[98] *On Liénard's differential equation.* pp. 149–153 of Wave Motion and Vibration Theory. New York, McGraw-Hill. (Amer. Math. Soc. Proceedings of Symposia in Applied Math., vol. 5.)

Part II

Papers in Algebraic Geometry

On the Complex Structures of a Class of Simply-Connected Manifolds

Aldo Andreotti

1. A topological variety of even dimension $2n$ has an *analytic complex structure*, if for every point there exists a neighborhood homeomorphic to an open subset of the complex (Euclidean) space of n dimensions, such that the correspondence which arises between local coordinates in the common part of two overlapping neighborhoods is analytic and bi-regular.

The definition itself causes us to consider as equivalent two structures of the same variety if in the neighborhood of every point the change from one system of local coordinates to the other is analytic and bi-regular.

Among topological varieties admitting a complex structure we have the Riemannian varieties of algebraic non-singular irreducible manifolds of a complex projective space. The natural complex structure in the neighborhood of each point is given by the equations which represent parametrically the branch through this point. In other words, this is the structure induced on the variety by the complex structure of the surrounding space.

An analytic complex structure will be said to be *algebraic* if there exists an analytic homeomorphism which carries the given variety into an algebraic (irreducible non-singular) manifold of a complex projective space, such that the structure becomes equivalent to the natural structure on the algebraic model.

In a recent paper Hirzebruch [11] has considered three types of topological varieties: the complex projective plane $P^{(2)}$, the Cartesian product of two spheres $S^2 \times S^2$, and the sum of the second kind (as defined by Rueff [18]) of two projective planes $P^{(2)} + P^{(2)}$. While for the projective plane the sole algebraic structure which it has occurred to him to consider is the natural one, for the two other types of varieties he has given an infinite number of different algebraic structures.

Our object is to re-examine the models studied by Hirzebruch; to place them in relation to the classification given by C. Segre of rational normal ruled surfaces of a complex projective space; and finally, to prove that the structures considered by Hirzebruch are the only possible algebraic structures for the given topological models with the 3-genus $P_3 \neq 25, 28$.†

In a second part of the paper we shall develop some applications to algebraic geometry of the above results. Among other things we shall give a complete classification of non-singular rational surfaces without exceptional curves of the first kind with respect to the group of birational transformations without exceptions.‡

I wish to acknowledge my gratitude to Professor Heinz Hopf for having called my attention to this problem and for his encouragement and helpful criticism.

1. The algebraic structures of Hirzebruch

2. The models of algebraic surfaces studied by Hirzebruch are defined as follows:

Let $P^{(n)}$ be the projective complex space of n dimensions. In the Cartesian product $P^{(2)} \times P^{(1)}$ of the projective plane (x_0, x_1, x_2) and the projective line (y_1, y_2) we consider the surfaces Σ_n defined by the equations

$$x_1 y_1^n - x_2 y_2^n = 0 \quad (n \geqq 0) \tag{1}$$

for every n.

For these models Hirzebruch has proved (a) that they are analytically distinct, and (b) that for n even, they are homeomorphic to $S^2 \times S^2$; for n odd to the sum of the second kind $P^{(2)} + P^{(2)}$.

Let us consider anew these surfaces, and let us prove first of all that *the surface Σ_n in Segre's variety, image of $P^{(2)} \times P^{(1)}$, is obtained by unexceptional birational projection of a ruled surface ϕ_{n+2} of $P^{(n+3)}$ with a directrix line.*

† It has privately been brought to my attention that F. Hirzebruch and K. Kodaira have proved the following fact: Let n be *odd* and M^n a compact Kähler manifold (with n complex dimensions) which is differentiably homeomorphic with the projective space $P^{(n)}$; then M^n is analytic (complex) equivalent with $P^{(n)}$. [*Added in proof.* I have realized, reading the proofs, that the condition for the plurigenera in the case of $P^{(2)}$ is actually superfluous; therefore the result of Hirzebruch and Kodaira can be extended also to the case of $n = 2$ (see the final remark at the end of § 6).]

‡ This classification could be deduced from a paper of G. Vaccaro, which, however, to be fully rigorous, would bring us into a very detailed analysis of Cremonian transformations. For this reason we have not made use of Vaccaro's results[27].

PROOF. Segre's variety mentioned above is defined [20] in $P^{(5)}$, where we assume as homogeneous coordinates z_{ik} $(i=0,1,2;\ k=1,2)$, by the equations

$$z_{ik}=x_i y_k$$

or by the three quadratic equations

$$(2)\qquad \begin{cases} z_{01}z_{12}=z_{02}z_{11}, \\ z_{01}z_{22}=z_{02}z_{21}, \\ z_{11}z_{22}=z_{12}z_{21}. \end{cases}$$

On this variety, Σ_n is given by the six equations

$$(3)\qquad z_{1h}z_{l1}^n=z_{2h}z_{l2}^n \quad (h=1,2;\ l=0,1,2).$$

We consider now in $P^{(n+3)}$ $(x_0, x_1, \ldots, x_{n+3})$ the line C_1:

$$x_0=1, \quad x_1=\lambda, \quad x_2=\ldots=x_{n+3}=0;$$

and the rational normal curve C_{n+1} in $x_0=x_1=0$:

$$x_0=x_1=0, \quad x_2=1, \quad x_3=\lambda, \quad \ldots, \quad x_{n+3}=\lambda^{n+1}.$$

These two curves are related projectively if we make them correspond one-to-one to the same value of the parameter λ. The ruled surface ϕ_{n+2} described by the lines joining corresponding points is given by the parametric equations

$$x_0=1, \quad x_1=\lambda, \quad x_2=\mu, \quad x_3=\mu\lambda, \quad \ldots, \quad x_{n+2}=\mu\lambda^n, \quad x_{n+3}=\mu\lambda^{n+1};$$

or by the equations

$$\begin{vmatrix} x_0 & x_2 & x_3 & x_4 & \cdots & x_{n+2} \\ x_1 & x_3 & x_4 & x_5 & \cdots & x_{n+3} \end{vmatrix}=0.$$

The subspace $x_0=x_1=x_2=x_3=x_{n+2}=x_{n+3}=0$ does not meet ϕ. We now project ϕ onto $x_4=\ldots=x_{n+1}=0$. The latter is a space of five dimensions. We superimpose it on the space $P^{(5)}$ of the Segre variety by putting

$$\begin{cases} x_0=z_{02}, & x_3=z_{11}, \\ x_1=z_{01}, & x_{n+2}=z_{22}, \\ x_2=z_{12}, & x_{n+3}=z_{21}. \end{cases}$$

The projection of ϕ,

$$x_0=1, \quad x_1=\lambda, \quad x_2=\mu, \quad x_3=\mu\lambda, \quad x_{n+2}=\mu\lambda^n, \quad x_{n+3}=\mu\lambda^{n+1},$$

is contained in the hyperquadrics (2) and in the hypersurfaces (3). Therefore it coincides with Σ_n. In addition, Σ_n has no singularities and the projection of ϕ on Σ_n is one-to-one without exception, as we can verify by the equations.

3. The surfaces ϕ_{n+2} that we have considered in the preceding section are special cases of non-singular irreducible surfaces of order r in a $P^{(r+1)}$.

These surfaces, according to a theorem of Del Pezzo,† are ruled rational surfaces (with the exception of the Veronese surface) and have been classified by C. Segre [3b]. Every non-singular rational normal ruled surface of $P^{(r+1)}$ can be obtained by joining with straight lines corresponding points in a projectivity established between two rational normal curves of orders m and $r-m$ contained in two spaces without common points. Therefore each one of these models $\phi(m, r-m)$ has, with a convenient choice of coordinates, the equations

$$x_0 = 1, \quad x_1 = \lambda, \quad \ldots, \quad x_m = \lambda^m,$$

$$x_{m+1} = \mu, \quad x_{m+2} = \mu\lambda, \quad \ldots, \quad x_{r+1} = \mu\lambda^{r-m}.$$

The surfaces Σ_n considered earlier are therefore birationally equivalent, without exceptions, to surfaces of the type $\phi(1, n+1)$. The significance of these models is emphasized by the fact that:

with respect to the group of birational transformations without exceptions the non-singular rational normal surfaces are fully represented by those surfaces with a rectilinear directrix.

In fact, let us consider one of these surfaces, $\phi(1, n+1)$, with the rectilinear directrix d: let a be a directrix of order $n+1$; let g be a generator.

Since $a+g$ is a hyperplane section of $\phi(1, n+1)$, we have

$$[a+g, a+g] = [a, a] + 2 = n+2, \quad \text{that is,} \quad [a, a] = n.$$

The linear system $|a+sg|$ has the virtual grade

$$[a+sg, a+sg] = n + 2s \quad (s \geqq 1);$$

moreover, the system is irreducible without neutral pairs because it contains partially the system of hyperplane sections. The genus of the generic curve of $|a+sg|$ is equal to the virtual genus of $a+sg$, that is, it is zero. We have, therefore, a linear system of rational curves. Its characteristic series is complete, and therefore its dimension is $n+2s+1$ and it has no base points, no fundamental curves, and no neutral pairs.

Taking the projective image of it, we have a surface ϕ of order $n+2s$ in $P^{(n+2s+1)}$, birationally equivalent without exceptions to $\phi(1, n+1)$. This surface is a ruled rational normal surface of type $\phi(s, n+s)$ (the generators of $\phi(1, n+1)$ change into generators of ϕ).

† See, for instance, [3a].

In fact, the directrices d and a change into directrices of orders $[d, a+sg]=s$, $[a, a+sg]=n+s$, respectively.

Since s can be any integer number, this establishes the result.

REMARK. From the above proof we see more precisely that every $\phi(m, r-m)$ is birationally equivalent without exceptions to

$$\phi(1, r-2m+1) \quad (r \geqq 2m).$$

Moreover, we have the theorem: *Of the ruled rational normal surfaces with a rectilinear directrix, no two are analytically equivalent.*

This follows from the result of Hirzebruch on the analytical non-equivalence of the models Σ_n and from the proven identity of these models with the above-mentioned surfaces.

Here is another proof. We first note that an analytical correspondence between two surfaces $\phi(1, n+1)$, $\phi(1, m+1)$ $(n \neq m)$ is necessarily birational and without exceptions, because of a theorem of Hurwitz-Severi-Chow [21, 7].

Let us now consider the rectilinear directrix d of $\phi(1, n+1)$. Its virtual grade is $[d, d] = -n$. In fact, a hyperplane through it cuts the surface in $n+1$ generators g, so that

$$[d+(n+1)g, d+(n+1)g]=n+2,$$

or

$$[d, d]+2(n+1)=n+2,$$

and so

$$[d, d] = -n.$$

The curves d and g give a base for linear equivalence on $\phi(1, n+1)$ [23a].

Let δ be the irreducible algebraic curve of $\phi(1, n+1)$ which corresponds to the rectilinear directrix of $\phi(1, m+1)$. The virtual grade of δ is $[\delta, \delta] = -m$, for reasons analogous to those given above. Let $\delta \equiv \alpha g + \beta d$. Because δ is irreducible and of negative virtual grade, it cannot be contained as a total curve in an infinite linear system. Therefore we must have $\alpha = 0$, $\beta = 1$, and $m = -[\delta, \delta] = -[d, d] = n$, which contradicts the hypothesis $m \neq n$.

4. From these premises we pass on to the demonstration that the only allowable algebraic structures with $P_3 \neq 25$ and 28 on the projective plane $P^{(2)}$, on the product of two spheres $S^2 \times S^2$, and on the sum $P^{(2)} + P^{(2)}$ of two projective planes, are those considered by Hirzebruch.

Let us start with the projective plane. We must prove that:

An irreducible, non-singular algebraic surface F with the 3-genus $P_3 \neq 28$, differentiably homeomorphic to the projective plane $P^{(2)}$, is birationally equivalent, without exceptions, to $P^{(2)}$.†

† This theorem is practically contained in a paper of Severi [22].

We recall that the projective plane $P^{(2)}$ has Betti numbers alternately equal to 1 and 0:

$$p^0=1, \quad p^1=0, \quad p^2=1 \quad p^3=0, \quad p^4=1,$$

and also that it has no torsion [28, 17, 2].

The same will hold for F, and therefore for both F and $P^{(2)}$ we have

$$q=\text{irregularity}=\tfrac{1}{2}p^1=0,$$

$$\rho+\rho_0=p^2=1$$

(ρ = base number, ρ_0 = number of transcendental 2-cycles), and so,

$$\rho=1, \quad \rho_0=0.$$

Moreover, from a theorem of Hodge [13] the geometric genus $p_g=0$ and also the arithmetic genus $p_a=p_g-q=0$.

Let D be an algebraic curve, minimal base on F; let K be a canonical curve of F. We will have $[D, D]=1$, and because K is an immersion character in $F\times F$ of the diagonal [6, 12, 19, 29], in the same way we will have†

$$K\equiv -3D \quad [4].$$

The virtual genus of D is 0; in fact, $3D+K$ is the 0-curve, and therefore the virtual genus of $3D$ is 1; if x is the virtual genus of D, we have

$$3x+3-2=1; \quad x=0.$$

Now if D is an arithmetically effective curve [23 b], it can be represented by an effective curve necessarily irreducible, because every irreducible component of D is a positive multiple of D.

Because F is regular, the characteristic series of D is complete and $|D|$ is a net of curves, all irreducible, of virtual grade 1 and virtual genus 0.

The linear system $|D|$ has no base point (each curve being irreducible and the characteristic series being a g_1^1). It also has no neutral pairs ($[D, D]=1$) and no fundamental curves.

The projective image of F constructed by means of $|D|$ is a projective plane $P^{(2)}$ and the correspondence is one-to-one, without exceptions.

Let us now suppose that D is not arithmetically effective. If by $i(D)$ we indicate the index of specialty of D, that is,

$$i(D)=\dim|K-D|+1,$$

we must have $i(D)>2$, that is, $|-4D|$ must be effective.

† By direct reasoning, we have $I+4=\Sigma(-1)^i p^i=3$; from Moether's formula $I+p^{(1)}=12p_a+9$, we get $p^{(1)}=10$, $[K, K]=9$, $K\equiv\pm 3D$. By changing the sign of D, if necessary, we may suppose $K\equiv -3D$.

Let C be a hyperplane section of F, and let us suppose that $C \equiv \delta D$; δ is the order of D: $\delta = [C, D]$.

Because $|-4D|$ is effective, the order of $-4D$ is positive, that is, $[C, -4D] = -4\delta > 0$, therefore $\delta < 0$.

Let λ be any integer > 3; we wish to calculate the virtual dimension of the linear system $|-\lambda D|$.

The virtual grade of $-\lambda D$ is $n_\lambda = [-\lambda D, -\lambda D] = \lambda^2$; the virtual genus of $-\lambda D$ is $\pi_\lambda = \frac{1}{2}[-\lambda D, -\lambda D + K] + 1 = \frac{1}{2}\lambda(\lambda+3)+1$; the index of specialty of $-\lambda D$ is $i(-\lambda D) = 0$, because $K + \lambda D = (\lambda - 3)D$ has the order $(\lambda - 3)\delta < 0$. Therefore the virtual dimension of $|-\lambda D|$ is

$$r_\lambda = n_\lambda - \pi_\lambda - i(-\lambda D) + 1 = \lambda^2 - \tfrac{1}{2}(\lambda(\lambda+3)) - 1 + 1 = \tfrac{1}{2}\lambda(\lambda-3) > 0.$$

Then $|-\lambda D|$ is effective and for $\lambda = 3k$ $(k = 2, 3, \ldots)$ we have for the plurigenera P_k of F the relations

$$P_k \geqq \tfrac{9}{2}k(k-1) + 1 \quad (k = 2, 3, \ldots).$$

In particular $P_2 \geqq 10$, $P_3 \geqq 28$.

We note that the bicanonical curves are virtually connected [10]. In fact, an effective curve of the system $|-6D|$ either is irreducible or, if reducible, two of its true components αD, βD, which are complementary, have positive order, and therefore $\alpha < 0$, $\beta < 0$; their intersection number is therefore $[\alpha D, \beta D] = \alpha\beta > 0$. From the theorem on the regularity of the adjoint system, given in the most comprehensive form by Franchetta [9],† we must have $P_3 = \pi_6 = 28$. This is the case we excluded in the hypothesis.

5. We pass on to the case of the product $S^2 \times S^2$. We shall prove that:

An irreducible non-singular algebraic surface F with the 3-genus $P_3 \neq 25$, differentiably homeomorphic to a non-singular quadric, is birationally equivalent, without exceptions, to a rational normal ruled surface $\phi(1, n+1)$, of even order ($n \equiv 0 \bmod 2$).

We recall that a non-singular quadric Q has the Betti numbers

$$p^0 = 1, \quad p^1 = 0, \quad p^2 = 2, \quad p^3 = 0, \quad p^4 = 1,$$

and has no torsion.

The same will hold for F, so that we will have for both Q and F, with the same notations as in the preceding section,

$$q = 0, \quad \rho + \rho_0 = 2,$$

and also, because of a theorem of Hodge [13],

$$p_g = 0.$$

† Note that F, being regular, cannot contain irrational pencils.

From a theorem of Lefschetz [14]

$$\rho_0 = 0 \quad \text{and therefore} \quad \rho = 2.$$

Let A and B be the curves of a minimal base on F. We can suppose that

$$[A, A] = 0, \quad [A, B] = 1, \quad [B, B] = 0$$

if we assume A and B to be the cycles corresponding to two generators of different systems of Q.

If K is the canonical curve of F, for the same reason given in the preceding section, we have†

$$K \equiv -2A - 2B.$$

From this it follows that the virtual genus of both A and B is 0.

If A and B are both arithmetically effective curves, they can be represented by two effective curves. We shall first discuss this case.

Let us suppose that

$$\text{order of } A \geqq \text{order of } B.$$

Let l_0 be the maximum integer $l \geqq 0$ such that $|A - lB|$ is effective. The curves $D = A - l_0 B$ and B form a minimal base and we have $[D, D] = -2l_0$.

We consider now the linear system

$$|D + (2l_0 + 1)B| = |A + (l_0 + 1)B| = |C|.$$

Its virtual grade is $2l_0 + 2$ and its dimension (evaluated by means of the Riemann-Roch theorem)‡ is $\geqq 2l_0 + 3$.

The curves of $|B|$ form at least a pencil. The curves of this system cannot have a fixed component. In fact, let us suppose

$$B \equiv B_1 + B_2,$$

with B_1 the fixed part and B_2 variable.

Let us suppose that

$$B_1 = \alpha_1 A + \beta_1 B,$$
$$B_2 = \alpha_2 A + \beta_2 B.$$

By addition it follows that

$$\alpha_1 + \alpha_2 = 0, \quad \beta_1 + \beta_2 = 1.$$

We cannot have $\alpha_1 = 0$ because B_1 cannot be a multiple of B unless it coincides with B, which is impossible. Therefore $\alpha_1 \neq 0$, $\alpha_2 \neq 0$.

† In this case it is $I + 4 = 4$, $p^{(1)} = 9$, $[K, K] = 8$; if $K \equiv \alpha A + \beta B$, $\alpha\beta = 4$ and $\alpha = \beta = \pm 2$ or $\alpha = \pm 1$, $\beta = \pm 4$. The second case is excluded because if π_B is the virtual genus of B, we would have $2\pi_B - 2 = \pm 1$.

‡ The virtual genus of $\lambda A + \mu B$ is $\lambda\mu - \lambda - \mu + 1$, the virtual grade, $2\lambda\mu$, the dimension of $|\lambda A + \mu B|$ is $\geqq \lambda\mu + \lambda + \mu$, if the curve $\lambda A + \mu B$ is nonspecial.

Let us suppose $\alpha_1 > 0$. Put $\alpha_1 = a$. We have $\alpha_2 = -a$. Because B_2 is effective, certainly $\beta_2 > 0$; put $\beta_2 = b+1$; we have $\beta_1 = -b$.

The second of the preceding equations can be written now:

$$B_2 = -aA + (b+1)B, \quad \text{with} \quad a > 0, \quad b \geqq 0.$$

From this it follow that $[B_2, B_2] = -2a(b+1)$, which is absurd since $[B_2, B_2] \geqq 0$ for the reason that B_2 is variable in a pencil without fixed components.

We must have therefore $\alpha_1 < 0$, and if we put $\alpha_2 = a$, we have $\alpha_1 = -a$. Moreover, $\beta_1 > 0$, and if we put $\beta_1 = b+1$, we have $\beta_2 = -b$, and it follows that

$$B_2 = aA - bB, \quad \text{with} \quad a > 0, \quad b \geqq 0.$$

Since $[B_2, B_2] \geqq 0$, we find $b = 0$, which is equally absurd because B_2, which is a part of B, cannot have an order which is greater than the order of A.

In conclusion, B is a total curve of a pencil without fixed components, and because $[B, B] = 0$, the pencil has no base points. Moreover, it contains no reducible curves. In fact, if we had $B = B_1 + B_2$ (B_1 and B_2 being parts of B), we would have

$$B_1 = \alpha A + \beta B.$$

But $[B, B_1] = \alpha = 0$, and therefore $\beta = 1$, that is, $B_1 = B$, and B_2 is lacking.

Thus B defines a pencil with no base point, the curves of which are all irreducible.

From this is follows that D is also an irreducible curve. In fact, $[B, D] = 1$ and the points of a component of D are in one-to-one correspondence with the curves of the pencil $|B|$. D cannot contain other parts, because each of these parts would coincide with some curves of $|B|$, and this contradicts the definition of D.

We shall now prove that the linear system $|C|$ has no neutral pairs.

Let P be a point of F. Among the curves of $|C|$ through P there are those curves composed of a B through P, of D, and of $2l_0$ curves B which do not contain P, if P is not on D. If P is on D, a curve of $|C|$ can be obtained by adding to D $2l_0 + 1$ curves B which do not contain P. It is sufficient to prove that the system $|C|$ cuts linear series without neutral pairs on every curve B and on curve D.

Since $[D, C] = 1$, the series cut on D is either a g_1^0 or a g_1^1. The first instance is impossible because in this case D would be a fundamental curve for $|C|$ and the residual system of D would have dimension

$\geqq 2l_0+2$, but actually it is composed of $2l_0+1$ curves of $|B|$ and its dimension cannot exceed $2l_0+1$.

Our assertion is proved for the curve D.

Let us now consider a curve B. Since $[B, C]=1$, the series cut on B is either a g_1^0 or a g_1^1. The first instance is impossible because the g_1^0 would consist of the point cut by D on B. This would be a base point for $|C|$, and $|C|$ would cut on D a g_1^0, which is contrary to the fact proved above.

In conclusion, $|C|$ has no neutral pairs; hence it cannot have fundamental curves or base points, nor can it be composed of the curves of a pencil.

The projective image of F by means of $|C|$ is therefore birationally equivalent, without exceptions, to F.

The generic curve C is irreducible and non-singular, and therefore its virtual genus is equal to its effective genus. The curves C are then rational and the dimension of $|C|$ is therefore exactly equal to $2l_0+3$.

The image we obtain of F is an algebraic surface, non-singular, of order $2l_0+2$ of a $P^{(2l_0+3)}$. From this our theorem follows immediately if we recall § 3. More directly, we can remark that the image we obtain of F is a ruled surface; its generators are the images of the curves B; this ruled surface has a rectilinear directrix, the image of D. The image of F is therefore a $\phi(1, 2l_0+1)$ of a $P^{(2l_0+3)}$, as we wished to prove.

Let us now suppose that one of the two curves A, B (A, for example) is not arithmetically effective. If by $i(A)$ we indicate the index of specialty of A, that is, $i(A)=\dim|K-A|+1$, we must have $i(A)>1$, that is, $|-3A-2B|$ must be effective.

Let C be a hyperplane section of a non-singular model of F and let us suppose that $C\equiv\beta A+\alpha B$; $\alpha=[C, A]$ is the order of A, $\beta=[C, B]$ that of B.

Because $[C, C]=2\alpha\beta$ is the order of F, $2\alpha\beta>0$, so that we have two possibilities:

1. $\alpha>0, \quad \beta>0,$
2. $\alpha<0, \quad \beta<0.$

In addition, $|-3A-2B|$ being effective, the order of its curves is positive, that is, $-3\alpha-2\beta\geqq 0$, which excludes the possibility of the first case.

Let us examine the second case. Let λ, μ be any two integers >2. We wish to calculate the virtual dimension of the system $|-\lambda A-\mu B|$.

The virtual grade of $-\lambda A-\mu B$ is

$$n_{\lambda\mu}=[-\lambda A-\mu B, -\lambda A-\mu B]=2\lambda\mu;$$

the virtual genus of $-\lambda A - \mu B$ is

$$\pi_{\lambda\mu} = \tfrac{1}{2}[-\lambda A - \mu B, -\lambda A - \mu B + K] + 1 = \lambda\mu + \lambda + \mu + 1;$$

the specialty index is $i_{\lambda\mu} = 0$, because the curves of $|K - (-\lambda A - \mu B)|$ have the order $(\lambda - 2)\alpha + (\mu - 2)\beta < 0$. Therefore the virtual dimension of $|-\lambda A - \mu B|$ is

$$r_{\lambda\mu} = n_{\lambda\mu} - \pi_{\lambda\mu} + P_a - i_{\lambda\mu} + 1 = \lambda\mu - \lambda - \mu > 0.$$

Hence $|-\lambda A - \mu B|$ is effective and for $\lambda = \mu = 2k$ $(k = 2, 3, \ldots)$ we have for the plurigenera P_k of F the relations

$$P_k \geqq 4k(k-1) + 1.$$

In particular $P_2 \geqq 9$, $P_3 \geqq 25$.

The bicanonical curves are effective, and since the linear genus of F is $p^{(1)} = 9$, they are virtually connected, because of a theorem of Franchetta [9]. The 3-canonical system is regular [9], therefore $P_3 = \pi_{4,4} = 25$. This is the case which we excluded in the hypothesis.

6. We come finally to the case of the sum of the second kind of two projective planes $P^{(2)} + P^{(2)}$. We must prove that:

An irreducible non-singular algebraic surface F, with the 3-genus $P_3 \neq 25$, differentiably homeomorphic to the ruled surface $\phi(1, 2)$ of $P^{(4)}$, is birationally equivalent without exceptions to a rational normal ruled surface $\phi(1, n+1)$ of odd order ($n \equiv 1 \bmod 2$).

The surface $\phi(1, 2)$ has Betti numbers

$$p^0 = 1, \quad p^1 = 0, \quad p^2 = 2, \quad p^3 = 0, \quad p^4 = 1,$$

and has no torsion. So for both F and $\phi(1, 2)$ we have $q = 0$, $\rho + \rho_0 = 2$, and also, because of the above-mentioned theorem of Hodge, $p_g = 0$. From this, as in the preceding section, we deduce $\rho_0 = 0$, $\rho = 2$.

Let A be the rectilinear directrix of $\phi(1, 2)$, and let B be a generator. We have $[A, A] = -1$, $[A, B] = 1$, $[B, B] = 0$, and A and B form a minimal base on ϕ.† Let K be a canonical curve of ϕ; we have $K \equiv -2A - 3B$, as we can easily verify.‡

† See §3 and [23a].

‡ For instance, we may observe that $\phi(1, 2)$ can be regarded as the projective image of conics in a projective plane passing through a given point O; O becomes A; the lines r through O the curves B, the canonical curve $-3r$, $-3B - 3A$, and to it we must add A in order to have the canonical curve of $\phi(1, 2)$. We can also refer to the projection of $\phi(1, 2)$ in $P^{(3)}$. It is a cubic primal with a double line which derives from the projection of a conic on $\phi(1, 2)$. On this projection the quadrics cut curves $C \equiv 2(A + 2B)$, the adjoint system $|C'|$ is given by the pencil B, and

$$K \equiv C' - C \equiv B - 2(A + 2B) \equiv -2A - 3B.$$

Also on F, if we indicate by A, B and K two curves of a minimal base and a canonical curve, we can assume that

$$[A,A]=-1, \quad [A,B]=1, \quad [B,B]=0$$

and we must have† $\qquad K\equiv -2A-3B.$

From this it follows that the curves A, B have virtual genus 0. If the specialty indices of A and B are zero, A and B are arithmetically effective curves and they can be represented by two effective curves. Moreover, from the Riemann-Roch theorem, we find that B is a total curve of a pencil (at least). We shall first discuss this case.

The curves of this system $|B|$ cannot have a fixed part. In fact, let $B=B_1+B_2$, B_1 being the fixed part and B_2 variable. Suppose that

$$B_1=\alpha_1 A+\beta_1 B,$$
$$B_2=\alpha_2 A+\beta_2 B.$$

We see that $\qquad \alpha_1+\alpha_2=0, \quad \beta_1+\beta_2=1.$

Certainly $\alpha_1 \neq 0$, and therefore $\alpha_2 \neq 0$, because B_1 is not a multiple of B.

If $\alpha_1>0$, putting $\alpha_1=a$, we have $\alpha_2=-a$. Because B_2 is effective, $\beta_2>0$, and putting $\beta_2=b+1$, we have $\beta_1=-b$. From the second of the above equivalences

$$B_2=-aA+(b+1)B, \quad \text{with} \quad a>0, \quad b\geqq 0.$$

From this it follows that $[B_2,B_2]=-a(a+2[b+1])$, and this is absurd because $[B_2,B_2]\geqq 0$.

We must have therefore $\alpha_1<0$, and putting $\alpha_2=a>0$, we have $\alpha_1=-a$. Consequently $\beta_1>0$, and putting $\beta_1=b+1$, we have $\beta_2=-b$. The second of the preceding equivalences gives

$$B_2=aA-bB.$$

From this it follows that $[B_2,B_2]=-a(a+2b)$, and this is absurd because $[B_2,B_2]\geqq 0$.

Thus B is a total curve of a pencil without a fixed part; and it is without base points, because $[B,B]=0$.

This pencil does not contain reducible curves. In fact, if $B=B_1+B_2$ and $B_1=\alpha A+\beta B$, since $[B,B_1]=0$, we must have $\alpha=0$ and $\beta=1$.

† We have $[K,K]=8$; let $K\equiv\alpha A+\beta B$; we may suppose $\alpha\leqq 0$; so from $\alpha(2\beta-\alpha)=8$, we obtain $\alpha=-2$, $\beta=-3$ or $\alpha=-4$, $\beta=-3$. The second case has to be excluded. In fact, if B is arithmetically effective, dim $|B|\geqq 2$. As in the discussion which follows, we see that $|B|$ has no fixed part, so $|B|$ is compound with couples of curves taken in a pencil of rational curves without base points, because the virtual genus of B is -1. This is impossible because if $B=B_1+B_2$, $B_1=\alpha A+\beta B$, since $[B_1,B]=0$, $\alpha=0$, $\beta=1$. If B is not arithmetically effective, we get again, as in the discussion below, $P_3=25$.

Thus $|B|$ is a pencil without base points and without reducible curves.

Now let l_0 be the maximum integer $l \geqq 0$ such that $|A - lB|$ is effective. The curves $D = A - l_0 B$ and B give again a minimal base, and we have $[D, D] = -2l_0 - 1$.

Let us consider the linear system

$$|D + (2l_0 + 2)B| = |A + (l_0 + 2)B| = |C|.$$

Its virtual grade is $2l_0 + 3$, and its dimension is $\geqq 2l_0 + 4$.

The curve D is irreducible because $[B, D] = 1$, for the same reason given in the preceding section. Moreover, $[D, C] = 1$, $[B, C] = 1$, and the reasoning of the preceding section holds without change. Therefore, if we consider the projective image of F by means of $|C|$, we have a surface in one-to-one correspondence, without exceptions, with F, which lies in a $P^{(2l_0+4)}$, and has order $2l_0 + 3$.

Thus the theorem is proved because of the results of § 3. More precisely, the image of F is $\phi(1, 2l_0 + 2)$, the image of D being a rectilinear directrix and the images of the curves B the generators.

Let us now suppose that the specialty index of A *or* B is not zero.

Let C be a hyperplane section of a non-singular model of F and let us suppose that $C \equiv \beta A + \alpha B$. The order of A is $[C, A] = \alpha - \beta$, that of B is $[C, B] = \beta$. Moreover, the order of the considered model of F is $[C, C] = \beta(2\alpha - \beta) > 0$. We have therefore two possibilities:

1. $\beta > 0, \quad 2\alpha - \beta > 0.$
2. $\beta < 0, \quad 2\alpha - \beta < 0.$

Let us first suppose that the specialty index of A is not zero, that is, that the system $|K - A| = |-3A - 3B|$ is effective. The order of its curves is therefore not negative: $-3\alpha \geqq 0$, that is, $\alpha \leqq 0$. This makes the first case impossible, because $2\alpha > \beta > 0$.

In the second case we shall calculate the virtual dimension r_k of the pluricanonical systems $|-2kA - 3kB|$ $(k = 2, 3, \ldots)$. The virtual grade of the curve $-2kA - 3kB$ is $n_k = 8k^2$; the virtual genus is

$$\pi_k = 4k(k+1) + 1;$$

the specialty index is $i_k = 0$, because the curves of

$$|K - (-2kA - 3kB)| = |2(k-1)A + 3(k-1)B|$$

have the order $(k-1)\{2\alpha + \beta\} < 0$, since $\beta < 0$ and $2\alpha < \beta < 0$. Therefore $r_k = n_k - \pi_k + p_a - i_k + 1 = 4k(k-1) > 0$. Hence we have for the plurigenera the relations

$$P_k \geqq 4k(k-1) + 1 \quad (k = 2, 3, \ldots).$$

In particular, for $k=2, 3$, we have $P_2 \geqq 9$, $P_3 \geqq 25$.

Similarly, if the specialty index of B is not zero, that is, if

$$|K-B| = |-2A-4B|$$

is effective, we must have $-2(\alpha+\beta) \geqq 0$, or $\alpha+\beta \leqq 0$. This again makes the first case impossible and we return to the second case, examined above.

In both instances the bicanonical curves are effective, and because the linear genus of F is $p^{(1)}=9$, they are virtually connected and the adjoint system, that is, the 3-canonical system, is regular [9]. Therefore $P_3 = \pi_2 = 25$; this is the case which we excluded in the hypothesis.

(REMARK. [*Added in proof.*] In the theorem of § 4 the condition $P_3 \neq 28$ can be omitted. Suppose, in fact, there exists an algebraic surface F differentiably homeomorphic to $P^{(2)}$, and with the 3-genus $P_3 = 28$. By expanding a point of F into an exceptional curve of the first kind, we get a new algebraic surface $\tilde{F}$ which is differentiably homeomorphic to $\phi(1, 2)$. The 3-genus of $\tilde{F}$ remains $P_3 = 28$, but, because of the preceding theorem, the 3-genus of $\tilde{F}$ can only have the values 25 and 0. So F cannot exist.

Since any Kähler manifold differentiably homeomorphic to $P^{(2)}$ is algebraic (because of a well-known theorem of Kodaira) we see that *only the natural complex structure on* $P^{(2)}$ *can carry a Kähler metric.*)

2. SOME GEOMETRIC APPLICATIONS

7. Let us begin with the proof of some preliminary lemmas.

LEMMA 1. *Let F be an irreducible non-singular algebraic surface of $P^{(r)}$, containing a pencil*† $|C|$ *without base points, of rational irreducible curves. If a curve C is reducible, at least one of the components is an exceptional curve of the first kind.*

Let K be a canonical curve of F. We have:

$$[C, C+K] = -2, \quad \text{and since} \quad [C, C] = 0,$$

$$[C, K] = -2.$$

Now let us suppose $C = \sum_1^h \mu_r \theta_r$ with the θ irreducible. We see that

$$[C, K] = \sum \mu_r [\theta_r, K] = \sum \mu_r \{[\theta_r, \theta_r + K] - [\theta_r, \theta_r]\}.$$

So if we call ρ_r the virtual genus of the component θ_r, we obtain

$$(1) \qquad -2 = \sum \mu_r \{2\rho_r - 2 - [\theta_r, \theta_r]\}.$$

† Rational or not.

Now $\rho_r \geqq 0$; and for $[\theta_r, \theta_r]$ we have

$$[C, \theta_r] = 0 = \textstyle\sum_1^h \mu_s[\theta_r, \theta_s],$$

that is, $$-\mu_r[\theta_r, \theta_r] = \textstyle\sum_{s \neq r} \mu_s[\theta_r, \theta_s].$$

Since $C = \sum \mu_r \theta_r$ is connected, the sum in the above equation is > 0; therefore $[\theta_r, \theta_r] \leqq -1$.

From this it follows that each bracket in the above equation (1) is $\geqq 0$ unless $\rho_r = 0$ and $[\theta_r, \theta_r] = -1$. For at least one r we must therefore have $\rho_r = 0$, $[\theta_r, \theta_r] = -1$, because the sum in equation (1) is negative.

LEMMA 2. *Under the hypothesis for F in the preceding lemma, let $|C|$ be a net of rational curves, irreducible and of virtual grade $[C, C] = 1$. If a curve C is reducible, at least one of its components is exceptional of the first kind.*

Let C be a reducible curve of $|C|$ and $C = C_1 + C_2$, C_1 and C_2 being proper parts of C. We have $[C, C_1] \geqq 0$ and $[C, C_2] \geqq 0$. Since

$$[C, C_1 + C_2] = [C, C_1] + [C, C_2] = 1,$$

we must have, for instance, $[C, C_1] = 0$, $[C, C_2] = 1$. If C_2 is reducible, reasoning in the same way as above for C, and so on, we find one of the irreducible components θ_1 of $C = \sum_{1r}^h \mu_r \theta_r$ (θ_r being irreducible) such that $[C, \theta_1] = 1$ and $[C, \theta_r] = 0$ if $r > 1$. Certainly $\mu_1 = 1$. So we can write

$$\textstyle\sum_1^h \mu_s[\theta_r, \theta_s] = \begin{cases} 1 & \text{if} \quad r = 1, \\ 0 & \text{if} \quad r > 1, \end{cases}$$

and therefore

$$-[\theta_1, \theta_1] + 1 = \textstyle\sum_{s>1} \mu_s[\theta_1, \theta_s],$$

$$-\mu_r[\theta_r, \theta_r] = \textstyle\sum_{s \neq r} \mu_s[\theta_r, \theta_s] \quad (r > 1).$$

We also have

$$-3 = \{2\rho_1 - 2 - [\theta_1, \theta_1]\} + \textstyle\sum_{r>1} \mu_r\{2\rho_r - 2 - [\theta_r, \theta_r]\}.$$

But $$[\theta_1, \theta_1] \leqq 0; \quad [\theta_r, \theta_r] \leqq -1 \quad (r > 1),$$

and therefore there exists an $r > 1$ such that $\rho_r = 0$, $[\theta_r, \theta_r] = -1$, and this proves the lemma.

LEMMA 3. *A non-singular rational surface F of $P^{(r)}$ always contains an infinite system of non-singular rational irreducible curves.*

For our purpose we shall make use of some arguments developed by Castelnuovo [5].

(1) Let F be a rational surface of order n of the projective complex

space S_r without singularities and without exceptional curves of the first kind.

We can suppose

(a) That F does not already have rational hyperplane sections (this excludes the possibility that F is ruled) and that $r \geqq 4$. In fact, in ordinary space a non-singular cubic surface contains exceptional curves of the first kind, its lines (a plane through a line a of the surface cuts the surface residually to a in a conic c variable in a pencil without base points, because F is non-singular; and we have

$$[a+c, a+c] = [a,a] + 2[a,c] = [a,a]+4 = 3; \quad [a,a] = -1).$$

(b) That F is normal, that is, that the system $|C|$ of the hyperplane sections is complete. If this is not so, we can always take the projective image of F by means of the system $|C|$ made complete. This implies, F being regular, the completeness of the characteristic series of $|C|$.

(c) That F does not contain lines. If this is not so, it is sufficient to take the image of F by means of the complete system $|2C|$.

(2) First of all, let us consider the case in which the hyperplane sections of F are elliptic. Then F is a surface of order r of S_r.

From a theorem of Del Pezzo [16]†, it appears that:

if $r > 9$, F is ruled, which for us is excluded;

if $r \leqq 9$, F can be represented on the plane by the system of cubics through $9-r$ base points, or by the system of the quartics with two double base points.

Therefore if $r = 9$, the surface is the image of the system of all the cubics of a plane. The image of the lines of the plane give a system of rational curves as sought.

If $r < 9$, the images of the lines through one of the base points give a system of conics on F. (We may note that, with the exception of the case of the representative system of quartics, the base points give lines of F, which we have excluded in (c) above.)

In that which follows, we can thus exclude the case in which F has hyperplane sections of genus 1.

(3) Let $|K|$ indicate the impure canonical system of F, certainly virtual since $p_g = 0$. Let us consider the adjoint system of the system of hyperplane sections of F: $|C'| = |C+K|$.

The virtual dimension r' of $|C'|$ is given by

$$r' = [C', C'] - \tfrac{1}{2}[C', C'+K] - 1 + 1,$$

† See also [8].

since the specialty index of $|C'|$ is zero (the curves $-C$ are not effective). Then

$$r' = \tfrac{1}{2}[C', C'] - \tfrac{1}{2}[C', K]$$
$$= \tfrac{1}{2}[C', C] = p - 1 > 0,$$

where $p > 1$ is the genus of the hyperplane sections of F.

The system $|C'|$ cuts canonical groups on the generic C. Not only that, but it cuts the complete canonical series g^{p-1}_{2p-2} because $|C'|$ cannot contain $|C|$, otherwise the impure canonical system would be effective.

From this it follows that $|C'|$ has no base points (nor, therefore, fixed components). In fact, if A were a base point for $|C'|$, the irreducible curves C through A, of genus p, would have a fixed point on the canonical series.

(4) Let us examine the possibility that $|C'|$ is reducible. Since the curves C' have the order $2p-2$ and $|C'|$ has the dimension $p-1$ and no fixed components, $|C'|$ is composed of groups of $p-1$ conics of a pencil. In this case, the lemma is proved.

We note that this possibility is present if, and only if, all curves of $|C|$ are hyperelliptic. In fact, if $|C'|$ is reducible, the pencil of conics cuts on any C a g^1_2 and the curves C are hyperelliptic. Conversely, let us suppose that the curves C are hyperelliptic. Take any point A on F. The locus of the conjugate point A' of A in the g^1_2 over the curves C through A is contained in all the curves C' through A. It is therefore a curve γ. Thus $|C'|$ is composed of the curves of a pencil, which is a pencil of conics.

In what follows, we can therefore exclude the case in which F has hyperelliptic sections and in which $|C'|$ is thus reducible. In particular, we may suppose that the genus p of the hyperplane sections is $p \geqq 3$.

(5) We wish now to examine the possibility that $|C'|$ is composed of an involution I. It is impossible that the generic C of $|C|$ should contain couples of conjugate points of I, because C would be hyperelliptic.

Any C of $|C|$ which contains two conjugate points of I is hyperelliptic and therefore contains an infinity of conjugate couples of I. I is composed of an involution I' of the second order.

The system $|C|$ contains ∞^{r-1} hyperelliptic curves. Moreover, the lines joining the conjugate couples of I' are such that any two of them impose three conditions on the hyperplanes which contain them. They intersect each other and therefore they pass through the same point O. The system of hyperplane sections through O is thus made up of hyperelliptic curves.

In conclusion, the fact that $|C'|$ is composed of an involution leads to the existence of a linear system ∞^{r-1} of hyperelliptic sections. And vice versa, as can be verified.

Let us put aside this case for the moment.

(6) We wish to prove that, if we exclude the case in which $|C'|$ is composed of an involution, $|C'|$ has no fundamental curves.

Let us suppose that the curves C' through a given point A of F contain, in consequence, a curve a. Let C_A be the generic C through A. The curves C' through A cut C_A residually to A counted with multiplicity one, in a further $2p-3$ points, all variable; otherwise the curves C_A would be hyperelliptic and we should find ourselves back in the case discussed in § (5) above.

From this it follows that A is simple for the curves C' which contain it.

The curve a must contain A, otherwise it would cut C_A in a point different from A and common to all canonical groups of C_A through A, so that C_A would be hyperelliptic.

The curve a cuts the curves C_A in the point A only. Therefore a is a line, and this is excluded.

(7) Let us now prove that $|C'|$ has no neutral pairs.

If, indeed, the curves C' through A contain another point B, the curves C through A and B form a linear system $|C_{AB}|$ (∞^2 at least), without fixed parts (because the line AB is not, by hypothesis, on F). The generic curve $|C_{AB}|$ is irreducible, without multiple points, and hyperelliptic because AB is a neutral pair of its canonical series.

Let P be a generic point of F. The curves C' through P must all contain a second point, the conjugate point of P in the g_2^1 which exists on an irreducible C_{AB} of genus p through P.

It follows that $|C'|$ is composed of an involution. We are therefore again in the case discussed in § (5) above.

(8) If we thus exclude the case of § (5), $|C'|$ is a system which will give an image ϕ of F birationally equivalent, without exceptions, to F.

We want to prove that ϕ, if it contains lines, is a ruled surface.

In fact, if D is a curve of F which is mapped on a line of ϕ, D is an irreducible non-singular rational curve such that $[C',D]=1$. If d denotes the order of D, and δ its virtual grade, we have

$$1=[C,D]+[K,D]=d+[-2-\delta]$$

$$=d-\delta-2$$

or

$$\delta=d-3.$$

But $d \geqq 2$, therefore $\delta \geqq -1$. We cannot have $\delta = -1$ because F has no exceptional curves of the first kind. Hence $\delta \geqq 0$, and D belongs to a system of dimension $\delta + 1 \geqq 1$. So ϕ is ruled.

(9) Now the adjunction process extinguishes itself on F because F is rational. If therefore we take the projective images of F by means of the successive adjoints of $|C|$, we arrive at a surface F of S_r for which the lemma is immediate or which contains a linear system ∞^{r-1} of hyperelliptic sections.

There remains only this last instance to be examined.

The system of hyperelliptic sections is obtained by cutting F with the hyperplanes through a point O of S_r.

We distinguish two cases, according to whether O is on F or not.

In the second case the projection of F from O is a surface ϕ of S_{r-1}, which has rational hyperplane sections. Moreover, ϕ is normal. In fact, if f is the mapping of F onto ϕ, $fF = \phi$, and if Γ_0 is a curve of ϕ linearly equivalent to a hyperplane section Γ, then $f^{-1}\Gamma_0 = C_0$ is linearly equivalent on F to $f^{-1}\Gamma = C$; that is, C_0 is a curve of $|C|$. Such a curve is composed of the involution I_m, of order m, existing on F, of which ϕ is the image. Hence the hyperplane of C_0 passes through O because it contains the lines belonging to the groups of I_m on C_0. Γ_0 is therefore one of the hyperplane sections of ϕ.

Consequently, the surface ϕ is either

(a) a non-singular rational normal ruled surface of order $r-2$ in S_{r-1};

(b) a rational normal cone of order $r-2$ in S_{r-1}; or

(c) the Veronese surface of S_5.

(a) In this case the generators g of ϕ give on F a pencil of plane curves $f^{-1}g$. This pencil has no base points. The curves $f^{-1}g$ belong therefore to a pencil of irreducible non-singular plane curves.

Let D indicate the generic section of ϕ by a hyperplane through g, residual to g. D is irreducible and non-singular and varies in a linear system without base points. The curves $f^{-1}D$ also vary in a system without base points; they are therefore non-singular. Moreover, they are irreducible, lying on rational normal cones of order $r-3$.†

† Consider the image of the cone by means of the system cut by quadrics through the vertex; it is a non-singular rational normal ruled surface of order $3(r-3)$. Let d be the image of the vertex and g that of a generator. We have $[d, d] = -(r-3)$. If L_0 is the image of a curve of the cone which does not pass through the vertex and which is reducible, $L_0 = L_1 + L_2$, we have

$$L_i \equiv \alpha_i d + \beta_i g \quad \text{and} \quad [L_i, g] = \alpha_i > 0, \quad [L_i, d] = -\alpha_i(r-3) + \beta_i = 0 \quad (i = 0, 1, 2).$$

Then

$$[L_i, L_2] = \alpha_1\beta_2 + \alpha_2\beta_1 - (r-3)\,\alpha_1\alpha_2 = (r-3)\,\alpha_1\alpha_2 > 0.$$

Let p_1, p_2 be the genera respectively of $f^{-1}g, f^{-1}D$. If $p_1 = 0$ or $p_2 = 0$, there is nothing to prove.

Let us suppose $p_1 \geqq 1$, $p_2 \geqq 1$. We cannot have $p_1 = 1$ because $f^{-1}g$ would be a plane cubic, but the order of such a curve is m, which is an even number since I_m is composed of an involution of the second order. Therefore $p_1 > 1$. The curves C' through a canonical group of $f^{-1}D$, through the m points common to $f^{-1}g, f^{-1}D$ and through $p_1 - 1$ generic points of $f^{-1}g$, form a pencil. It cannot contain $f^{-1}g$ as a fixed component, otherwise there would exist a curve C' containing $f^{-1}g + f^{-1}D$, which is impossible (see § (3) above). The pencil therefore cuts a g^1_2 on $f^{-1}g$, and this curve must be a plane hyperelliptic non-singular curve, which is impossible.

(b) Let V be the vertex of the cone ϕ, $|g|$ the system of generators, and f the mapping of F onto ϕ. The curves $f^{-1}g$ are plane curves of a pencil without fixed components because the line OV is not on F. The pencil cannot have base points, which would consequently be multiple points for F, because a hyperplane section through OV cuts F in at least two curves of the pencil. The curve $f^{-1}g$ cuts OV in a variable group and OV would belong to F, which is excluded.

(c) Let g be the system of the conics on the Veronese surface ϕ, f the mapping of F on ϕ. The curves $f^{-1}g$ are a net of curves without base points, therefore the generic $f^{-1}g$ is non-singular and also irreducible because it lies on a quadric cone.

Let p_1 be the genus of $f^{-1}g$. If $p_1 = 0$, there is nothing to prove. If $p_1 = 1$, $f^{-1}g$ are elliptic quartics,† $m = 2$, and the involution I_2 represents F on a double plane with a branch curve of the fourth order not consisting of four lines of a pencil. (The plane is the image of the Veronese surface by means of the system $|g|$.)

A double tangent g to the branch quartic is the 'projection' of two conics C_1, C_2, of the surface F of the same virtual grade δ (because they are changed one into the other by the birational transformation which generates I_2), for which we have

$$[C_1 + C_2, C_1 + C_2] = 2\delta + 4 = 2.$$

† A curve Γ_m of a quadric cone which does not pass through the vertex and which cuts the generators in m points is the intersection of the cone with a surface of order m. In fact, the surfaces of order m cut on the cone a linear system $|C_m|$ of curves which is complete because the characteristic series is complete. Let d be the image of a generator over the model of the cone that we obtain by means of the system cut out by the quadrics through the vertex. If C is the image of a plane section of the cone, we have $C \equiv 2g + d$. If Γ_m denotes also its image on the non-singular model of the cone, we have $\Gamma_m \equiv \alpha g + \beta d$; $[\alpha g + \beta d, g] = \beta = m$; $[\alpha g + \beta d, d] = \alpha - 2\beta = 0$, therefore $\Gamma_m \equiv 2mg + md \equiv m(2g + d) \equiv mC$. Hence on the cone Γ_m is a curve of $|C_m|$.

Therefore they have the virtual grade $\delta = -1$, and they are thus exceptional of the first kind, which is excluded.

If $p_1 > 1$, certainly $m > 2$. We recognize, as in (a) above, that these curves are hyperelliptic. On the other hand, they are on a quadric cone and do not pass through the vertex. Therefore each one is a complete intersection of the cone with a surface of order m. The surfaces of order $m-2$ cut on the curve the canonical series which cannot be compound, and this is a contradiction.

(10) Let us suppose that O in on F. The projection of F from O is a surface ϕ, normal and with rational sections, therefore ϕ belongs to types (a), (b) and (c) of § (9) above.

We exclude the case in which ϕ is a cone, as in § (9). Moreover, the point O gives rise by projection to a line d on ϕ. That makes it impossible for ϕ to be the Veronese surface. It is therefore a rational normal ruled surface of order $r-2$ of S_{r-1}.

Let $|g|$ be the system of generators of ϕ; let $|D|$ be the residual system of $|g|$ with respect to the system of hyperplane sections; and let f be the mapping of F onto ϕ.

If d is a directrix line on ϕ, we have $[d, d] = -(r-4)$, and because $D \equiv d + (r-4)g$, we have $[D, d] = 0$. The system of the curves $f^{-1}D$ cannot therefore have base points (the only possible one would be O, which would lead to $[D, d] > 0$). The generic curve $f^{-1}D$ lies on a rational normal cone, and is therefore irreducible and non-singular.

The system of curves $f^{-1}g$ is a system of plane curves with the ordinary simple base point O and therefore its curves are generically irreducible and non-singular.

Let p_1 be the genus of the generic $f^{-1}g$ and p_2 that of the generic $f^{-1}D$. If $p_1 = 0$ or $p_2 = 0$, there is nothing to prove. Let us suppose $p_1 \geqq 1, p_2 \geqq 1$.

If $p_1 > 1$, as in (a) of § (9) above, we have a contradiction.

We have only to examine the case in which $p_1 = 1$, $p_2 \geqq 1$. This implies that the curves $f^{-1}g$ are plane elliptic cubics with the base point O. Therefore $m = 2$, and F is of order $m(r-2)+1 = 2(r-2)+1$. Moreover, there is a hyperplane section of ϕ consisting of d and $r-3$ generators g; a hyperplane section of F can therefore be broken into $r-3$ curves $f^{-1}g$, so that we must have $r = 6$.† Thus F is a surface of

† If x is the order of the curve cut on F by the plane Od ($x = 0$ if the curve is lacking), we have $x + 3(r-3) = 2(r-2)+1$, so that $r = 6 - x$ and the order of F is 9, 7, 5 according as $x = 0$, 1, 2 ($r \geqq 4$). The second case has to be excluded; the third implies that the S_3 through an $f^{-1}g$ cut F in a pencil of conics; there remains only the first case, discussed in the text above.

order 9 of S_6 with hyperplane sections of genus 4; the curves $f^{-1}D$ are of genus $p_2 = 2$.

F is doubly mapped onto ϕ with a branch curve Δ which cuts the generators in four points and the curves D in six points. Therefore $\Delta \equiv 4d + 6g$. Because $[\Delta, d] = -2$, Δ contains the line d and a residual part $\Delta_1 \equiv 3d + 6g$ (this means that the cubics $f^{-1}g$ all have inflection at O). The curve Δ_1 is a curve of the 9th order $\equiv 3D$. It does not meet d because $[\Delta_1, d] = 0$, and d cannot be doubly subtracted from Δ, because Δ meets the generic generator in four distinct points. The curve Δ_1 is irreducible and non-singular. In fact, the presence on Δ_1 of a multiple point A would imply that $f^{-1}A$ is multiple for the curve $f^{-1}g$ which contains it, and also for the curves $f^{-1}D$ which contain it; these two implications are contradictory.

The curve Δ_1 is therefore a curve of genus 4 (which on the quadric cone, image of $|D|$, is a canonical sextic not passing through the vertex). Hence there exist irreducible curves D which touch Δ_1 at three points (they correspond to the triple tangent planes to the sextic which are not tangent planes of the cone). Let D_1 be one such curve; the curve $f^{-1}D_1$ is composed of two twisted cubics C_1, C_2, one of which, if both are not exceptional of the first kind, must vary in an infinite linear system.

If d is a generator, $|f^{-1}g|$ has no base points and we can repeat with slight changes the reasoning of (a) in § (9) above.

8. *A theorem of M. Noether.* As an illustration of the preceding premises, and more precisely as a consequence of the theorems of §§ 5 and 6, and Lemma 1 of the preceding section, we will prove a well-known theorem of M. Noether:

Each irreducible algebraic surface with a linear pencil of rational curves can be birationally transformed into a rational ruled surface so that the generators are the images of the curves of the pencil.

We can refer to a non-singular model of F. If we expand the possible base points of the pencil by introducing a suitable group of exceptional curves of the first kind [1], we may assume that on F the pencil has no base points. Finally, if we eliminate the exceptional curves of the first kind which are contained in some curves of the pencil, we obtain a model of F on which the pencil is still without base points and whose curves are all irreducible because of Lemma 1. It is sufficient to prove the theorem for this model.

We note that the 3-genus of F is $P_3 = 0$. In fact, if $P_3 \geqq 1$, because of the adjunction formula, it would follow that the genus of the curves of the pencil is $\geqq 1$. Moreover, F is a fibre bundle [26] with base space a 2-sphere and whose fibres are also 2-spheres. The group is the projective group on the complex projective line.

In fact, mark a point on the 2-sphere which is the base space. To this point corresponds as a fibre a curve of the pencil. We can find three analytic branches with origins in three different points of the fibre, which cut in one point each of the fibres in a convenient neighborhood of the considered one. This bunch of fibres is therefore homeomorphic to the product of a 2-cell and the complex sphere, in such a way that the correspondence induced on the fibre is a projective one. From this the assertion follows.

Now, by virtue of a theorem of Steenrod [25, 26], the only possible types of sphere bundles are those we have considered in §§5 and 6, and because of the theorems we have proved there, F is birationally equivalent, without exceptions, to a rational normal ruled surface (non-singular). From this the assertion follows, because the only pencil without base points (or the only two in the case of a quadric) is the pencil of the generators.

Remark. The theorem we have proved is both somewhat more and less wide than the original theorem of Noether. It is less wide in relation to the ground field in which we obtain the required transformation. We have supposed the ground field to be the complex field, but the theorem holds in every algebraically closed field of characteristic 0 and also in more general fields. On the other hand, our theorem is more comprehensive because we have proved that every abstract rational ruled surface [24], without exceptional curves of the first kind contained in the pencil, is always birationally equivalent, without exceptions, to a projective rational normal ruled surface with a directrix line. This is a particular case of a theorem due to Maroni [15].

9. *The classification of rational surfaces with respect to birational transformations without exceptions.* Let us consider the totality of non-singular rational surfaces without exceptional curves of the first kind. With respect to the group of all birational transformations, they stand in a single class; on the contrary, with respect to birational transformations without exceptions they stand in infinitely many classes. This is a consequence of the following theorem (first stated by G. Vaccaro [27]):

Every non-singular rational surface F without exceptional curves of the first kind is birationally equivalent, without exceptions, either to the projective plane or to a rational normal ruled surface of even order with a directrix line.

Let us consider on F a complete and infinite linear system $|C|$ of irreducible rational non-singular curves. Among all possible choices we fix a system of minimum order. All this is possible because of Lemma 3 in § 7.

The dimension of $|C|$ is certainly <3. Otherwise let us consider the image ϕ of the complete system $|C|$. It is a surface with rational sections and therefore ruled, rational and normal (possibly a cone), or the Veronese surface. The system $|C|$ has no base points and is not composed of an involution; hence the correspondence between F and ϕ in the sense $F \rightarrow \phi$ is strictly one-valued. The system of generators of ϕ, or that of the conics of ϕ when ϕ is the Veronese surface, is mapped onto F by a system of non-singular rational curves whose order is less than the order of the curves C because it is partially contained in $|C|$.

So $|C|$ is either a net or a pencil.

In the first case, because of the completeness of the characteristic series, we must have $[C, C] = 1$. Moreover, since F does not contain exceptional curves of the first kind, no curve C of $|C|$ can be reducible (Lemma 2, § 7). Consequently $|C|$ gives an unexceptional birational transformation of F on the projective plane.

In the second case, we must have $[C, C] = 0$, that is, the pencil has no base points. Moreover, no curve of the pencil is reducible, because of Lemma 1, § 7. F is therefore an abstract ruled surface without reducible generators, and as such (§ 8, Remark) it is birationally equivalent, without exceptions, to a rational normal ruled surface with a directrix line.

The order of this ruled surface cannot be odd, because with the directrix line and a convenient number of generators, we can construct a curve of virtual genus 0 and virtual grade -1, that is, an exceptional curve of the first kind. So the theorem is proved.

Remark. The models we have considered in the statement of the theorem are distinct with respect to birational transformations without exceptions, as follows from § 3.

Turin, Italy

References

[1] S. F. Barber and O. Zariski, *Reducible exceptional curves of the first kind*, Amer. J. Math., 57 (1935), pp. 119–141.

[2] A. Bassi, *Sulla riemanniana dell' S_n proietivo*, Rend. Circ. Mat. Palermo, 56 (1932), pp. 228–237.

[3] E. Bertini, Introduzione alla geometria degli iperspazi, Principato, Messina, 1923: (a) p. 398; (b) p. 362.

[4] G. Castelnuovo, *Ricerche generali sopra i sistemi lineari di curve piane*, Memorie Scelte, Zanichelli, Bologna, 1937, pp. 137–186.

[5] G. Castelnuovo, *Sulle superficie di genere zero*, Memerie della Società dei XL, s. 3, 10 (1894), pp. 103–123; or in Memorie Scelte, Zanichelli, Bologna, 1937, pp. 307–332.

[6] S. S. Chern, *Characteristic classes of Hermitian manifolds*, Ann. of Math., 47 (1946), pp. 85–121.

[7] W. L. Chow, *On compact complex analytic varieties*, Amer. J. Math., 71 (1949), pp. 893–914.

[8] F. Conforto, *Le superficie razionali, nelle lezioni del Prof. F. Enriques*, Zanichelli, Bologna, 1945, p. 313.

[9] A. Franchetta, *Sui sistemi pluricanonici di une superficie algebrica*, Rend. Mat., s. 5, 8 (1949), pp. 423–440.

[10] ——, *Sulle curve riducibili appartenenti ad una superficie algebrica*, Rend. Mat., s. 5, 8 (1949), pp. 378–398.

[11] F. Hirzebruch, *Über eine Klasse von einfach-zusammenhangenden komplexen Mannigfaltigkeiten*, Math. Ann., 124 (1951), pp. 77–86.

[12] W. V. D. Hodge, *The characteristic classes on algebraic varieties*, Proc. London Math. Soc., s. 3, 1 (1951), pp. 138–151.

[13] ——, The theory and applications of harmonic integrals, Cambridge University Press, 1952, second edition, p. 224.

[14] S. Lefschetz, L'analysis situs et la géométrie algébrique, Gauthier-Villars, Paris, 1924, p. 82.

[15] A. Maroni, *Sulle rigate astratte*, Rend. Mat., s. 5, 7 (1948), pp. 236–242.

[16] P. del Pezzo, *Sulle superficie dell' n^{mo} ordine immerse nello spazio a n dimensioni*, Rend. Circ. Mat. Palermo, 1 (1887), pp. 241–255.

[17] G. de Rham, *Sur l'analysis situs des variétés a n dimensions*, J. Math. Pures Appl., s. 9, 10 (1931), pp. 115–200.

[18] M. Rueff, *Beiträge zur Untersuchung der Abbildungen von Mannigfaltigkeiten*, Compositio Math., 6 (1938), pp. 161–202.

[19] B. Segre, *Nuovi metodi e risultati...*, Annali di Mat., s. 4, 35 (1953), pp. 1–128.

[20] C. Segre, *Sulle varietà che rappresentano le coppie di punti di due piani o spasi*, Rend. Circ. Mat. Palermo, 5 (1891), pp. 192–204.

[21] F. Severi, *Alcune proprietà fondamentali dell'insieme dei punti singolari di una funzione analitica di più variabili*, Memorie dell'Accademia d'Italia, 3 (1932), pp. 5–20.

[22] ——, *Caratterizzazione topologica delle superficie razionali e delle rigate*, Viertel Jahrsschrift Nat. Ges., Zurich, 85 (1940), pp. 51–60.

[23] ——, *Serie, sistemi di equivalenza..., a cura di Conforto e Martinelli*, Cremonese, Roma, 1942: (a) p. 245; (b) p. 395.

[24] ——, *Sulla classificazione delle rigate algebriche*, Rend. Circ. Mat. Palermo, s. 5, 2 (1941), pp. 1–32.

[25] N. E. Steenrod, *Classification of sphere bundles*, Ann. of Math., 45 (1944), pp. 294–311.

[26] ——, The topology of fibre bundles, Princeton University Press, 1951, pp. 135–137.

[27] G. Vaccaro, *Le Superficie Razionali...*, Rend. Lincei, s. 8, 4 (1948), pp. 549–551.

[28] B. L. van der Waerden, *Topologische Bergründung des Kalküls der Abzählender Geometrie*, Math. Ann., 102 (1930), pp. 337–362.

[29] E. Vesentini, *Classi caratteristiche e varietà covarianti d'immersione*, Rend. Lincei, s. 3, 16 (1954), pp. 199–204.

On Kähler Manifolds with Vanishing Canonical Class

Eugenio Calabi

THE purpose of this note is to show certain properties of compact Kähler manifolds whose canonical class vanishes; an application of the results obtained leads to a partial classification of all such manifolds, reducing it to the problem of classifying a more restricted class of compact Kähler manifolds, namely, those whose canonical class vanishes and whose first Betti number is zero. A complete classification of manifolds of the latter type is claimed by Severi [9] in the case of algebraic surfaces.

We shall first recall that the canonical class of an algebraic manifold M_n of dimension n is the homology class of the divisor of any non-zero meromorphic n-uple differential. This definition is not suitable for generalization to compact, complex manifolds, since it is not known whether one can always find a non-trivial meromorphic n-uple differential in all manifolds of this wider class. An alternate definition which allows such a generalization is due to Chern [3]. Let M_n be a compact, n-dimensional complex manifold, in which let g be any real, differentiable, positive-valued density of weight 1; g is expressible in any coordinate domain $\{U;(z)\}$ $((z)=(z^1,z^2,\ldots,z^n))$ by means of a positive, differentiable function $g_U(z,\bar{z})$, so that, if $\{V;(z')\}$ is another coordinate domain, then at every point of $U\cap V$

$$g_V(z',\bar{z}')=g_U(z,\bar{z})\left|\frac{\partial(z)}{\partial(z')}\right|^2, \tag{1}$$

where $\partial(z)/\partial(z')$ is the Jacobian of the coordinated transformation. It follows from (1) that

$$\log g_V=\log g_U+2\mathrm{Re}\log((\partial(z)/\partial(z')),$$

and hence the tensor components defined in each coordinate domain $\{U;(z)\}$ by

$$R_{\alpha\beta^*}(z,\bar{z})=\frac{\partial^2(\log g_U(z,\bar{z}))}{\partial z^\alpha\partial\bar{z}^\beta}\qquad(\alpha,\beta=1,2,\ldots,n) \tag{2}$$

define the same hermitian tensor over all of M_n; consequently the exterior form†

(3) $$\Sigma = -\tfrac{1}{2}dCd(\log g) = \sqrt{(-1)}\, R_{\alpha\beta^*} dz^\alpha \wedge d\bar{z}^\beta$$

is a closed, real-valued exterior differential form of type $(1, 1)$‡ defined over M_n. If a different density g' is chosen, then g'/g is a positive, differentiable scalar; hence

(4) $$\Sigma' - \Sigma = -\tfrac{1}{2}dCd\left(\log \frac{g'}{g}\right)$$

is a closed form cohomologous to zero. Thus the cohomology class of Σ depends only on the complex analytic structure of M_n, and not on the particular density chosen. It is shown by Chern [3] that whenever a non-trivial meromorphic n-uple differential exists in M_n, then the dual of its divisor cycle is cohomologous to $(1/2\pi)\Sigma$. Thus for any compact, complex manifold one can define the canonical class (or first Chern class) to be the cohomology class of $(1/2\pi)\Sigma$.

From this point on we shall assume that the complex manifold M_n admits an infinitely differentiable, positive-definite Kähler metric (for a definition, see [2])

(5) $$ds^2 = 2g_{\alpha\beta^*}(z, \bar{z})\, dz^\alpha d\bar{z}^\beta;$$

then one can take the density g to be the determinant of the matrix $(g_{\alpha\beta^*})$, i.e. the volume distribution with respect to a local coordinate system. One of the properties of a Kähler metric which we shall use here is that the tensor $R_{\alpha\beta^*}$ defined by (2) from the volume element g is equal to the Ricci tensor of the metric (cf. [1], p. 789); we shall therefore call the form Σ defined by (3) the Ricci form relative to the metric. The property which is characteristic of and was originally used to define a Kähler metric is that the exterior form ω associated with (5) and defined by

(5a) $$\omega = \sqrt{(-1)}\, g_{\alpha\beta^*} dz^\alpha \wedge d\bar{z}^\beta$$

is, like the Ricci form, closed, differentiable, of type $(1, 1)$, and real-valued (see [6]); we shall call ω the principal form of the metric. Since $\omega^n/n!$ is the volume element of M_n,§ its integral over the cycle carried by M_n is positive, so that, as a closed differential form of type (n, n), ω^n cannot be cohomologous to zero; therefore ω belongs to a non-trivial, real cohomology class (see [4]), which we shall call the principal class.

† The notation here is the same as in Kodaira ([8], pp. 95 et seq.).

‡ An exterior form of type (p, q) $(0 \leq p,\ q \leq n)$ is an exterior form of degree $p+q$, which is p-fold linear in $dz^1, \ldots, dz^n$ and q-fold linear in $d\bar{z}^1, \ldots, d\bar{z}^n$.

§ An exponent written over an exterior form denotes its power with respect to exterior multiplication.

The topology of compact, Kähler manifolds from the standpoint of homology theory has been studied extensively by Hodge [4], who generalized to this class of manifolds the homology properties with respect to real coefficients that Lefschetz [7] had discovered for algebraic varieties. In particular we shall use the following facts. All holomorphic and all antiholomorphic differentials in M_n are harmonic; those of degree 1 (Picard differentials of the first kind and their conjugates) for a complete set of representatives of cohomology classes with complex coefficients for dimension 1; thus the first Betti number B_1 of M_n is twice the irregularity g_1, or number of linearly independent Picard differentials of the first kind in M_n. If $\alpha_1, \ldots, \alpha_{g_1}$ form a basis for these differentials, we call the Albanese variety A_M of M_n the complex torus of (real) dimension $2g_1$ whose periods are those of the integrals $\oint \alpha_i$ $(i = 1, \ldots, g_1)$ over the 1-cycles of M_n; we have also a complex analytic mapping of M_n into A_M, which we shall call the Jacobi map, defined up to a translation in A_M, described as follows: if $p_0 \in M_n$ and $a_0 \in A_M$ are fixed points and if p ranges over M_n,

$$J(p) = a_0 + \left(\int_{p_0}^{p} \alpha_i \right)_{i=1}^{g_1} \quad \text{(modulo periods of } \oint \alpha_i\text{)}.$$

The map J induces the natural isomorphism of the translation invariant simple differentials of A_M with the Picard differentials of M_n. As in algebraic geometry we shall call M_n *regular*, if $B_1 = 2g_1 = 0$.

We shall now postulate a general result about compact, Kähler manifolds. While the proof of this result is not complete, we feel justified in assuming the truth of the statement for several concurrent, intuitive reasons. A sketch of a heuristic argument for the proposition is given in the Appendix.

PROPOSITION 1. *Let M_n be a compact, complex manifold admitting an infinitely differentiable Kähler metric with principal form ω and Ricci form Σ. If Σ' is any closed, real-valued, infinitely differentiable form of type* (1, 1) *and cohomologous to Σ, then there exists a unique Kähler metric with principal form ω' cohomologous to ω and Ricci form equal to Σ'. This metric is always infinitely differentiable; it is real analytic, if Σ' is analytic.*

We shall now assume that the canonical class of M_n is zero, or, equivalently, that the Ricci form Σ is cohomologous to zero. We shall call M_n with this restriction, for short, a *special Kähler manifold*, because (see Chern [3]) this property is equivalent to the reducibility of the structural group of the tangent bundle from the unitary to the special (or unimodular) unitary group.

Examples of special Kähler manifolds of dimension n are given: (a) by the n-dimensional complex tori (group manifolds of compact, Abelian, complex Lie groups); (b) by the non-singular hypersurfaces of order $n+2$ in a complex projective $(n+1)$-space; the latter examples are, for $n \geqq 2$, also regular; for $n=1$ of course the only possible examples are algebraic curves of genus 1.

The following theorem describes the first characteristic property of special Kähler manifolds:

THEOREM 1. *If M_n is a special Kähler manifold of dimension n and with irregularity g_1, then $0 \leqq g_1 \leqq n$ (equality holding only for a complex torus); more precisely the Jacobi map J of M_n into its Albanese variety A_M:*

(i) *is a map onto;*

(ii) *it is the projection map of a complex analytic fibre bundle with base space A_M and bundle space M_n;*

(iii) *the structural group of the bundle is finite and Abelian;*

(iv) *the fibre is a connected, special Kähler manifold.*

PROOF. Since Σ is cohomologous to zero, we can set $\Sigma' = 0$; then, by applying Proposition 1, we can replace the given Kähler metric by another one with the same principal class, for which the Ricci tensor vanishes identically; this new metric is unique and real analytic. All of the statements that will follow will be referred to this metric.

An immediate application of the theorems of Bochner ([1], Theorems 1, 2, 5, 6) to the case of vanishing Ricci tensor leads to the following statement: a vector function in M_n cannot be harmonic, nor satisfy Killing's equation, nor have holomorphic or antiholomorphic contravariant components, unless it is a parallel vector field (of course if it is parallel, then it trivially satisfies all three other conditions simultaneously). In other words, a real vector function in M_n is harmonic (i.e. the real part of a Picard differential of the first kind), if and only if, its contravariant components are the sum of a holomorphic and an antiholomorphic vector, or if, and only if, it satisfies Killing's equation (i.e. the vector generates a one-parameter group of complex analytic, isometric transformations of M_n). It follows from this that the mapping J induces a one-to-one linear correspondence between the parallel vector fields of M_n and those of A_M; hence the group G' of translations of A_M is locally isomorphic under a map induced by J to the component G of the identity in the group of analytic (and isometric) transformations of M_n. This shows that J maps M_n onto A_M; in the special case where $g_1 = n$, the map is locally one-to-one and without singularities, because of the properties of the group of translations of A_M; therefore M_n is a compact covering of A_M under J, whence it is itself a complex torus, i.e. its own Albanese variety, so that J becomes the identity map.

The inverse image $J^{-1}(a)$ of each point $a \in A_M$ is a complex analytic subvariety of M_n. If U is a geodesically convex, sufficiently small neighborhood of a, the set $\tilde{U}$ of translations of A_M mapping a into U is a neighborhood of the identity in the translation group G', constituting a local group which under the local isomorphism induced by J carries each point of $J^{-1}(a)$ into an analytic cross-section of $J^{-1}(U)$; the image of each element of $\tilde{U}$ under the local isomorphism maps $J^{-1}(a)$ analytically and isometrically onto $J^{-1}(a')$ for some $a' \in U$. One verifies easily that each cross-section over U thus defined is analytically isometric with U and intersects orthogonally the subvariety $J^{-1}(a')$ for each $a' \in U$. Thus $J^{-1}(U)$ is analytically isometric with $U \times J^{-1}(a)$. Since each local cross-section as constructed above is uniquely determined by one of its points but for its extent, we have established the complex fibre bundle structure of M_n over A_M. Furthermore, each component of a fibre is a complex submanifold without singularities, because of the local metric product structure of the bundle space.

It is clear that J induces a homomorphism of G onto G', and that the kernel consists of the analytic isometries of M_n which are analytically deformable to the identity, and map each $J^{-1}(a)$ for each $a \in A_M$ into itself. But it is well known that the group of all isometries of a compact Riemannian space is a compact Lie group; thus G is a compact, connected Lie group, locally isomorphic to the compact, Abelian group G', so that G is itself Abelian. The kernel K of the homomorphism of G onto G' is thus a discrete, and hence finite, Abelian subgroup of G. It is easy to see that K is the structural group of the fibre bundle of M_n over A_M, so that the third contention of the theorem is proved.

We have already verified that each component of a fibre is a complex submanifold of M_n without singularities; from the local metric product structure we see that it is totally geodesic, so that its Ricci curvature tensor is the tensor induced from the Ricci tensor of the ambient space. It follows that the Ricci curvature of each component of a fibre vanishes identically, so that it is a special Kähler manifold. In order to prove that each fibre is connected, we identify all points of M_n lying on the same component of any fibre; we obtain thus a complex manifold A', over which M_n is fibered by connected submanifolds, and which is in turn fibered over A_M by fibres consisting of a finite number of points, so that the composition of the two fibre projections is J. The manifold A' thus obtained is a finite covering of A_M; therefore it is a complex torus, in which the group of periods of Picard integrals of the first kind is a subgroup of finite index of the corresponding one for A_M. This would imply, if the covering of A_M by A' were not trivial, that the group of periods of the integrals in M_n would be a proper subgroup of the group of periods of A_M, which would contradict the definition of the Albanese variety. Theorem 1 is thereby proved.

The next theorem is a new interpretation of the first, but it is perhaps more helpful in describing the structure of special Kähler manifolds.

THEOREM 2. *If M is a special Kähler manifold, then there exists a finite covering manifold $\tilde{M}$ of M, which is a direct product of its own Albanese variety $A_{\tilde{M}}$ with a special Kähler manifold Y, which is regular, and possibly a point. The relative covering group of $\tilde{M}$ over M is solvable.*

PROOF. We construct the principal bundle (cf. Steenrod, [10], pp. 35–9) associated with the fibre bundle $\{M, A_M, J\}$; since the structural group of the bundle is finite, the principal bundle is given by a finite, necessarily Abelian, covering A_1 of A_M. Denote the fibre of M by Y_0 and set $M_0 = M$ and $M_1 = Y_0 \times A_1$. Then the principal map P: $M_1 \to M_0$ defines a finite, Abelian covering of M_0 by M_1. The complex torus A_1 appearing as a direct factor of M_1 is not necessarily the Albanese variety of M_1, since M_1 may have some Picard differentials of the first kind other than those induced from M under the covering map P; it is obvious that the additional differentials appearing in M_1 are precisely those induced from its other direct factor, namely, Y_0. Thus the Albanese variety of M_1 is the direct product of A_1 with the Albanese variety of Y_0.

Since it is clear that M_1 is again a special Kähler manifold, the process may be repeated with M_0 replaced by its covering manifold M_1, thus obtaining a special Kähler manifold M_2 as a finite, Abelian covering of M_1, and continue inductively. At the k'th step one obtains a special Kähler manifold M_k, which is a finite, Abelian covering of M_{k-1}, such that the Albanese variety of M_k is the direct product of the corresponding finite covering of the Albanese variety of M_{k-1} with the Albanese variety of the fiber Y_{k-1} of M_{k-1} under J. Since

$$g_1(M_k) - g_1(M_{k-1}) = \dim(Y_{k-1}) - \dim(Y_k) = g_1(Y_{k-1}),$$

either the process stops when the fibre Y_k is reduced to a point, in which case the covering manifold M_k is a complex torus, or one reaches a stage when the irregularity of Y_k is zero, in which case the next inductive step will yield as covering manifold $M_{k+1} = Y_k \times A_{k+1}$ with the properties we set out to prove. Repeated application of the process will then yield only the identity covering. Thus we have obtained a finite covering manifold of M, which is the direct product of a complex torus (its own Albanese variety) with a special Kähler manifold that is regular. The relative covering group is solvable, because it has a normal series consisting of finite, Abelian groups. This completes the proof of Theorem 2.

By means of the result just established one could completely classify all special Kähler manifolds in terms of those that are regular. We shall describe briefly how this can be done.

Let A be a complex torus and G a finite soluble group of complex analytic transformations of A such that no element of G except the identity leaves any point of A fixed (for a discussion of these crystallographic groups, cf. H. Weyl [11] and H. Hopf [5]). Let Y be a regular, special Kähler manifold (this may even be one point) and (G_0, ϕ) a (possibly trivial) representation of G as a group of analytic transformations of Y. We then define a group G' of transformations of $Y \times A$, naturally isomorphic to G as follows: if $x \in G$ and x' is the corresponding element of G', and if $y \in Y$ and $a \in A$, then

$$x'(y, a) = (\phi(x)(y), x(a)).$$

It is clear that no element of G' other than the identity leaves any point of $M_0 \times A$ fixed; the identification of points in $y \times A$ equivalent under G' produces a Kähler manifold, which is obviously special. As a result of Theorem 2 we know that every special Kähler manifold can be obtained in this way, so that the whole class of special Kähler manifolds can be contructed from a classification of the regular ones plus the family of all compact Kähler manifolds which are metrically flat.

It might be interesting to know immediately, given a special Kähler manifold M_n with its Ricci flat metric, what the dimension is of the complex torus $A_{\tilde{M}}$ and the complementary dimension of the regular manifold Y such that $Y \times A_{\tilde{M}}$ is a finite covering of M_n. It is clear that an upper bound for the dimension of $A_{\tilde{M}}$ is given by maximum number m such that at each point of M_n there exist m linearly independent (over complex coefficients) parallel vector fields; however, this upper bound is not the best possible, at least for $n \geqq 3$, since one can construct a compact, regular, 3-dimensional Kähler manifold, with Riemann curvature identically zero, admitting a complex torus as a finite covering. Another unanswered question in this connection is whether a special Kähler manifold with no local parallel vector fields with respect to its Ricci flat metric is not only regular, but also simply connected.

One interesting application of the above results is the following one, suggested to the author by G. Washnitzer. Let M_n be a special Kähler manifold with an arbitrary (i.e not necessarily Ricci flat) Kähler metric. We consider on M_n the density g' defined by

$$g' = g \exp(2\Lambda G\Sigma), \tag{6}$$

where g is the volume density, Λ is algebraic operator on exterior p-forms ϕ defined by $\Lambda\phi = (-1)^p * (\omega \wedge * \phi)$ (i.e. the negative of the one so denoted by Kodaira [8]), G is the Green's operator as in de Rham [8], and Σ is the Ricci form (3). From the well-known identities in the exterior calculus in Kähler manifolds, and from the assumption that the closed form Σ is cohomologous to 0, we obtain the identity for the functional expression $g'_U(z, \bar{z})$ of g' in a complex coordinate system $\{U; (z)\}$

$$dCd(\log g'_U(z, \bar{z})) = 0.$$

Consequently the density g' is locally equal to the squared absolute value of a complex analytic, nowhere vanishing, n-uple differential Θ, uniquely determined up to constant factors of modulus unity, i.e. a generalization of a Prym differential. The analytical continuations of Θ over closed paths depend on the homology class of the latter only, and hence furnish a multiplicative representation of the first homology group of M_n. We remark that, if M_n is regular, then Θ is single-valued; the same is true if M_n is a complex torus, for then Θ becomes, with a constant factor, the exterior product of a base for the Picard differentials of the first kind.

A direct application of Theorem 2, in connection with the considerations just made, produces the following result.

Corollary to Theorem 2. *Let M_n be a special Kähler manifold and let the smallest finite covering manifold $\tilde{M}_n$ of M_n that is a product of a complex torus with a regular, special Kähler manifold be k-sheeted. Then there exists a divisor d of k such that the m-genera of M_n $(m = 1, 2, \ldots)$ are* 1 *or* 0 *according as to whether d divides m or not.*

Proof. We recall that the m-genus of a complex manifold is the maximum number of linearly independent single-valued, everywhere holomorphic densities of weight m; for $m = 1$ it is also called the geometric genus. If M_n is itself the product of a regular, special Kähler manifold with a complex torus, then $k = d = 1$; in this case we have already verified that there is a single-valued, holomorphic, n-uple differential Θ with no zeros. Each mth tensor power of Θ $(m=1, 2, \ldots)$ is a holomorphic density of weight m; it is unique up to constant factors, since its divisor vanishes; thus the m-genera of M_n are all 1.

In the general case there exists a k-sheeted covering $\tilde{M}_n$ of M_n, which is a product of a regular, special Kähler manifold with a complex torus. The real density g' defined by (6) on M_n induces the corresponding one of $\tilde{M}_n$. Hence the generalized Prym differential Θ on M_n has multiplicative periods corresponding to a representation of the relative covering group of M_n by $\tilde{M}_n$, which is finite and of order k; thus the multiplicative periods of Θ are the totality of dth roots of unity for some divisor d of k. For each positive integer m the density Θ^m is an m-uple, holomorphic Prym density on M_n, obviously unique up to a constant factor, since its divisor vanishes, and single-valued, if and only if m is a multiple of d. This concludes the proof.

Appendix. We shall give here a heuristic argument to make the statement of Proposition 1 seem plausible. The proof of the uniqueness of the Kähler metric with prescribed Ricci form (subject to the necessary restrictions stated) will be rigorous; the existence part will rely on the assumption of the existence of a solution of an integro-differential equation.

We remark first that, if M_n has two distinct metrics whose principal forms ω and ω' belong to the same (principal) class, then there exists a real-valued scalar Φ defined over all of M_n and unique up to an additive real constant, such that

$$\omega' - \omega = dCd\Phi; \tag{7}$$

in fact, it is enough to set $\Phi = \Lambda G(\omega' - \omega)$, where Λ and G are taken relative to either (or any other) Kähler metric, and verify that Φ is a solution. If Φ' is any other solution, then $dCd(\Phi - \Phi') = 0$, meaning that the difference between the two solutions of (7) is the sum of an everywhere holomorphic with an everywhere antiholomorphic scalar, i.e. a constant. Since the conjugate of a solution is clearly a solution, it follows that the imaginary part of any solution of (7) is a constant, so that its real part will do.

We now prove that, if the principal forms ω and ω' are distinct, but cohomologous, then their corresponding Ricci forms Σ and Σ' are distinct. In fact, suppose that $\Sigma = \Sigma'$; from equation (4) it follows that g'/g is the real part of an everywhere holomorphic scalar; therefore it is a constant, namely, the ratio of the volumes of M_n as calculated by each of the two metrics. But the volume elements are given respectively by the forms $\omega^n/n!$ and $\omega'^n/n!$, which are clearly cohomologous, so that their integrals over M_n are equal. Thus we have $\omega'^n = \omega^n$.

Using the solution Φ of equation (6), we obtain

$$
\begin{aligned}
(8) \qquad 0 = *(\omega'^n - \omega^n) &= *[dCd\Phi \wedge (\textstyle\sum_{k=0}^{n-1} \omega^k \wedge \omega'^{n-k-1})] \\
&= \textstyle\sum_{k=0}^{n-1} *(\omega^k \wedge \omega'^{n-k-1} \wedge dCd\Phi).
\end{aligned}
$$

Each term in the last summation on the above equation can be written in the form

$$H_{(k)}^{\alpha\beta^*} \frac{\partial^2 \Phi}{\partial z^\alpha \partial \bar{z}^\beta},$$

where the coefficient matrices $H_{(k)}^{\alpha\beta^*}$ are, for each value of k $(0 \leqq k \leqq n-1)$, Hermitian contravariant tensors; using the local orthogonal developments of the two metrics, one can easily show that each tensor $H_{(k)}^{\alpha\beta^*}$ is positive-definite. Thus the tensor $H^{\alpha\beta^*} = \sum_{k=0}^{n-1} H_{(k)}^{\alpha\beta^*}$ is positive-definite, and the scalar Φ satisfies the equation

$$(9) \qquad H^{\alpha\beta^*} \partial^2\Phi / \partial z^\alpha \partial \bar{z}^\beta = 0;$$

these two facts imply that, since M_n is compact, Φ is a constant, because no solution of (9) can achieve a maximum value in the interior of its domain of existence, unless it is a constant (cf. Bochner and Yano [2], pp. 26–31). From this it follows trivially that $\omega' - \omega = 0$.

We shall now derive the integro-differential equation whose solution yields the Kähler metric with the prescribed Ricci form Σ', subject to the conditions stated in Proposition 1. Assume that for each such form Σ' there existed a corresponding Kähler metric with principal form ω'; suppose further that for each continuous system $\Sigma(t)$ of Ricci forms depending differentiably on a real parameter t $(0 \leqq t \leqq 1)$ the corresponding principal forms $\omega(t)$ depended also differentiably on t. Then, by using equations (3) and (4) we could compute the relationship between the derivatives of $\Sigma(t)$ and $\omega(t)$ with respect to t. In fact

$$
\begin{aligned}
\frac{\partial \Sigma(t)}{\partial t} &= \lim_{h\to 0} -\frac{1}{2h} dCd\left(\log \frac{g(t+h)}{g(t)}\right) = -\tfrac{1}{2} dCd\Lambda_t\left(\frac{\partial \omega(t)}{\partial t}\right) \\
&= -\tfrac{1}{2}\Delta_t\left(\frac{\partial \omega(t)}{\partial t}\right),
\end{aligned}
$$

where the subscript t under the operators Λ and Δ indicates that they are the ones associated with the metric corresponding to $\omega(t)$. Since both of the derivatives with respect to t of $\omega(t)$ and $\Sigma(t)$ are closed forms cohomologous to zero, using the Green operator G_t relative to $\omega(t)$, we obtain the integro-differential equation

$$\frac{\partial\omega(t)}{\partial t} = -2G_t\left(\frac{\partial\Sigma(t)}{\partial t}\right). \tag{10}$$

It can be shown analytically that, subject to sufficient conditions of differentiability of the function $\partial\Sigma(t)/\partial t$, the integro-differential equation above has a solution $\omega(t)$ with initial value given by the principal form $\omega(0)$ of an arbitrary, infinitely differentiable Kähler metric. In particular, on letting the initial values correspond to the Kähler metric assumed in the statement of the Proposition, and choosing $\partial\Sigma(t)/\partial t$ to be the form $\Sigma' - \Sigma$, the solution $\omega(t)$ of equation (10) such that $\omega(0) = \omega$ formally gives the principal form of the Kähler metric whose Ricci form $\Sigma(t)$ satisfies

$$\Sigma(t) = \Sigma + t(\Sigma' - \Sigma); \tag{11}$$

for $t = 1$ the solution becomes the metric whose existence is asserted.

The first essential step in rigorizing this argument consists in showing that, as t varies, equation (10) preserves its elliptic character. In fact the corresponding equation for the volume element $\omega^n(t)/n!$ is given by

$$\begin{aligned}\frac{\partial(\omega^n(t))}{\partial t} &= n\omega^{n-1}(t)\frac{\partial\omega(t)}{\partial t} = \omega^n(t)\,\Lambda_t\left(\frac{\partial\omega(t)}{\partial t}\right)\\ &= -2\omega^n(t)\,\Lambda_t G_t(\Sigma' - \Sigma).\end{aligned} \tag{12}$$

Remembering again that the operator ΛG acting on closed forms of type (1, 1) and cohomologous to zero gives the same scalar modulo additive constants independently of the metric, and that the volume of M_n under any of the metrics considered is constant, we have the explicit solution of equation (12):

$$\omega^n(t) = \omega^n(0)\exp[-2\Lambda G(\Sigma' - \Sigma)]/H\{\exp[-2\Lambda G(\Sigma' - \Sigma)]\}, \tag{13}$$

where the operators Λ, G and H (harmonic projection) are taken relative to the given metric $\omega = \omega(0)$. This fact indicates that, since the determinant $g(t)$ of the metric remains positive for all values of t, the solutions of equation (10) can be continued locally with respect to t.† The proof of existence of the solution $\omega(t)$ of (10) for

$$\partial\Sigma/\partial t = \Sigma' - \Sigma(0)$$

† There seems to be some question as to whether the interval in t for which one can solve for $\omega(t)$ is unbounded. This essential gap in the proof of Proposition 1 makes the results of this paper depend on the conjecture that a compact Kähler manifold admits a Kähler metric with any assigned, positive, differentiable volume element.

can then be completed by evaluating formally from it $\partial^2\omega(t)/\partial t^2$ and obtaining a uniform *a priori* estimate of the latter expression in terms of $\omega(t)$ and $\partial\omega(t)/\partial t$, so that one can then apply the Cauchy-Peano method of polygonal approximations to a solution.

LOUISIANA STATE UNIVERSITY

REFERENCES

[1] S. BOCHNER, *Vector fields and Ricci curvature*, Bull. Amer. Math. Soc., 52 (1946), pp. 776–797.

[2] —— and K. YANO, Curvature and Betti numbers, Annals of Mathematics Studies, no. 32, Princeton, 1953.

[3] SHIING-SHEN CHERN, *Characteristic classes of Hermitian manifolds*, Ann. of Math., 47 (1946), pp. 85–121.

[4] W. V. D. HODGE, Theory and applications of harmonic integrals, Cambridge University Press, 1941.

[5] HEINZ HOPF, *Zum Clifford-Kleinschen Raumproblem*, Math. Ann., 95 (1925), pp. 313–339.

[6] E. KÄHLER, *Über eine bemerkenswerte Hermitesche Metrik*, Abh. Math. Sem. Hamburg Univ., 9 (1933), pp. 173–186.

[7] S. LEFSCHETZ, L'analysis situs et la géométrie algébrique, Paris, Gauthier-Villars, 1950.

[8] G. DE RHAM and K. KODAIRA, Harmonic integrals, Institute for Advanced Study Seminar, 1950.

[9] F. SEVERI, *Sulle superficie algebriche con curva canonica d'ordine zero*, Atti Ist. Veneto, 68 (1908–9), pp. 31–32.

[10] N. E. STEENROD, The topology of fibre bundles, Princeton University Press, 1951.

[11] H. WEYL, Group theory, Lecture Notes, Princeton University, Spring, 1949.

Quotient d'un Espace Analytique par un Groupe d'Automorphismes

Henri Cartan

LE quotient d'une variété analytique complexe X par un groupe proprement discontinu d'automorphismes G n'est pas, en général, une variété analytique complexe, à cause de la présence possible de points fixes dans les transformations de G. Récemment, diverses extensions de la notion de variété analytique complexe ont été proposées (cf. [2, 3, 4]). Nous prouvons ici que si X est un 'espace analytique', il en est de même de X/G (Théorème 4). Dans ce but, nous étudions d'abord le quotient d'un espace numérique complexe C^n par un groupe linéaire fini; dans la mesure où les résultats ont un caractère purement algébrique, ils sont établis pour un corps quelconque K au lieu du corps C. Les démonstrations sont de nature élémentaire.

Revenant au cas d'un groupe proprement discontinu G d'automorphismes d'une variété analytique complexe X, nous démontrons, sous certaines hypothèses, que l'espace analytique X/G, supposé compact, peut être réalisé comme variété algébrique V dans un espace projectif; V peut avoir des singularités, mais est 'normale' en tout point (au sens de Zariski). Un théorème d'immersion de ce type a été récemment démontré dans le cas où le groupe G n'a pas de points fixes [8, 9]. W. L. Baily [1] a aussi annoncé un théorème d'immersion lorsque G possède des points fixes; mais ici, nous précisons que l'immersion peut être obtenue à l'aide de 'séries de Poincaré' d'un poids convenable.

Le présent article constitue une mise au point de développements donnés dans mon Séminaire de l'Ecole Normale en 1953–54 [4]. De nombreuses discussions avec J. P. Serre m'ont apporté une aide efficace, dont je tiens à le remercier.

1. Groupe linéaire fini opérant dans K^n

Soit K un corps commutatif. On notera $GL(n, K)$ le groupe linéaire à n variables; $S = K[x_1, \ldots, x_n]$ l'algèbre des polynômes; $F = K[[x_1, \ldots, x_n]]$ l'algèbre des séries formelles; si de plus K est valué complet (non discret), on notera $H = K\{x_1, \ldots, x_n\}$ la sous-algèbre de F formée des fonctions holomorphes (i.e. séries entières en $x_1, \ldots, x_n$ qui convergent dans un voisinage de l'origine).

Soit G un sous-groupe *fini* de $GL(n, K)$. Notons S^G (resp. F^G, H^G) la sous-algèbre des éléments de S (resp. F, H) *invariants* par G.

PROPOSITION 1. *L'algèbre S^G est engendrée par un nombre fini d'éléments.*

C'est le classique 'théorème des invariants'. En voici une démonstration simple, qui ne suppose pas que K soit de caractéristique 0, et est valable pour tout *anneau noethérien* K (commutatif avec élément unité). Les générateurs $x_1, \ldots, x_n$ de l'algèbre S sont entiers algébriques sur S^G; il existe donc une sous-algèbre A de S^G, engendrée par un nombre fini d'éléments, et telle que $x_1, \ldots, x_n$ soient entiers sur A; S est ainsi un A-module de type fini. Comme A est noethérien, le sous-A-module S^G de S est de type fini. Alors un système fini de générateurs de S^G (comme A-module) et un système fini de générateurs de A (comme K-algèbre) engendrent S^G comme K-algèbre.

Choisissons une fois pour toutes un système de q polynômes homogènes $Q_i(x) \in S^G$, qui engendre l'algèbre S^G; soit $d_i > 0$ le degré de Q_i. Un polynôme $R(y_1, \ldots, y_q)$ sera dit *isobare* de poids p si

$$R(k^{d_1}y_1, \ldots, k^{d_q}y_q) = k^p R(y_1, \ldots, y_q)$$

pour $k \in K$. Tout élément de S^G, homogène de degré p, s'écrit $R(Q_1, \ldots, Q_q)$, où R est isobare de poids p. Donc *tout élément de F^G s'exprime comme série formelle en les Q_i* (cf. Théorème 1 ci-dessous).

PROPOSITION 2. *Soient $f_i \in F^G$ $(i = 1, \ldots, q)$ des séries formelles telles que $\omega(f_i - Q_i) > d_i$ (en notant $\omega(f)$ l'ordre* d'une série formelle f). *Alors tout élément de F^G s'exprime comme série formelle en les f_i.*

Cette proposition va résulter de la suivante:

PROPOSITION 3. *Soient $f_i \in F^G$ des séries formelles telles que*

$$\omega(f_i - Q_i) > d_i.$$

Il existe q séries formelles $F_i(y_1, \ldots, y_q)$ telles que:

(1) $F_i(Q_1, \ldots, Q_q) = f_i$;

(2) *si $L_i(y_1, \ldots, y_q)$ désigne la composante homogène de degré un de $F_i(y_1, \ldots, y_q)$, les formes linéaires L_i sont linéairement indépendantes.*

La Proposition 3 entraîne que *les Q_i s'expriment comme séries formelles en les f_i*, d'où la Proposition 2.

Démonstration de la Proposition 3. Disons que le système de générateurs homogènes Q_i est *irréductible* si tout polynôme isobare $R(y_1, \ldots, y_q)$ tel que $R(Q_1, \ldots, Q_q) = 0$ est d'ordre $\geqq 2$. Il est immédiat que tout système de générateurs contient un système irréductible de générateurs; appelons-les $Q_1, \ldots, Q_r$ $(r \leqq q)$. Prenons r séries formelles $F_i(y_1, \ldots, y_r)$ $(i = 1, \ldots, r)$ telles que $f_i = F_i(Q_1, \ldots, Q_r)$; soit L_i la composante homogène de degré un de F_i. Par hypothèse, la série formelle $F_i(Q_1, \ldots, Q_r) - Q_i$ est d'ordre $> d_i$, $i \leqq r$; donc, pour tout $d \leqq d_i$, l'ensemble des termes de poids d de $L_i(y_1, \ldots, y_r) - y_i$ est nul, sinon le système $(Q_1, \ldots, Q_r)$ ne serait pas irréductible. Ainsi la forme linéaire $L_i(y_1, \ldots, y_r) - y_i$ ne contient aucune des variables y_j telles que $d_j \leqq d_i$. Si on range les y_i dans l'ordre des d_i croissants, la matrice des formes L_i est triangulaire, avec 1 dans la diagonale principale. Pour $i > r$, choissons des séries formelles $g_i(y_1, \ldots, y_r)$ telles que

$$f_i = Q_i + g_i(Q_1, \ldots, Q_r).$$

En prenant $F_i(y_1, \ldots, y_q) = y_i + g_i(y_1, \ldots, y_r)$ pour $i > r$, la condition (2) de la Proposition 3 sera satisfaite; ce qui prouve la proposition.

Pour chaque point $a = (a_1, \ldots, a_n)$ de K^n, on notera $G(a)$ le *groupe d'isotropie* de a: sous-groupe des éléments de G laissant fixe a.

Lemme 1. *Pour tout entier r, et tout polynôme $R(x_1, \ldots, x_n)$ invariant par $G(a)$, il existe un polynôme $Q \in S^G$ tel que l'ordre de $R - Q$ au point a soit $> r$.*

Démonstration. Prenons un polynôme $U(x)$ tel que $U(x) - 1$ soit d'ordre $> r$ au point a, et que $U(x)$ soit d'ordre $> r$ aux points $sa \neq a$ $(s \in G)$. Le produit V des transformés de U par $G(a)$ jouit des mêmes propriétés. Le polynôme $R - RV$ est d'ordre $> r$ en a, et RV est d'ordre $> r$ aux points $sa \neq a$. Comme RV est invariant par $G(a)$, l'ensemble G_a des classes $s \cdot G(a)$ (où $s \in G$) opère dans RV; la somme Q des transformés de RV par G_a répond à la question.

Appliquons la Proposition 2 au point a et au groupe $G(a)$, et tenons compte de Lemme 1; on obtient:

Théorème 1. *Soit $a = (a_1, \ldots, a_n) \in K^n$. Toute série formelle en les $x_k - a_k$ $(k = 1, \ldots, n)$, invariante par le groupe d'isotropie $G(a)$, s'exprime comme série formelle en les $Q_i(x) - Q_i(a)$, en désignant par (Q_i) un système fini de générateurs homogènes de l'algèbre S^G.*

2. L'espace quotient K^n/G comme variété algébrique affine

L'application ψ: $x \to (Q_i(x))$ de K^n dans K^q passe au quotient suivant G, et définit une application ϕ de K^n/G dans K^q. L'application ϕ est *biunivoque*, car les polynômes G-invariants séparent les points de K^n/G: en effet, si $x' \in K^n$ et $x'' \in K^n$ ne sont pas congrus modulo G, il existe un polynôme R égal à 1 aux points sx' ($s \in G$) et à 0 aux points sx''; le produit des transformés de R par G sépare la classe de x' et celle de x''.

PROPOSITION 4. *Si le corps K est algébriquement clos, l'application ϕ applique biunivoquement K^n/G sur la variété algébrique $V \subset K^q$, lieu des zéros de l'idéal $I \subset K[y_1, \ldots, y_q]$ formé des polynômes $R(y_1, \ldots, y_q)$ tels que $R(Q_1, \ldots, Q_q) = 0$.*

DÉMONSTRATION. Tout revient à prouver que tout point de V est image d'un point de K^n. D'après le 'théorème des zéros' de Hilbert, les points de V correspondent aux idéaux maximaux de l'algèbre $A = K[y_1, \ldots, y_q]/I$ (anneau de la variété affine V); or ψ définit un isomorphisme de A sur S^G. Il suffit donc de montrer que tout idéal maximal J de S^G est induit par au moins un idéal maximal de S. Cela résulte d'un théorème de Krull,† puisque tout élément de S est entier sur S^G. Directement: l'idéal de S engendré par J ne contient pas 1; car si on avait $1 = \sum_i a_i u_i$, $a_i \in S$, $u_i \in J$, on aurait $1 = \prod_{s \in G}(\sum_i (s \cdot a_i) u_i)$, et le second membre appartiendrait à J, ce qui est absurde.

Jusqu'à la fin de ce numéro, nous supposerons que le corps K est *algébriquement clos*. Ce n'est pas une restriction essentielle, car le groupe G se prolonge en un groupe linéaire opérant dans $\bar{K}^n$ ($\bar{K}$: clôture algébrique de K). La variété V qui réalise K^n/G est *normale* au sens affine: l'anneau de V, isomorphe à S^G, est un *anneau d'intégrité intégralement clos*. En effet, il est évident que si un groupe G opère dans un anneau d'intégrité intégralement clos (ici, l'anneau S), le sous-anneau des invariants de G est un anneau d'intégrité intégralement clos.

Il est classique que si une variété algébrique affine V est normale, alors, pour tout point $b \in V$, l'*anneau local* de V (anneau des fractions rationnelles dont le dénominateur est $\neq 0$ en b) est un anneau d'intégrité intégralement clos; la réciproque est vraie si K est algébriquement clos. De plus, un théorème de Zariski ([13], Théorème 2) affirme que si l'anneau local d'une variété algébrique V, en un point

† [10], Satz 1. Plus généralement, si B est un sous-anneau d'un anneau commutatif A, et si tout élément de A est entier sur B, tout idéal maximal de B est induit par un idéal maximal de A, même si A n'est pas un anneau d'intégrité (théorème de Cohen-Seidenberg [6]).

$b \in V$, est un anneau d'intégrité intégralement clos, il en est de même de son *complété* (anneau induit sur V par les séries formelles de l'espace ambiant au point b). Dans notre cas particulier, cela résulte directement du Théorème 1, qui dit que l'anneau local complété est isomorphe au sous-anneau des éléments $G(a)$-invariants de l'anneau des séries formelles en les $x_k - a_k$, anneau qui est intégralement clos.

3. L'espace K^n/G comme variété analytique

Supposons désormais que le corps K soit *valué complet*, non discret.

THÉORÈME 2. *Soit (Q_i) un système fini de polynômes homogènes engendrant l'algèbre S^G. Alors tout élément de H^G s'exprime comme fonction homomorphe des Q_i.*

DÉMONSTRATION. Soit H' la sous-algèbre des éléments de H qui s'expriment comme fonctions holomorphes des Q_i. Les éléments $x_1, \ldots, x_n$ sont entiers sur S^G, donc sur H'; par application du Vorbereitungssatz de Weierstrass, on voit facilement que tout élément de H s'exprime comme polynôme en $x_1, \ldots, x_n$ à coefficients dans H'; donc H est un H'-module de type fini. Ceci entraîne que H' est *fermé* dans H pour la topologie définie par les puissances de l'idéal maximal de H (idéal des fonctions holomorphes nulles à l'origine).† Or le développement d'une fonction holomorphe en série de polynômes homogènes montre que S^G est *dense* dans H^G pour cette topologie; *a fortiori*, H' est dense dans H^G, d'où $H' = H^G$.

A partir du Théorème 2, on obtient des énoncés analogues aux Propositions 2 et 3 et au Théorème 1. En particulier:

PROPOSITION 3 bis. *Soient $f_i \in H^G$ tels que $\omega(f_i - Q_i) > d_i$. Il existe q fonctions holomorphes $F_i(y_1, \ldots, y_q)$ telles que:*

(1) *$F_i(Q_1, \ldots, Q_q) = f_i$;*

(2) *si $L_i(y_1, \ldots, y_q)$ désigne la composante homogène de degré un de $F_i(y_1, \ldots, y_q)$, les formes linéaires L_i sont linéairement indépendantes. Les Q_i s'expriment donc comme fonctions holomorphes des f_i.* (Même démonstration que pour la Proposition 3.)

THÉORÈME 1 bis. *Soit $a = (a_1, \ldots, a_n) \in K^n$. Toute fonction holomorphe en les $x_k - a_k$ $(k = 1, \ldots, n)$, invariante par le groupe d'isotropie*

† Soient K un corps, A une K-algèbre locale (noethérienne), M l'idéal maximal de A, A' une sous-K-algèbre locale de A, d'idéal maximal $M \cap A' = M'$. Si A est un A'-module de type fini, la M-topologie de A (i.e. la topologie définie par les puissances de M) induit sur A' la M'-topologie de A', et A' est *fermé* dans A (cf. [4], Exp. VIII bis). Cela résulte d'un théorème de Krull: soit E un module de type fini sur un anneau noethérien A', et F un sous-module de E; pout tout idéal I' de A', il existe un entier n tel que $F \cap (I'^n E) \subset I'F$.

$G(a)$, s'exprime comme fonction holomorphe en les $Q_i(x)-Q_i(a)$. (Même démonstration que pour le Théorème 1.)

Supposons maintenant, pour simplifier l'exposition, que K soit le *corps C des nombres complexes.* La Proposition 4 est applicable. De plus:

PROPOSITION 4 bis. *L'application ϕ est un homéomorphisme de l'espace quotient C^n/G sur la variété algébrique V.*

DÉMONSTRATION. Soit $x \to tx$ l'homothétie de rapport $t > 0$ dans l'espace C^n/G; et soit $\sigma(t)$ la transformation $(y_i) \to (t^{d_i} y_i)$ de V en elle-même. On a $\phi(tx) = \sigma(t) . \phi(x)$. Soit U un voisinage compact de l'origine 0 dans C^n/G; $\phi(U)$ est un *voisinage* de 0 dans V, sinon, utilisant le groupe des homothéties, et sachant que ϕ applique C^n/G *sur* V, on trouverait une suite de points de C^n/G ayant une limite $\neq 0$ et dont les transformés par ϕ tendraient vers $0 \in V$, ce qui est absurde. Comme U est compact et ϕ biunivoque, ϕ est un homéomorphisme de U sur $\phi(U)$. Utilisant à nouveau le groupe des homothéties, on voit que ϕ est bicontinue en chaque point de C^n/G.

On a vu (fin du nº 2) que l'anneau local complété de V en chacun de ses points b est un anneau d'intégrité intégralement clos. L'*anneau local holomorphe* en b (anneau des fonctions induites sur V, au point b, par les fonctions holomorphes de l'espace ambiant) est aussi intégralement clos, d'après un raisonnement classique. Ici, cela résulte directement du Théorème 1 bis, qui définit un isomorphisme de l'anneau local holomorphe au point $b = \psi(a)$, sur le sous-anneau des fonctions holomorphes en $a \in C^n$ et invariantes par le groupe $G(a)$.

Ainsi V est *H-normale* (normale au sens 'holomorphe') en chacun de ses points. Prenons alors des $f_i \in H^G$ telles que $\omega(f_i - Q_i) > d_i$, et soient des $F_i(y_1, \ldots, y_q)$ holomorphes, comme dans la Proposition 3 bis. Il existe dans C^q un voisinage ouvert U' de 0 tel que les F_i soient holomorphes dans U' et définissent un homéomorphisme de U' sur un ouvert U de C^q, la transformation réciproque étant holomorphe. On en déduit:

THÉORÈME 3. *Soient $f_i \in H^G$ telles que $\omega(f_i - Q_i) > d_i$. Il existe dans C^n un voisinage ouvert A de l'origine, stable par G, dans lequel les f_i sont holomorphes, et qui jouit de la propriété suivante: l'application f de A dans C^q, définie par $x \to (f_i(x))$, induit un homéomorphisme de A/G sur un sous-ensemble analytique*† *W d'un ouvert $U \subset C^q$; W est H-normal*

† Nous disons qu'un ensemble W de points d'un ouvert U de C^q (ou plus généralement d'une variété analytique complexe U) est un *sous-ensemble analytique* s'il est fermé dans U et peut, au voisinage de chacun de ses points b, être défini par l'annulation d'un système fini de fonctions holomorphes au point b.

en chacun de ses points; pour chaque point $a \in A$, l'application f définit un isomorphisme de l'anneau local holomorphe de W au point $f(a)$, sur l'anneau des fonctions holomorphes au point $a \in C^n$ et invariantes par le groupe d'isotropie $G(a)$.

4. Espaces analytiques et groupes d'automorphismes

Le Théorème 3 va permettre d'étudier le quotient d'une variété analytique complexe par un groupe d'automorphismes satisfaisant à des conditions locales convenables. Auparavant, il faut élargir la notion de variété analytique complexe (cf. [2, 3, 4]).

Un *espace analytique* (complexe) sera, par définition, un espace topologique séparé X muni de la donnée, en chaque point $x \in X$, d'un sous-anneau $\mathscr{H}_x$ de l'anneau $\mathscr{C}_x$ des germes de fonctions continues au point x (à valeurs complexes), de manière que soit satisfaite la condition suivante: pour chaque point $a \in X$, il existe un voisinage ouvert A de a dans X, et un *isomorphisme* f de A sur un sous-ensemble analytique† V d'un ouvert U d'un espace C^q. Le mot 'isomorphisme' signifie que, pour chaque $x \in A$, l'application $\alpha \to \alpha \circ f$ est un isomorphisme de l' 'anneau local homomorphe' de V au point $f(x)$, sur l'anneau $\mathscr{H}_x$.

Pour tout ouvert A de X, les fonctions continues dans A et appartenant à $\mathscr{H}_x$ pour tout $x \in A$, forment un anneau: l'anneau des *fonctions holomorphes dans* A. Les éléments de $\mathscr{H}_x$ s'appellent les *fonctions holomorphes au point* x (ce sont des germes de fonctions).

L'espace analytique X sera dit *normal* en un point a si V est H-normal au point $f(a)$, ou en d'autres termes si $\mathscr{H}_a$ est un anneau d'intégrité intégralement clos.‡ Si X est normal en chacun de ses points, X s'appelle un *espace analytique normal.*

Les points $a \in X$ qui possèdent un voisinage ouvert *isomorphe* à un ouvert d'un espace C^n sont dits *réguliers.* Si tous les points de X sont réguliers, X n'est autre qu'une variété analytique complexe, au sens usuel.

Soit X un espace analytique, muni des anneaux $\mathscr{H}_x$. Soit R une relation d'équivalence dans X, et notons p l'application de X sur l'espace quotient $Y = X/R$ supposé séparé. Attachons à chaque point $y \in Y$ un sous-anneau $\mathscr{K}_y$ de $\mathscr{C}_y$ comme suit: pour U ouvert $\subset Y$, soit $\mathscr{K}_U$ l'anneau des fonctions α continues (complexes) dans U telles que

† See note on p. 95.

‡ Il résulte d'un théorème d'Oka ([11], 'lemme fondamental') que l'ensemble des points de X en lesquels X n'est pas H-normal est un sous-ensemble analytique (voir [4, Exposé X]). En particulier, si X est H-normal en un point a, X est H-normal aux points assez voisins de a.

$\alpha \circ p$ soit 'holomorphe' dans l'ouvert $p^{-1}(U)$. Par définition, $\mathscr{H}_y$ est la limite inductive ('direct limit') des anneaux $\mathscr{H}_U$ quand U parcourt l'ensemble des ouverts contenant y. La question peut alors se poser de savoir si Y, muni des anneaux $\mathscr{H}_y$, est un espace analytique; on la résoudra dans un cas particulier (ci-dessous, Théorème 4).

Définition. Soit X un espace topologique séparé. Un groupe G d'homéomorphismes de X sera dit *proprement discontinu* s'il satisfait aux deux conditions suivantes:

(a) si $x' \in X$ et $x'' \in X$ ne sont pas congrus modulo G, il existe un voisinage A' de x' et un voisinage A'' de x'' tels que, pour tout $s \in G$, sA' et A'' soient disjoints;

(b) pour tout $a \in X$, le groupe d'isotropie $G(a)$ est *fini*; et il existe un voisinage ouvert A de a, stable par $G(a)$, tel que les relations

$$s \in G, \quad x \in A, \quad sx \in A$$

entraînent $s \in G(a)$.

La condition (a) exprime que l'espace quotient $Y = X/G$ est séparé. La condition (b) implique que l'image de A dans Y est isomorphe au quotient de A par le groupe d'isotropie $G(a)$, qui est fini.

Théorème 4. *Soit X un espace analytique, et soit G un groupe proprement discontinu d'automorphismes de X* (respectant la structure d'espace analytique). *Alors;* (1) *l'espace quotient $Y = X/G$, muni des anneaux $\mathscr{H}_y$ comme ci-dessus, est un espace analytique;* (2) *si X est normal, Y est normal.*

Démonstration. Soit $b \in Y$, $b = p(a)$, $a \in X$. Pour étudier l'espace Y au voisinage de b, on prend un voisinage ouvert A de a, stable par le groupe fini $G(a)$; si A est assez petit, alors pour tout $x \in A$, l'application $\alpha \to \alpha \circ p$ est un isomorphisme de l'anneau $\mathscr{H}_{p(x)}$ sur l'anneau des fonctions holomorphes au point x et invariantes par $G(x)$: cela résulte de la condition (b). On va montrer que $A/G(a)$ est un espace analytique; ceci établira la première partie de l'énoncé; la seconde en résultera, car si X est normal en a, l'anneau $\mathscr{H}_{p(a)}$ est un anneau d'intégrité intégralement clos, donc Y est normal au point $b = p(a)$.

La démonstration du fait que $A/G(a)$ est un espace analytique est facile dans le cas où a est un point *régulier* de X; on peut alors supposer que X est un ouvert de C^n, a étant à l'origine. Par une transformation holomorphe sur les coordonnées $x_1, \ldots, x_n$, on peut se ramener au cas où le groupe fini $G(a)$ est *linéaire*: en effet, pour $s \in G(a)$, soit s' la transformation linéaire tangente à s; la transformation

$$x \to \sigma x = (1/r) \Sigma_{s \in G} s'^{-1} sx$$

(où r désigne l'ordre du groupe $G(a)$) est tangente à l'identité, et

$\sigma s = s'\sigma$ pour $s \in G(a)$. La transformation σ transforme $x_1, \ldots, x_n$ en un système de coordonnées locales, au voisinage de l'origine, sur lequel $G(a)$ opère linéairement. On peut alors appliquer à ce groupe linéaire les résultats du n° 3, qui montrent que le quotient par $G(a)$ est un espace analytique normal.

Il reste à examiner le cas où le point a n'est pas régulier. Voici une méthode dont l'idée est due à J. P. Serre:

LEMME 2. *Soient X un espace analytique, $a \in X$, G un groupe fini d'automorphismes de X laissant fixe a. Il existe un ouvert A contenant a, stable par G, un isomorphisme f de A sur un sous-ensemble analytique E d'un ouvert B d'un espace C^r (avec $f(a) = 0$), et un groupe linéaire fini Γ d'automorphismes de C^r, isomorphe à G, tel que B et E soient stables par Γ et que f transforme le groupe G dans le groupe Γ' d'automorphismes de E induit par Γ.*

Admettons d'abord le Lemme 2, et déduisons-en que A/G est un espace analytique. Il suffit de prouver que E/Γ' est un espace analytique. Or, d'après le n° 3, on a un isomorphisme ϕ de C^r/Γ sur un sous-ensemble algébrique V d'un espace C^q, et ϕ est induit par une application polynomiale ψ: $C^r \to C^q$, d'image V, telle que $\psi(0) = 0$. Pour $x \in E$, soit $\mathcal{H}_x$ l'anneau des fonctions holomorphes (dans C^r) au point x, et $\mathcal{I}_x$ l'idéal des fonctions nulles sur E. Soit $\Gamma(x)$ le groupe d'isotropie au point x. Soit $F = \phi(E/\Gamma') = \psi(E)$. On a vu au n° 3 que ψ définit un isomorphisme de l'anneau local holomorphe $\mathcal{K}_{\psi(x)}$ de V au point $\psi(x)$, sur le sous-anneau $(\mathcal{H}_x)^{\Gamma(x)}$ des éléments $\Gamma(x)$-invariants de $\mathcal{H}_x$; dans cet isomorphisme, l'idéal $\mathcal{J}_{\psi(x)}$ des éléments de $\mathcal{K}_{\psi(x)}$ nuls sur F correspond à $(\mathcal{I}_x)^{\Gamma(x)}$. Ceci prouve que F est un sous-ensemble analytique au voisinage de $\psi(x)$, car E, au voisinage de x, est l'ensemble des zéros communs aux fonctions de $(\mathcal{I}_x)^{\Gamma(x)}$. De plus, ψ définit un isomorphisme de l'anneau local holomorphe $\mathcal{K}_{\psi(x)}/\mathcal{J}_{\psi(x)}$ de F sur l'anneau $(\mathcal{H}_x)^{\Gamma(x)}/(\mathcal{I}_x)^{\Gamma(x)}$. Ce dernier est canoniquement isomorphe à $(\mathcal{H}_x/\mathcal{I}_x)^{\Gamma(x)}$, comme on le voit en faisant une moyenne. Finalement, ϕ est un *isomorphisme* de E/Γ' sur l'espace analytique F, et par suite E/Γ' est un espace analytique. Ainsi s'achève la démonstration du Théorème 4.

Il reste à démontrer le Lemme 2. Prenons d'abord un ouvert A contenant a, stable par G, et un isomorphisme g de A sur un sous-ensemble analytique F d'un ouvert U d'un espace C^m, avec $g(a) = 0$. A chaque $s \in G$ associons un exemplaire U_s de U; alors G opère sur le produit $\prod_{s \in G} U_s = B$ en permutant ses facteurs. Ces opérations sont induites par un groupe Γ, isomorphe à G, de transformations linéaires de l'espace ambiant C^{km} (k désignant l'ordre du groupe G): en fait, les transformations de Γ sont des permutations sur les coordonnées de C^{km}. Soit $f: A \to C^{km}$ l'application définie par

$$f(x) = (g(s_1 x), \ldots, g(s_k x)),$$

où $s_1, \ldots, s_k$ désignent les éléments de G. L'image E de f est un sous-ensemble analytique de B, stable par Γ; f est un isomorphisme de A sur E (comme espaces analytiques) et transforme les automorphismes de A dans ceux de E induits par Γ. Ceci démontre le Lemme 2.

5. Un théorème d'immersion

Le Théorème 4 s'applique notamment au cas où X est un *domaine borné* de C^n, et G un groupe *discret* d'automorphismes de X. Il est classique† que G est alors proprement discontinu (au sens du n° 4). L'espace quotient X/G est donc un *espace analytique normal* (résultat bien connu dans le cas $n=1$). On va montrer que, dans ce cas, l'espace analytique X/G, lorsqu'il est *compact*, peut être réalisé comme *sous-ensemble algébrique V d'un espace projectif*, V étant *normal* en chacun de ses points. Cette réalisation peut être obtenue au moyen de *séries de Poincaré* d'un poids convenable.

D'une manière générale, considérons un groupe G, proprement discontinu, d'automorphismes d'une variété analytique complexe X. L'espace $Y = X/G$ n'est pas supposé compact. Supposons donné un *facteur d'automorphie*, c'est-à-dire, pour chaque $s \in G$, une fonction holomorphe $J_s(x)$ dans X, partout $\neq 0$, et telle que

$$J_{st}(x) = J_s(tx)\, J_t(x) \quad \text{pour} \quad x \in K,\ s \in G,\ t \in G.$$

Attachons à chaque point $a \in X$ un entier $q(a)$ tel que $(J_s(x))^{q(a)} = 1$ pour tout $s \in G(a)$ (groupe d'isotropie de a) et tout x assez voisin de a; par exemple, il suffirait de prendre pour $q(a)$ l'ordre du groupe $G(a)$. On peut supposer que l'entier $q(a)$ ne dépend que de la classe $\bar{a}$ de a modulo G; on le notera aussi $q(\bar{a})$.

Supposons attaché à chaque entier m assez grand un espace vectoriel L_m de *formes automorphes de poids m*, c'est-à-dire de fonctions f holomorphes dans X et telles que

$$f(sx)\,(J_s(x))^m = f(x) \quad \text{pour} \quad x \in X,\ s \in G.$$

Et supposons vérifiées les trois conditions suivantes:

(i) *pour tout couple (x, x') de points de X, non congrus modulo G, il existe un entier $m(x, x')$ tel que, pour tout $m \geqq m(x, x')$ et multiple de $q(x)$ et $q(x')$, L_m contienne une fonction f satisfaisant à $f(x) = a$, $f(x') = b$ (a et b nombres complexes arbitraires)*;

(ii) *pour tout $x_0 \in X$ et tout entier d, il existe un entier $n(x_0, d)$ jouissant*

† Voir par exemple [12, Chap. x].

de la propriété suivante: si $m \geqq n(x_0, d)$ *est multiple de* $q(x_0)$, *et si* $h(x)$ *est holomorphe au voisinage de* x_0 *et invariante par* $G(x_0)$, *il existe une* $f \in L_m$ *telle que l'ordre de* $f-h$ *au point* x_0 *soit* $>d$;

(iii) *le produit d'une fonction de* L_m *et d'une fonction de* $L_{m'}$ *est dans* $L_{m+m'}$.

EXEMPLE. Supposons que X soit un *domaine borné* de C^n, et que $J_s(x)$ désigne la valeur, au point $x \in X$, du jacobien complexe de la transformation $x \to sx$. Prenons pour $q(a)$ le plus petit des entiers q tels que $(J_s(a))^q = 1$ pour tout $s \in G(a)$. Soit P_m ($m \geqq 2$) l'espace vectoriel des séries de Poincaré $\sum_{s \in G} \phi(s)\,(J_s(x))^m$, où ϕ est un *polynôme*; et prenons pour L_m l'espace vectoriel engendré par les produits finis $f_1 \cdots f_k$, avec $f_i \in P_{m_i}$ et $\sum_i m_i = m$. Chaque fonction de L_m est une série de Poincaré† de poids m. Alors on démontre assez facilement‡ que les conditions (i) et (ii) sont satisfaites; pour (iii), c'est évident.

Revenons au cas général. Sous les hypothèses (i), (ii), (iii) nous démontrerons au nº 6 les trois propositions que voici:

PROPOSITION 5. *Etant donnés arbitrairement* y_1 *et* $y_2 \in Y$, *il existe un ouvert* V *contenant* y_1 *et* y_2 *et jouissant de la propriété suivante: pour tout multiple* m *de* $q(y_1)$ *et* $q(y_2)$, *assez grand, il existe une* $f \in L_m$ *telle que* $f(x) \neq 0$ *pour tout* $x \in p^{-1}(V)$. (On note p l'application de X sur $Y = X/G$.)

PROPOSITION 6. *Pour tout couple* (y_1, y_2) *de points distincts de* Y, *il existe deux voisinages* V_1 *et* V_2 *de* y_1 *et* y_2 *respectivement, jouissant de la propriété suivante: pour tout* m *assez grand et multiple de* $q(y_1)$ *et* $q(y_2)$, *il existe deux fonctions* g *et* $h \in L_m$ *telles que* $g(x) \neq 0$ *et* $|h(x)/g(x)| < 1$ *pour* $x \in p^{-1}(V_1)$, $h(x) \neq 0$ *et* $|g(x)/h(x)| < 1$ *pour* $x \in p^{-1}(V_2)$.

PROPOSITION 7. *Tout point* $y_0 \in Y$ *possède un voisinage ouvert* W *jouissant de la propriété suivante: pour tout multiple assez grand de* $q(y_0)$, *il existe un système fini de fonctions de* L_m, *dont l'une est* $\neq 0$ *en tout point de* $p^{-1}(W)$ *et dont les rapports mutuels définissent un isomorphisme de* W *sur un sous-ensemble analytique normal d'un ouvert de l'espace projectif.*

Ces propositions étant admises, considérons un *compact* K de $Y = X/G$. Il résulte facilement des Propositions 5, 6, 7 que, pour tout entier m assez grand et multiple d'un entier $q(K)$ (à savoir le plus petit commun multiple des $q(y)$ pour $y \in K$), il existe un système *fini* de fonctions de L_m dont les rapports mutuels définissent un *homéomorphisme* Φ de K sur un compact d'un espace projectif P; et ceci de manière que, pour tout point y *intérieur* à K, Φ soit un *isomorphisme*

† Nous appelons 'série de Poincaré' toute série $f(x) = \sum_{s \in G} \phi(x)\,(J_s(x))^m$ qui converge uniformément sur tout compact de X, et où ϕ est holomorphe dans X, non nécessairement bornée. Une telle $f(x)$ est évidemment une forme automorphe de poids m.

‡ Voir [7], et l'Exposé I de [4].

d'un voisinage ouvert de y sur un sous-ensemble analytique normal (dans un ouvert de P). En particulier, si Y est compact, $\Phi(Y)$ est un sous-ensemble analytique de l'espace projectif P. D'après un théorème classique de Chow [5], $\Phi(Y)$ est un ensemble *algébrique*. Il est clair que toute composante irréductible de $\Phi(Y)$ est ouverte et fermée; donc si X est connexe, $\Phi(Y)$ est irréductible. En conclusion:

THÉORÈME 5. *Soit G un groupe proprement discontinu d'automorphismes d'une variété analytique complexe X, tel que l'espace quotient $Y = X/G$ soit compact. Supposons donnés un facteur d'automorphie $J_s(x)$ et des espaces vectoriels L_m de formes automorphes de poids m, vérifiant* (i), (ii) *et* (iii). *Alors, pour tout entier m assez grand et tel que $(J_s(x))^m = 1$ pour tout $x \in X$ et tout $s \in G(x)$, il existe un système fini de fonctions de L_m dont les rapports mutuels définissent un isomorphisme de l'espace analytique normal X/G sur un sous-ensemble algébrique de l'espace projectif, normal en chacun de ses points, et irréductible si X est connexe.*

REMARQUE. Il est connu† que si X/G est compact, l'espace vectoriel de *toutes* les formes automorphes de poids m est de dimension finie.

6. Démonstration des Propositions 5, 6, 7

DÉMONSTRATION‡ DE LA PROPOSITION 5. Soient $x_1 \in X$, $x_2 \in X$ tels que $p(x_1) = y_1$, $p(x_2) = y_2$. Soit q le plus petit commun multiple de $q(y_1)$ et $q(y_2)$. D'après (i), il existe deux multiples consécutifs de q, soient m' et $m'' = m' + q$, et deux fonctions $f' \in L_{m'}$ et $f'' \in L_{m''}$, telles que $f'(x_i) = f''(x_i) = 1$ $(i = 1, 2)$. Soit U un ouvert de X contenant x_1 et x_2, tel que $f'(x) \neq 0$ et $f''(x) \neq 0$ pour $x \in U$. Soit V l'image de U dans Y. Si m est un multiple de q au moins égal à $(m'm'')/q$, il existe des entiers positifs a' et a'' tels que $m = a'm' + a''m''$; la fonction $f = (f')^{a'}(f'')^{a''}$ appartient à L_m et est $\neq 0$ en tout point de $p^{-1}(V)$.

DÉMONSTRATION DE LA PROPOSITION 6. Soient y_1 et y_2 deux points distincts de Y: soient $x_1 \in X$, $x_2 \in X$, tels que $p(x_1) = y_1$, $p(x_2) = y_2$. Soit q le plus petit commun multiple de $q(y_1)$ et $q(y_2)$. D'après (i), il existe un entier m_0, multiple de q, une $f_1 \in L_{m_0}$ telle que $f_1(x_1) = 1$, $f_1(x_2) = 0$, et une $f_2 \in L_{m_0}$ telle que $f_2(x_1) = 0$, $f_2(x_2) = 1$. Soit U_1 un ouvert contenant x_1 et tel que $|f_1(x)| > 1/2$ et $|f_2(x)| < 1/2$ pour $x \in U_1$; et soit U_2 un ouvert contenant x_2 et tel que $|f_1(x)| < 1/2$ et $|f_2(x)| > 1/2$ pour $x \in U_2$. On peut choisir U_1 et U_2 assez petits pour que $V_1 = p(U_1)$ et $V_2 = p(U_2)$ soient contenus dans l'ouvert V de la Proposition 5. D'après la Proposition 5, pour tout multiple m assez grand de q, il existe une $f \in L_{m-m_0}$ telle que $f(x) \neq 0$ pour tout $x \in p^{-1}(V_1) \cup p^{-1}(V_2)$. Alors $f_1 f = g$ et $f_2 f = h$ satisfont aux conditions de la Proposition 6.

† Voir par exemple l'Exposé II de J. P. Serre dans [4].

‡ Cf. [7].

Démonstration de la Proposition 7. Soit $x_0 \in X$ tel que $p(x_0) = y_0$. D'après le n° 4, on peut choisir dans X, au voisinage de x_0, des coordonnées locales (nulles en x_0) sur lesquelles le groupe d'isotropie $G(x_0)$ opère linéairement. Soit $(Q_i(x))$ un système fini de polynômes homogènes par rapport à ces coordonnées, et engendrant l'anneau des polynômes invariants par $G(x_0)$. Soit d_i le degré de Q_i. Soit m_0 un multiple de $q(x_0)$, au moins égal à tous les entiers $n(x_0, d_i)$ de la condition (ii). D'après (ii), il existe, pour chaque i, une $f_i \in L_{m_0}$ telle que l'ordre de $f_i - Q_i$ au point x_0 soit $> d_i$. D'après le Théorème 3, il existe un ouvert U contenant x_0, stable par $G(x_0)$, tel que les f_i, restreintes à U, induisent un homéomorphisme de $W = p(U)$ sur un sous-ensemble analytique normal N d'un ouvert de l'espace numérique C^r (r désignant le nombre des f_i). On peut de plus choisir U assez petit pour qu'il existe, pour tout multiple m assez grand de $q(x_0)$, une $g \in L_{m-m_0}$ et une $f \in L_m$ qui soient $\neq 0$ en tout point de $p^{-1}(W)$ (cf. Proposition 5). Alors, pour chaque tel m, les fonctions gf_i ($1 \leqq i \leqq r$) et f sont dans L_m, et leurs rapports mutuels définissent un isomorphisme de W sur un sous-ensemble analytique normal d'un ouvert de l'espace projectif.

Paris

References

[1] W. L. Baily, *On the quotient of an analytic manifold by a group of analytic homeomorphisms*, Proc. Nat. Acad. Sci., U.S.A., 40 (1954), pp. 804–808.

[2] H. Behnke und K. Stein, *Modifikation komplexer Mannigfaltigkeiten und Riemannscher Gebiete*, Math. Ann., 124 (1951), pp. 1–16.

[3] H. Cartan, Séminaire E.N.S. 1951–52, Exposé xiii.

[4] H. Cartan, Séminaire E.N.S. 1953–54.

[5] W. L. Chow, *On compact analytic varieties*, Amer. J. Math., 71 (1949), pp. 49–50.

[6] I. S. Cohen and A. Seidenberg, *Prime ideals and integral independence*, Bull. Amer. Math. Soc., 52 (1946), pp. 252–261.

[7] M. Hervé, *Sur les fonctions fuchsiennes de deux variables complexes*, Ann. Ecole Norm., 69 (1952), pp. 277–302.

[8] J. Igusa, *On the structure of a certain class of Kaehler varieties*, Amer. J. Math., 76 (1954), pp. 669–678; cf. Theorem 3.

[9] K. Kodaira, *On Kähler varieties of restricted type*, Proc. Nat. Acad. Sci. U.S.A., 40 (1954), pp. 313–316.

[10] W. Krull, *Beiträge zur Arithmetik kommutativer Integritätsbereiche*, III, Math. Zeit., 42 (1937), pp. 745–766.

[11] K. Oka, *Sur les fonctions analytiques de plusieurs variables*, VIII, J. Math. Soc. Japan, 3 (1951), pp. 204–278.

[12] C. L. Siegel, Analytic functions of several complex variables, Princeton, 1948–49.

[13] O. Zariski, *Sur la normalité analytique des variétés normales*, Ann. Institut Fourier, 2 (1950), pp. 161–164.

On a Generalization of Kähler Geometry

Shiing-shen Chern

1. Introduction

A Kähler manifold is a complex Hermitian manifold, whose Hermitian metric

$$(1)\qquad ds^2=\textstyle\sum_{1\leqq\alpha,\beta\leqq m} g_{\alpha\beta}(z^1,\ldots,z^m;\bar z^1,\ldots,\bar z^m)\,dz^\alpha d\bar z^\beta \quad (\bar g_{\alpha\beta}=g_{\beta\alpha}),$$

has the property that the corresponding exterior differential form

$$(2)\qquad \Omega=\textstyle\sum_{1\leqq\alpha,\beta\leqq m} g_{\alpha\beta}\,dz^\alpha\wedge d\bar z^\beta$$

is closed. The importance of Kähler manifolds lies in the fact that they include as special cases the non-singular algebraic varieties over the complex field.

So far the most effective tool for the study of the homology properties of compact Kähler manifolds is Hodge's theory of harmonic integrals or harmonic differential forms.† The notion of a harmonic differential form is defined on any orientable Riemann manifold, and can be briefly introduced as follows: The Riemann metric allows us to define the star operator $*$, which transforms a differential form of degree p into one of degree $n-p$, n being the dimension of the manifold. From the operator $*$ and the exterior differentiation operator d we introduce the operators

$$(3)\qquad \begin{cases}\delta=(-1)^{np+n+1}*d*,\\ \Delta=d\delta+\delta d.\end{cases}$$

If the manifold is compact, as we shall assume from now on, a differential form η is called harmonic, if $\Delta\eta=0$.

In the case of a complex manifold it will be convenient to consider complex-valued differential forms. The star operator can be extended in an obvious way to such differential forms. For its definition we

† Various accounts of this study are now in existence; cf. [3], [4], [5], [8], [9]. The numbers refer to the Bibliography at the end of this paper.

follow the convention of Weil,† without repeating the details. We only mention that we can define an operator $\bar{*}$ by

$$\bar{*}\eta = *\bar{\eta}. \tag{4}$$

The operator δ is then extended to complex-valued differential forms by the definition

$$\delta = (-1)^{np+n+1}\bar{*}d\bar{*}. \tag{5}$$

By means of this we define Δ by the second equation of (3). For a Kähler manifold we introduce furthermore the operators

$$\begin{cases} L\eta = \Omega \wedge \eta, \\ \Lambda = \bar{*}L\bar{*}. \end{cases} \tag{6}$$

A differential form η on a Kähler manifold is called effective or primitime, if $\Lambda\eta = 0$.

The notion of a primitive harmonic form is a formulation, in terms of cohomology, of the effective cycles of Lefschetz on an algebraic variety.‡ Lefschetz proved that on a complex algebraic variety every cycle is homologous, with respect to rational coefficients, to a linear combination of effective cycles and the intersection cycles, by linear spaces of the ambient projective space, of effective cycles of higher dimension. This result can be expressed in terms of harmonic differential forms by the following decomposition theorem of Hodge:

Every harmonic form ω of degree p on a compact Kähler manifold of (complex) dimension m can be written in a unique way in the form

$$\omega = \Sigma L^k \omega_k, \tag{7}$$

where the summation is extended over the following range of k:

$$\max(0, p-m) \leqq k \leqq q = [\tfrac{1}{2}p],$$

and where ω_k is a primitive harmonic form of degree $p-2k$, completely determined by ω.

The existing proofs of this theorem depend on the establishment of various identities between the operators introduced above. We attempt to give in this paper what seems to be a better understanding of this theorem by generalizing it and proving it in an entirely different way.

It is well known§ that the existence of a positive definite Hermitian metric on a complex manifold allows us to define a connection with the unitary group and that the Kähler property $d\Omega = 0$ is equivalent

† Since we are dealing with real manifolds, our δ operator differs from Weil's in sign.

‡ [6] or [5], p. 182.

§ [1], p. 112.

to the absence of torsion of this connection. Our contention is that the latter condition accounts more for the homology properties of Kähler manifolds than the analytically simpler condition $d\Omega = 0$.

Utilizing this idea, we generalize the Kähler property as follows: Let M be a real differentiable manifold of dimension n. Suppose that the structural group of its tangent bundle, which is the general linear group $GL(n, R)$ in n real variables, can be reduced, in the sense of fiber bundles, to a subgroup G of the rotation group $R(n) \subset GL(n, R)$. It will be proved in § 2 that a connection can be defined, with the group G. In general, the torsion tensor of this connection does not vanish. The vanishing of torsion of this connection is then a natural generalization of the Kähler property.

On the other hand, the group G acts on the tangent vector space V of M at a point and also on its dual space V^*. This induces a linear representation of G in the exterior q^{th} power $\Lambda^q(V^*)$ of V^*, which can also be described as the representation of G into the space of all anti-symmetric covariant tensors of order q. If $G = GL(n, R)$, it is well known that this representation is irreducible. However, if G is a proper subgroup of $R(n)$, it is possible that this representation is reducible. When this is the case, suppose W be an invariant subspace of this representation. Since $G \subset R(n)$, there is an inner product defined in $\Lambda^q(V^*)$, and the subspace W' in $\Lambda^q(V^*)$ orthogonal to W is also invariant. The invariance property of W allows us to introduce the notion of a differential form of degree q and type W, as one which assigns to every point $x \in M$ an element of $W(x) \subset \Lambda^q(V^*(x))$. Similarly, we can define an operator P_W on differential forms of degree q, its projection in W. With these preparations we can state our decomposition theorem:

Let M be a compact differentiable manifold of dimension n, which has the following properties: (1) *The structural group of its tangent bundle can be reduced to a subgroup G of the rotation group $R(n)$ in n variables.* (2) *There is a connection with the group G, whose torsion tensor vanishes. Let $W \subset \Lambda^q(V^*)$ be an invariant subspace of $\Lambda^q(V^*)$ under the action of G, and let P_W be the projection of an exterior differential form of degree q into W. Then*

$$P_W \Delta = \Delta P_W. \tag{8}$$

It follows that if $W_1, \ldots, W_k$ are irreducible invariant subspaces of $\Lambda^q(V^*)$ under the action of G and if η is a harmonic form of degree q, then $P_{W_1}\eta, \ldots, P_{W_k}\eta$ are harmonic. Moreover, if η is a form of degree q and type W, then $\Delta\eta$ is also a form of degree q and type W.

When $n=2m$ is even and $G=U(m)$ is the unitary group in m complex variables considered as a subgroup of $GL(n,R)$, our notion of a manifold having a G-connection without torsion includes that of a Kähler manifold. As will be shown in § 4, it also includes a generalization of Kähler manifold studied by A. Lichnerowicz,† namely, an orientable even-dimensional Riemann manifold with the property that there exists an exterior quadratic differential form, everywhere of the highest rank, whose covariant derivative is zero. To derive Hodge's decomposition theorem from ours it remains to solve the following algebraic problem: Let V_1^*, V_2^* be two m-dimensional complex vector spaces and V^* their direct sum. Let $U(m)$ act on V^* such that it acts on V_1^* in the usual way but on V_2^* by the conjugate-complex transformation. This induces a representation of $U(m)$ into the group of linear transformations of $\Lambda^q(V^*)$. Our problem is to decompose this representation into its irreducible parts. It will be shown in § 4 that the summands in (7) correspond to the irreducible parts of the representation. In this sense the Hodge decomposition theorem cannot be further improved.

To illustrate that the scope of our theorem goes beyond Kähler geometry, we consider in § 5 the case that $G=R(s)\times R(n-s)$ $(0<s<n)$ is the direct product of two rotation groups of dimensions s and $n-s$ respectively. As is well known, the reduction of the structural group of the tangent bundle to G is equivalent to the existence of a continuous field of oriented s-dimensional linear spaces over M. The existence of a G-connection without torsion means the existence of such a field with the further property that its linear spaces are parallel with respect to a Riemann metric. Our decomposition theorem shows that the cohomology classes of M can be given a bi-degree. In particular, it follows that the s-dimensional Betti number of such a manifold is $\geqq 1$.‡

2. G-connection in a tangent bundle; torsion tensor

We shall derive in this section the basic notions and formulas for a G-connection in the tangent bundle of a real n-dimensional differentiable manifold M. Since the results are local, M will not be assumed to be compact.

The tangent bundle of M has as structural group the general linear group $GL(n,R)$ in n real variables. We consider $GL(n,R)$ to be the group of all $n\times n$ real non-singular matrices. Let G be a closed sub-

† [7].

‡ I was first informed of this theorem by Dr T. J. Willmore.

group of $GL(n, R)$. By a G-structure in M we mean a covering of M by coordinate neighborhoods U_α and, to each U_α, a set of n linearly independent Pfaffian forms θ^i_α in U_α,† such that, when $U_\alpha \cap U_\beta \neq 0$, we have

$$\theta^i_\alpha = \textstyle\sum_j g^i_{\alpha\beta,j}(x)\,\theta^j_\beta \quad (x \in U_\alpha \cap U_\beta), \tag{9}$$

where $g_{\alpha\beta}(x) = (g^i_{\alpha\beta,j}(x)) \in G$ and the mapping $U_\alpha \cap U_\beta \to G$ defined by $x \to g_{\alpha\beta}(x)$ is differentiable. Let θ_α and θ_β denote respectively the one-rowed matrices, whose elements are θ^i_α and θ^k_β. Then equation (9) can be abbreviated in the matrix form

$$\theta_\alpha = \theta_\beta g_{\alpha\beta}(x). \tag{9a}$$

The 'coordinate transformations' $g_{\alpha\beta}(x)$ define a principal fiber bundle $p: B_G \to M$ with the structural group G. We recall that B_G is the union of the sets $\{U_\alpha \times G\}$, under the identification

$$y_\beta = g_{\alpha\beta}(x)\,y_\alpha, \tag{10}$$

where $\quad (x, y_\alpha) \in U_\alpha \times G, \quad (x, y_\beta) \in U_\beta \times G, \quad x \in U_\alpha \cap U_\beta.$

From (9a) and (10) it follows that the one-rowed matrix of Pfaffian forms

$$\omega = \theta_\alpha y_\alpha = \theta_\beta y_\beta \tag{11}$$

is globally defined in B_G. The elements ω^i in $\omega = (\omega^1, \ldots, \omega^n)$ are clearly linearly independent. By exterior differentiation we find

$$d\omega = -\omega \wedge y_\alpha^{-1} dy_\alpha + d\theta_\alpha y_\alpha, \tag{12}$$

where $y_\alpha^{-1} dy_\alpha$ is a matrix of left-invariant Pfaffian forms of G. If π^ρ $(1 \leqq \rho \leqq r)$ is a set of linearly independent left-invariant Pfaffian forms of G, we can write (12) in the form

$$d\omega^i = -\textstyle\sum_{\rho,k} a^i_{\rho k}\,\omega^k \wedge \pi^\rho + \frac{1}{2}\sum_{j,k} c^i_{jk}(b)\,\omega^j \wedge \omega^k, \tag{13}$$

where $a^i_{\rho k}$ are constants and $c^i_{jk}(b)$ are functions in B_G satisfying the conditions

$$c^i_{jk}(b) + c^i_{kj}(b) = 0 \quad (b \in B_G). \tag{14}$$

The constants $a^i_{\rho k}$ in (13) have a simple geometrical meaning. In fact, we can regard G as acting on an n-dimensional vector space with the coordinates (ξ^i), according to the equations

$$\xi^i \to \xi'^i = \textstyle\sum_j y^i_j \xi^j \quad ((y^i_j) \in G). \tag{15}$$

† Unless otherwise stated, we agree on the following ranges of indices:

$$1 \leqq i, j, k, l \leqq n, \quad 1 \leqq \rho, \sigma, \tau \leqq r.$$

In this section we use α, β, γ to index the neighborhoods. When we discuss almost complex manifolds of dimension $2m$, we suppose $1 \leqq \alpha, \beta, \gamma \leqq m$.

Then the infinitesimal transformations

$$X_\rho = \sum_{i,k} a^i_{\rho k} \xi^k \frac{\partial}{\partial \xi^i} \tag{16}$$

are linearly independent and generate the group G. It follows that their commutators satisfy equations of the form

$$[X_\rho, X_\sigma] = \sum_\tau \gamma^\tau_{\rho\sigma} X_\tau, \quad \gamma^\tau_{\rho\sigma} = -\gamma^\tau_{\sigma\rho}, \tag{17}$$

where $\gamma^\tau_{\rho\sigma}$ are the constants of structure of G. When the expressions (16) are substituted into (17), we get

$$\sum_i (a^i_{\rho k} a^j_{\sigma i} - a^i_{\sigma k} a^j_{\rho i}) = \sum_\tau \gamma^\tau_{\rho\sigma} a^j_{\tau k}. \tag{18}$$

The equations (12) or (13) are derived with reference to a representation of ω in $p^{-1}(U_\alpha)$ and are therefore local in character. To put this in a different way, the forms π^ρ are not globally defined in B_G. The permissible transformation on π^ρ is given by

$$\pi^\rho \to \pi'^\rho = \pi^\rho + \sum_k b^\rho_k \omega^k. \tag{19}$$

Under this transformation the equations (13) preserve the same form, with new coefficients c'^i_{jk} given by

$$c'^i_{jk} = c^i_{jk} + \sum_\rho (-a^i_{\rho k} b^\rho_j + a^i_{\rho j} b^\rho_k). \tag{20}$$

The $\frac{1}{2}n^2(n-1)$ expressions

$$A^i_{jk} = \sum_\rho (-a^i_{\rho k} b^\rho_j + a^i_{\rho j} b^\rho_k) \tag{21}$$

are linear and homogeneous in b^ρ_k, with constant coefficients. We say that the group G has the property (C), if there are nr linearly independent ones among them, that is, if $A^i_{jk} = 0$ implies $b^\rho_k = 0$.

If the group G has the property (C), we can define a connection† in the bundle $p'\colon B_{G'} \to M$, where G' is the group of non-homogeneous linear transformations on n variables with G as the homogeneous part and the bundle is obtained from $p\colon B_G \to M$ by enlarging the group from G to G'. To define such a connection it suffices to determine in B_G a set of forms π^ρ satisfying (13). Suppose $A^{i'}_{j'k'}$ be a subset of nr linearly independent expressions among the A^i_{jk}, and suppose $A^{i''}_{j''k''}$ be its complementary set. Then there exists one, and only one, set of forms π^ρ in B_G satisfying the equations (13) together with the conditions $c^{i'}_{j'k'} = 0$. These forms, together with ω^i, define a connection in the bundle $p'\colon B_{G'} \to M$. By abus du langage, we call such a connection a G-connection in the tangent bundle.

The bundles $p\colon B_G \to M$ and $p'\colon B_{G'} \to M$ are G'-equivalent in the sense of bundles. From the point of view of connections it is, however,

† [2].

necessary to consider the second bundle. The curvature form of our connection is a tensorial quadratic differential form in M, of type $ad(G')$ and with values in the Lie algebra $L(G')$ of G'. Since the Lie algebra $L(G)$ of G is a subalgebra of $L(G')$, there is a natural projection of $L(G')$ into the quotient space $L(G')/L(G)$. The image of the curvature form under this projection will be called the torsion form or the torsion tensor. If the forms π^ρ in (13) define a G-connection, the vanishing of the torsion form is expressed analytically by the conditions

$$c^{i''}_{j''k''}=0. \tag{22}$$

We proceed to derive the analytical formulas for the theory of a G-connection without torsion in the tangent bundle. In general we will consider such formulas in B_G. The fact that the G-connection has no torsion simplifies (13) into the form

$$d\omega^i=\textstyle\sum_{\rho,k} a^i_{\rho k}\pi^\rho \wedge \omega^k. \tag{23}$$

By taking the exterior derivative of (23) and using (18), we get

$$\textstyle\sum_{\rho,k} a^i_{\rho k}\Pi^\rho \wedge \omega^k=0, \tag{24}$$

where we put

$$\Pi^\rho=d\pi^\rho+\tfrac{1}{2}\textstyle\sum_{\sigma,\tau}\gamma^\rho_{\sigma\tau}\pi^\sigma\wedge\pi^\tau. \tag{25}$$

For a fixed value of k we multiply the above equation by

$$\omega^1\wedge\ldots\wedge\omega^{k-1}\wedge\omega^{k+1}\wedge\ldots\wedge\omega^n,$$

getting

$$\textstyle\sum_\rho a^i_{\rho k}\Pi^\rho\wedge\omega^1\wedge\ldots\wedge\omega^n=0,$$

or

$$\textstyle\sum_\rho a^i_{\rho k}\Pi^\rho\equiv 0, \quad \operatorname{mod}\omega^j.$$

Since the infinitesimal transformations X_ρ are linearly independent, this implies that

$$\Pi^\rho\equiv 0, \quad \operatorname{mod}\omega^j.$$

It follows that Π^ρ is of the form

$$\Pi^\rho=\textstyle\sum_j \phi^\rho_j\wedge\omega^j,$$

where ϕ^ρ_j are Pfaffian forms. Substituting these expressions into (24), we get

$$\textstyle\sum_{\rho,j,k}(a^i_{\rho k}\phi^\rho_j-a^i_{\rho j}\phi^\rho_k)\wedge\omega^j\wedge\omega^k=0.$$

It follows that

$$\textstyle\sum_\rho(a^i_{\rho k}\phi^\rho_j-a^i_{\rho j}\phi^\rho_k)\equiv 0, \quad \operatorname{mod}\omega^l.$$

Since G has the property (C), the above equations imply that

$$\phi^\rho_j\equiv 0, \quad \operatorname{mod}\omega^k.$$

In other words, we have

(26) $$\Pi^\rho = \tfrac{1}{2}\textstyle\sum_{j,k} R^\rho_{jk}\omega^j \wedge \omega^k, \quad R^\rho_{jk} + R^\rho_{kj} = 0.$$

These are essentially the curvature form of the G-connection.

It will be convenient to introduce the quantities

(27) $$S^i_{jkl} = \textstyle\sum_\rho a^i_{\rho j} R^\rho_{kl}.$$

By substituting (26) into (24) and equating to zero the coefficients of the resulting cubic differential form, we get

(28) $$S^i_{jkl} + S^i_{klj} + S^i_{ljk} = 0.$$

We now consider a differential form of degree q in B_G which belong to the base manifold M, that is, which is the dual image of a differential form in M under the projection p. Such a differential form can be written as

(29) $$\eta = \frac{1}{q!}\textstyle\sum_{i_1,\ldots,i_q} P_{i_1\ldots i_q}\,\omega^{i_1}\wedge\ldots\wedge\omega^{i_q},$$

where $P_{i_1\ldots i_q}$ can be supposed to be anti-symmetric in any two of its indices. In order that η belongs to M, it is necessary that $d\eta$ has the same property. By using (23), we see that this implies the relations

(30) $$dP_{i_1\ldots i_q} + \textstyle\sum_{s=1}^q \sum_\rho P_{i_1\ldots i_{s-1} j\, i_{s+1}\ldots i_q} a^j_{\rho i_s}\pi^\rho = \sum_l P_{i_1\ldots i_q|l}\omega^l.$$

Exterior differentiation of this equation gives

$$\textstyle\sum_l (dP_{i_1\ldots i_q|l} + \sum_{s=1}^q \sum_{\rho,j} P_{i_1\ldots i_{s-1} j\, i_{s+1}\ldots i_q|l}\, a^j_{\rho i_s}\pi^\rho + \sum_{\rho,j} P_{i_1\ldots i_q|j}\, a^j_{\rho l}\pi^\rho)\wedge\omega^l$$
$$= \tfrac{1}{2}\textstyle\sum_{s=1}^q \sum_{j\,k,l} P_{i_1\ldots i_{s-1} j\, i_{s+1}\ldots i_q} S^j_{i_s kl}\omega^k\wedge\omega^l.$$

This allows us to put

(31) $$dP_{i_1\ldots i_q|l} + \textstyle\sum_{s=1}^q \sum_{\rho,j} P_{i_1\ldots i_{s-1} j\, i_{s+1}\ldots i_q|l}\, a^j_{\rho i_s}\pi^\rho$$
$$+ \textstyle\sum_{\rho,j} P_{i_1\ldots i_q|j}\, a^j_{\rho l}\pi^\rho = \sum_k P_{i_1\ldots i_q|l|k}\omega^k.$$

Substituting this into the last equation and equating to zero the coefficient of $\omega^k\wedge\omega^l$, we get

(32) $$P_{i_1\ldots i_q|l|k} - P_{i_1\ldots i_q|k|l} = \textstyle\sum_{s=1}^q \sum_j P_{i_1\ldots i_{s-1} j\, i_{s+1}\ldots i_q} S^j_{i_s kl}.$$

These equations are usually known as the interchange formulas.

An important case of a G-connection is when G is a subgroup of the orthogonal group $O(n)$ in n variables. In this case we shall lower the superscripts of our symbols and use subscripts throughout.

We first remark that *such a group G always has the property* (C). Since the infinitesimal transformations X_ρ leave invariant the quadratic form

$$(\xi^1)^2 + \ldots + (\xi^n)^2,$$

we have

$$a_{i\rho k} + a_{k\rho i} = 0. \tag{33}$$

Suppose that $\qquad \sum_\rho (-a_{i\rho k} b_j^\rho + a_{i\rho j} b_k^\rho) = 0.$

Permuting this equation cyclically in i, j, k, we get

$$\sum_\rho (-a_{j\rho i} b_k^\rho + a_{j\rho k} b_i^\rho) = 0,$$

$$\sum_\rho (-a_{k\rho j} b_i^\rho + a_{k\rho i} b_j^\rho) = 0.$$

By subtracting the first equation from the sum of the last two equations, we find

$$\sum_\rho a_{j\rho k} b_i^\rho = 0.$$

But these equations imply $b_i^\rho = 0$. This proves that G has the property (C).

When G is a subgroup of $O(n)$, there are some symmetry properties of S_{ijkl} which will be useful later on. From the second equation of (26) and (33), we have

$$S_{ijkl} = -S_{jikl} = -S_{ijlk}. \tag{34}$$

It is well known that these relations and (28) imply

$$S_{ijkl} = S_{klij}. \tag{35}$$

3. Proof of the decomposition theorem

We are now ready to give a proof of the decomposition theorem as stated in the Introduction. Since $G \subset R(n)$, the G-structure on M defines an orientation on M by the condition $\omega_1 \wedge \ldots \wedge \omega_n > 0$ and a Riemann metric on M by

$$ds^2 = \omega_1^2 + \ldots + \omega_n^2. \tag{36}$$

Relative to these the operators in (3) are defined. Our first problem is to compute $\Delta\eta$, with η given by (29).

This is a routine computation, and we shall only give the relevant formulas. First of all we have

$$d\eta = \frac{1}{q!} \sum_{i_1, \ldots, i_q, j} P_{i_1 \ldots i_q | j}\, \omega_j \wedge \omega_{i_1} \wedge \ldots \wedge \omega_{i_q}. \tag{37}$$

To make the coefficients anti-symmetric, we can write

$$d\eta = \frac{(-1)^q}{(q+1)!} \sum_{i_1, \ldots, i_{q+1}} (P_{i_1 \ldots i_q | i_{q+1}} - P_{i_{q+1} i_2 \ldots i_q | i_1} - \ldots - P_{i_1 \ldots i_{q-1} i_{q+1} | i_q})\, \omega_{i_1} \wedge \ldots \wedge \omega_{i_{q+1}}. \tag{38}$$

We also have, by definition,

$$(39)\qquad *\eta = \frac{1}{q!\,(n-q)!}\sum_{i_1,\dots i_n}\epsilon_{i_1\dots i_q i_{q+1}\dots i_n} P_{i_1\dots i_q}\omega_{i_{q+1}}\wedge\dots\wedge\omega_{i_n},$$

where $\epsilon_{i_1\dots i_n}$ is equal to $+1$ or -1, according as $i_1, \dots, i_n$ form an even or odd permutation of $1, \dots, n$, and is otherwise equal to zero. Using (3), we find

$$(40)\qquad \delta\eta = \frac{(-1)^q}{(q-1)!}\sum_{i_1,\dots,i_{q-1},j} P_{i_1\dots i_{q-1}j|j}\,\omega_{i_1}\wedge\dots\wedge\omega_{i_{q-1}}.$$

Further computation gives

$$(41)\qquad \begin{cases} -(q-1)!\,d\delta\eta = \sum_{i_1,\dots,i_q,j} P_{i_1\dots i_{q-1}j|j|i_q}\,\omega_{i_1}\wedge\dots\wedge\omega_{i_q}, \\ -(q-1)!\,\delta d\eta = \dfrac{1}{q}\sum_{i_1,\dots,i_q,j} P_{i_1\dots i_q|j|j}\,\omega_{i_1}\wedge\dots\wedge\omega_{i_q} \\ \qquad\qquad - \sum_{i_1,\dots,i_q,j} P_{i_1\dots i_{q-1}j|i_q|j}\,\omega_{i_1}\wedge\dots\wedge\omega_{i_q}. \end{cases}$$

By using (32), we get the following fundamental formula:

$$(42)\qquad \begin{aligned} -(q-1)!\,\Delta\eta = {} & \frac{1}{q}\sum_{i_1,\dots,i_q,j} P_{i_1\dots i_q|j|j}\,\omega_{i_1}\wedge\dots\wedge\omega_{i_q} \\ & + \sum_{i_1,\dots,i_q,k,j} P_{i_1\dots i_{q-1}k} S_{kji_qj}\,\omega_{i_1}\wedge\dots\wedge\omega_{i_q} \\ & - (q-1)\sum_{i_1,\dots,i_q,k,j} P_{i_1\dots i_{q-2}kj} S_{ki_{q-1}ji_q}\,\omega_{i_1}\wedge\dots\wedge\omega_{i_q}. \end{aligned}$$

The disadvantage of this formula is that the coefficients are not anti-symmetric in their indices. The following artifice is used to anti-symmetrize the coefficients: Let $\epsilon(i_1\dots i_q; j_1\dots j_q)$ denote the number which is equal to $+1$ or -1 according as $j_1, \dots, j_q$ form an even or odd permutation of $i_1, \dots, i_q$, and is otherwise equal to zero. We define

$$(43)\qquad \begin{aligned} & S(i_1\dots i_q, j_1\dots j_q;\ k_1\dots k_q, l_1\dots l_q) \\ & \quad = \Sigma\epsilon(i_1\dots i_q;\ r_1\dots r_{q-1}g)\,\epsilon(j_1\dots j_q;\ r_1\dots r_{q-1}h) \\ & \qquad \times \epsilon(k_1\dots k_q;\ s_1\dots s_{q-1}u)\,\epsilon(l_1\dots l_q;\ s_1\dots s_{q-1}r)\,S_{ghuv} \\ & \qquad\qquad (1 \leqq r, s, g, h, u, v \leqq n), \end{aligned}$$

where all the indices run from 1 to n and the summation is over all the repeated ones. To shorten our notation we write the symbol on the left-hand side also as $S((i)(j); (k)(l))$. It is easily seen that these symbols have the following properties:

(1) They are anti-symmetric in any two indices of each of the sets $i_1, \dots, i_q; j_1, \dots, j_q; k_1, \dots, k_q; l_1, \dots, l_q$.

(2) $S((i)(j); (k)(l)) = -S((j)(i); (k)(l))$.

(3) $S((i)(j); (k)(l)) = -S((i)(j); (l)(k))$.

(4) $S((i)(j); (k)(l)) = \ \ S((k)(l); (i)(j))$.

From these quantities we define

$$(44) \qquad S((i)(k)) = S(i_1 \dots i_q, k_1 \dots k_q) = \frac{1}{q!} \sum_{j_1, \dots j_q} S(i_1 \dots i_q, j_1 \dots j_q; k_1 \dots k_q, j_1 \dots j_q).$$

Then $S((i)(k))$ are anti-symmetric in the indices of each one of the sets $i_1, \dots, i_q$ and $k_1, \dots, k_q$, and

$$(45) \qquad S((i)(k)) = S((k)(i)).$$

It turns out that $S(i_1 \dots i_q; k_1 \dots k_q)$ are the quantities which occur in the expression for $\Delta\eta$. In fact, we find

$$(46) \qquad -\Delta\eta = \frac{1}{q!} \sum_{i_1, \dots, i_q, j} P_{i_1 \dots i_q | j | j} \omega_{i_1} \wedge \dots \wedge \omega_{i_q} + \frac{1}{(q!(n-q)!)^2} \sum_{i,k} P_{i_1 \dots i_q} S(i_1 \dots i_q, k_1 \dots k_q) \omega_{k_1} \wedge \dots \wedge \omega_{k_q}.$$

The exterior q^{th} power $\Lambda^q(V^*(x))$ of the space of covectors $V^*(x)$ at $x \in M$ of is dimension $N = \binom{n}{q}$ and has as base

$$(47) \qquad \omega_{i_1} \wedge \dots \wedge \omega_{i_q} \quad (1 \leqq i_1 < \dots < i_q \leqq n).$$

In $\Lambda^q(V^*(x))$ an inner product is defined by

$$(48) \qquad (\eta, \eta) = \frac{1}{q!} \sum_{i_1, \dots i_q,} P^2_{i_1 \dots i_q}.$$

Under the action of G through its linear representation this inner product remains invariant. If W_1 is an invariant subspace of $\Lambda^q(V^*(x))$ under G, its orthogonal space W_2 is also invariant. There exist therefore base vectors $\Phi_1, \dots, \Phi_N \in \Lambda^q(V^*(x))$, which are related to the base (47) by an orthogonal transformation

$$(49) \qquad \omega_{i_1} \wedge \dots \wedge \omega_{i_q} = \sum_{\lambda=1}^{N} g_{i_1 \dots i_q, \lambda} \Phi_\lambda,$$

such that $\Phi_1, \dots, \Phi_h$ and $\Phi_{h+1}, \dots, \Phi_N$ span W_1 and W_2 respectively. We can assume $g_{i_1 \dots i_q, \lambda}$ to be defined for all $i_1, \dots, i_q$ and anti-symmetric in any two of its first q indices. Then equations (49) can be solved for Φ_λ, giving

$$(50) \qquad \Phi_\lambda = \frac{1}{q!} \sum_{i_1, \dots i_q,} g_{i_1 \dots i_q, \lambda} \omega_{i_1} \wedge \dots \wedge \omega_{i_q}.$$

From now on till the end of this section we shall agree on the following ranges of indices:

$$(51) \qquad 1 \leqq A, B, C \leqq h, \quad h+1 \leqq \alpha, \beta, \gamma \leqq N.$$

We shall find the condition that W_1 and W_2 are invariant under G. For this purpose we compute the exterior derivative $d\Phi_A$ and find

(52) $$q!\, d\Phi_A = q \sum_{i_1,\ldots,i_q,l,\rho} g_{i_1\ldots i_{q-1}l,A}\, a_{l\rho i_q} \pi_\rho \wedge \omega_{i_1} \wedge \ldots \wedge \omega_{i_q}$$
$$= q \sum_{i_1,\ldots,i_{q-1},l,m,\lambda,\rho} g_{i_1\ldots i_{q-1}l,A}\, g_{i_1\ldots i_{q-1}m,\lambda}\, a_{l\rho m} \pi_\rho \wedge \Phi_\lambda.$$

It follows that the invariance of W_1 under G implies

(53) $$\sum_{i_1,\ldots,i_{q-1},l,m} g_{i_1\ldots i_{q-1}l,A}\, g_{i_1\ldots i_{q-1}m,\alpha}\, a_{l\rho m} = 0$$

or

(54) $$\sum_{i_1,\ldots,i_{q-1},l,m} g_{i_1\ldots i_{q-1}l,A}\, g_{i_1\ldots i_{q-1}m,\alpha}\, S_{lmjk} = 0.$$

As to be expected, this relation is symmetric in A and α.

Our theorem will be established if we prove that the condition that η is of type W_1 implies that $\Delta\eta$ is of type W_1. Suppose therefore that

(55) $$\eta = \sum_A P_A \Phi_A = \frac{1}{q!} \sum_{i_1,\ldots,i_q,A} P_A g_{i_1\ldots i_q,A}\, \omega_{i_1} \wedge \ldots \wedge \omega_{i_q}.$$

By (46) we find

(56) $$-\Delta\eta = \frac{1}{q!} \Sigma P_{A|j|j}\, g_{i_1\ldots i_q,A}\, \omega_{i_1} \wedge \ldots \wedge \omega_{i_q}$$
$$+ \frac{1}{(q!\,(q-1)!)^2} \Sigma P_A g_{i_1\ldots i_q,A}\, S(i_1\ldots i_q,\, k_1\ldots k_q)\, \omega_{k_1} \wedge \ldots \wedge \omega_{k}$$
$$= \Sigma P_{A|j|j} \Phi_A + \frac{1}{(q!\,(q-1)!)^2} \Sigma P_A g_{i_1\ldots i_q,A}\, g_{k_1\ldots k_q,\lambda}\, S(i_1\ldots i_q,\, k_1\ldots k_q)\, \Phi_\lambda.$$

It suffices to prove that

(57) $$\sum_{i_1,\ldots,i_q,k_1,\ldots,k_q} S(i_1\ldots i_q, k_1\ldots k_q)\, g_{i_1\ldots i_q,A}\, g_{k_1\ldots k_q,\alpha} = 0.$$

For this purpose we consider the quantities introduced in (43), and put

(58) $$(q!)^4 R_{\kappa\lambda\mu\nu} = \sum_{i_1,\ldots,l_q} g_{i_1\ldots i_q,\kappa}\, g_{j_1\ldots j_q,\lambda}\, g_{k_1\ldots k_q,\mu}\, g_{l_1\ldots l_q,\nu}$$
$$\times S(i_1\ldots i_q,\, j_1\ldots j_q;\, k_1\ldots k_q,\, l_1\ldots l_q),$$

where the indices of $R_{\kappa\lambda\mu\nu}$ have the ranges

(59) $$1 \leqq \kappa, \lambda, \mu, \nu \leqq N.$$

We also put

(60) $$R_{\kappa\mu} = \sum_\lambda R_{\kappa\lambda\mu\lambda}.$$

Because of similar properties of

$$S(i_1\ldots i_q,\, j_1\ldots j_q;\, k_1\ldots k_q,\, l_1\ldots l_q),$$

$R_{\kappa\lambda\mu\nu}$ has the properties

(61) $$R_{\kappa\lambda\mu\nu} = -R_{\lambda\kappa\mu\nu} = -R_{\kappa\lambda\nu\mu},$$
$$R_{\kappa\lambda\mu\nu} = R_{\mu\nu\kappa\lambda}.$$

From (54) it follows that, on remembering the ranges of indices as agreed upon in (51),

(62) $$R_{A\alpha\mu\nu}=0.$$

From this we find

$$R_{A\alpha}=\sum_\lambda R_{A\lambda\alpha\lambda}=\sum_B R_{AB\alpha B}+\sum_\beta R_{A\beta\alpha\beta}=0.$$

But this is exactly the equation (57) to be proved. Thus the proof of our decomposition theorem is complete.

4. The case of the unitary group

As discussed in the Introduction, our decomposition theorem reduces the proof of (7) to a purely algebraic problem. The latter has been solved in the theory of representations of the unitary group. To be precise, the problem can be formulated as follows:

Let L be a complex vector space of dimension $2m$, which is a direct sum of two complex vector spaces V, $\bar{V}$ of dimension m. Let ω_α, $\bar{\omega}_\alpha$ be base vectors of V, $\bar{V}$ respectively. The equations

(63) $$\begin{cases}\omega_\alpha\to\omega'_\alpha=\sum_\beta u_{\alpha\beta}\omega_\beta,\\ \bar{\omega}_\alpha\to\bar{\omega}'_\alpha=\sum_\beta \bar{u}_{\alpha\beta}\bar{\omega}_\beta,\end{cases}$$

where $(u_{\alpha\beta})$ is a unitary matrix, define a linear mapping of L, which maps the vector $\sum_\alpha(f_\alpha\omega_\alpha+g_\alpha\bar{\omega}_\alpha)$ into the vector $\sum_\alpha(f_\alpha\omega'_\alpha+g_\alpha\bar{\omega}'_\alpha)$. The linear mappings so obtained, for all $(u_{\alpha\beta})\in U(m)$, define a representation of $U(m)$. It induces a linear representation of $U(m)$ in the exterior power $\Lambda^r(L)$. Our problem is to decompose this representation into its irreducible parts.

As base vectors of $\Lambda^r(L)$ we can take

(64) $$\omega_{\alpha_1}\wedge\ldots\wedge\omega_{\alpha_p}\wedge\bar{\omega}_{\beta_1}\wedge\ldots\wedge\bar{\omega}_{\beta_q}$$
$$(p+q=r,\ 1\leqq\alpha_1<\ldots<\alpha_p\leqq m,\ 1\leqq\beta_1<\ldots<\beta_q\leqq m).$$

For fixed values of p, q these vectors clearly span an invariant subspace of $\Lambda^r(L)$, to be denoted by $\Lambda^{p,q}(L)$. An element of $\Lambda^{p,q}(L)$ is said to be bi-homogeneous with the bi-degree (p,q). Such an element can be written in the form

(65) $$\sum_{\alpha,\beta}P_{\alpha_1\ldots\alpha_p\beta_1\ldots\beta_q}\omega_{\alpha_1}\wedge\ldots\wedge\omega_{\alpha_p}\wedge\bar{\omega}_{\beta_1}\wedge\ldots\wedge\bar{\omega}_{\beta_q},$$

where we can suppose the coefficients to be anti-symmetric in the α's and β's separately. For $p\geqq 1$, $q\geqq 1$, the linear subspace in $\Lambda^{p,q}(L)$, defined by the equation

(66) $$\sum_\gamma P_{\alpha_1\ldots\alpha_{p-1}\gamma\beta_1\ldots\beta_{q-1}\gamma}=0,$$

is an invariant subspace.

We wish to remark that the following theorem is true: *For $p+q \leqq m$ the representation of $U(m)$ in the linear subspace* (66) *of $\Lambda^{p,q}(L)$ is irreducible.*

This can be verified by a computation of the character of the representation. In fact, we easily show that in the notation of H. Weyl† this is the representation of signature $(\underbrace{1 \ldots 1}_{p} 0 \ldots 0 \underbrace{-1 \ldots -1}_{q})$.

Moreover, if we consider the maximal Abelian subgroup of all diagonal matrices

$$\begin{pmatrix} \epsilon_1 & & 0 \\ & \ddots & \\ 0 & & \epsilon_n \end{pmatrix} \tag{67}$$

of $U(m)$ and introduce the integers

$$\begin{cases} l_1 = m, \ldots, l_p = m-p+1, \; l_{p+1} = m-p-1, \ldots, \\ \qquad l_{m-q} = q, \; l_{m-q+1} = q-2, \ldots, l_m = -1, \end{cases} \tag{68}$$

this representation has the character

$$\chi = \frac{|\epsilon^{l_1} \ldots \epsilon^{l_m}|}{|\epsilon^{m-1} \ldots \epsilon^0|}, \tag{69}$$

where

$$|\epsilon^{l_1} \ldots \epsilon^{l_m}| = \begin{vmatrix} \epsilon_1^{l_1} & \ldots & \epsilon_1^{l_m} \\ \ldots & \ldots & \ldots \\ \epsilon_m^{l_1} & \ldots & \epsilon_m^{l_m} \end{vmatrix}, \quad |\epsilon^{m-1} \ldots \epsilon^0| = \begin{vmatrix} \epsilon_1^{m-1} & \ldots & \epsilon_1^0 \\ \ldots & \ldots & \ldots \\ \epsilon_m^{m-1} & \ldots & \epsilon_m^0 \end{vmatrix}. \tag{70}$$

As is well known, condition (66) characterizes the primitive elements. Hence all these add to the remark that each summand in (7) corresponds to an irreducible representation of $U(m)$, so that in this sense Hodge's decomposition theorem cannot be improved.

We shall show briefly that our considerations include also as a particular case a generalization of Kähler geometry, which has been studied by A. Lichnerowicz.‡ This is the geometry on a compact even-dimensional manifold on which there are given an exterior quadratic differential form Ω of highest rank and a Riemann metric with the property that the covariant derivative of Ω vanishes. Lichnerowicz proved that if such is the case the Riemann metric can be so modified that we can suppose

$$\begin{cases} ds^2 = \sum_\alpha \theta_\alpha^2 + \sum_{\alpha'} \theta_{\alpha'}^2, \\ \Omega = \sum_\alpha \theta_\alpha \wedge \theta_{\alpha'}, \end{cases} \tag{71}$$

† [10], in particular, pp. 198–201. ‡ [7].

where $\alpha' = \alpha + m$, etc., and where θ_α, $\theta_{\alpha'}$ are linearly independent linear differential forms. By putting

$$\omega_\alpha = \frac{1}{\sqrt{2}}(\theta_\alpha + i\theta_{\alpha'}), \quad \overline{\omega}_\alpha = \frac{1}{\sqrt{2}}(\theta_\alpha - i\theta_{\alpha'}), \tag{72}$$

we can also write

$$\begin{cases} ds^2 = 2\sum_\alpha \omega_\alpha \overline{\omega}_\alpha, \\ \Omega = i\sum_\alpha \omega_\alpha \wedge \overline{\omega}_\alpha. \end{cases} \tag{73}$$

These forms define an almost complex structure on the manifold, and the group of the bundle is reduced to $U(m)$. By following the general discussions in § 2, we see that a connection can be defined in the bundle, with the group $U(m)$. Without going into details, we state that the forms $\omega_{\alpha\beta}$ which define the connection are characterized by the conditions

$$\begin{cases} d\omega_\alpha = \sum_\beta \omega_\beta \wedge \omega_{\beta\alpha} + \Omega_\alpha, \\ \Omega_\alpha = \sum_{\beta,\gamma} (A_{\alpha\beta\gamma}\omega_\beta \wedge \omega_\gamma + B_{\alpha\beta\gamma}\overline{\omega}_\beta \wedge \overline{\omega}_\gamma) \\ \qquad\qquad (A_{\alpha\beta\gamma} + A_{\alpha\gamma\beta} = 0,\ B_{\alpha\beta\gamma} + B_{\alpha\gamma\beta} = 0), \end{cases} \tag{74}$$

and

$$\omega_{\alpha\beta} + \overline{\omega}_{\beta\alpha} = 0. \tag{75}$$

It is possible to express these equations in the real form. For this purpose we write

$$\omega_{\alpha\beta} = \phi_{\alpha\beta} + i\psi_{\alpha\beta}, \tag{76}$$

where $\phi_{\alpha\beta}$, $\psi_{\alpha\beta}$ are real. Conditions (75) are equivalent to the conditions

$$\begin{cases} \phi_{\alpha\beta} + \phi_{\beta\alpha} = 0, \\ \psi_{\alpha\beta} - \psi_{\beta\alpha} = 0, \end{cases} \tag{77}$$

and equations (74) will then take the real form

$$\begin{cases} d\theta_\alpha = \sum_\beta (\theta_\beta \wedge \phi_{\beta\alpha} - \theta_{\beta'} \wedge \psi_{\beta\alpha}) + \Theta_\alpha, \\ d\theta_{\alpha'} = \sum_\beta (\theta_\beta \wedge \psi_{\beta\alpha} + \theta_{\beta'} \wedge \phi_{\beta\alpha}) + \Theta_{\alpha'}, \end{cases} \tag{78}$$

where Θ_α, $\Theta_{\alpha'}$ are defined by

$$\begin{cases} \Omega_\alpha = \dfrac{1}{\sqrt{2}}(\Theta_\alpha + i\Theta_{\alpha'}), \\ \overline{\Omega}_\alpha = \dfrac{1}{\sqrt{2}}(\Theta_\alpha - i\Theta_{\alpha'}). \end{cases} \tag{79}$$

On the other hand, the ds^2 in (71) defines a Riemann metric on the manifold. To this structure with the orthogonal group there always exists a connection without torsion, the parallelism of Levi-Civita.

The latter will be defined by the forms $\theta_{AB} = -\theta_{BA}$ $(A, B = 1, \ldots, 2m)$, satisfying the equations

$$(80) \qquad \begin{cases} d\theta_\alpha = \sum_\beta (\theta_\beta \wedge \theta_{\beta\alpha} + \theta_{\beta'} \wedge \theta_{\beta'\alpha}), \\ d\theta_{\alpha'} = \sum_\beta (\theta_\beta \wedge \theta_{\beta\alpha'} + \theta_{\beta'} \wedge \theta_{\beta'\alpha'}). \end{cases}$$

In terms of this connection we express the condition that the covariant derivative of Ω is zero. This gives

$$(81) \qquad \theta_{\alpha'\beta} = \theta_{\beta'\alpha}, \quad \theta_{\alpha'\beta'} = \theta_{\alpha\beta}.$$

Equations (78) are therefore satisfied, if we put

$$(82) \qquad \phi_{\beta\alpha} = \theta_{\beta\alpha}, \quad \psi_{\beta\alpha} = -\theta_{\beta'\alpha}, \quad \Theta_\alpha = 0, \quad \Theta_{\alpha'} = 0.$$

Since the forms $\phi_{\beta\alpha}$, $\psi_{\beta\alpha}$, Θ_α, $\Theta_{\alpha'}$ in (78) are completely determined by their symmetry properties (77) and the form of Ω_α, it follows that the condition of Lichnerowicz is equivalent to saying that the group of the tangent bundle can be reduced to $U(m)$ in such a way that the resulting connection has no torsion. Our decomposition theorem applies therefore to this case and gives a decomposition of harmonic forms identical with Hodge's theorem for Kähler manifolds.

As an illustration to derive topological consequences let us prove the following theorem: *If a compact manifold of dimension $2m$ has a $U(m)$-connection without torsion, then its odd-dimensional Betti numbers are even.*

On the complex-valued differential forms η of degree r we define the operator

$$(83) \qquad C = \sum_{p+q=r} i^{p-q} P_{p,q},$$

where $P_{p,q}\eta$ is the bihomogeneous component of η of bidegree (p, q). If η is harmonic, $P_{p,q}\eta$ is harmonic, and the same is true of $C\eta$. Moreover, this operator C has the following properties: (1) $C^2 = (-1)^r$; (2) C is a real operator, that is, it maps a real form into a real form. If r is odd, C defines a linear mapping on the space of real harmonic forms of degree r, such that $C^2 = -1$. It follows that the dimension of this vector space, that is, the r-dimensional Betti number, must be even.

5. Riemann manifolds with a field of parallel oriented linear spaces

We first prove the theorem:

Let $G = R(s) \times R(n-s) \subset R(n)$ be the product of two rotation groups in s and $n-s$ variables respectively, $0 < s < n$. The existence, on a manifold of dimension n, of a G-connection without torsion is equivalent to that *of a Riemann metric and a continuous field of oriented s-dimensional linear spaces which are parallel with respect to the Riemann metric.*

We adopt in this section the following ranges of indices:

$$1 \leqq \alpha, \beta, \gamma \leqq s, \quad s+1 \leqq a, b, c \leqq n. \tag{84}$$

Following the general method in § 2, we define, in the corresponding principal bundle, a uniquely determined set of forms $\omega_\alpha, \omega_a, \pi_{\alpha\beta} = -\pi_{\beta\alpha}, \pi_{ab} = -\pi_{ba}, \omega_{a\alpha}$, such that $\omega_\alpha, \omega_a, \pi_{\beta\alpha}, \pi_{ba}$ are linearly independent and

$$\omega_{\alpha a} = \sum_\beta A_{\alpha a \beta} \omega_\beta + \sum_b B_{\alpha a b} \omega_b, \tag{85}$$

which satisfy the conditions

$$\begin{cases} d\omega_\alpha = \sum_\beta \omega_\beta \wedge \pi_{\beta\alpha} - \sum_a \omega_a \wedge \omega_{\alpha a}, \\ d\omega_a = \sum_\alpha \omega_\alpha \wedge \omega_{\alpha a} + \sum_b \omega_b \wedge \pi_{ba}. \end{cases} \tag{86}$$

The forms $\pi_{\beta\alpha}, \pi_{ba}$ define a G-connection in the bundle. It is without torsion, if and only if

$$\omega_{\alpha a} = 0. \tag{87}$$

Let e_α, e_a be tangent vectors which are dual to the covectors ω_β, ω_b. The s-dimensional linear space spanned by e_α defines a parallel field, if and only if (87) is fulfilled. This proves our theorem.

Suppose from now on that we have a compact manifold with such a G-structure without torsion. A differential form of degree r can be written in the form

$$\eta = \sum_{p+q=r} \sum_{\alpha, a} P_{\alpha_1 \ldots \alpha_p a_1 \ldots a_q} \omega_{\alpha_1} \wedge \ldots \wedge \omega_{\alpha_p} \wedge \omega_{a_1} \wedge \ldots \wedge \omega_{a_q}, \tag{88}$$

where the coefficients $P_{\alpha_1 \ldots \alpha_p a_1 \ldots a_q}$ are supposed to be anti-symmetric in the α's and in the a's. For fixed p, q we define the operator

$$P_{p,q} \eta = \sum_{\alpha, a} P_{\alpha_1 \ldots \alpha_p a_1 \ldots a_q} \omega_{\alpha_1} \wedge \ldots \wedge \omega_{\alpha_p} \wedge \omega_{a_1} \wedge \ldots \wedge \omega_{a_q}. \tag{89}$$

Then we have

$$\eta = \sum_{p+q=r} P_{p,q} \eta. \tag{90}$$

Each of these summands is said to be bihomogeneous of bidegree p, q. According to our decomposition theorem, $P_{p,q}$ commutes with the operator Δ. It follows that if η is harmonic, then each of the summands in (90) is harmonic. Let $B^{p,q}$ be the number of linearly independent bihomogeneous harmonic forms of bidegree p, q. Then the r-dimensional Betti number of the manifold is given by

$$B^r = \sum_{p+q=r} B^{p,q}. \tag{91}$$

It can be shown that each of the summands in (90) corresponds to an irreducible representation of G. In this sense the decomposition of a harmonic form η given by (90) cannot be improved.

Let

$$\Omega_1 = \omega_1 \wedge \ldots \wedge \omega_s, \tag{92}$$

and let Φ be a bihomogeneous differential form of bidegree $0, q$. We wish to prove the following lemma:

The differential form Φ is harmonic, if and only if $\Omega_1 \wedge \Phi$ is harmonic.

To prove this lemma we remark that the exterior derivative of a bihomogeneous differential form η of bidegree (p, q) is a sum of two bihomogeneous differential forms, of bidegrees $(p+1, q)$ and $(p, q+1)$ respectively. We call them $d_1\eta$ and $d_2\eta$, so that

$$d = d_1 + d_2. \tag{93}$$

We write

$$\Phi = \frac{1}{q!} \Sigma_{a_1, \ldots, a_q} P_{a_1 \ldots a_q} \omega_{a_1} \wedge \ldots \wedge \omega_{a_q}, \tag{94}$$

where the coefficients $P_{a_1 \ldots a_q}$ are supposed to be anti-symmetric in their indices. We define an operator $*_2$ by

$$*_2 \Phi = \frac{1}{q!\,(n-s-q)!} \Sigma_{a_1, \ldots, a_{n-s}} \epsilon_{a_1 \ldots a_{n-s}} P_{a_1 \ldots a_q} \omega_{a_{q+1}} \wedge \ldots \wedge \omega_{a_s}. \tag{95}$$

Then we have

$$\begin{cases} * \Phi = \pm \Omega_1 \wedge *_2 \Phi, \\ *(\Omega_1 \wedge \Phi) = \pm *_2 \Phi. \end{cases} \tag{96}$$

It follows immediately from definition that, for a form of type $(0, q)$, the conditions $d_1 \Phi = 0$ and $d_1(*_2 \Phi) = 0$ are equivalent.

To prove our lemma, suppose Φ be harmonic:

$$d\Phi = 0, \quad d * \Phi = 0.$$

From the first equation follow

$$d_1 \Phi = 0, \quad d_1(*_2 \Phi) = 0.$$

From the second equation we get

$$d(\Omega_1 \wedge *_2 \Phi) = 0 \quad \text{or} \quad d_2(*_2 \Phi) = 0.$$

It follows that

$$d(\Omega_1 \wedge \Phi) = 0,$$
$$d * (\Omega_1 \wedge \Phi) = \pm d(*_2 \Phi) = \pm d_1(*_2 \Phi) \pm d_2(*_2 \Phi) = 0.$$

Hence $\Omega_1 \wedge \Phi$ is harmonic.

Conversely, suppose $\Omega_1 \wedge \Phi$ be harmonic:

$$d(\Omega_1 \wedge \Phi) = 0, \quad d * (\Omega_1 \wedge \Phi) = 0.$$

From the first equation we get

$$d_2 \Phi = 0.$$

From the second equation we get

$$d(*_2 \Phi) = 0,$$

which gives $$d_1(*_2 \Phi) = 0, \quad d_1 \Phi = 0.$$

It follows that

$$d\Phi = 0, \quad d * \Phi = \pm d(\Omega_1 \wedge *_2 \Phi) = 0.$$

Therefore Φ is harmonic.

From our lemma we get the following equalities:

$$B^{0,q} = B^{s,q}, \quad B^{p,0} = B^{p,n-s}. \tag{97}$$

In particular, we have

$$B^s \geqq B^{s,0} = B^{0,0} = 1, \tag{98}$$

which is the result stated at the end of the Introduction.

UNIVERSITY OF CHICAGO

REFERENCES

[1] S. CHERN, *Characteristic classes of Hermitian manifolds*, Ann. of Math., 47 (1946), pp. 85–121.

[2] ——, *Topics in differential geometry*, mimeographed notes, Princeton, 1951.

[3] P. R. GARABEDIAN and D. C. SPENCER, *A complex tensor calculus for Kähler manifolds*, Acta Math., 89 (1953), pp. 279–331.

[4] H. GUGGENHEIMER, *Über komplex-analytische Mannigfaltigkeiten mit Kählerscher Metrik*, Comment. Math. Helv., 25 (1951), pp. 257–297.

[5] W. V. D. HODGE, The theory and application of harmonic integrals, Cambridge University Press, 1941.

[6] S. LEFSCHETZ, L'analysis situs et la géometrie algébrique, Paris, Gauthier-Villars, 1950.

[7] A. LICHNEROWICZ, Généralisations de la géométrie kählerienne globale, Colloque de géométrie différentielle, Louvain, 1951, pp. 99–122.

[8] A. WEIL, *Sur la théorie des formes différentielles attachées à une variété analytique complexe*, Comment. Math. Helv., 20 (1947), pp. 110–116.

[9] ——, Theorie der Kählerschen Mannigfaltigkeiten, Göttingen, 1953.

[10] H. WEYL, The classical groups, Princeton, 1939.

On the Projective Embedding of Homogeneous Varieties

Wei-Liang Chow

IN an article [5] elsewhere in this volume, Weil has shown that an idea of Lefschetz on the projective embedding of an Abelian variety over the complex field, which seemingly depends upon the use of theta functions ([1], pp. 368–9), can actually be extended to the case of an abstract Abelian variety over a field of arbitrary characteristic. In this note we shall show that this idea can be further extended to get a projective embedding not only of an arbitrary group variety, but also of any homogeneous variety. We shall say that a variety V can be embedded in a projective space or has a projective embedding, if there is an everywhere biregular birational transformation of V onto a (not necessarily complete) variety contained in a projective space. A group variety G is said to act on a variety V, if there is a subvariety T in $G \times V \times V$, such that for any point a in G, the cycle

$$T_a = pr_{23}(T \cdot (a \times V \times V))$$

is defined and is an everywhere biregular birational transformation of V onto itself, and that for any two points a and b in G, we have the relation $T_{ab} = T_a T_b$; G is said to act transitively on V, if for any two points p and q in V, there exists a point a in G such that $p = T_a(q)$. A variety is said to be *homogeneous* if there is a group variety which acts transitively on it; it is clear that a group variety is homogeneous, since it acts transitively on itself by left translation. A field is said to be a field of definition for a homogeneous variety V, if it is such for the varieties V and G as well as for the subvariety T in $G \times V \times V$.

For the sake of convenience, we shall avail ourselves of the use of topological terminology by introducing the Zariski topology on an algebraic variety V, in which the closed subsets are bunches of subvarieties in V. If k is a field of definition for V, then we shall say that a subset W in V is k-closed if the bunch W is normally algebraic over k, and that a subset is k-open if it is the complement of a k-closed subset. A subset in a projective space is evidently a variety if and only if it is an open subset in a complete variety, and it is defined over a field k if and only if it is a k-open subset. We do not know whether any variety, not necessarily one embedded in a projective space, is also an open subset in a complete variety, but the following Lemma 1 offers in a sense a substitute for this property. This simple lemma is significant for our purpose for the following reason. One of the main difficulties in extending Weil's proof to our more general case is the fact that, for an incomplete variety V, the linear system of all functions f on V which are multiples of a given divisor X in V has in general an infinite dimension and hence does not possess a finite base; since the embedding mapping in the Weil proof is obtained from a finite base of a suitably chosen linear system, one sees readily that the lack of such a base is a serious obstacle. In case V is an open subset in a complete variety $\bar{V}$, any divisor X is the restriction to V of a divisor $\bar{X}$ in $\bar{V}$ and any function f is the restriction of a function $\bar{f}$ on $\bar{V}$; it is then natural to consider only those functions f such that $\bar{f}$ is a multiple of $\bar{X}$ on $\bar{V}$. As we shall see later, the complete variety V' in Lemma 1 below can be used for a similar purpose in the general case.

LEMMA 1. *Let V be a variety and k be a field of definition for V; then there exist a complete normal subvariety V' in a projective space and a birational correspondence F between V' and V, both defined over a purely inseparable extension k' of k, such that F is defined at every point in a k'-open subset W in V' and only at such points, and the image of W under F is V; furthermore, the inverse image of every point in V under F is a closed subset in V'.*

PROOF. Let $V_1, \ldots, V_t$ be the representatives of the variety V and let $B_1, \ldots, B_t$ be the frontiers of $V_1, \ldots, V_t$ respectively; without any loss of generality, we can consider each V_i to be a complete subvariety in a projective space, so that the product variety $V_1 \times \ldots \times V_t$ can also be considered as a complete subvariety in a projective space. Let $x_1, \ldots, x_t$ be a system of corresponding generic points of $V_1, \ldots, V_t$ respectively over k, and let V' be the locus of the point $x_1 \times \ldots \times x_t$ over k; then V' is a complete subvariety in $V_1 \times \ldots \times V_t$, defined over k, and the projection from V' to V_i defines a birational correspondence F_i between V' and $V_i - B_i$, defined over k, which is defined at every point in the k-open subset $W_i = V' \cap (V_1 \times \ldots \times V_{i-1} \times V_i - B_i \times V_{i+1} \times \ldots \times V_t)$; furthermore, it is clear that for any point y in $V_i - B_i$, its inverse image under F_i is the closed subset $V' \cap (V_1 \times \ldots \times V_{i-1} \times y \times V_{i+1} \times \ldots \times V_t)$ in V'. The system $(F_1, \ldots, F_t)$ then defines a birational correspondence F between V' and V, defined over k, which is defined at every point in the k-open subset $W = \bigcup_{i=1}^{t} W_i$ and only at these points; since $V_1, \ldots, V_t$ constitute a system of representatives of V, one sees readily that the image of W under F is V. Finally, if V' is not already normal (in the absolute sense), we can replace it by a derived normal model of it over the perfect closure of k and replace F by the induced correspondence between this derived normal model and V. This proves the lemma.

We further observe that W coincides with V' in case V is complete; however, we have no particular use here for this fact.

If X' is a divisor in V', we define $F(X') = pr_1(F \cdot (V \times X'))$; if X is a divisor in V, then $pr_2(F \cdot (X \times V'))$ is a divisor in W and we define $F^{-1}(X)$ to be the divisor in V' which contracts to the divisor $pr_2(F \cdot (X \times V'))$ in W. It can be easily shown that there exists a k'-closed subset C in W, such that F is biregular at every point in $W - C$; we set $D = (V' - W) + C$, so that D is also a k'-closed subset in V'. Since the image set of C under F cannot contain any variety of dimension $r-1$ (r being the dimension of V) which is not singular on V and since V' has no singular subvariety of dimension $r-1$, we have the relation $F(F^{-1}(X)) = X$ for every divisor X in V; while for any divisor X' in V', the divisor $X' - F^{-1}(F(X'))$ contains only components which lie in D. Similarly, if f is a function on V and $f' = f \cdot F$ is the function induced by f on V', then we have the relation $(f) = F((f'))$, while the divisor $(f') - F^{-1}((f))$ contains only components which lie in D. If K is an extension of k' such that the divisor (f) is rational over K, then the divisor $F^{-1}((f))$ is also rational over K; since the divisor $(f') - F^{-1}((f))$, having only components which lie in the k'-closed subset D, is rational over $\bar{k}$, it follows that the divisor (f') is rational over $\bar{K}$. Since V' is a complete variety, there is a constant element c such that the function cf' is defined over $\bar{K}$, and hence the function cf is also defined over $\bar{K}$. We shall find this remark useful presently.

Let V be a homogeneous variety and G be the group variety that acts on V; let V' be a complete normal subvariety in a projective space with the property stated in Lemma 1, and let F be the corresponding birational correspondence between V' and V. Consider a function f on V and a divisor X in V, and let K be a field of definition for V, V' and F such that X is rational over K and f is defined over K; let u be a generic point of G over K, and let f'_u be the function induced on V' by the function $f \cdot T_u^{-1}$. We shall say that the function f is a *strong multiple* of the divisor X (relative to F) and we shall write $(f) \succ\succ X$, if we have the relation $(f'_u) \succ F^{-1}(X_u)$ on V'. We observe that, on account of the relations $X_u = F(F^{-1}(X_u))$ and $(f) = F((f'))$ mentioned before, the relation $(f'_u) \succ F^{-1}(X_u)$ implies the relation $(f \cdot T_u^{-1}) \succ X_u$

and hence the relation $(f) \succ X$, so that a strong multiple of X is also a multiple of X in the usual sense. Also, it is clear that the definition of a strong multiple is independent of the choice of the generic point u of G over K. We shall denote by $L(X)$ the set of all functions on V which are strong multiples of the divisor $-X$; it is easily seen that $L(X)$ is a vector space of finite dimension over the field of constants. In fact, the dimension of $L(X)$ cannot exceed the dimension of the linear system $|F^{-1}(X_u)|$ on V'. For, if $f_1, \ldots, f_d$ is a set of linearly independent functions in $L(X)$, and if we denote by $f'_1, \ldots, f'_d$ the functions induced on V' by the functions $f_1 \cdot T_u^{-1}, \ldots, f_d \cdot T_u^{-1}$ respectively, then $f'_1, \ldots, f'_d$ are evidently linearly independent functions which are multiples of the divisor $F^{-1}(X_u)$, and the number of such functions cannot exceed the dimension of $|F^{-1}(X_u)|$. We shall denote by $|X|$ the set of all positive divisors in V which has the form $(f)+X$, where f is any function in $L(X)$, and we shall call this set the complete linear system determined by X. We shall say that a divisor X in V is *strongly equivalent to zero* (relative to F), and we shall write $X \approx 0$, if there exist a function f on V such that we have both the relations $(f) \succ\succ -X$ and $(1/f) \succ\succ X$, and we shall say that a divisor Y is *strongly equivalent to* X (relative to F) if $X - Y \approx 0$. It is easily seen that we have the relation $X \approx 0$ if and only if there exist a function f on V such that $(f'_u) = F^{-1}(X_u)$, where f'_u is the function induced on V' by the function $f \cdot T_u^{-1}$ and u is a generic point of G over a field of definition for V, V', F and f over which X is rational. The complete linear system $|X|$ contains all positive divisors which are strongly equivalent to X, but it may contain also other 'partial' divisors in case V is not complete. It can be easily seen that the relations $X \approx Y$ and $|X| = |Y|$ imply each other, and that both imply that $L(X)$ and $L(Y)$ have the same dimension.

LEMMA 2. *Let X be a divisor in V, and let K be a field of definition for V, V' and F, such that X is rational over K; then the relation $X \approx 0$ holds if and only if the relation $F^{-1}(X_u) \sim 0$ holds for every generic point u of G over K.*

PROOF. That the condition is necessary follows immediately from what we have just said above, so that we need only to prove the converse statement. Let f' be a function on V', rational over $K(u)$, such that $(f') = F^{-1}(X_u)$, and let f be the function induced by f' on V; if we set $g = f \cdot T_u$, then g is defined over $K(u)$ and we have the relation $(g) = X$. According to a remark we made before, there is a constant element c such that the function $h = cg$ is defined over $\bar{K}$; if h'_u is the function induced on V' by the function $h \cdot T_u^{-1} = cg \cdot T^{-1} = cf$, then we have evidently the relation $h'_u = cf'$ and hence $(h'_u) = (f') = F^{-1}(X_u)$. Since u is a generic point of G over $\bar{K}$, this proves our lemma.

We shall say that a linear system X on a homogeneous variety V is *ample*, if it has no fixed component and satisfies the two conditions (A) and (B) stated in Weil [5]. Similarly as in the case of a complete variety, we shall show that the existence of an ample linear system $|X|$ on V implies the existence of a projective embedding of V, whereby it is already sufficient to assume the condition (B) for a generic point of V. In fact, the condition (A) shows that a base of $L(X)$ will define a rational transformation T of V into a variety U (of the same dimension as V) in a projective space, such that T is defined at every point in V and is one-to-one between the points of V and its image U_0 under T; and the validity of condition (B) at a generic point shows that T is a birational transformation. Let U' be a derived (absolutely) normal model of U, and let R be the birational transformation of U onto U', so that $T' = RT$ is the induced birational transformation of V into U' and $T'' = T'F$ is the induced birational transformation of V' onto U'; let U'_0 be the image of U_0 under R. Since the image of any point in V under T' consists of at most a finite number of points and since V is non-singular, it follows from ([3], Chap. VI, Theorem 13), that T' is defined at every point in V; on the other hand, since the image of any point in U'_0 under T'^{-1} consists of a single point in V and since U' is normal, it follows, from [6], Main Theorem, that T'^{-1} is defined at every point in U'_0. Thus T' is an everywhere biregular birational transformation of V onto U'_0, and it remains to show that U'_0 is an open subset in U'. Since the image of any point in U'_0 under T'^{-1} consists of a single point in V and since the image of a point in V under F^{-1} is a closed subset in V' which is contained in W, it follows that the image of any point in U'_0 under T''^{-1} consists of a non-empty closed subset in V' which does not intersect $V' - W$ and a closed subset in $V' - W$; since U' is normal, it follows from Zariski's Connectedness Theorem ([7], p. 6), that the image of any point in U'_0 under T''^{-1} does not intersect $V' - W$, and hence $U' - U'_0$ is the image of $V' - W$ under T''. Since $V' - W$ is a closed subset in V', $U' - U'_0$ must be a closed subset in U' and hence U'_0 must be an open subset in U'.

In order to prove the possibility of a projective embedding of a homogeneous variety V, it is therefore sufficient (and necessary) to show the existence of an ample complete system of divisors in V. In order to do this, we observe that Lefschetz's idea, as formulated by Weil, can be carried over to our more general case with no essential change, provided we have the relation $3X \approx X_a + X_b + X_{a^{-1}b^{-1}}$ for any divisor X in V and any two points a and b in G. For this, we prove the following lemma, which is known to be true for Abelian varieties (Weil [4], Theorem 30, Corollary 2):

LEMMA 3. *If X is any divisor in a homogeneous variety V, and if a and b are any two points in the group variety G which acts on V, then we have the relation $X_{ab} - X_a - X_b + X \approx 0$.*

PROOF. We shall use the following result from the theory of Picard varieties: For any complete normal variety M in a projective space, there exists an Abelian variety P, called the Picard variety of M, and a rational homomorphism of $G_a(M)$ onto P, called the canonical homomorphism, which has the group $G_l(M)$ as its kernel; here $G_a(M)$ is the group of all divisors in M which are algebraically equivalent to zero and $G_l(M)$ is the group of all divisors in M which are linearly equivalent to zero. For the proof of this fact we refer to Matsusaka [2] as well as a forthcoming paper of ours on Picard varieties over arbitrary ground fields. Let P be the Picard variety of V', and let Φ be the canonical homomorphism of $G_a(V')$ onto P; let K be a field of definition for V, V', F, P and such that the divisor X and the points a and b are all rational over K, and let u and v be independent generic points of G over K. The divisor $F^{-1}(X_{uv}) - F^{-1}(X_v)$ is rational over $K(u, v)$, and is contained in $G_a(V')$; hence the point $\Phi(F^{-1}(X_{uv}) - F^{-1}(X_u))$ is rational over $K(u, v)$; therefore the correspondence $v \to \Phi(F^{-1}(X_{uv}) - F^{-1}(X_u))$ defines a rational transformation ϕ of G into P, defined over $K(u)$, which carries the unit element in G into the unit element in P. According to ([4], § 19), the rational transformation ϕ is a homomorphism of G into P; furthermore, for any point c in G such that uc is a generic point of G over K, we have the relation $\phi(c) = \Phi(F^{-1}(X_{uc}) - F^{-1}(X_u))$. Since the points uab, ua, ub evidently are generic points of G over K, we have then the relation

$$\begin{aligned}\Phi(F^{-1}(X_{uab}) - F^{-1}(X_u)) = \phi(ab) &= \phi(a) + \phi(b) \\ &= \Phi(F^{-1}(X_{ua}) - F^{-1}(X_u)) + \Phi(F^{-1}(X_{ub}) - F^{-1}(X_u)) \\ &= \Phi(F^{-1}(X_{ua}) + F^{-1}(X_{ub}) - 2F^{-1}(X_u)).\end{aligned}$$

This means that we have the relation

$$F^{-1}(X_{uab}) - F^{-1}(X_u) \sim F^{-1}(X_{ua}) + F^{-1}(X_{ub}) - 2F^{-1}(X_u)$$

and hence $F^{-1}(X_{uab}) - F^{-1}(X_{ua}) - F^{-1}(X_{ub}) + F^{-1}(X_u) \sim 0$; according to Lemma 2, this implies the relation $X_{ab} - X_a - X_b + X \approx 0$.

If we apply the above lemma to the pairs of points (a, b), (ab, b^{-1}) and $(a, a^{-1}b^{-1})$, we obtain the following relations:

$$X \approx X_a + X_b - X_{ab},$$
$$X \approx X_{ab} + X_{b^{-1}} - X_a,$$
$$X \approx X_a + X_{a^{-1}b^{-1}} - X_{b^{-1}},$$

which, when added together, gives the relation $3X \approx X_a + X_b + X_{a^{-1}b^{-1}}$. From here on we can proceed exactly as in Weil [5], except for a few obvious changes here and there; there is no need to repeat the proof here.

THEOREM. *Any homogeneous variety has a projective embedding.*

JOHNS HOPKINS UNIVERSITY

REFERENCES

[1] S. LEFSCHETZ, *On certain numerical invariants of algebraic varieties, with application to abelian varieties,* Trans. Amer. Math. Soc., 22 (1921), pp. 327–482.

[2] T. MATSUSAKA, *On the algebraic construction of the Picard variety*, Jap. J. Math., 21 (1951), pp. 217–235.

[3] A. WEIL, Foundations of algebraic geometry, New York, 1946.

[4] ——, *Variétés abeliennes et courbes algébriques*, Act. Sci. et Ind., no. 1064, Paris, 1948.

[5] ——, *On the projective embedding of abelian varieties*, this volume, pp. 177–181.

[6] O. ZARISKI, *Foundations of a general theory of birational correspondence,* Trans. Amer. Math. Soc., 53 (1943), pp. 490–542.

[7] ——, *Theory of applications of holomorphic functions on algebraic varieties over arbitrary ground fields*, Memoirs Amer. Math. Soc., no. 5 (1951).

Various Classes of Harmonic Forms

G. F. D. Duff

1. Introduction

ON a closed Riemannian manifold there is no distinction between harmonic forms and harmonic fields. Consequently the relationship of both these classes of differential forms to the homology structure of the manifold is contained in Hodge's theorem, which states that the number of linearly independent harmonic fields of degree p is equal to the pth Betti number of the manifold. Since the dual of a harmonic field is again a harmonic field, the Betti numbers of complementary dimensions must be equal—a well-known conclusion of the Poincaré duality theorem.

On a Riemannian manifold with boundary, a harmonic form is not necessarily a harmonic field. We shall consider these two classes of differential forms, and also a third and intermediate class, that of closed (or co-closed) harmonic forms. The theory of boundary value problems for such forms or tensors is closely connected with the relative homology theory of a manifold M with boundary B. In particular, we shall see several instances wherein the Lefschetz duality theorem $R_p(M,B)=R_{N-p}(M)$ plays an important role. Here N is the dimension of the manifold.

The boundary-value problems which we discuss will not be proved here from the beginning, our intention being rather to consider the relationships among the various problems for the different classes of harmonic forms. We shall derive a number of boundary-value theorems for special classes of harmonic forms as special cases of three general boundary-value theorems for harmonic forms.

The following notations are included for the convenience of the reader [2, 7]. An exterior differential form of degree p $(0 \leqq p \leqq N)$, represented in a given coordinate neighborhood by the expression

$$\phi = \phi_p = \phi_{(i_1 \ldots i_p)} dx^{i_1} \wedge \ldots \wedge dx^{i_p},$$

is constructed by multiplying components of a totally skew-symmetric tensor $\phi_{i_1 \ldots i_p}$ of rank p by the exterior differential product $dx^{i_1} \wedge \ldots \wedge dx^{i_p}$

of degree p and contracting over all sets of indices $i_1 < i_2 < \ldots < i_p$. The differential or exterior derivative is defined by

$$d\phi = (d\phi_{(i_1 \ldots i_p)}) \wedge dx^{i_1} \wedge \ldots \wedge dx^{i_p},$$

where the $d\phi_{i_1 \ldots i_p}$ is the 'scalar' total differential. Upon multiplying out the differentials a form of degree $p+1$ is found. If $d\phi = 0$, ϕ is said to be closed, and if there exists a $(p-1)$-form θ_{p-1}, such that $\phi = d\theta$ ϕ is said to be derived. Since $d \cdot d\phi \equiv 0$ in all cases, a derived form is necessarily closed.

Denoting by bC the boundary of a $(p+1)$ dimensional chain $C = C_{p+1}$, we have the general form of Stokes's theorem

$$\int_{C_{p+1}} d\phi = \int_{bC_{p+1}} \phi.$$

From this formula it is easy to show that the integrals of closed forms over homologous cycles are equal; their common value is called the period of ϕ over cycles of the homology class. A derived form necessarily has zero periods. It has been proved by G. de Rham that in a closed manifold there exists a closed form having $R_p(M)$ assigned periods on independent p-cycles, and that a closed form with vanishing periods on all cycles is derived. Analogous theorems for a manifold with boundary have been proved and will be quoted when necessary.

The preceding developments are independent of any Riemannian metric. However, we now introduce such a metric, based upon a covariant and symmetric tensor a_{ik}. We suppose that a_{ik} is positive-definite, that is, that $ds^2 = a_{ik} dx^i dx^k > 0$ for $\sum_i (dx^i)^2 \neq 0$. We also assume that $M+B$ is compact and C^∞, that B is C^∞ in $M+B$, and that a_{ik} is C^∞ in the local coordinate systems.

The adjoint, or dual, of the form ϕ is the form of degree $(N-p)$, denoted by $*\phi$, where

$$(*\phi)_{j_1 \ldots j_{N-p}} = e_{(i_1 \ldots i_p) j_1 \ldots j_{N-p}} \phi^{(i_1 \ldots i_p)};$$

$e_{i_1 \ldots i_N}$ being the 'unit' alternating N-tensor density. The relation $**\phi = (-1)^{Np+p} \phi$ holds. Exterior multiplication of ϕ by its dual $*\phi$ leads to an N-form $\phi \wedge *\phi$ which when integrated over the manifold defines a positive definite norm $N(\phi) = (\phi, \phi) = \int_M \phi \wedge *\phi$, and corresponding symmetric scalar product $(\phi, \psi) = (\psi, \phi)$.

The co-differential $\delta\phi$ of degree $(p-1)$ is defined as

$$\delta\phi = (-1)^{Np+N+1} * d * \phi,$$

and satisfies the operator relation $\delta\delta\phi \equiv 0$. A form is co-closed or co-derived according as $\delta\phi = 0$ or $\phi = \delta\psi$. From Stokes's theorem we may derive Green's formula

$$(d\phi, \psi) - (\phi, \delta\psi) = \int_B \phi \wedge * \psi,$$

where $\phi = \phi_{p-1}$, $\psi = \psi_p$ and the brackets indicate scalar products over M. Thus δ is the formal differential adjoint of d.

We now introduce the Laplacian operator $\Delta = d\delta + \delta d$. A harmonic form ϕ satisfies, by definition, the linear second-order equation $\Delta\phi = 0$. In Euclidean space the components of a harmonic form are harmonic functions in the usual sense. The equations satisfied by harmonic fields are $d\phi = 0$, $\delta\phi = 0$, which imply at once that a harmonic field is a harmonic form. Green's formula for harmonic forms may be written

$$(d\phi, d\psi) + (\delta\phi, \delta\psi) - (\phi, \Delta\psi) = \int_B (\phi \wedge * d\psi - \delta\psi \wedge * \phi).$$

On the boundary manifold B there is an induced form of degree p which we denote by $t\phi$. Letting x^N denote a coordinate normal to B, we see that $t\phi$ consists of those terms of ϕ not containing dx^N. The remaining, or normal part, contains dx^N as a factor. Dualizing interchanges the t and n operations; thus $*t = n*$ and $*n = t*$.

Finally, we remark that a harmonic form on M is C^∞ [7].

2. Harmonic forms

The following boundary-value theorems for harmonic forms have been proved elsewhere [2, 3]. We state them here as a starting point for our comparisons of the various special cases. In the Dirichlet problem the assigned data are values of the components of ϕ on the boundary. For brevity we state once for all that all such data are assumed sufficiently smooth (Hölder continuous).

THEOREM I. *There exists a harmonic form ϕ having assigned boundary values of $t\phi$ and $n\phi$.*

The solution is known to be unique if the metric tensor a_{ik} is analytic, and in other cases it can be proved that the number of linearly independent eigensolutions of the Dirichlet problem is finite. From Green's formula we see that the Dirichlet integral $(d\phi, d\phi) + (\delta\phi, \delta\phi)$ of any harmonic form ϕ with $t\phi = 0$, $n\phi = 0$ vanishes. That is, ϕ must be a harmonic field, a result analogous to the situation in a closed space.

We state next the mixed boundary-value theorem for harmonic forms, in which $t\phi$, $t\delta\phi$ are assigned. The reader may verify that the total number of components assigned is equal to $\binom{N}{p}$.

THEOREM II. *There exists a harmonic form ϕ having assigned values of $t\phi$ and $t\delta\phi$ if and only if these values satisfy*

$$\int_B \delta\phi \wedge * \rho = 0,$$

for every harmonic form ρ with $t\rho = 0$, $t\delta\rho = 0$.

From Green's formula we see that $\Delta\rho = 0$, $t\rho = 0$ and $t\delta\rho = 0$ imply $d\rho = 0$, $\delta\rho = 0$. Thus the independent conditions satisfied by ρ are $d\rho = 0$, $\delta\rho = 0$ and $t\rho = 0$, so that ρ is a harmonic field with vanishing tangential part on B. We shall see below that the number of these linearly independent eigenforms ρ is precisely the relative Betti numbers $R_p(M, B)$. Dual to Theorem II is a problem in which $n\phi$ and $nd\phi$ are assigned. However, this dual problem for p-forms is equivalent to the original one for $(N-p)$-forms and so need not be stated separately. The number of linearly independent eigenforms of this problem is $R_{N-p}(M, B) = R_p(M)$, the number of independent absolute p-cycles.

Finally, we have the Neumann problem in which only values of the components of $d\phi$ and $\delta\phi$ are given.

THEOREM III. *There exists a harmonic form ϕ having assigned values of $nd\phi$ and $t\delta\phi$ if and only if*

$$\int_B (\tau \wedge * d\phi - \delta\phi \wedge * \tau) = 0$$

for every harmonic field τ which is C^∞ in M.

This problem differs from the two preceding inasmuch as the number of linearly independent eigensolutions is infinite. This we will demonstrate in § 4 by showing that infinitely many linearly independent harmonic fields can be constructed on a manifold with boundary. Green's formula again shows that an eigenform of this problem is necessarily a harmonic field.

Proofs of these results have been given by the integral equation method of Poincaré and Fredholm. However, since certain tangential derivatives of components on the boundary are present in the problem, the singularity of the kernels in the integral equation is of a higher order than in the classical Poincaré-Fredholm theory. This necessitates the use of principal values to define the integrals, and the theory of Giraud [5] to solve the resulting singular integral equations.

3. Co-closed harmonic forms

The simplest formal analogue of the scalar Laplacian $\Delta = \operatorname{div} \operatorname{grad}$ is the operator δd, which for $p \geqq 1$ has somewhat weaker properties than the Laplacian $\Delta = d\delta + \delta d$. However, a co-closed harmonic form satisfies both $\delta d\phi = 0$ and $\delta\phi = 0$. The boundary-value problems appropriate to this pair of differential equations may be derived as special cases of Theorems II and III above in view of the following

LEMMA. *If $\Delta\phi = 0$ in M and $t\delta\phi = 0$ on B, then $\delta\phi \equiv 0$ in M.*

To prove this we use Green's formula to calculate first

$$N(d\delta\phi) = (d\delta\phi, d\delta\phi) = (\delta\phi, \delta d\delta\phi) + \int_B \delta\phi \wedge * d\delta\phi.$$

The first term on the right vanishes since $\delta d\delta\phi = \delta\Delta\phi = 0$. The second term contains only tangential components of $\delta\phi$ in the first factor of the integrand; these are zero by hypothesis. Thus $N(d\delta\phi) = 0$ which implies that $d\delta\phi = 0$. Now consider

$$N(\delta\phi) = (\delta\phi, \delta\phi) = (\phi, d\delta\phi) - \int_B \delta\phi \wedge * \phi$$

by Green's formula. Since each term on the right contains a vanishing factor, we see that $N(\delta\phi) = 0$ so $\delta\phi = 0$ in M as stated.

In the Dirichlet problem for co-closed harmonic forms the values of $t\phi$ are assigned.

THEOREM IV. *There exists a co-closed harmonic form ϕ having assigned values of $t\phi$.*

To establish this we construct the harmonic form of Theorem II having the given values of $t\phi$ and zero values for $t\delta\phi$. Since the orthogonality condition in Theorem II is satisfied if $t\delta\phi = 0$, such a harmonic form exists. Applying the Lemma, we see that it is co-closed throughout M. Clearly the eigenforms of Theorem II are also eigenforms of this problem.

The Neumann problem for co-closed harmonic forms is a consequence of Theorem III, in which we take $t\delta\phi = 0$, and apply the Lemma. However, the orthogonality condition can be expressed in a more convenient form in this case by means of the concept of an 'admissible' tangential boundary value, due to Tucker [8]. Since, for any closed form ϕ we have

$$d_B t\phi = t d\phi = 0,$$

where d_B is the differential operator in the boundary B, we see that the tangential boundary values $t\phi$ of ϕ must be closed in B. If, moreover, C_p is a p-cycle of B which bounds in M, we have $C_p = bR_{p+1}$ for some relative cycle R_{p+1} of M (mod B) and

$$\int_{C_p} t\phi = \int_{C_p} \phi = \int_{R_{p+1}} d\phi = 0,$$

by Stokes's theorem. Thus $t\phi$ has zero period on all such cycles C_p of B which are homologous to zero in M. Conversely given any form θ_p, defined on B, closed, and having zero periods on these cycles, there exists a form ϕ in M, closed ($d\phi = 0$), and with $t\phi = \theta$ [1]. Such a form θ is therefore said to be admissible as the tangential boundary value of a form closed in M.

If $t\delta\phi = 0$, the orthogonality condition of Theorem III becomes

$$\int_B \tau \wedge * d\phi = 0, \quad d\tau = 0, \quad \delta\tau = 0.$$

In the Neumann problem for co-closed harmonic forms we require $d * d\phi \equiv \pm * (\delta d\phi) = 0$, so the tangential boundary values $t * d\phi$ must be admissible. That is, we take $t * d\phi = t\psi$, for some closed form ψ defined in M. Thus

$$\int_B \tau \wedge * d\phi = \int_B \tau \wedge \psi = \pm \int_B \psi \wedge * (*\tau) = \pm (d\psi, *\tau) \mp (\psi, \delta * \tau) = 0,$$

and the condition is satisfied. Therefore we have

THEOREM V. *There exists a co-closed harmonic form ϕ with assigned admissible boundary values of $t * d\phi$.*

The solution is unique up to an additive harmonic field only.

Since the condition of admissibility for $t * d\phi$ is necessary if the differential equation $\delta d\phi = 0$ is to hold, we see that the solution of Theorem III is co-closed if and only if $t\delta\phi = 0$ and $t * d\phi$ is admissible.

The existence of p-tensor potentials ($\delta d\phi = 0$) satisfying the boundary conditions of Theorems IV and V was conjectured by Tucker in his early work on this subject.

4. Harmonic fields

The theorem of this section is not new, but the proof of existence is considerably simpler [4]. From the results to be found here we shall, in the next section, derive alternative statements for the theorems of § 3.

In the Dirichlet problem for harmonic fields ϕ,

$$d\phi = 0, \quad \delta\phi = 0,$$

values of $t\phi$, necessarily admissible, are assigned.

THEOREM VI. *There exists a unique harmonic field ϕ_p having assigned admissible values of $t\phi$, and assigned periods ν_i on $R_p(M, B)$ given independent relative p-cycles.*

Since the values of $t\phi$ may differ from zero the relative periods are not relative homology invariants; hence the boundaries of the relative cycles must be $p-1$-dimensional point sets assigned in advance. The uniqueness follows from the fact that if $t\phi=0$ and the relative periods ν_i all vanish, then ϕ is derived, $\phi=d\theta$, say, where $t\theta=0$ ([1], Theorem 6). Thus

$$N(\phi)=(\phi,d\theta)=(\delta\phi,\theta)+\int_B \theta\wedge*\phi=0,$$

and ϕ vanishes identically.

To construct the solution in the non-homogeneous case, we start with a closed form ρ having the assigned boundary values and periods. The existence of such a form ρ is established in ([1], Theorem 4). From the Kodaira or de Rham orthogonal decomposition formulae [6, 7] we can express ρ_p as a sum

$$\rho_p=d\sigma_{p-1}+\chi_p,$$

where χ_p is a harmonic field. Now let ϕ^2_{p-1} be a solution of Theorem IV with $t\phi^2_{p-1}=t\sigma_{p-1}$. Then consider the form

$$\phi=d\phi^2_{p-1}+\chi_p,$$

which is a harmonic field since

$$\delta d\phi^2=0.$$

Since $$td\phi^2=d_Bt\phi^2=d_Bt\sigma=td\sigma,$$

we see that $t\phi=t(d\sigma_{p-1}+\chi_p)=t\rho$ which is the assigned boundary value. Since also

$$\int_{R^i_p} d\phi^2_{p-1}=\int_{bR^i_p}\phi^2_{p-1}=\int_{bR^i_p}\sigma_{p-1}\int_{R^i_p}=d\sigma_{p-1},$$

we see that ϕ has the same relative periods as ρ and therefore constitutes the required solution.

As a first Corollary we see that if $t\phi=0$ the above harmonic field is an eigenform of Theorem II. Since the $R_p(M,B)$ relative periods ν_i can be chosen independently, we have the

COROLLARY I. *The dimension of the eigenspace of Theorem II:*

$$d\rho=0,\quad \delta\rho=0,\quad t\rho=0$$

is the relative Betti number $R_p(M,B)$.

We state next, as a Corollary of Theorem VI, a result which is a kind of dual to that theorem.

COROLLARY II. *There exists a unique harmonic field ϕ having assigned admissible values of $t*\phi$, and assigned periods μ_i on $R_p(M)$ independent absolute p-cycles.*

The uniqueness follows at once, for if the absolute periods are zero, ϕ is a derived form, $\phi = d\xi$, say. If then $t * \phi$ is zero, we have

$$N(\phi) = (\phi, d\xi) = (\delta\phi, \xi) + \int_B \xi \wedge * \phi = 0,$$

and ϕ vanishes identically.

To show that the solution exists, we take an $N-p$-form ϕ'_{N-p} which is a solution of Theorem VI with the assigned boundary values $t * \phi$. We require to adjust the absolute periods of $* \phi'_{N-p}$. Now the period matrix on the absolute p-cycles of the duals of the

$$R_{N-p}(M, B) = R_p(M)$$

independent eigenforms of degree $N-p$ for Theorem VI is non-singular. Indeed, if any linear combination of these forms had vanishing absolute periods it would, by the above demonstration of uniqueness, vanish identically, in contradiction to the property of linear independence. We may therefore find a form ρ_{N-p}, such that $t\rho_{N-p} = 0$, $d\rho_{N-p} = 0$, $\delta\rho_{N-p} = 0$, whose dual $* \rho_{N-p}$ has suitable periods on the absolute p-cycles. Adding this form to ϕ'_{N-p} we obtain a form whose dual is the required solution.

COROLLARY III. *There exists a unique co-derived harmonic form ϕ having given admissible values of $t\phi$.*

The differential equations in this case are $d\phi = 0$ and $\phi = \delta\chi$ for some form χ. The uniqueness is immediate, since if $t\phi = 0$, we have

$$N(\phi) = (\phi, \delta\chi) = (d\phi, \chi) - \int_B \phi \wedge * \chi = 0.$$

To demonstrate that a co-derived harmonic field exists, we note that the duals of the eigenforms of Theorem VI have a non-singular period matrix on the absolute cycles of dimension $N-p$. We may therefore add to a solution ϕ_1 of Theorem VI having the assigned values of $t\phi$, an eigenform ρ so chosen that the absolute periods of the dual of the resulting harmonic field $\phi_1 + \rho$ vanish. Since this dual form is then derived, the form $\phi = \phi_1 + \rho$ is co-derived, as required.

Theorem VI was also conjectured by Tucker, and was proved in [4]. By analogy with the case $N = 2$, $p = 1$, when a harmonic field and its dual may be represented as the differentials of the real and imaginary parts of a function of a complex variable, the second Corollary above has been called the Neumann problem for harmonic fields. This is because the Dirichlet problem for a harmonic function $u(x, y)$ is in a certain sense equivalent to the Neumann problem for the harmonic conjugate function $v(x, y)$. However, in the higher dimensional cases it seems better to regard the boundary value problems as equivalent and then the choice of relative or absolute periods to specify the solution uniquely is seen as a consequence of the Lefschetz duality theorem.

5. Co-derived harmonic forms

From the first Corollary of the preceding section we can deduce alternative statements of the results of § 3.

THEOREM IVa. *There exists a unique co-derived harmonic form ϕ having assigned values of $t\phi$.*

Thus the differential equations to be satisfied are $\delta d\phi = 0$, $\phi = \delta\chi$. The uniqueness can be proved in two stages as follows [4]. If $t\phi = 0$, then

$$N(d\phi) = (d\phi, d\phi) = (\phi, \delta d\phi) + \int_B \phi \wedge * d\phi = 0,$$

so that ϕ is closed. Next,

$$N(\phi) = (\phi, \delta\chi) = (d\phi, \chi) - \int_B \phi \wedge * \chi = 0,$$

showing that ϕ vanishes identically, and establishing the uniqueness of the solution.

To construct the solution, let ϕ_1 be a solution of Theorem IV with the assigned boundary values $t\phi$. Then $*\phi_1$ is closed. To ϕ_1 we add a suitable eigenform ρ_1 (the existence of which is assured by the work of § 4), whose dual has as absolute periods on the $N-p$ cycles the negatives of the periods of ϕ_1. Thus $*\phi_1 + *\rho_1$ is derived, so $\phi = \phi_1 + \rho_1$ is co-derived as required. A special case of this result has been proved in [4].

The Neumann problem for co-derived harmonic forms can be solved in an exactly similar way. Taking ϕ_1 as a solution of Theorem V with assigned values of $t * d\phi$, we add to ϕ_1 a harmonic field ρ_2 such that $*(\phi_1 + \rho_2)$ has zero periods on the absolute cycles of dimension $N-p$. Thus $*(\phi_1 + \rho_2)$ is derived, and $\phi = \phi_1 + \rho_2$ is co-derived.

THEOREM Va. *There exists a co-derived harmonic form ϕ having assigned admissible values for $t * d\phi$.*

The solution is not unique, for any co-derived harmonic field can be added to it.

In conclusion, it may be remarked that since our results have all flowed from the theorems for harmonic forms, these might be regarded as the basic boundary-value theorems for all classes of harmonic forms. However, the connections with the homology properties of the manifold are most evident when we turn to the special cases.

UNIVERSITY OF TORONTO

References

[1] G. F. D. Duff, *Differential forms in manifolds with boundary*, Ann. of Math., 56 (1952), pp. 115–127.

[2] ——, *Boundary value problems associated with the tensor Laplace equation*, Canadian J. Math., 5 (1953), pp. 57–80.

[3] ——, *A tensor boundary value problem of mixed type*, Canadian J. Math., 6 (1954), pp. 427–440.

[4] —— and D. C. Spencer, *Harmonic tensors on Riemannian manifolds with boundary*, Ann. of Math., 56 (1952), pp. 128–156.

[5] G. Giraud, *Équations et systèmes d'équations où figurent des valeurs principales d'intégrales*, C.R. Acad. Sci., Paris, 204 (1937), pp. 628–630.

[6] K. Kodaira, *Harmonic fields in Riemannian manifolds*, Ann. of Math., 50 (1949), pp. 587–665.

[7] G. de Rham and K. Kodaira, *Harmonic integrals*, Mimeographed lectures, Institute for Advanced Study, 1950.

[8] A. W. Tucker, *A boundary-value theorem for harmonic tensors*, Bull. Amer. Math. Soc., 47 (1941), p. 714.

On the Variation of Almost-Complex Structure

*K. Kodaira and D. C. Spencer**

1. Introduction

LET M be an almost-complex manifold of class C^∞, and let Φ be the exterior algebra of the complex-valued differential forms of class C^∞ on M. We denote by $\bar{\partial}$ the anti-derivation of degree $+1$ of Φ which maps elements of Φ^0 (functions) into forms of type $(0, 1)$ and which satisfies the commutativity relation $d\bar{\partial} + \bar{\partial}d = 0$, where d is the exterior differential of Φ. Given an arbitrary Hermitian metric on M, let $\mathfrak{d}$ be the adjoint of $\bar{\partial}$ in the sense that $(\bar{\partial}\phi, \psi) = (\phi, \mathfrak{d}\psi)$ for forms $\phi, \psi \in \Phi$ with compact supports. We introduce as Laplacian the operator $\Delta = 2(\mathfrak{d}\bar{\partial} + \bar{\partial}\mathfrak{d})$ and show that the almost-complex structure of M is integrable if $\Delta = \delta d + d\delta$, where δ is the adjoint of d. Denote by $\mathscr{H}^{r,s}$ the complex vector space of norm-finite harmonic forms of type (r, s) (forms annihilated by $\bar{\partial}$ and $\mathfrak{d}$). If M is compact, $\mathscr{H}^{r,s}$ is finite-dimensional; if, in addition, M is complex, $\mathscr{H}^{r,s}$ is isomorphic to the $\bar{\partial}$-cohomology of forms of type (r, s) (theorem of Dolbeault).

Now let M be a compact manifold of class C^∞ which carries an almost-complex structure $\mathscr{S}(t)$ which depends in a C^∞ manner on a point t in a domain of a Euclidean space. We construct a metric which is Hermitian with respect to the structure $\mathscr{S}(t)$ and depends on t in a C^∞ manner, and consider the corresponding space $\mathscr{H}^{r,s}(t)$. Writing $h^{r,s}(t) = \dim \mathscr{H}^{r,s}(t)$, we investigate the integer $h^{r,s}(t)$ in its dependence on t and show that it is upper semi-continuous; that is

$$\limsup_{t \to t_0} h^{r,s}(t) \leqq h^{r,s}(t_0).$$

Moreover, if $\mathscr{S}(t)$ is a Kähler structure for each t, then $h^{r,s}(t)$ is continuous, therefore a constant.

* This work was supported by a research project at Princeton University sponsored by the Office of Ordnance Research, U.S. Army.

2. Almost-complex structure

Let M be a real differentiable manifold of class C^∞ and dimension m, and let $T=T(M)$ be the sheaf (faisceau) of germs of C^∞ cross-sections of the tangent bundle of M, $A=A(M)$ the de Rham sheaf of germs of exterior differential forms of class C^∞. Write $CT=T\otimes_R C$, $CA=A\otimes_R C$, where R and C denote respectively the real and complex numbers. For each point $x\in M$, $(CT)_x$ is a left $(CA)^0_x$-module which is m-dimensional and free. The manifold M has a complex almost-product structure if and only if there exists an endomorphism P of CT satisfying: (1) $P^2=P$; (2) image P_x is a free submodule of $(CT)_x$ for each $x\in M$. We therefore have a direct sum decomposition $CT=P(CT)\oplus Q(CT)$, where Q is the projection of CT onto the kernel of P. The manifold has an almost-complex structure if and only if it has a complex almost-product structure with a conjugate-linear involutory isomorphism $u\to\bar{u}$ (conjugation) of CT which maps $P(CT)$ onto $Q(CT)$. If M is almost-complex, its real dimension m is even, $m=2n$.

Assume that M has an almost-complex structure, and let $\Phi=\Gamma(CA,M)$. Thus Φ is the exterior algebra of $\Phi^1=\Gamma((CA)^1,M)$, and we denote the symmetric algebra of Φ^1 by Ψ. The direct sum decomposition of CT induces corresponding decompositions of Φ and Ψ, namely,

$$\Phi=\Sigma_{r,s}\Pi_{r,s}\Phi,\quad \Psi=\Sigma_{r,s}\Pi_{r,s}\Psi,$$

where $\Pi_{r,s}$ denotes projection onto the subspace composed of elements of type (r,s). A form is of type (r,s) if and only if its contraction with $r+s$ tangent vectors at a point $p\in M$ vanishes unless r of these vectors belong to $P(CT)$, s of them to $Q(CT)$.

Let ∂ be the anti-derivation of degree $+1$ of Φ which coincides on Φ^0 with $\Pi_{1,0}d$ and which satisfies the relation $d\partial+\partial d=0$. We have the decomposition $d=d_1+d_2+\bar{d}_1+\bar{d}_2$, where

$$d_1=\Sigma_{r,s}\Pi_{r+1,s}\,d\,\Pi_{r,s},\quad d_2=\Sigma_{r,s}\Pi_{r+2,s-1}\,d\,\Pi_{r,s},$$

and it is easy to verify that

$$\partial=2d_2+d_1-\bar{d}_2.$$

We denote the conjugate of ∂ by $\bar\partial$ ($\bar\partial=2\bar{d}_2+\bar{d}_1-d_2$), and we have the decomposition

$$d=\partial+\bar\partial.$$

The almost-complex structure of M is said to be (completely) integrable if and only if $\partial^2=\partial\cdot\partial=0$.

An element $\gamma \in \prod_{1,1} \Psi^r$ which is positive-definite at each point defines an Hermitian metric on M. Such a metric always exists, and we assume that one has been chosen. Then, in terms of this metric, we have the duality isomorphism $*$: $\Phi^r \to \Phi^{2n-r}$, and the Hermitian character of the metric is expressed by the formula

$$* \prod_{r,s} = \prod_{n-s,n-r} *.$$

Further, if we let

$$(\phi, \psi) = \int_M \phi \wedge * \overline{\psi}, \quad \| \phi \| = \sqrt{[(\phi, \phi)]},$$

then

$$(\prod_{r,s} \phi, \psi) = (\phi, \prod_{r,s} \psi).$$

Now let

$$\mathfrak{d} = - * \partial *.$$

Writing $\delta = - * d *$, $\delta_k = - * d_k *$ $(k=1, 2)$, we have the decomposition $\delta = \delta_1 + \delta_2 + \overline{\delta}_1 + \overline{\delta}_2$, where

$$\delta_1 = \Sigma_{r,s} \prod_{r,s-1} \delta \prod_{r,s}, \quad \delta_2 = \Sigma_{r,s} \prod_{r+1,s-2} \delta \prod_{r,s},$$

and

$$\mathfrak{d} = 2\delta_2 + \delta_1 - \overline{\delta}_2.$$

Therefore

$$\delta = \mathfrak{d} + \overline{\mathfrak{d}},$$

where

$$\overline{\mathfrak{d}} = 2\overline{\delta}_2 + \overline{\delta}_1 - \delta_2.$$

Let Φ_c be the subspace of Φ composed of forms with compact supports. For ϕ, $\psi \in \Phi$ we have $(d\phi, \psi) = (\phi, \delta\psi)$, and it follows that a similar formula is valid for $\overline{d}_k$, δ_k, namely,

$$(\overline{d}_k \phi, \psi) = (\phi, \delta_k \psi).$$

Hence

$$(\overline{\partial} \phi, \psi) = (\phi, \mathfrak{d} \psi), \quad \phi, \psi \in \Phi_c.$$

If we write $\Delta = 2(\mathfrak{d}\overline{\partial} + \overline{\partial}\mathfrak{d})$, we therefore have

$$(\Delta \phi, \psi) = (\phi, \Delta \psi), \quad \phi, \psi \in \Phi_c.$$

We say that the Hermitian metric γ is a Kähler metric if and only if $\Delta = 2(\mathfrak{d}\overline{\partial} + \overline{\partial}\mathfrak{d}) = \delta d + d\delta$. If a Kähler metric exists on an almost-complex manifold, we say that the manifold is a Kähler manifold or that the almost-complex structure is Kählerian.

THEOREM 2.1. *An almost-complex Kähler manifold possesses an integrable structure.*

PROOF. Let M be an almost-complex manifold with a Kähler metric. In terms of this metric, $\Delta = 2(\mathfrak{d}\overline{\partial} + \overline{\partial}\mathfrak{d})$ is a real operator, that is,

$$\mathfrak{d}\overline{\partial} + \overline{\partial}\mathfrak{d} = \overline{\mathfrak{d}}\partial + \partial\overline{\mathfrak{d}},$$

and we have

$$\begin{aligned}\delta d+d\delta &= (\mathfrak{b}+\bar{\mathfrak{b}})(\partial+\bar\partial)+(\partial+\bar\partial)(\mathfrak{b}+\bar{\mathfrak{b}})\\ &= (\mathfrak{b}\bar\partial+\bar\partial\mathfrak{b})+(\bar{\mathfrak{b}}\partial+\partial\bar{\mathfrak{b}})+\{(\mathfrak{b}\partial+\partial\mathfrak{b})+(\bar{\mathfrak{b}}\bar\partial+\bar\partial\bar{\mathfrak{b}})\}\\ &= \delta d+d\delta+(\mathfrak{b}\partial+\partial\mathfrak{b})+\bar{\mathfrak{b}}\bar\partial+\bar\partial\bar{\mathfrak{b}}.\end{aligned}$$

Therefore

$$(\mathfrak{b}\partial+\partial\mathfrak{b})+(\bar{\mathfrak{b}}\bar\partial+\bar\partial\bar{\mathfrak{b}})=0. \tag{2.1}$$

Substituting for ∂ and $\mathfrak{b}$ in terms of d_k, δ_k $(k=1,2)$, we obtain

$$\begin{aligned}\mathfrak{b}\partial+\partial\mathfrak{b} = &-2\{(\delta_2\bar d_2+\bar d_2\delta_2)+(\bar\delta_2 d_2+d_2\bar\delta_2)\}+\{\delta_1 d_1+d_1\delta_1\}\\ &-\{(\delta_1\bar d_2+\bar d_2\delta_1)+(d_1\bar\delta_2+\bar\delta_2 d_1)\}\\ &+2\{(\delta_1 d_2+d_2\delta_1)+(d_1\delta_2+\delta_2 d_1)\}\\ &+4\{\delta_2 d_2+d_2\delta_2\}+\{\bar\delta_2\bar d_2+\bar d_2\bar\delta_2\}.\end{aligned} \tag{2.2}$$

We say that an operator on an almost-complex manifold is of type (μ,ν) if it maps a form of type (r,s) into a form of type $(r+\mu,s+\nu)$. The bracketed terms on the right side of (2.2) are grouped according to type:

$$\begin{array}{cl}(\delta_2\bar d_2+\bar d_2\delta_2)+(\bar\delta_2 d_2+d_2\bar\delta_2) & \text{is of type } (0,0);\\ (\delta_1 d_1+d_1\delta_1) & \text{of type } (1,-1);\\ (\delta_1\bar d_2+\bar d_2\delta_1)+(d_1\bar\delta_2+\bar\delta_2 d_1) & \text{of type } (-1,1);\\ (\delta_1 d_2+d_2\delta_1)+(d_1\delta_2+\delta_2 d_1) & \text{of type } (2,-2);\\ (\delta_2 d_2+d_2\delta_2) & \text{of type } (3,-3);\\ (\bar\delta_2\bar d_2+\bar d_2\bar\delta_2) & \text{of type } (-3,3).\end{array}$$

Since the terms of type $(0,0)$ in the left side of (2.1) must equal zero, we obtain

$$(\delta_2\bar d_2+\bar d_2\delta_2)+(\bar\delta_2 d_2+d_2\bar\delta_2)=0. \tag{2.3}$$

Given an arbitrary form $\phi\in\Phi$ and point $x\in M$, let $\psi\in\Phi_c$ be such that the support of $\phi-\psi$ vanishes in some neighborhood of x. Since $\psi\in\Phi_c$, we have by (2.3)

$$\begin{aligned}0&=((\delta_2\bar d_2+\bar d_2\delta_2)\psi,\psi)+((\bar\delta_2 d_2+d_2\bar\delta_2)\psi,\psi)\\ &=(d_2\psi,d_2\psi)+(\bar\delta_2\psi,\bar\delta_2\psi)+(\bar d_2\psi,\bar d_2\psi)+(\delta_2\psi,\delta_2\psi).\end{aligned}$$

Therefore $d_2\psi$ vanishes everywhere and $d_2\phi=d_2\psi=0$ at x. Since x is an arbitrary point of M, we see that d_2 maps Φ into zero, that is, $\partial=d_1$ and

$$d=d_1+\bar d_1.$$

Finally, the vanishing of d^2 implies $d_1^2=\partial^2=0$ and it follows that the almost-complex structure of M is integrable.

3. Potential theory

Assume that M is an almost-complex manifold with Hermitian metric. Let D: $\Phi \to \Phi \oplus \Phi$ be defined by $D(\phi) = \bar{\partial}\phi \oplus \mathfrak{d}\phi$, and let $\mathscr{D}$: $\Phi \oplus \Phi \to \Phi$ be defined by $\mathscr{D}(\phi \oplus \psi) = \mathfrak{d}\phi + \bar{\partial}\psi$. Then the composite map $2\mathscr{D}D$: $\Phi \to \Phi$ coincides with Δ.

We define a norm in the space $\Phi \oplus \Phi$ by setting

$$\| \phi \oplus \psi \|^2 = \| \phi \|^2 + \| \psi \|^2.$$

We denote the Hilbert space of norm-finite differential forms on M by $\mathscr{A}$: if $\phi \in \mathscr{A}$ there exists a sequence $\{\phi_\mu\}$, $\phi_\mu \in \Phi$, $\| \phi_\mu \| < \infty$, such that $\| \phi - \phi_\mu \| \to 0$. We say that a form ϕ belongs to (the closure of) the domain of D if and only if there is a sequence $\{\phi_\mu\}$, $\phi_\mu \in \Phi$, $\| \phi_\mu \| < \infty$, $\| D\phi_\mu \| < \infty$, such that $\| \phi - \phi_\mu \|$ and $\| D(\phi_\mu - \phi_\nu) \|$ tend to zero $(\mu, \nu \to \infty)$. If we impose the additional restriction that $\phi_\mu \in \Phi_c$ for all μ, we say that ϕ belongs to the domain of the operator D^c. The domains of the operators $\mathscr{D}$, $\mathscr{D}^c$ are then defined in a similar manner to those of D, D^c. If a form ϕ belongs to the domain of an operator such as D, we shall write $\phi \in D$.

A form ϕ belongs to the domain of $\Delta = 2\mathscr{D}D$ if and only if ϕ belongs to the domain of D and $D\phi$ belongs to the domain of $\mathscr{D}$. We introduce also the operators $\Delta^c = 2\mathscr{D}D^c$, ${}^c\Delta = 2\mathscr{D}^c D$, and we denote the domain of Δ^c by $\mathfrak{G}$, that of ${}^c\Delta$ by $\mathfrak{N}$. Let

$$\mathscr{H}^c = \{\phi \mid \phi \in D^c, \bar{\partial}\phi = \mathfrak{d}\phi = 0\}, \quad \mathscr{H} = \{\phi \mid \phi \in D, \bar{\partial}\phi = \mathfrak{d}\phi = 0\}.$$

Given a constant $\sigma > 0$, let D_σ: $\Phi \to \Phi \oplus \Phi \oplus \Phi$ be defined by $D_\sigma(\phi) = \bar{\partial}\phi \oplus \mathfrak{d}\phi \oplus \sqrt{(\tfrac{1}{2}\sigma)}\,\phi$, and let $\Delta_\sigma = \Delta + \sigma$, where σ denotes the trivial operation of multiplication by σ.

THEOREM 3.1. *For each $\sigma > 0$ there exist linear isomorphisms G_σ: $\mathscr{A} \to \mathfrak{G}$ and N_σ: $\mathscr{A} \to \mathfrak{N}$, where $\Delta_\sigma G_\sigma \phi = \phi$, $\Delta_\sigma N_\sigma \phi = \phi$, $\phi \in \mathscr{A}$. As operators on $\mathscr{A}$, G_σ and N_σ are symmetric and*

$$\| D_\sigma G_\sigma \phi \| \leq \| \phi \| / \sqrt{(2\sigma)}, \quad \| D_\sigma N_\sigma \phi \| \leq \| \phi \| / \sqrt{(2\sigma)}.$$

Moreover

$$\mathscr{H}^c = \{\phi \mid \phi \in \mathscr{A}, \phi - \sigma G_\sigma \phi = \Delta G_\sigma \phi = 0\},$$

$$\mathscr{H} = \{\phi \mid \phi \in \mathscr{A}, \phi - \sigma N_\sigma \phi = \Delta N_\sigma \phi = 0\}.$$

A proof is to be found in [5].

Given $\beta \in \mathscr{A}$, let $E_\sigma(\phi) = E_\sigma(\phi; \beta) = \| D\phi \|^2 + \tfrac{1}{2}\sigma \| \phi - (\beta/\sigma) \|^2$; then $G_\sigma \beta$, $N_\sigma \beta$ are the unique elements which minimize $E_\sigma(\phi)$ in D^c, D respectively. If we write $e_\sigma = \inf E_\sigma(\phi)$, $\phi \in D$, we have the inequality (cf. [5])

$$\| D_\sigma(\phi - \psi) \| \leqq \sqrt{[E_\sigma(\phi) - e_\sigma]} + \sqrt{[E_\sigma(\psi) - e_\sigma]}. \tag{3.1}$$

Assume that M is compact; then the spaces $\mathfrak{G}$, $\mathfrak{N}$ coincide with $\mathscr{A}$ and $G_\sigma = N_\sigma$. In this case G_σ is a completely continuous operator and $\mathscr{H} = \mathscr{H}^c$ is therefore a finite-dimensional complex vector space. Let $\mathscr{H}^{r,s} = \prod_{r,s} \mathscr{H}$, $h^{r,s} = \dim \mathscr{H}^{r,s}$. If M is complex, $\mathscr{H}^{r,s}$ is isomorphic to the $\bar{\partial}$-cohomology of forms of type (r, s) (after a theorem of P. Dolbeault) and is thus independent of the particular choice of Hermitian metric.

4. Variation of almost-complex structure

Let M be a compact manifold of class C^∞ with an almost-complex structure $\mathscr{S}(t)$ which depends in a C^∞ manner on a point $t = (t^1, \ldots, t^m)$ which varies on a domain E of real Euclidean m-space. Given an arbitrary point $t_0 \in E$, let γ_0 be an Hermitian metric compatible with the structure $\mathscr{S}(t_0)$. Then there is a neighborhood of t_0 on E such that $\gamma(t) = \prod_{1,1}(t)\gamma_0$ is an Hermitian metric on M for the structure $\mathscr{S}(t)$ provided that t varies in the neighborhood. Let $\bar{\partial}(t)$, $\mathfrak{d}(t)$, $G_\sigma(t)$, $*(t)$ and $\|\phi\|_t$ correspond to $\mathscr{S}(t)$ with the metric $\gamma(t)$, and let $h^{r,s}(t) = \dim \mathscr{H}^{r,s}(t)$. We observe, however, that the space $\mathscr{A}$ does not depend on t.

THEOREM 4.1. *There exists a neighborhood U on E, $t_0 \in U$, such that $h^{r,s}(t) \leqq h^{r,s}(t_0)$ for $t \in U$.*

PROOF. Given an arbitrary point $x_0 \in M$ and a positive integer ν, there exists a neighborhood W of x_0 on M, a neighborhood V of t_0 on E, and a form $\pi = \pi(x, y, t)$ on $M \times W \times V$ which satisfies the following conditions:

(1) π is C^∞ except for $x = y$.

(2) Let $r = r(x, y, t)$ be the geodesic distance of the points x, y with respect to the metric $\gamma(t)$; then $r^{2(n-1)} \cdot \pi$ is of class C^{n-1} in $W \times W \times V$.

(3) For fixed $y \in W$, $\Delta_\sigma(t)\pi = p$ where p is of class C^ν in $M \times W \times V$.

(4) Given $\phi \in \Phi$, let

$$(\Pi\phi)(x, t) = \int_W \bar{\pi}(x, y, t) \wedge *(t)\,\phi(y);$$

$$(P\phi)(x, t) = \int_W \bar{p}(x, y, t) \wedge *(t)\,\phi(y).$$

The parametrix π can be constructed along the lines of the method used by Kodaira [2], provided that the power series employed by him are replaced by partial sums whose orders depend on the given integer ν.

Let Π', P' be the adjoint operators satisfying $(\Pi'\phi, \psi)_t = (\phi, \Pi\psi)_t$, $(P'\phi, \psi)_t = (\phi, P\psi)_t$. Let X, Y be neighborhoods of x_0 on M, $\bar{X} \subset Y$, $\bar{Y} \subset W$, let ρ be a function of class C^∞ on M which is equal to 1 in X, equal to 0 in $M - Y$, and write $\tilde{\pi}(x, y, t) = \rho(x)\pi(x, y, t)$. Given $\beta \in \mathscr{A}$,

β independent of t, we have by Green's formula (ν being chosen sufficiently large) that

$$G_\sigma(t)\,\beta = \tilde{\Pi}'\beta - \tilde{P}'G_\sigma(t)\,\beta,$$

where $\tilde{P} = \Delta_\sigma(t)\,\tilde{\Pi}$ and where the formula is to be interpreted in the sense that the norm over X of the difference of the two sides of this equation vanishes. It follows that

$$\left\| \frac{\partial}{\partial t} G_\sigma(t)\,\beta \right\|_{t_0} \leqq c \cdot \| \beta \|_{t_0}, \tag{4.1}$$

where the constant c is independent of β and $t \in V$.

We denote by $|t' - t''|$ the Euclidean distance between t' and t''. Then we infer from (4.1) the following

LEMMA 4.1. *There is a number c independent of β and $t', t'' \in V$ such that*

$$\| G_s(t')\,\beta - G_s(t'')\,\beta \|_{t_0} \leqq c \cdot \| \beta \|_{t_0} \cdot |t' - t''|.$$

Now let $\{\phi_j(t)\}$ be a complete orthonormal set of eigenforms of the operator $G_\sigma(t)$ of type (r, s); that is,

$$G_\sigma(t)\,\phi_j(t) = \lambda_j(t)\,\phi_j(t), \quad 1/s \geqq \lambda_1(t) \geqq \lambda_2(t) \geqq \ldots > 0,$$

where $(\phi_i(t), \phi_j(t))_t = \delta_{ij}$ and where each ϕ_j is of type (r, s).

LEMMA 4.2. *If there is a sequence $\{t_\mu\}$, $t_\mu \to t_0$ $(\mu \to \infty)$, such that $\lambda_j(t_\mu) \to \lambda > 0$ for $j = k+1, \ldots, k+h$, then there exist h eigenforms ϕ_j of type (r, s) of $G_\sigma(t_0)$, $j = k+1, \ldots, k+h$, such that*

$$G_\sigma(t_0)\,\phi_j = \lambda \phi_j, \quad (\phi_j, \phi_j)_{t_0} = \delta_{ij}.$$

PROOF. For all but a finite number of μ, $\| \phi_j(t_\mu) \|_{t_0} \leqq 2$ and therefore, since $G_\sigma(t_0)$ is completely continuous, we can find a subsequence, which we again denote by $\{t_\mu\}$, such that $G_\sigma(t_0)\,\phi_j(t_\mu)$ converges in norm to a limit ψ_j $(j = k+1, \ldots, k+h)$. By Lemma 4.1,

$$\| G_\sigma(t_\mu)\,\phi_j(t_\mu) - G_\sigma(t_0)\,\phi_j(t_\mu) \|_{t_0} \to 0$$

and hence

$$\| \lambda_j(t_\mu)\,\phi_j(t_\mu) - \psi_j \|_{t_0} \to 0;$$

that is,

$$\| \phi_j(t_\mu) - \lambda^{-1}\psi_j \|_{t_0} \to 0.$$

Let $\phi_j = \lambda^{-1}\psi_j$ $(j = k+1, \ldots, k+h)$. Then $\| \phi_j(t_\mu) - \phi_j \|_{t_0} \to 0$ where

$$\| \phi_j \|_{t_0} = \lim \| \phi_j(t_\mu) \|_{t_0} = \lim \| \phi_j(t_\mu) \|_{t_\mu} = 1.$$

Now $G_\sigma(t_0)\,\phi_j = \lim G_\sigma(t_0)\,\phi_j(t_\mu) = \psi_j = \lambda \phi_j$ where, for $i \neq j$,

$$(\phi_i, \phi_j)_{t_0} = \lim\, (\phi_i(t_\mu), \phi_j(t_\mu))_{t_0}.$$

Let $\omega_i = \phi_i(t_\mu)$; then for suitable choice oi a real number θ, we have

$$\begin{aligned} 2\,|(\omega_i, \omega_j)_{t_0}| &= 2(\omega_i, e^{i\theta}\omega_j)_{t_0} \\ &= \|\omega_i + e^{i\theta}\omega_j\|^2_{t_0} - \|\omega_i\|^2_{t_0} - \|\omega_j\|^2_{t_0} \\ &= \|\omega_i + e^{i\theta}\omega_j\|^2_{t_\mu} - \|\omega_i\|^2_{t_\mu} - \|\omega_j\|^2_{t_\mu} + 0(|t_\mu - t_0|) = 0(|t_\mu - t_0|). \end{aligned}$$

Hence $(\phi_i, \phi_j)_{t_0} = 0$, and this completes the proof of Lemma 4.2.

Let $M(\lambda, t)$ be the dimension of the space of eigenforms of $G_\sigma(t)$ of type (r, s) which belong to the eigenvalue $\lambda > 0$. It follows from Lemma 4.2 that $M(\lambda, t) \leqq M(\lambda, t_0)$ for $|t - t_0| < \epsilon$. In particular, taking $\lambda = 1/\sigma$, we obtain Theorem 4.1.

If $\mathscr{S}(t)$ is a Kähler structure for each t, then

$$\textstyle\sum_{r+s=p} h^{r,s}(t)$$

is equal to the pth Betti number of the manifold M and we have in this case:

THEOREM 4.2. *If $\mathscr{S}(t)$ is a Kähler structure for each t, $h^{r,s}(t)$ is equal to a constant.*

5. Variation of complex line bundles

Let M be a (fixed) compact-complex manifold, $F(t)$ a complex line bundle over M depending in a C^∞ manner on t, and let $\mathscr{H}^{r,s}(M, F(t))$ be the space of harmonic forms on M with coefficients in $F(t)$. Then, by the same method as above, we can prove that $\dim \mathscr{H}^{r,s}(M, F(t))$ is an upper semi-continuous function of t, while

$$\mathscr{H}^{r,s}(M, F(t)) \cong H^s(M, \Omega^r(F(t))),$$

where $\Omega^r(F(t))$ is the sheaf of germs of holomorphic r-forms with coefficients in $F(t)$. Hence we obtain

THEOREM 5.1. *The dimension* $\dim H^s(M, \Omega^r(F(t)))$ *is an upper semi-continuous function in t.*

As an example of applications, we derive from the above result the following theorem:†

Let V be an algebraic manifold and let F be a complex line bundle over V. The Euler characteristic $\chi(V, \Omega^r(F))$ depends only on the characteristic class $c(F)$ of F, where $\Omega^r(F)$ denotes the sheaf over V of germs of holomorphic r-forms with coefficients in F. Let $\mathfrak{P}$ be the Picard variety, i.e., the additive group consisting of all line bundles P over V with $c(P) = 0$. Then it can be shown‡ that there exists on the

† This is an immediate consequence of a deep theorem of F. Hirzebruch [1] to the effect that $\chi(V, \Omega^r(F))$ can be represented by the Todd polynomial in $c(F)$. A different proof of this result was given earlier by K. Kodaira [3].

‡ This construction will be given in a forthcoming paper by K. Kodáira.

product manifold $V \times \mathfrak{P}$ a complex line bundle Ξ such that the restriction $\Xi_{V \times P}$ of Ξ to $V \times P$ (regarded as a bundle over V in an obvious manner) coincides with P. Take sufficiently ample bundles E over V and $\mathscr{E}$ over $\mathfrak{P}$ and form the sum

$$F = \Xi + E + \mathscr{E} \quad \text{over} \quad V \times \mathfrak{P},$$

where E and $\mathscr{E}$ are to be regarded as bundles over V in an obvious way. Letting $r_{V \times P}$ be the restriction map of $\Omega(F)$ to $V \times P$, we can prove that the map

$$r^*_{V \times P}\colon\ H^0(V \times \mathfrak{P}, \Omega^r(F)) \to H^0(V, \Omega^r(P+E))$$

is *onto*. Considering $P+E$ as a bundle depending analytically on a point $P \in \mathfrak{P}$, we therefore infer that $\dim H^0(V, \Omega^r(P+E))$ is a *lower* semi-continuous function in P, while the above theorem asserts that $\dim H^0(V, \Omega^r(P+E))$ is *upper* semi-continuous in P. Hence we obtain

$$\dim H^0(V, \Omega^r(P+E)) = \dim H^0(V, \Omega^r(E)).$$

This proves that $\dim H^0(V, \Omega^r(E))$ is determined uniquely by the characteristic class $c(E)$ of E, provided that E is sufficiently ample.

Let S be a general hypersurface section of V of sufficiently high order and let $E = F + [S]$, where $[S]$ denotes the line bundle over V defined by S. We have

$$\chi^s(V, F) = \chi^s(V, E) - \chi^s(S, E_S) - \chi^{s-1}(S, F_S),$$

and, since E is sufficiently ample,

$$\chi^s(V, E) = \dim H^0(V, \Omega^s(E)).$$

Hence we obtain

$$\chi^s(V, F) = \dim H^0(V, \Omega^s(E)) - \chi^s(S, F_S + [S]_S) - \chi^{s-1}(S, F_S).$$

Now, applying induction on the dimension of the variety V, we conclude from this that $\chi^s(V, F)$ is determined uniquely by $c(F)$.

We note that $\dim H^s(M, \Omega^r(F(t)))$ is not necessarily a constant even in case M is an algebraic variety. As an example we consider an Abelian variety A of dimension 2 (i.e. a complex torus of complex dimension 2 whose period matrix is a Riemann matrix). Let $\mathfrak{P}$ be the Picard variety attached to A, i.e. the group consisting of complex line bundles F over A which are topologically trivial [4].

The Picard variety $\mathfrak{P}$ is also an Abelian variety and each member $F \in \mathfrak{P}$ can be written as $F = F(t)$, where t denotes a point on the complex torus $\mathfrak{P}$ and where $F(t)$ depends analytically on t.

Now we have

$$\dim H^0(A, \Omega^0(F(t))) = \begin{cases} 1 & \text{if} \quad t = 0, \\ 0 & \text{otherwise.} \end{cases}$$

Since the Euler characteristic $\chi(A, F(t))$ depends only on the characteristic class of $F(t)$, we have

$$\sum_{s=0}^{2}(-1)^s \dim H^s(A, \Omega^0(F(t))) = \chi(A, 0) = 0,$$

while by Serre's duality theorem,

$$\dim H^2(A, \Omega^0(F(t))) = \dim H^0(A, \Omega^0(-F(t))).$$

Consequently we obtain

$$\dim H^1(A, \Omega^0(F(t))) = \begin{cases} 2 & \text{if } t=0, \\ 0 & \text{otherwise.} \end{cases}$$

As an example of a non-algebraic complex manifold, we consider the Hopf manifold defined as follows: Let C_2 be the space of two complex variables (z_1, z_2) and let $\mathscr{B} = C_2 - (0, 0)$. Moreover, let $\Delta = \{T^m \mid m = 0, 1, 2, \ldots\}$ be the discontinuous group of analytic automorphisms of $\mathscr{B}$ generated by

$$T: (z_1, z_2) \to (2z_1, 2z_2).$$

The factor space $V = \mathscr{B}/\Delta$ is obviously a compact complex manifold which is called the Hopf manifold. We note that, for an arbitrary complex manifold M, the group $\mathfrak{P}$ consisting of all complex line bundles over M which are topologically trivial may be called the *Picard variety attached to M*. We have the canonical isomorphism†

$$\mathfrak{P} \cong \frac{H^1(M, \Omega)}{i^* H^1(M, Z)},$$

where Ω is the sheaf over M of germs of holomorphic functions, Z is the integers, and where i is the map: $k \to 2ik$ of Z into Ω. For the Hopf manifold $V = \mathscr{B}/\Delta$, we have

$$H^1(V, Z) \cong Z,$$

and‡

$$H^1(V, \Omega) \cong C.$$

Therefore the Picard variety $\mathfrak{P}$ attached to V is given by

$$\mathfrak{P} \cong C/2\pi i Z \cong C^*.$$

Moreover, for each number $t \in C^*$ we can construct the corresponding line bundle $F(t) \in \mathfrak{P}$ explicitly in the following manner: Let

$$\tilde{\Delta} = \{\tilde{T}^m \mid m = 0, 1, 2, \ldots\}$$

be the discontinuous group of analytic automorphisms of the product space $\mathscr{B} \times C$ generated by

$$\tilde{T}: (z_1, z_2, \zeta) \to (2z_1, 2z_2, t \cdot \zeta).$$

† The proof of this fact is contained in [4], p. 871.

‡ This result has been proved by A. Borel.

Then the factor space $F(t) = \mathscr{B} \times C/\tilde{\Delta}$

forms, in an obvious manner, a complex line bundle over $V = \mathscr{B}/\Delta$. Moreover, we have

$$F(s) \pm F(t) = F(st^{\pm 1}).$$

Clearly $F(t)$ depends analytically on t.

Now it is easy to compute $\dim H^0(V, \Omega^0(F(t)))$. Clearly each holomorphic section $\phi \in H^0(V, \Omega^0(F(t)))$ corresponds one-to-one to a holomorphic function $\tilde{\phi} = \tilde{\phi}(z_1, z_2)$ on $\mathscr{B}$ satisfying

$$\tilde{\phi}(2z_1, 2z_2) = t\tilde{\phi}(z_1, z_2), \tag{5.1}$$

while, by a theorem of Hartogs, an arbitrary holomorphic function on $\mathscr{B}$ must be holomorphic on the whole of $C_2 \supset \mathscr{B}$. Hence we can write

$$\tilde{\phi}(z_1, z_2) = \sum_{m,n=0} c_{mn} z_1^m z_2^n,$$

and therefore the equation (5.1) is reduced to

$$2^{m+n} c_{mn} = t \cdot c_{mn}.$$

This shows that (5.1) has non-trivial solutions $\tilde{\phi}$ if and only if $t = 2^l$ $(l = 0, 1, 2, \ldots)$, and, if $t = 2^l$, the number of linearly independent solutions $\tilde{\phi}$ of (5.1) is equal to $l+1$. Thus we obtain

$$\dim H^0(V, \Omega(F(t))) = \begin{cases} l+1 & \text{if} \quad t = 2^l \ (= 0, 1, 2, \ldots), \\ 0 & \text{otherwise.} \end{cases}$$

Added in proof. We indicate another application of the upper semi-continuity principle (details are contained in a forthcoming Princeton thesis of J. Kohn). Let M be a compact complex manifold and introduce the differential operator $\{(\tau - \sqrt{-1})\partial + (\tau + \sqrt{-1})\bar{\partial}\}/2$ where τ is an arbitrary but fixed complex number. It is obvious that $d_\tau^2 = 0$, $d_{\bar{\tau}} = \bar{d}_\tau$. In terms of any given Hermitian metric on M we can introduce the adjoint operator $\delta_{\bar{\tau}} = \bar{\delta}_\tau = -*d_{\bar{\tau}}*$ and form the Laplacian $\Delta_\tau = 2(\delta_{\bar{\tau}} d_\tau + d_\tau \delta_{\bar{\tau}})$ which, for $\tau = \sqrt{-1}$, coincides with the Laplacian $2(\mathfrak{d}\bar{\partial} + \bar{\partial}\mathfrak{d})$ introduced in §2. It may be verified that the Poincaré lemma holds for d_τ: if ϕ is of positive degree and $d_\tau \phi = 0$ in the neighborhood of some point of M, then there exists locally a form ψ such that $\phi = d_\tau \psi$. Moreover, in case $\tau \neq \pm\sqrt{-1}$, a 0-form (function) f satisfies $d_\tau f = 0$ if and only if f is a constant. Let $\mathscr{H}_\tau^p$ be the space of harmonic forms of degree p on M (forms annihilated by Δ_τ) and let $h_\tau^p = \dim \mathscr{H}_\tau^p$; then $\mathscr{H}_\tau^p$ is isomorphic to the d_τ-cohomology of forms of degree p and is thus independent of the choice of the Hermitian metric. For $\tau \neq \pm\sqrt{-1}$, one verifies readily that $\mathscr{H}_\tau^p$ is isomorphic to $H^p(M, C)$ and hence h_τ^p is equal to the

p^{th} Betti number b^p of M. On the other hand, for $\tau=\sqrt{-1}$, $h_\tau^p = \sum_{r+s=p} h^{r,s}$. An argument similar to that given in §4 shows that h_τ^p is upper semi-continuous from which we conclude that

$$b^p \leqq \sum_{r+s=p} h^{r,s}.$$

This is a recent result of A. Frölicher (Proc. Nat. Acad. Sci., *U.S.A.*, 41 (1955), pp. 641–644).

PRINCETON UNIVERSITY

REFERENCES

[1] F. HIRZEBRUCH, *Arithmetic genera and the theorem of Riemann-Roch for algebraic varieties*, Proc. Nat. Acad. Sci., U.S.A., 40 (1954), pp. 110–114.

[2] K. KODAIRA, *Harmonic fields in Riemannian manifolds (generalized potential theory)*, Ann. of Math., 50 (1949), pp. 587–665.

[3] ——, *Some results in the transcendental theory of algebraic varieties*, Ann. of Math., 59 (1954), pp. 86–134.

[4] —— and D. C. SPENCER, *Groups of complex line bundles over compact Kähler varieties*, Proc. Nat. Acad. Sci., U.S.A., 39 (1953), pp. 868–872.

[5] D. C. SPENCER, *Potential theory and almost-complex manifolds*, Lectures on Functions of a Complex Variable, ed. Wilfred Kaplan *et al.*, The University of Michigan Press, Ann Arbor, 1955.

Commutative Algebraic Group Varieties

Maxwell Rosenlicht

---

1. Among the many fields enriched by Lefschetz is that of Abelian varieties. These constitute one much studied extreme in the set of all algebraic group varieties, another consisting of the set of all linear groups, i.e. group varieties that are biregularly isomorphic to algebraic matric groups. It is our object here to discuss a number of results in the theory of arbitrary algebraic group varieties, and especially of commutative group varieties. Many of the results given here are not new.† The proofs, which appear here only in the merest outline, will be given in detail in subsequent papers.

Before proceeding with the general theory we make a few remarks about the invariant differential forms and Lie algebra associated with any given group variety. These can be constructed in a manner entirely analogous to that used in the theory of Lie groups and most of the usual relations can be shown to hold. In the case of characteristic zero the Lie algebra mechanism functions more or less normally and one can apply it to get most of what one would expect. For example, the adjoint representation of a group variety is a rational homomorphism onto a linear group with kernel equal to the center of the original group. However, if the characteristic is $p \neq 0$, the Lie theory is of very limited use, so we neglect it entirely in what follows.

2. Suppose that the group variety G *operates rationally* on the variety V. This means that we have given a rational map of $G \times V$ into V such that if the point $g \times p$ of $G \times V$ maps into $g(p) \in V$, then (1) if k is a field of definition for G, V and the rational map in question, then $g(p)$ is defined whenever p is a generic point of V over $k(g)$, (2) if p is a generic point of V over k and $g_1, g_2 \in G$, then

$$g_1(g_2(p)) = g_1 g_2(p),$$

† At the time of the conference we learned of I. Barsotti's independent work which overlaps heavily with the material below. Some of these results had previously been published by S. Nakano.

and (3) if e is the identity element of G, then $e(p)=p$. If $g(p)$ is always defined, we may say that G operates *regularly* on V. As a first result one has the following:

If G operates rationally on V, then there exists a variety $\bar{V}$ and a rational map $\tau\colon V \to \bar{V}$ such that if k is any algebraically closed field of definition for everything in question and p is a generic point of V over k, then $\tau(p)$ is generic for $\bar{V}$ over k and $k(\tau(p))$ consists precisely of all elements of $k(p)$ left fixed by all the automorphisms of $k(p)$ induced by points of G that are rational over k. Furthermore, these properties characterize $\bar{V}$ and τ to within a birational transformation of $\bar{V}$.

If G, V, $\bar{V}$, τ, k are as above, and if q is a generic point of $\bar{V}$ over k, it is easy to verify that $\tau^{-1}(q)$ consists of the orbit $G(p)$ of a certain generic point p of V over k. In the case where V, G are replaced by a group variety G and an algebraic subgroup H we can let H act on G according to the law $h(g)=gh^{-1}$ to get a sharpened version of the preceding result:

If G is a group variety and H an algebraic subgroup, then there exists a variety (which we denote by G/H) and a rational map τ of G onto G/H such that (1) the points of G/H are in natural 1-1 correspondence (under τ) with the left cosets of G modulo H and (2) the map $g \times aH \to gaH$ makes G operate regularly and transitively on G/H. If k is any algebraically closed field of definition for G and H, then we may take G/H, τ and the action of G on G/H to be defined over k. In this case, if p is a generic point of G over k, then $k(p)$ is separably generated over $k(\tau(p))$. Furthermore, these properties characterize G/H and τ to within a biregular transformation of G/H.

If H is a normal algebraic subgroup of the group variety G then G/H is also a group variety. G/H and the map $\tau\colon G \to G/H$ are characterized to within a biregular isomorphism of G/H by the following properties: (1) G/H is a group variety, (2) τ is a rational homomorphism of G onto G/H with kernel H, (3) if k is a field of definition for G, H, G/H and τ, and if p is a generic point of G over k, then $k(p)$ is separably generated over $k(\tau(p))$.

By suitably modifying a few definitions, we can extend the above results to reducible group varieties. The same will be true of much of what follows.

The preceding results enable one to prove that the elementary algebraic theorems on groups, such as the homomorphism theorems and the Jordan-Hölder-Schreier theorem, can be extended to group varieties in such a way that they hold from the point of view of algebraic geometry also, i.e. the group isomorphisms are given by algebraic correspondences, etc. Without this many of our results would be of limited significance.

If the group variety G acts on V, if $\bar{V}$ is the variety of orbits on V, as above, with τ the natural rational map from V to $\bar{V}$, then a rational map σ from $\bar{V}$ to V is called a *cross-section* if $\tau\sigma = 1$ on $\bar{V}$. The situation is particularly simple if such a cross-section σ exists in the case where G operates regularly on V in such a manner that for any $p \in V$, the natural map of $G \to G(p)$ is birational. In this case, let k be a field of definition for G, V, $\bar{V}$, τ, σ and let p be a generic point of V over k. Then $q = \tau(p)$ is generic for $\bar{V}$ over k, so $\sigma(q)$ is defined and $\sigma(q) \in G(p)$. Hence there exists an element $g \in G$ such that $p = g(\sigma(q))$. One easily shows that $k(p) = k(g, \sigma(q))$, and a dimension argument shows that g is generic for G over $k(q)$. Hence the map of $G \times \bar{V}$ into V defined by mapping $g \times q$ into p is birational.

Define G_a to be the group of the affine line with the group composition $(x) \circ (x') = (x + x')$, and let G_m be the group of the affine line with one point deleted and the group composition $(x) \circ (x') = (xx')$. G_a and G_m are the only linear groups of dimension one. Define an irreducible group variety to be *solvable* if it contains a normal chain of algebraic subgroups going down to $\{e\}$ such that the factor group of two successive groups in the chain is always either G_a or G_m. A detailed analysis for the case of groups of dimension one and then induction on the dimension of the group yields the following:

There exists a cross-section whenever a solvable group variety G acts on a variety V.

An easy consequence is that if G is any group variety and H a solvable algebraic subgroup of G, then G can be considered a principal fiber bundle with base space G/H and fiber H. Repeated application shows that any solvable group variety is *rational*, i.e. has a rational function field. Using this one can prove:

A solvable group variety is linear.

3. *If G is a commutative complex Lie group, then we can write $G = C^{\nu_1} \times G_1$, where C is the additive group of complex numbers, ν_1 an integer, and G_1 a subgroup of G that contains a subgroup $(C^*)^{\nu_2}$, where C^* is the multiplicative group of non-zero complex numbers and ν_2 an integer, such that $G_1/(C^*)^{\nu_2}$ is compact, i.e. a complex multitorus.*

This result is easily proved by passing to the universal covering group of G, which is $C^{\dim G}$. Of course, if our universal domain is the complex number field, we have $C = G_a$ and $C^* = G_m$. Unfortunately, the analogous result for group varieties (when we consider algebraic rather than analytic structure) is not valid, even in the case of the complex number field, since, for example, the theory of generalized Jacobian varieties shows the existence of group varieties G of dimension greater than one that contain exactly *one* irreducible algebraic subgroup, and that subgroup a G_a. However, a somewhat similar result will be seen to be true.

Let V be an abstract variety. Then there exists a projective model V' of V and a subvariety W of V' such that, k being any field of definition for everything present, any valuation v of $k(V)$ over k has a center on V if and only if its center on V' is not contained in W.

This is an easily proved lemma that helps in the proof of the following fundamental fact:

Let the group variety G operate regularly on the non-complete variety V, and let k be an algebraically closed field of definition for G, V and the action of G on V. Then there exists a local ring $\mathfrak{o}$, properly between k and $k(V)$, whose quotient field is $k(V)$, such that $\mathfrak{o}$ is invariant under all the automorphisms of $k(V)$ over k induced by elements of G that are rational over k.

If G, V, k, $\mathfrak{o}$ are as above, and if we restrict ourselves to points of G that are rational over k, we get a normal algebraic subgroup H of G by defining H to consist of all elements $h \in G$ satisfying $h(x) - x \in \mathfrak{m}$ for each $x \in \mathfrak{o}$, $\mathfrak{m}$ being the maximal ideal of $\mathfrak{o}$. H is then a group of automorphisms of the vector space $\mathfrak{m}^i/\mathfrak{m}^{i+1}$ over the field $\mathfrak{o}/\mathfrak{m}$, for any integer $i \geqq 0$, and G/H becomes a group of k-automorphisms of the function field $\mathfrak{o}/\mathfrak{m}$. If we let $V = G =$ a non-complete group variety we are led to:

A non-complete group variety of dimension greater than one contains a non-complete group variety of strictly smaller dimension.

The next result is easily proved:

If the group variety G contains a complete algebraic subgroup H such that G/H is complete, then G is complete.

We now get the main structure theorem for commutative group varieties:

If G is a commutative group variety then G contains an irreducible linear subgroup H such that G/H is an Abelian variety.

To prove this, note that it suffices to prove the result with the word 'solvable' replacing the word 'linear'. Since the result is trivial if G is complete or $\dim G = 1$ we may assume the contrary and use induction on $\dim G$. Let G_1 be a minimal non-complete algebraic subgroup of G. Then G_1 is either a G_a or a G_m and G/G_1 satisfies the theorem by our induction hypothesis. Hence G satisfies the theorem. Note that H

may be characterized as the greatest linear, or solvable, or rational, irreducible algebraic subgroup of G.

We remark that it is probable that a refinement of the method used here will produce an elementary proof of the last theorem without the restriction of commutativity of G, a result that has been proved recently by Chevalley by using the theory of the Picard and Albanese variety.†

4. This section is devoted to a few results on commutative linear groups. We define an *affine group* to be a group variety which is bi-regularly equivalent to an affine space. The first result is:

If G is a commutative linear group variety, then we can write G, in one and only one way, as a direct product $G = G_1 \times G_2$, where G_1 is a direct product of G_m's and G_2 is an affine group. In the case of characteristic zero, G_2 is the direct product of G_a's but this need not be the case for characteristic $p \neq 0$.

The structure of a commutative affine group can be very complicated in the case of characteristic $p \neq 0$, but one easily shows, at any rate, that any irreducible algebraic subgroup of an affine group is also affine, and that any factor group of an affine group is affine. We already know that any affine group is solvable, so that one can develop the theory of these groups by induction on the dimension. The case of dimension one is trivial, the only affine group being G_a. For the rest, we confine ourselves to the case of dimension two, where things are particularly simple: One can show that any affine group of dimension two over a field of characteristic $p \neq 0$ is either $G_a \times G_a$ or the group with law of composition

$$(x, y) \circ (x', y') = (x + x', y + y' + \textstyle\sum_{i=0}^{n} a_i(F(x, x'))^{p^i}),$$

where $a_0, \ldots, a_n$ are non-zero constants and

$$F(x, x') = \frac{1}{p}((x + x')^p - x^p - (x')^p).$$

Finally, any affine group of dimension two of the latter type is isogenous to the group with composition

$$(x, y) \circ (x', y') = (x + x', y + y' + F(x, x')).$$

5. Let H be an irreducible commutative linear group, A an Abelian variety. The group variety G is said to be an *extension of H by A* if H is a normal algebraic subgroup of G such that $G/H = A$. Here the essential problem is to find all G for given H and A. This problem

† Since the conference both the author and Barsotti have obtained such elementary proofs of Chevalley's theorem.

falls into the cadre of fiber bundle questions, since G is a principal fiber bundle over A with group H. The theory of principal fiber bundles with group G_m is known to lead one to the theory of the Picard variety; that of principal fiber bundles with group G_a leads one to the theory of simple differentials of second kind. In the latter case we arrive naturally at the following results:

Let the commutative group variety G be an extension of the direct product of a certain number of G_a's by the Abelian variety A. Then if $\dim G > 2 \dim A$, *we can write G as the direct product of G_a by another extension of a direct product of G_a's by A; furthermore the number* $2 \dim A$ *is minimal here. In the case of characteristic $p \neq 0$, G is isogenous to the direct product of A and a certain number of G_a's.*

We remark that the last statement is false for characteristic zero. The first statement implies that, given an Abelian variety A, there exists a unique commutative extension A' of $(G_a)^{\dim A}$ by A with the property that any commutative extension of a product of G_a's by A is 'contained in' (in a certain sense) the product of A' by a certain number of G_a's. In the case of complex numbers, A' is analytically, but not algebraically, the group $(G_m)^{2 \dim A}$.

The theory of generalized Jacobian varieties, in addition to giving the first non-trivial examples of extensions of linear groups by Abelian varieties, plays a crucial role in the proofs of the foregoing statements.

NORTHWESTERN UNIVERSITY

On the Symbol of Virtual Intersection of Algebraic Varieties

Francesco Severi

A DEBATE which took place (on September 10th, 1954) in the Symposium of algebraic geometry in Amsterdam, during my lecture on 'Problèmes résolus et problèmes nouveaux dans la théorie des séries et des systèmes d'équivalence', convinced me of the desirability of illustrating once again the value of the symbol of virtual intersection of the algebraic varieties (effective or virtual), introduced and used since 1904 in my works, especially in the papers quoted below. As a result of this debate, I realized that some of my listeners were not quite aware of these researches of mine, which had constituted the basis of my lecture [2, 3, 4, 6].

The following reflections do not make any change in my preceding definitions and conclusions, but they condense them in a systematic synthesis.

I dedicate these reflections to the celebration of the 70th birthday of my good and great friend Solomon Lefschetz, who, by a fortunate coincidence, presided that day over the session of the Symposium.

I do so even more willingly, because the concepts of virtual varieties and virtual intersections, which I introduced half a century ago into algebraic geometry, established more intimate relations between this branch of geometry and topology. Since that time the reciprocal repercussions of the two branches of study have been remarkable; to these repercussions the work of Lefschetz has made significant contributions.

1. We deal with pure varieties, lying upon an algebraic irreducible non-singular variety, M_r, of dimension r, and we shall always suppose these varieties to be either irreducible, or with *simple* components.

If $h_1, h_2, \ldots, h_s$ are the dimensions of s such varieties on M_r, we shall suppose that the inequality $h_1 + h_2 + \ldots + h_s \geqq (s-1)r$ holds; therefore the case may arise where the set of the common elements of those varieties has the *regular* or *normal dimension*

$$l = h_1 + \ldots + h_s - (s-1)r,$$

or else the dimension of at least one component of the intersection of this is *irregular* or *abnormal*, i.e., $>l$, or finally those varieties have no common point.

Indeed, by means of elementary theorems of the theory of analytic functions, and by using an analytic parametrization of the neighborhood of a (simple) point P of M_r, taken to be a common point to those varieties, we conclude at once that, if such a point P exists, the set of common points of those varieties certainly has a dimension $\geqq l$, and that P belongs to at least one component of the intersection, of dimension not less than l.

2. We consider first the case of two (pure) varieties V, W of dimensions h, k, with $l=h+k-r=0$.

Let us define the *virtual set* VW to within an algebraic equivalence.†

If the points common to V, W (when they exist) are all *simple* intersections,‡ then as virtual set VW, we take the effective set of such intersections (each taken with coefficient $+1$). If V, W have no common point, we assume by definition $VW \equiv 0$.

Now let V, W be in a quite arbitrary mutual situation (possibly with an infinity of intersections, or also, if possible, one of them contained completely or partially in the other, or in one of its components). In order to define VW, we first prove that it is possible to find, in an infinity of ways, a pure ∞^h variety V', such that $V+V'$ belongs, on M_r, to an irreducible algebraic system of dimension >0, $\{V+V'\}$, and besides both V' and the generic variety H of $\{V+V'\}$ have only simple intersections with W. We then put by definition

$$VW \equiv HW - V'W.$$

† We must remember that I have called (in 1905) *algebraically equivalent* two (pure) effective varieties A, B, of equal dimension, upon M_r (and I have symbolized this fact with $A \equiv B$) if A and B are (total) varieties of some irreducible algebraic system lying on M_r, or if they can be so reduced by adding to them one variety having their common dimension. On this subject, cf. [1]. In this paper there are also references to former works of the author, where this notion has been introduced for the first time. The attribute of 'total', referring to a variety A, denotes, according to the concepts and terminology of E. Noether-van der Waerden, that A is a *specialization* of the generic variety of a given irreducible algebraic system. Our definition of a total variety is rather different.

‡ An intersection P of two varieties of complementary dimensions h, $k(h+k=r)$ is called *simple* (of zero dimension) if the two varieties go through P simply, without any common tangent. A simple intersection is necessarily isolated; therefore, when there are only simple intersections, these are certainly finite in number. The same conceptual process which leads us to define in any case the group VW, and which we are going now to indicate, leads also to the precise notion of *multiplicity of intersection* of V, W at one of their isolated intersections. In any simple intersection the multiplicity of intersection is 1. Cf. particularly [6] and the following §9.

According to this definition, every point of the set VW has a coefficient $+1$ or -1.

If we change V' and H to $\bar{V}'$ and $\bar{H} \equiv V + \bar{V}'$ respectively, the set VW remains algebraically equivalent to itself (therefore it is defined only to within algebraic equivalence); this follows from the equivalence

$$HW - V'W \equiv \bar{H}W - \bar{V}'W,$$

because, as

$$H + \bar{V}' \equiv \bar{H} + V',$$

the two effective sets $(H + \bar{V}')\,W$, $(\bar{H} + V')\,W$, formed by simple intersections, contain the same number of points (with coefficients $+1$) and are therefore algebraically equivalent.

We denote by the symbol $[VW]$ the *order* of the virtual set VW, that is, the *algebraic* number (independent of the variable elements of the definition) of its points.

The topological comparisons with the properties of the Kronecker index between cycles are spontaneous!

3. Let us suppose $l = h + k - r > 0$. We define the *virtual* ∞^l *variety*, VW. If the common points of V, W are distributed upon components of regular dimension l, and if, in the generic point P of each one of these components, V, W have a *simple intersection of dimension* l, namely, they pass simply through P and their linear tangent spaces S_h, S_k intersect along an S_l (one also says, in such a case, that *that component is simple*, although it may happen that some of its particular points are not simple intersections), then we represent by the symbol VW the variety which is the sum of those effective components, each taken with the coefficient $+1$.

If there does not exist any common point we put $VW \equiv 0$.†

In any other more intricate case we can extend the procedure of § 2, with the only variation that the difference

$$HW - V'W$$

now represents a virtual variety with components of dimension l, some with the coefficient $+1$ and others with the coefficient -1. (We do not exclude the possibility that some of the preceding symbols may be equivalent to 0.) The symbol VW remains algebraically equivalent to itself if V', H are changed to $\bar{V}'$, $\bar{H}$ respectively, under the stated conditions, because the (all simple) components of the two effective ∞^l varieties $(H + \bar{V}')\,W$, $(\bar{H} + V')\,W$ belong, each one as a

† As we have remarked (§ 1), if there exists a common point, then $l \geqq h + k - r$ always.

total variety, upon W or upon every component of W, to the same irreducible algebraic system.†

4. From the definition above, it follows immediately that *the symbol VW is distributive with respect to addition*, both to the left and to the right, because the symbol has such a property when it represents simple effective intersections of regular dimensions. Hence if V splits up into the sum $V^1 + V^2$, we have

$$VW \equiv V^1W + V^2W;$$

likewise, if $W = W^1 + W^2$, it follows

$$VW \equiv VW^1 + VW^2.$$

This suggests that *the symbol VW is commutative*, that is, $VW \equiv WV$. In fact, if

$$H \equiv V + V',$$

the symbol HW satisfies the equivalence

$$HW \equiv VW + V'W,$$

and likewise for WH

$$WH \equiv WV + WV';$$

but, as we are dealing with symbols of effective varieties with simple components, we have

$$HW = WH; \quad V'W = WV',$$

and so it follows that $VW \equiv WV$.

5. The definition of the virtual ∞^l variety VWZ, is also immediate, where the dimensions h_1, h_2, h_3 of the three varieties V, W, Z having only simple components satisfy the inequality $h_1 + h_2 + h_3 \geqq 2r$, and we assume $l = h_1 + h_2 + h_3 - 2r$. In fact, from what precedes,

$$VW \equiv A - B,$$

where A, B are certain effective $\infty^{h_1+h_2-r}$ varieties, having only simple components. Therefore, after having defined VWZ as the set $(VW)Z$, we have

$$VWZ \equiv AZ - BZ \equiv C - D,$$

where C, D are two convenient effective ∞^l varieties with simple components.

† For: Two algebraically equivalent effective varieties A, B of dimension h cut out two algebraically equivalent varieties, upon any irreducible variety C of dimension k, when $h+k \geqq r$ and the intersections are regular. In fact, A, B can always be joined (possibly with the addition of an h-dimensional variety) in an irreducible system Σ, to which they belong, by an irreducible ∞^1 system, Σ', containing none of the varieties of Σ passing through C (if any); and Σ' cuts out upon C an algebraic system of effective varieties which is in unirational correspondence with Σ', and so is irreducible.

It is then obvious that VWZ has the associative and the commutative properties; similarly for any number of varieties, in the case when their intersection exists normally, of dimension $\geqq 0$.

Moreover, if the pure varieties $V, W, \ldots, Z$ to which we are referring have multiple components, for instance, if

$$V = a_1 V^1 + \ldots + a_s V^s, \quad W = b_1 W^1 + \ldots + b_\sigma W^\sigma, \quad \ldots, \quad Z = c_1 Z^1 + \ldots + c_\tau Z^\tau$$

the symbol $VW \ldots Z$ is defined by the relation

$$VW \ldots Z \equiv \sum_{i,j,\ldots,\lambda} a_i b_j \ldots c_\lambda V^i W^j \ldots Z^\lambda.$$

6. The definitions of the virtual groups VW, VWZ, ... and the consequences deduced above, concern the chosen type of equivalence, which has until now been that of algebraic equivalence.

However, definitions and properties hold exactly in the same way for any type of equivalence, that is, for any choice of the class of equivalence zero, provided that certain conditions are satisfied. More precisely we require the conditions which make possible the construction of the varieties V' and H, as well as those that allow us to transfer the chosen equivalence relation in M_r onto the subvarieties of M_r, by means of intersections.

We can, for instance, refer to *rational equivalence*, taking as class zero, for any given dimension h, that of the differences between the pairs of ∞^h varieties belonging to a certain kind of elementary system.†

7. Concerning the definition of elementary system and of systems of rational equivalence, we warn the reader that the definitions contained in the works and lectures [4] of the author never attribute the symbols of virtual intersection to algebraic equivalence (this would imply the coincidence of the type of equivalence to be introduced there with that of algebraic equivalence).

Here are in fact the definitions concerning rational equivalence.

Let us take upon $M_r \, r-h$ $(h \leqq r-1)$ linear systems of hypersurfaces, $|A^1|, |A^2|, \ldots, |A^{r-h}|$, and suppose that they are generic and

† The set of these differences is not a group, but it becomes one if we consider all their linear combinations (Abelian minimum group containing all those differences within the Abelian group of the virtual varieties ∞^h lying on M_r). It is obvious that it is possible to construct in this case, under the stated conditions, the varieties V', H. The possibility of transferring by intersection the relation of rational equivalence within M_r is an obvious consequence of the properties possessed by linear systems of hypersurfaces upon M_r, according to which they give as intersections linear systems of hypersurfaces on a variety lying on M_r, and therefore elementary systems give elementary systems, and linear combinations of elementary systems give similar combinations.

generically placed, in the sense that the variety V_h common to $r-h$ hypersurfaces taken generically, one for each of the given systems, is pure and ∞^h. Let us consider the totality of the effective intersection $A^1A^2\ldots A^{r-h}$, each component of which has to be counted with its own multiplicity of intersection; then this totality completed, if necessary, with the ∞^h varieties (*specializations*), each of which is of accumulation for the V_h mentioned above, results algebraic (and even rational). This is what we shall call an *intersection system* or *elementary system* (a *series* for $h=0$).

It must be understood that, in this definition, possible ∞^h base varieties, appearing as *fixed components* of the above intersection variety, may, at one's will, be included or not in the variety describing the elementary system.

It is hardly necessary to warn the reader that each one of the specializations mentioned above has dimension h; nevertheless, it can be contained in a pure or impure variety, with components at least one of which has the dimension $>h$, common to $r-h$ *particular* A. In this case the specialization referred to may vary over one of the components of abnormal dimension, because, when we try to obtain it as a limit of the generic V_h, the limiting position constitutes the *functional equivalence* on the component of abnormal dimension, according to the concepts of p. 17 of the 'Lezioni' [4].

Further complements may be added to the definition of elementary systems (e.g. transferring the idea in the virtual field).† But they become useless because finally everything is included in the more general definition of systems of equivalence at which one arrives, as we shall recall at the end of this section.

As the set of the elementary systems does not constitute a group, it is necessary to pass (as the author did in 1934) to the *systems of rational equivalence*, each of which in fact consist of all possible linear combinations (with integer positive, negative, or zero coefficients) of the varieties taking one from each member of a given set of elementary systems.

According to the fundamental theorem of p. 84 of the 'Lezioni' referred to in [4], the set formed by the elementary systems of kind h becomes closed with respect to the operations of addition and subtraction, when, instead of referring to elementary systems of effective varieties, we consider elementary systems of virtual varieties. *The set thus obtained then coincides with the one of all the systems of (rational) equivalences of kind h.*

† As it is done in Lectures [4], pp. 70 and 75.

8. We now make a further remark on the values of the symbols of virtual intersection of several hypersurfaces of M_r.

Let $A^1, A^2, \ldots, A^{r-h}$ be $r-h$ $(h \leqq r-2)$ such hypersurfaces. The value of the symbol $A^1A^2 \ldots A^{r-h}$ with respect to rational equivalence has already been implied by the general considerations of § 6; but we may also establish it directly, as follows, in agreement with the concepts of § 2.†

First we suppose the hypersurfaces A to be irreducible, and assume $h = r-2$. We can, in an infinity of ways, determine upon M_r a linear system $|B|$ of hypersurfaces, such that, if $|C| = |B + A^2|$, the generic hypersurfaces B, C of the systems mentioned above cut irreducible varieties on A^1. Then we shall assume

$$A^1A^2 = A^1C - A^1B,$$

where A^1B, A^1C denote the generic ∞^{r-2} varieties of the linear systems cut out upon A^1 by $|B|$, $|C|$. If we let A^1C, A^1B vary in the respective systems, the difference $A^1C - A^1B$ describes on A^1, that is, on M_r, an elementary virtual system. The same occurs if B varies under the said conditions, because, if $|B^1|$ and $|C^1|$ are systems analogous to $|B|$, $|C|$, we have $A^2 \equiv C - B \equiv C^1 - B^1$. Therefore A^1A^2 is defined only for its variability in a virtual linear system with simple components (therefore only for rational equivalence). Further, with simple considerations analogous to those of § 4, one finds that $A^1A^2 \equiv A^2A^1$.

Let us proceed by supposing $h = r-3$. We can, in an infinity of ways, determine a linear system $|D|$ such that, assuming

$$|E| = |D + A^3|,$$

the generic hypersurfaces D, E cut out on each one of the irreducible components of $H = A^1C$, $K = A^1B$ of the elementary system A^1A^2 an irreducible ∞^{r-3} variety. We assume then

$$A^1A^2A^3 = (H-K)E - (H-K)D = (H-K)(E-D).$$

The ∞^{r-3} variety defined by the right-hand side of the above equality varies, as is now obvious, in an elementary virtual system (with simple components) independent of the choice of $|D|$, under the given conditions; moreover the symbol $A^1A^2A^3$ is associative and commutative.

We can continue with the same procedure; and, on applying it inductively, we reach the conclusion that, *however complicated or abnormal or non-existent the intersection of the hypersurfaces A is in the effective domain, we obtain for the symbol $A^1A^2 \ldots A^{r-h}$ a precise meaning*

† What follows is essentially included in §41 of Lectures[4], although in a less concise form.

so that it always denotes a virtual variety with simple components, of dimension h, mobile in an elementary system.

This is but a natural extension of the very concept which half a century ago led the author to the introduction in [3] of the notion of the characteristic virtual series on a curve of a surface, independently of the variability of the curve in a continuous system.

The meaning of $A^1A^2 \ldots A^{r-h}$, when the hypersurfaces A can vary in an arbitrary way, and also have multiple components, is now clear. $A^1A^2 \ldots A^{2-h}$ is given by an equality identical with the one stated explicitly in § 5 for the algebraic equivalence.

Finally, if we remember that every pure variety ∞^h can be obtained in a linear space as a complete intersection of $r-h$ suitable virtual hypersurfaces (cf. 'Lezioni' [4], p. 62), we deduce in an autonomous way the *meaning of the symbol of virtual intersection of several varieties of any dimension with respect to rational equivalence.*

9. The general concept, constituting the foundation of the process by which we arrive at the notion of virtual intersection of several varieties, is the same one (as we said in footnote ‡ on p. 158) as that which led the author to the precise notion of multiplicity of intersection, remaining strictly in the algebraic domain, by the use of a process the author has called *dynamic.*†

The definition of multiplicity of intersection and of functional equivalence which we obtain, is, we believe, both clear and elegant.

After having defined the symbol VW, for instance, only for an algebraic equivalence, with the relation

$$VW \equiv HW - V'W,$$

let us make H tend to $V+V'$ ($H \to V+V'$) with the only condition, in the case $l=0$, that the sets HW, $V'W$ contain distinct simple points, certainly at the limit; and, in the case $l>0$, that the components of the varieties HW, $V'W$ are simple, at the limit. If P is an isolated intersection of V, W ($l=0$), and V' does not pass through P, the number of positive points of the set of distinct points of the set $HW-V'W$, which tend to P, is the multiplicity of intersection of V, W in P. If N is a component of the effective intersection of V, W ($l>0$), the number of distinct positive components of the variety $HW-V'W$ tending to N, when V' does not contain N as a component, gives the multiplicity of N as a component of the effective intersection of the two varieties.

† See [6], where the previous treatment by van der Waerden, based on topology (intersections of cycles, after Lefschetz), also is quoted.

Finally, if Q is a component having abnormal dimension ($>l$) in the effective intersection of V, W, and V' does not contain Q, then the set of the components of the virtual variety VW, which can partly be positive and partly negative or all of the same sign, tending to lie upon Q, gives the functional equivalence of Q in the intersection of V, W, and the algebraic sum of the coefficients $+1$ and -1 of the above mentioned components gives the *numerative equivalence* of Q in that intersection.†

10. We conclude this work with some further remarks. When two virtual or effective varieties A, B (pure and ∞^h) are rationally equivalent (so that they totally belong to some rational system of virtual varieties), they are always joined by an infinity of rational ∞^1 systems of virtual varieties.

Let Σ be one of these systems which contain totally A, B, and let $D=E-F$ be a variable virtual variety in Σ, equal to the difference of two effective varieties E, F. Then two alternatives are possible.

(a) The negative part $-F$ is missing in D. Then for $D \to A$ or $D \to B$ the limits of E are A, B, and therefore E describes a rational ∞^1 system of effective varieties containing A, B totally.

(b) The order of E is greater than the order of A and B. Then for $D \to A$ or $D \to B$ the limits of E are respectively $A+A_0$, $B+B_0$, where A_0, B_0 are limits of F and they are joined by the *rational* system of effective varieties described by F. Therefore:

Two rationally equivalent effective varieties A, B are either totally contained in a rational system of effective varieties or they are reduced to two varieties joined by a rational system by adjoining two effective varieties A_0, B_0 totally contained in a rational system of effective varieties.

If the case (a) occurs for every pair of varieties totally contained in a complete system of rationally equivalent effective varieties, then a variety of points representative of the system is such that any two of its points are always joined by some rational curve. *Does this imply that the variety is unirational?* Once before the author had the opportunity of asking this question which still remains unanswered. The conjecture made in the lecture of September 10th quoted at the beginning, namely the unirationality of the complete systems of effective

† The most recent paper [5] by the author on the multiplicities of intersection was published in 1950 for the celebration of the 70th birthday of Perron. In this paper we prove the equivalence of the definition of the author with the *static* one (which is more suited for fields where one cannot or will not use the continuity), given by A. Weil in his work [7]. Severi's paper deals also more fully with the multiplicity of intersection in particular points of a different type from that considered in the short remark of §9.

varieties rationally equivalent on M_r, i.e. of the *traces* in the effective domain of the virtual systems of rational equivalence (in particular, of the families of effective varieties of a given order and a given dimension lying in a projective space) is specified by the above remark.

ROME

REFERENCES

[1] F. SEVERI, *Il punto di vista gruppale nei vari tipi di equivalenza sulla varieta algebriche*, Comment. Math. Helv., 21 (1948), pp. 189–224.

[2] ——, Memorie scelte, Bologna, 1950, p.185.

[3] ——, *Osservazioni sui sistemi continui di curve tracciate sopra una superficie algebrica*, Atti Accad. Sci. Torino, 39 (1904), pp. 490–508.

[4] ——, *Serie, sistemi d'equivalenza e corrispondenze algebriche sulle varietà algebriche* (Lectures given in Rome at the Istituto Nazionale di Alta Matematica, collected by F. Conforto and E. Martinelli, Cremonese, Roma, 1942).

[5] ——, *Sulla molteplicità d'intersezione delle varietà algebriche ed analitiche e sopra una teoria geometrica dell'eliminazione*, Math. Zeit., 52 (1950), pp. 827–851.

[6] ——, *Ueber die Grundlagen der algebraischen Geometrie*, Abhandlungen aus dem Math. Seminar der Hamburgischen Universität (1933), pp. 335–364.

[7] A. WEIL, Foundations of algebraic geometry, Amer. Math. Soc. Colloquium Publications, 29, 1946.

Integral Closure of Modules and Complete Linear Systems

Ernst Snapper

Introduction

LET Σ be the field of rational functions of an irreducible, r-dimensional, algebraic variety V over an arbitrary groundfield k. Then, r is the degree of transcendency of Σ/k and Σ/k has a finite number of field generators, say $\Sigma = k(\sigma_1, \ldots, \sigma_m)$, where $\sigma_1, \ldots, \sigma_m \in \Sigma$. It is known that every linear system of V, which arises from its $(r-1)$-dimensional subvarieties, is contained in a complete linear system. Hence we ask ourselves the question 'What algebraic theorem concerning finitely generated field extensions is equivalent to the geometric theorem that a linear system is always contained in a complete linear system?'

Let L be a k-module of Σ, i.e. L is a subgroup of the additive group of Σ and is closed under multiplication by elements of k. We say that L has a finite number of k-generators $\alpha_1, \ldots, \alpha_n \in \Sigma$, if L consists of the linear combinations $c_1\alpha_1 + \ldots + c_n\alpha_n$, where $c_1, \ldots, c_n \in k$; we indicate this by $L = (\alpha_1, \ldots, \alpha_n)$. We introduce in § 1 the notion of *the integral closure* $|L|_i$ *of* L. This $|L|_i$ is again a k-module and the existence of complete linear systems is equivalent to the following theorem.

THEOREM. *If L has a finite number of k-generators, the same is true for* $|L|_i$.

The integral closure of fractional ideals of integrally closed rings occurs in [1], § 6. Although we find ourselves in the study of Σ/k in quite a different situation, many of the methods of [1] can be applied in our case.

Whenever we refer to 'the theorem', we mean the theorem stated in this introduction. Sections 2, 3 and 4 are devoted to the proof of the theorem, while in § 5 we discuss briefly why the theorem is equivalent to the existence of complete linear systems. The first four sections are self-contained and require, for their reading, no knowledge of algebraic geometry. The last three sections lean heavily on [3], §§ 17 and 18 and on [2], §§ 2–4.

1. The integral closure of a module

As in the introduction, Σ/k denotes a finitely generated field extension of degree of transcendency r. Whenever we say 'module' we mean 'k-module of Σ', and when we say that a module is finitely generated we mean that it has a finite number of k-generators.

The product LM of two modules L and M is the smallest module which contains all products $\lambda\mu$, where $\lambda \in L$ and $\mu \in M$; clearly, if L and M are finitely generated, so is LM. When $\sigma \in \Sigma$, we write σL for $(\sigma)L$ and hence the module σL consists of all the products $\sigma\lambda$, where $\lambda \in L$. Multiplication of modules is clearly commutative and associative; in particular, the powers L^m, for $m \geqq 1$, are well defined. We add to this the convention that $L^0 = k$.

As is customary, the sum (L, M) of two modules is defined as the smallest module which contains all the elements of L and M. Again, when L and M are finitely generated, so is (L, M) and we write (σ, L) instead of $((\sigma), L)$. Addition of modules is also commutative and associative, while multiplication and addition are combined by the law of distributivity. We note for the proof of Statement 1.1 that, consequently,

$$(\sigma, L)^m = (\sigma^m, L(\sigma, L)^{m-1}) \quad \text{for} \quad m \geqq 1;$$

namely,

$$(\sigma, L)^m = (\sigma^m, \sigma^{m-1}L, \ldots, L^m) = (\sigma^m, L(\sigma^{m-1}, \sigma^{m-2}L, \ldots, L^{m-1}) = (\sigma^m, L(\sigma, L)^{m-1}).$$

For § 2 we observe that, if we extend the notion of addition of modules in the usual way to infinitely many modules, we obtain that the ring $k[L]$ is equal to the infinite sum $(L^0, L, \ldots, L^m, \ldots)$.

Definition 1.1. *If L is a module, its integral closure $|L|_i$ consists of all elements $\sigma \in \Sigma$ for which there exists a non-zero, finitely generated module M such that $\sigma M \subset ML$.* (M depends of course on σ.)

Since k is finitely generated and $Lk = kL$, certainly $L \subset |L|_i$; this shows in particular that $|L|_i$ is not empty. Let $\sigma_1, \sigma_2 \in |L|_i$ and let the corresponding modules be M_1 and M_2. We then conclude from

$$\sigma_1 M_1 \subset M_1 L \quad \text{and} \quad \sigma_2 M_2 \subset M_2 L$$

that $$\sigma_1 M_1 M_2 \subset M_1 M_2 L \quad \text{and} \quad \sigma_2 M_1 M_2 \subset M_1 M_2 L$$

and hence that $$(\sigma_1 \pm \sigma_2) M_1 M_2 \subset M_1 M_2 L;$$

consequently, $|L|_i$ is an additive group. Finally, since $\sigma M \subset ML$ implies that $(c\sigma) M \subset ML$ for any $c \in k$, $|L|_i$ is closed under multiplication by elements of k and hence is a k-module.

Observe that $|k|_i$ is exactly the algebraic closure k' of k in Σ. Hence the theorem to be proven implies the well-known fact that, *since Σ/k is a finitely generated field extension*, k' has a finite field degree with respect to k. We would furthermore like to point out that $|L|_i$ is actually closed under multiplication by elements of k' and not just by elements of k. Namely, if $\alpha \in k'$ and $\sigma \in |L|_i$, there exist non-zero, finitely generated, modules M_1 and M_2, such that $\alpha M_1 \subset M_1$ and $\sigma M_2 \subset M_2 L$. We conclude that $\alpha\sigma M_1 M_2 \subset M_1 M_2 L$ and hence that $\alpha\sigma \in |L|_i$.

Observe that if L had been a subring R of Σ, instead of a module, and we had been dealing with R-modules instead of with k-modules, Definition 1.1 would have given rise to the usual integral closure of R in Σ. The following statement, which we will need later, bears out further the close relationship between our integral closure of modules and the usual integral closure of rings.

STATEMENT 1.1. *If L is finitely generated, an element of Σ belongs to $|L|_i$ if and only if it satisfies an equation*

$$x^m + a_1 x^{m-1} + \dots + a_{m-1}x + a_m = 0,$$

where $\quad m \geqq 1 \quad$ *and* $\quad a_j \in L^j \quad$ *for* $\quad j = 1, \dots, m.$

It follows that $|L|_i$ is contained in the usual integral closure of the ring $k[L]$ in Σ.

PROOF. Let $\sigma \in |L|_i$. Then there exists a module $M = (\alpha_1, \dots, \alpha_n)$, where not all $\alpha_1, \dots, \alpha_n$ are zero, such that $\sigma M \subset ML$. Hence

$$\sigma\alpha_j = \textstyle\sum_{h=1}^{n} \lambda_{jh}\alpha_h, \quad \text{where} \quad \lambda_{jh} \in L \quad \text{and} \quad j = 1, \dots, n,$$

which shows that σ is a characteristic root of the matrix (λ_{jh}). Since the characteristic equation of this matrix has the required form, the 'only if' part has been proved; we have not used, so far, that L is finitely generated. Now let σ be an element of Σ which satisfies an equation of the form described in Statement 1.1. Then

$$\sigma^m \in (\sigma^{m-1}L, \sigma^{m-2}L^2, \dots, L^m) = L(\sigma^{m-1}, \sigma^{m-2}L, \dots, L^{m-1}) = L(\sigma, L)^{m-1}.$$

Hence we conclude from the relation, observed earlier, that

$$(\sigma, L)^m = L(\sigma, L)^{m-1},$$

i.e., that $(\sigma, L)(\sigma, L)^{m-1} = (\sigma, L)^{m-1}L$. If $\sigma = 0$, of course $\sigma \in |L|_i$. If $\sigma \neq 0$, the module $(\sigma, L)^{m-1}$ is non-zero and finitely generated, *since L is finitely generated*; hence every element of (σ, L), in particular σ, belongs to $|L|_i$. The proof of Statement 1.1 is now complete.

We conclude this section with the remark that for any module L and $\sigma \in \Sigma$, $|\sigma L|_i = \sigma |L|_i$. This is an immediate consequence of Definition 1.1 and is used in the next section.

2. Preparation for the proof of the theorem

Let L be a module of our field extension Σ/k. If E is a field which contains Σ, the module defined by L in E, that is, the smallest k-module of E which contains L, is set-theoretically identical with L. However, the integral closure of L as a module of E (denoted by '$|L|_i$ in E') may be larger than the integral closure of L as a module of Σ (denoted for distinction by '$|L|_i$ in Σ'). Consequently, if we can prove that, for a certain choice of E, the first closure is finitely generated, so is the latter one and we are done. The following statement, although not necessary for the proof of the theorem, is convenient to have, since we will use as E a simple transcendental extension of Σ.

STATEMENT 2.1. *Let L be finitely generated. Then*

$$(|L|_i \text{ in } E) \cap \Sigma = |L|_i \text{ in } \Sigma.$$

If E is a purely transcendental extension of Σ, where the degree of transcendency of E/Σ may be any cardinal number, $|L|_i$ in $E = |L|_i$ in Σ.

PROOF. We know from Statement 1.1 that $\sigma \in (|L|_i \text{ in } E) \cap \Sigma$ if and only if $\sigma \in \Sigma$ and σ is the characteristic root of a matrix with elements in L, i.e. if and only if $\sigma \in |L|_i$ in Σ. If E is a purely transcendental extension of Σ, a characteristic root in E of a matrix with elements in L must necessarily belong to Σ, since in that case a polynomial with coefficients in Σ cannot have roots in E which do not already belong to Σ. Hence the statement is proved.

Let L be any module of Σ. Since Σ/k is a finitely generated field extension, we can find elements $\alpha_1, \ldots, \alpha_n \in \Sigma$. such that

$$k(\alpha_1, \ldots, \alpha_n, L) = \Sigma.$$

Denoting by M the module $(1, \alpha_1, \ldots, \alpha_n, L)$, we see that $k \subset M$ and that $k(M) = \Sigma$. Furthermore, we conclude from $L \subset M$ that $|L|_i \subset |M|_i$, and hence, if $|M|_i$ is finitely generated, so is $|L|_i$. This argument shows that, in order to prove the theorem, we can restrict ourselves to modules L which are such that $k \subset L$ and $k(L) = \Sigma$.

Let L then be a module of Σ which has the just mentioned two properties and let $E = \Sigma(t)$, where t is transcendental over Σ. We have seen earlier that all we have to show is that $|L|_i$ in E is finitely generated. Instead of studying the module L in E, we consider the module tL in E. Now $|tL|_i$ in $E = t(|L|_i$ in $E)$ and hence, if $|tL|_i$ in E is finitely generated (always over k), so is $|L|_i$ in E. Observe that tL has lost the property of containing k, but that tL contains t and hence is certainly not the zero-module. Furthermore, since $k(L) = \Sigma$

and $t \in tL$, $k(tL) = \Sigma(t) = E$. Finally, the additive decomposition $k[tL] = ((tL)^0, tL, \ldots, (tL)^m, \ldots)$ is *direct*, i.e. if $\gamma_j \in (tL)^{h_j}$ for $j = 1, \ldots, s$ and the exponents $h_1, \ldots, h_s$ are mutually distinct, $\gamma_1 + \ldots + \gamma_s$ can be zero only if each $\gamma_j = 0$. This follows immediately from the fact that $(tL)^h$ consists of the polynomials $t^h f(\alpha_1, \ldots, \alpha_n)$, where $f(\alpha_1, \ldots, \alpha_n)$ is a homogeneous polynomial of degree h of $k[\alpha_1, \ldots, \alpha_n]$ and $\alpha_1, \ldots, \alpha_n \in L$; it is at this point that we use that t is transcendental over Σ. Hence, we have to prove the theorem only for non-zero modules L of Σ which are such that (A) *the additive decomposition* $k[L] = (L^0, L, \ldots, L^m, \ldots)$ *is direct and* (B) $k(L) = \Sigma$; we refer to these last two italicized properties of L as respectively 'property A' and 'property B'.

3. Non-zero modules with property A

In this section we assume that L is a non-zero module of Σ *which possesses property* A. We say that a non-zero element $\sigma \in \Sigma$ is *homogeneous of degree m* if $\sigma = \lambda/\mu$, where $\lambda \in L^h$, $\mu \in L^j$ and $h - j = m$; here, h and j are any two non-negative integers and property A guarantees that the definition makes sense. It is clear that, if σ and σ' are homogeneous elements of degrees respectively m and m', $\sigma\sigma'$ is homogeneous of degree $m + m'$, while, if $m = m'$ and $\sigma \neq \sigma'$, $\sigma - \sigma'$ is also homogeneous of degree m. It follows that the homogeneous elements of fixed degree $m \geqq 0$, together with the zero element of Σ, form a module (as always over k) which we denote by H_m and that the elements of Σ which can be written as sums of homogeneous elements (not necessarily of the same degree) form a ring, denoted by H. Clearly, H is equal to the infinite sum $(H_0, H_1, \ldots, H_m, \ldots)$ and we now show that property A can be extended from $k[L]$ to H.

STATEMENT 3.1. *The additive decomposition* $H = (H_0, H_1, \ldots, H_m, \ldots)$ *is direct.*

PROOF. Let $\gamma_j \in H_{t_j}$ for $j = 1, \ldots, q$, where the indices t_j are mutually distinct. We assume furthermore that $\gamma_1 + \ldots + \gamma_q = 0$, and we have to show that consequently each $\gamma_j = 0$. If not all γ_j's are zero, let us first delete all those γ_j's which happen to be zero and then assume that in the remaining sum $\gamma_1 + \ldots + \gamma_s = 0$ no term is zero. Now $\gamma_j = \lambda_j/\mu_j$, where $\lambda_j \in L^{u_j}$ and $\mu_j \in L^{v_j}$ and $u_j - v_j = t_j$; we denote $v_1 + v_2 + \ldots + v_s$ by w. Since

$$0 = \gamma_1 + \ldots + \gamma_s = (1/\mu_1\mu_2 \cdots \mu_s) \textstyle\sum_{j=1}^{s} \lambda_j \mu_1 \cdots \mu_{j-1}\mu_{j+1} \cdots \mu_s,$$

we conclude that $\sum_{j=1}^{s} \lambda_j \mu_1 \cdots \mu_{j-1}\mu_{j+1} \cdots \mu_s = 0$.

The facts that $\lambda_j \mu_1 \ldots \mu_{j-1}\mu_{j+1} \cdots \mu_s \in L^{w+t_j}$ and that property A holds, imply that each $\lambda_j\mu_1 \dots \mu_{j-1}\mu_{j+1} \dots \mu_s = 0$ and hence that each $\gamma_j = 0$, which is a contradiction; the proof is now complete.

We now go over to the study of submodules of H which are closed under multiplication by all elements of the ring $k[L]$ and not just of the field k. Observe in this connection that $k[L] \subset H$.

STATEMENT 3.2. *Let the module L of Σ, which still satisfies property* A, *be finitely generated (over k). Let J be a submodule of H which is closed under multiplication by the elements of the ring $k[L]$ and is finitely generated over $k[L]$. Then the module $J \cap H_m$ is finitely generated over k for each $m \geqq 0$.*

PROOF. We are given that there exists a finite number of elements $z_1, \dots, z_s \in J$ such that, if $\gamma \in J$, $\gamma = f_1 z_1 + \dots + f_s z_s$, where $f_i \in k[L]$. Let $z_i = \alpha_{i1} + \dots + \alpha_{ih_i}$ be the decomposition of z_i into a sum of homogeneous elements; we do not claim that necessarily $\alpha_{ij} \in J$. Since $f_j \in k[L]$, we see that $\gamma = \sum \mu_{ij}\alpha_{ij}$, where $\mu_{ij} \in L^{t_{ij}}$. If γ is homogeneous of degree m, we conclude from Statement 3.1 that $\gamma = \sum \mu'_{ij}\alpha'_{ij}$, where we have deleted from the sum $\sum \mu_{ij}\alpha_{ij}$ all those terms whose degree is not exactly m. Then $\mu'_{ij} \in L^{t'_{ij}}$, where $t'_{ij} \leqq m$, which shows that $J \cap H_m$ is contained in the product of the k-module $(L^0, L, \dots, L^m)$ and the k-module which is generated by all the elements α_{ij}. Since both these modules are finitely generated over k (here we use the finite generation of L), so is $J \cap H_m$ and we are done.

Let us return to the case where L is an arbitrary module of Σ which possesses property A. Since products and quotients of homogeneous elements of degree zero are again homogeneous elements of degree zero, the module H_0 is a field which contains k. (Actually, Statement 3.4 implies that H_0 contains the whole algebraic closure of k in $k(L)$.)

STATEMENT 3.3. *Every non-zero element α of L is transcendental over H_0. Furthermore, $H_0[\alpha] = H_0[L] = H$ and $H_0(\alpha) = H_0(L) = k(L)$ and $H_m = \alpha^m H_0$. Finally, $H_0 = k(L_\alpha)$, where L_α denotes the module $(1/\alpha)L$.*

PROOF. If $\gamma_0\alpha^m + \gamma_1\alpha^{m-1} + \dots + \gamma_m = 0$, where $m \geqq 1$ and $\gamma_i \in H_0$, each term $\gamma_i\alpha^{m-i}$ is either zero or is homogeneous of degree $m-i$; hence we conclude from Statement 3.1 that each $\gamma_i = 0$, i.e. that α is transcendental over H_0. Observe that $H_0[\alpha] \subset H_0[L] \subset H$ is trivial. If $\gamma \in H$, decompose γ into homogeneous elements, say $\gamma = \mu_1 + \dots + \mu_s$, where μ_i is homogeneous of degree m_i. Then

$$\gamma = (\mu_1/\alpha^{m_1})\alpha^{m_1} + \dots + (\mu_s/\alpha^{m_s})\alpha^{m_s}$$

and consequently, since $\mu_i/\alpha^{m_i} \in H_0$, $\gamma \in H_0[\alpha]$ which shows that $H_0[\alpha] = H_0[L] = H$. All elements of L_α clearly belong to H_0 and hence

$k(L_\alpha) \subset H_0$. Conversely, if $\gamma \in H_0$, there exist elements $\lambda, \mu \in L^s$ for some $s \geqq 0$ such that $\gamma = \lambda/\mu$; it follows that $\gamma = (\lambda/\alpha^s)/(\mu/\alpha^s) \in k(L_\alpha)$ and hence that $H_0 = k(L_\alpha)$. We conclude from $H_0[\alpha] = H_0[L]$ that $H_0(\alpha) = H_0(L)$. Furthermore, $H_0(\alpha) = k(L_\alpha, \alpha)$ and, since $\alpha \in L$, $k(L_\alpha, \alpha) = k(L)$. Finally, observe that $\alpha^m H_0 \subset H_m$ is trivial; conversely, if $\gamma \in H_m$, then $\gamma = (\gamma/\alpha^m)\alpha^m$ and, since $\gamma/\alpha^m \in H_0$, $\gamma \in \alpha^m H_0$. This shows that $\alpha^m H_0 = H_m$ and Statement 3.3 is proved.

STATEMENT 3.4. *The field of quotients of the ring H is $k(L)$. The ring H is integrally closed in $k(L)$ and the field H_0 is algebraically closed in $k(L)$.*

PROOF. We know from Statement 3.3 that $H = H_0[\alpha]$ and that $k(L) = H_0(\alpha)$; hence $k(L)$ is the field of quotients of H. The remainder of Statement 3.4 follows from the fact that α is transcendental over H_0.

Let I denote the usual integral closure of the ring $k[L]$ in its field of quotients $k(L)$. We conclude from Statement 3.4 that $I \subset H$. It is well known that, if L is finitely generated (over k), I is finitely generated over $k[L]$. Hence we can then use I as the module J of Statement 3.2 and conclude that *the modules $I \cap H_m$ are finitely generated over k for $m \geqq 0$.*

4. Non-zero modules with properties A and B

All symbols have the same meaning as before. In particular, I is the usual integral closure of $k[L]$ in $k(L) = \Sigma$. Observe that $k[L^m] \subset k[L]$ and that all elements of L depend integrally on $k[L^m]$; hence I is also the usual integral closure of $k[L^m]$ in Σ.

The following statement, when restricted to the case $m = 1$, shows, in connection with the last italicized remark of the previous section, that the theorem is correct for the special type of finitely generated modules with which the present section is concerned. Consequently, as observed in the last sentence of § 2, our theorem is proved.

STATEMENT 4.1. *Let L be a finitely generated, non-zero module which possesses properties* A *and* B. *Then, for any $m \geqq 0$, $|L^m|_i = I \cap H_m$.*

PROOF. Let $\gamma \in |L^m|_i$. We know from Statement 1.1 that $|L^m|_i$ is contained in the usual integral closure of $k[L^m]$ in Σ, i.e. $|L^m|_i \subset I$; hence, in order to show that $|L^m|_i \subset I \cap H_m$, all we have to show is that $\gamma \in H_m$. Statement 1.1 tells us that $\gamma^h + a_1\gamma^{h-1} + \ldots + a_h = 0$, where $h \geqq 1$ and $a_j \in L^{mj}$ for $j = 1, \ldots, h$. If α is any non-zero element of L, we conclude that

$$(\gamma/\alpha^m)^h + (a_1/\alpha^m)(\gamma/\alpha^m)^{h-1} + (a_2/\alpha^{2m})(\gamma/\alpha^m)^{h-2} + \ldots + (a_h/\alpha^{hm}) = 0.$$

Since $a_j/\alpha^{jm} \in H_0$ and H_0 is algebraically closed in Σ, $\gamma/\alpha^m \in H_0$, and hence $\gamma \in H_m$. We have not used, so far, that L is finitely generated.

Conversely, let γ be a non-zero element of $I \cap H_m$ and let us prove that $\gamma \in |L^m|_i$. Hence we know that γ satisfies an equation

$$x^h + a_1 x^{h-1} + \ldots + a_h = 0,$$

where $h \geqq 1$ and $a_j \in k[L]$ for $j = 1, \ldots, h$, and also that $\gamma = \lambda/\mu$, where $\lambda \in L^s$, $\mu \in L^t$, $s - t = m$. We conclude that

$$\lambda^h + a_1 \mu \lambda^{h-1} + a_2 \mu^2 \lambda^{h-2} + \ldots + a_h \mu^h = 0.$$

In this expression, replace each a_j by the sum $\alpha_{j1} + \ldots + \alpha_{jn_j}$, where $\alpha_{j1} \in L^{u_{ji}}$ and $a_j = \alpha_{j1} + \ldots + \alpha_{jn_j}$. It follows then from Statement 3.1 and the fact that λ^h is homogeneous of degree sh, that

$$\lambda^h + \alpha_1 \mu \lambda^{h-1} + \alpha_2 \mu^2 \lambda^{h-2} + \ldots + \alpha_h \mu^h = 0,$$

where $\alpha_j \in L^{mj}$. We then conclude from

$$(\lambda/\mu)^h + \alpha_1 (\lambda/\mu)^{h-1} + \ldots + \alpha_h = 0,$$

and from Statement 1.1 that $\lambda/\mu \in |L^m|_i$ and we are done; this last conclusion uses that L is finitely generated.

5. Complete linear systems

As in the introduction, let V be an irreducible, r-dimensional variety over the groundfield k whose field of rational functions is our field Σ. We know how V selects, from among all the $(r-1)$-dimensional valuations of Σ/k, a subset $\mathfrak{B}$ whose elements are called the prime divisors of the first kind of V. These divisors arise from the irreducible, $(r-1)$-dimensional subvarieties of V where, unless V satisfies special conditions, several divisors of the first kind may correspond to the same subvariety of V. The knowledge of the set $\mathfrak{B}$ alone suffices to define the linear systems and complete linear systems of V. Namely, if α is a non-zero element of Σ, its pole divisor $P(\alpha)$, zero divisor $Z(\alpha)$ and its divisor $D(\alpha) = Z(\alpha) - P(\alpha)$ are defined as usual, *where we restrict ourselves completely to the valuations which occur in* $\mathfrak{B}$. If L is a non-zero, finitely generated module of Σ, every valuation reaches a minimum on L; hence we can speak of the pole divisor $P(L)$, zero divisor $Z(L)$ and divisor $D(L) = Z(L) - P(L)$ of L, where we restrict ourselves again to the valuations which occur in $\mathfrak{B}$. The linear system g without fixed divisor which arises from L consists of the divisors $D(\alpha) - D(L)$, where α runs through the non-zero elements of L. We denote by $|L|_v$ the module of Σ which consists of the zero element of Σ together with those non-zero elements α of Σ for which $D(\alpha) \geqq D(L)$; here, the inequality sign has the usual meaning for divisors. It is clear that $|L|_v$ is a module which contains L and which determines the complete linear system $|g|$ in which g is contained. Precisely,

$|g|$ consists of the divisors $D(\alpha)-D(L)$, where α runs through the non-zero elements of $|L|_v$. Consequently, the geometric theorem that a linear system is always contained in a complete linear system is equivalent to the statement that the modules $|L|_v$ are always finitely generated (over k).

Of course, since $|L|_i$ is constructed in a strictly invariant fashion, that is, without any choice of a set of divisors $\mathfrak{B}$, usually $|L|_i \neq |L|_v$. Only if $r=1$, i.e. if V is a curve, does $\mathfrak{B}$ not depend on the choice of our model V for the function field Σ. $\mathfrak{B}$ then *always* consists of all the valuations of Σ/k, no matter how V is chosen, and it is easy to show that in that case always $|L|_i = |L|_v$. In general, all we can say is the following.

STATEMENT 5.1. $|L|_i \subset |L|_v$.

PROOF. Let α be a non-zero element of $|L|_i$. There exists a finitely generated, non-zero module M such that $\alpha M \subset ML$. It follows that $D(\alpha)+D(M) \geqq D(M)+D(L)$ and hence that $D(\alpha) \geqq D(L)$. Consequently, $\alpha \in |L|_v$ and we are done.

We see from Statement 5.1 that the finite dimensionality of complete linear systems, that is, of the modules $|L|_v$, implies that integral closures of finitely generated modules are again finitely generated. Conversely, let us show how the finite generation of integral closures leads to the finite dimensionality of the modules $|L|_v$. Hereto, let g_m denote the linear system cut out on our V by the hypersurfaces of degree m, for $m \geqq 1$. It is well known that, in order to show that *all* complete linear systems of V are finite dimensional, we only have to prove the finite dimensionality of the complete linear systems $|g_m|$. Let $\alpha_1, \ldots, \alpha_n$ be elements of Σ such that $\alpha_1, \ldots, \alpha_n$ are the coordinates of a generic point of V. We consider, for each $m \geqq 1$, the module L^m, where $L=(1, \alpha_1, \ldots, \alpha_n)$. The linear system without fixed divisor which arises, in the manner described above, from L^m is exactly the system g_m. Hence, all we have to show is that the special modules $|L^m|_v$ are finitely generated for $m \geqq 1$. It follows easily from the proof of Theorem 1 of [2] and our § 4 that for these special modules again $|L|_v = |L|_i$ and hence we are done.

The arguments of this section can be carried out in a strictly algebraic fashion, in which the set of valuations $\mathfrak{B}$ is defined ring-theoretically and no reference is made to projective models of the function field Σ/k. This will be worked out elsewhere.

UNIVERSITY OF SOUTHERN CALIFORNIA
and
HARVARD UNIVERSITY

REFERENCES

[1] H. PRÜFER, *Untersuchungen über Teilbarkeitseigenschaften in Körpern*, J. Reine Angew. Math., 168 (1932), pp. 1–36.

[2] O. ZARISKI, *Complete linear systems on normal varieties and a generalization of a lemma of Enriques-Severi*, Ann. of Math., 55 (1952), pp. 552–592.

[3] ——, *Some results in the arithmetic theory of algebraic varieties*, Amer. J. Math., 61 (1939), pp. 249–294.

On the Projective Embedding of Abelian Varieties

André Weil

THE modern theory of Abelian varieties may be said to have originated from Lefschetz's Bordin prize memoir [1]. The abstract theory, developed more recently, consists largely in nothing else than the extension to arbitrary groundfields, not only of Lefschetz's results, but also, whenever possible, of his methods.

The purpose of the present note, respectfully dedicated to Lefschetz, is to give still another example of the application of his ideas to a problem of the abstract theory; we shall concern ourselves with the projective embedding of Abelian varieties, a question which has attracted a good deal of attention in recent years and to which Chow and Matsusaka have applied the method of associated forms ('Chow coordinates'). In the classical case, Lefschetz had given a solution ([1], pp. 368–369) which seemingly depended upon the use of theta-functions. It will be shown here that his idea can be extended very simply to the abstract case, giving a more complete result than those of Chow and of Matsusaka.

Following modern usage in the theory of complex-analytic varieties, we shall say that a complete linear system is *ample* if it has no fixed component and defines a one-to-one and everywhere biregular embedding of the variety into a projective space. On a complete non-singular abstract variety V, a complete linear system S is ample if and only if the following two conditions are satisfied:

(A) S *separates points on* V. This means that, given any two distinct points P, Q on V, there is a divisor in S going through P and not through Q; when that is so, S can have no 'base-points'.

(B) S *separates infinitely nearby points* in the sense that, given any point P on V and any tangent vector T to V at P, there is a divisor X in S going simply through P and transversal to T at P. This means that there is just one component of X going through P, that it has the coefficient 1 in X and has a simple point at P, and that the tangent linear variety to it at P is tranversal to T.

We shall say that a class of divisors (for linear equivalence) is *ample* if the complete linear system consisting of all the positive divisors in that class is ample.

To say that a variety is projectively embeddable is to say that there exists on it at least one ample class of divisors. For Abelian varieties over complex numbers, Lefschetz proved much more; he showed that a suitable multiple of the class determined by any Riemann form (satisfying Riemann's bilinear relations and inequalities) is ample. His proof will be extended to the corresponding theorem in the abstract case.

Let A be an Abelian variety of dimension n. If X is any divisor on A, we denote by X_a, as usual, the transform of X by the translation $x \rightarrow x+a$. Let the $W^{(i)}$ be finitely many subvarieties of A of dimension $n-1$, all going through O and with the following properties: (a) $\bigcap W^{(i)} = \{0\}$; (b) given any tangent vector T to A at O, there is a $W^{(i)}$ having a simple point at O and transversal to T at O. *Then the class determined on A by the divisor $3\sum W^{(i)}$ is ample.* In fact, by a known result ([3], no. 57, Theorem 30, Corollary 2), this class contains all the divisors

$$X = \sum(W^{(i)}_{u_i} + W^{(i)}_{v_i} + W^{(i)}_{-u_i-v_i}).$$

Let a and b be two distinct points of A. By our assumption (a), one at least of the $W^{(i)}$, say $W^{(1)}$, does not go through $b-a$. Take $u_1 = a$; take for v_1 and the u_i, v_i for $i \neq 1$ a set of independent generic points of A over $k(a, b)$, where k is a common field of definition of A and the $W^{(i)}$. Then X goes through a and not through b, i.e. it satisfies condition (A). Similarly, let T be a tangent vector to A at a; the translation $x \rightarrow x-a$ transforms it into a tangent vector T_0 to A at O. By our assumption (b), one at least of the $W^{(i)}$, say $W^{(1)}$, has at O a simple point and is transversal to T_0. Taking the u_i, v_i just as before, we get a divisor X which satisfies condition (B).

Now the existence of varieties $W^{(i)}$ such as we want them here is a trivial consequence of the following facts:

(a′) given any point a on A, there is a subvariety W of A of dimension $n-1$ going through O and not through a;

(b′) given any linear subvariety L of dimension $n-1$ of the space of tangent vectors to A at O, there is a subvariety W of A of dimension $n-1$ having O as a simple point and L as its tangent linear variety at O.

The latter is true for arbitrary varieties; it is a purely local property of simple points, is obvious for a variety in an affine space, and is therefore always true. As to the former, A being given as an abstract variety, let A' be one of its representatives; if k is a field of definition for A, and u a generic point of A over $k(a)$, $u+a$ is also generic on A over $k(a)$, and so both u and $u+a$ have representatives u', v' on A'. If W' is any subvariety of A' of dimension $n-1$ going through u' and not through v' (e.g. a component going through u' of the intersection of A' with any hyperplane going through u' and not through v'), it will be the representative on A' of a subvariety W of A such that W_{-u} satisfies (a').

This already shows that A is projectively embeddable. But we want to prove the following precise result:

THEOREM. *Let X be a positive divisor on A. In order that there may exist an integer $n>0$ such that the class of nX be ample on A, it is necessary and sufficient that X should be non-degenerate.*

Here, following an interesting recent paper by Morikawa [2], we say that a divisor X on A is *non-degenerate* if there are only finitely many points t of A such that $X_t \sim X$; otherwise it is called *degenerate*.

Assume first that X is non-degenerate. Write X as $X=\sum X^{(\nu)}$, where the $X^{(\nu)}$ are subvarieties of A of dimension $n-1$. The set γ of points a of A such that $X^{(\nu)}_a = X^{(\nu)}$ for all ν is contained in the set of those a for which $X_a \sim X$ and is therefore finite. Consider the divisor

$$Y=\sum(X^{(\nu)}_{u_\nu}+X^{(\nu)}_{v_\nu}+X^{(\nu)}_{-u_\nu-v_\nu}).$$

We see, as before, that $Y\sim 3X$. Let a and b be two points of A such that $b-a$ is not in γ; then there is an $X^{(\nu)}$, say $X^{(1)}$, and a point c in $X^{(1)}$ such that $c+(b-a)$ is not in $X^{(1)}$. Take $u_1=a-c$ and take for v_1 and the u_ν, v_ν for $\nu\neq 1$ a set of independent generic points of A over $K(a,b,c)$, where K is a common field of definition for A and the $X^{(\nu)}$; then the divisor Y goes through a and not through b. This shows that the complete linear system determined by $3X$ separates at any rate all pairs of points a, b such that $b-a$ is not in γ. Let $f_0=1, f_1, \ldots, f_N$ be a basis for the vector-space of all functions f on A such that $(f)\succ -3X$; these functions, taken as homogeneous coordinates, determine a mapping F of A into the projective space P^N in which two points a, b cannot have the same image unless $b-a$ is in γ. As γ is finite, this implies that the image $F(A)$ of A has the same dimension n as A itself. Therefore, if W is any subvariety of A of dimension $n-1$, its image $F(W)$ is not $F(A)$, so that there is a homogeneous polynomial $P(X_0, \ldots, X_N)$ which is 0 on $F(W)$ and not on $F(A)$. Let d be the degree of P, and put $g=P(f_0, \ldots, f_N)$; g is not identically 0 on A; we have $(g)\succ -3dX$; and W is a component of the positive divisor $Z=(g)+3dX$.

Now let $W^{(i)}$ be a system of finitely many subvarieties of A such as was used above in order to embed A projectively. We have just shown that to every $W^{(i)}$ one can find an integer d_i and a positive divisor $Z_i \sim 3d_iX$ having $W^{(i)}$ as one of its components. Put $d=\sum d_i$ and $Z=\sum Z_i$; Z is a positive divisor, linearly equivalent to $3dX$, having all the $W^{(i)}$ among its components. Putting $Z=\sum W^{(i)}+\sum Z^{(j)}$, and reasoning just as before, one sees that the complete linear system determined by the class of $3Z$, i.e. by that of $9dX$, is ample. This completes the proof of the sufficiency of the condition in our theorem.†

In order to prove the converse, we shall have to lean somewhat more heavily upon the results and notations of [3]. We need the following lemmas:

LEMMA 1. *Let W be a subvariety of A of dimension $n-1$. Let f be a mapping of a curve Γ into A, and λ its linear extension* ([3], no. 42) *to the Jacobian variety J of Γ. Then, if $d(\lambda, W)=0$* ([3], no. 44), *W is invariant by the translation λz for every z in J.*

Let k be a field of definition for A, W, Γ, f; let M and x be independent generic points of Γ and W over k. The assumption means (loc. cit.) that $x-f(M)$ is of dimension $<n$ over k; its locus W' over k is therefore of dimension $\leqq n-1$. But W' contains all points $y-f(P)$, where y is any point of W and P any point of Γ; hence it contains all the varieties $W_{-f(P)}$, and therefore, since these have the dimension $n-1$, it must coincide with everyone of them. This shows that W is invariant by the translation $f(P)-f(Q)$ when P, Q are any two points of Γ, and therefore by the whole subgroup of A generated by these translations; this subgroup consists precisely of the translations λz, with z in J.

LEMMA 2. *Let f, Γ, λ, J be as in Lemma* 1; *let X be a divisor on A, and t a point such that $X_t \sim X$. Then we have $\lambda'_X t=0$.*

In fact, we have $X_{u+t} \sim X_u$ for every u. Taking u such that $X_u \cdot f(\Gamma)$ and $X_{u+t} \cdot f(\Gamma)$ are defined, and calling ϕ a function on A such that $(\phi)=X_{u+t}-X_u$, one finds that the divisor $\overset{-1}{f}(X_{u+t}-X_u)$ is the divisor of the function $\phi \circ f$ on Γ and therefore is ~ 0 on Γ; the assertion follows by Theorem 23 of [3], no. 45.

Now let $\mathfrak{g}$ be the subgroup of A consisting of all points t such that $X_t \sim X$; and assume that X is degenerate, i.e. that $\mathfrak{g}$ is infinite. It could be shown that $\mathfrak{g}$ is a closed algebraic set, and therefore, being infinite, must contain an Abelian subvariety of A; but we do not wish

† A similar reasoning shows that the class of $3Z+mX_1$ is ample if $m \geqq 2$ and X_1 is any positive divisor. In particular, the class of nX is ample for every $n \geqq 9d+2$. Using a result of Morikawa ([2], Lemma 7; cf. ibid., Lemma 10), one can even see that, if X_1 is any positive divisor, the class of $3Z+2X+X_1$, i.e. that of $(9d+2)X+X_1$, is ample.

to prove this here, and therefore reason as follows. Among all Abelian subvarieties of A containing infinitely many points of $\mathfrak{g}$, let B be one of smallest dimension. If f, Γ, λ, J are as in Lemmas 1 and 2, Lemma 2 shows that $\mathfrak{g}$ is contained in the kernel N of λ'_X. It follows that N must contain B; in fact, if this were not so, the component of 0 in $N \cap B$ would be an Abelian subvariety B' of B, other than B, and therefore would contain only a finite number of points of $\mathfrak{g}$; as B' is of finite index in $N \cap B$, the same would then be true of $N \cap B$ and therefore of B since $\mathfrak{g} \subset N$.

Take now for f a non-constant mapping of a curve Γ into B; this is possible, since B contains infinitely many points and is therefore not of dimension 0. Then λ is a homomorphism of J into B; as B is contained in the kernel of λ'_X, we have $\lambda'_X \lambda = 0$ and therefore, by Theorem 31 of [3], no. 61, $d(\lambda, X) = 0$.

Now let X' be any positive divisor linearly equivalent to a multiple nX of X; then $d(\lambda, X') = 0$. As X' is positive, we can write it as $X' = \sum X_\nu$, where the X_ν are subvarieties of A of dimension $n-1$. By the definition of $d(\lambda, X)$ as an intersection number, we have $d(\lambda, X_\nu) \geqq 0$ for all ν. Since $d(\lambda, X') = 0$, this gives $d(\lambda, X_\nu) = 0$ for all ν. By Lemma 1, this shows that the X_ν are invariant by all translations λz, where z is any point of J. Since this is so for every component of every positive divisor $X' \sim nX$, it is clear that such divisors cannot separate two points λz, $\lambda z'$ on A, and therefore the class of nX cannot be ample. This completes the proof.

UNIVERSITY OF CHICAGO

REFERENCES

[1] S. LEFSCHETZ, *On certain numerical invariants of algebraic varieties, with application to abelian varieties*, Trans. Amer. Math. Soc., 22 (1921), 327–482.

[2] H. MORIKAWA, *On abelian varieties*, Nagoya Math. J., 6 (1953), pp. 151–170.

[3] A. WEIL, Variétés abéliennes et courbes algébriques, Act. Sci. Ind., no. 1064, Hermann et Cie, Paris, 1948.

The Connectedness Theorem for Birational Transformations †

Oscar Zariski

1. Introduction

In our memoir [4] on abstract holomorphic functions we have proved the so-called 'principle of degeneration' in abstract algebraic geometry, by establishing a general connectedness theorem for algebraic correspondences. It is desirable to establish the connectedness theorem without using the theory of holomorphic functions which we have developed in the above-cited memoir. In the classical case this theorem follows, for birational transformations, from very simple topological considerations, as was pointed out in our memoir ([4], footnote on p. 7). By Lefschetz's principle it follows that in the case of characteristic zero we have a proof of the connectedness theorem for birational transformations, independent of the theory of abstract holomorphic functions. In the present paper we propose to deal algebraically with the connectedness theorem for birational transformations, without making any use of our theory of abstract holomorphic functions. We shall prove the connectedness theorem, in the case of birational transformations, only for non-singular varieties (or, more precisely, for simple points, since the question is of a purely local character).

2. The connectedness theorem for birational transformations

Let V be a variety defined over a ground field k. We say that V/k *is connected*, or that V *is connected over* k, if V is not the union of two proper subvarieties which are defined over k and have no points in common.

Let V and V' be two irreducible varieties, of dimension r, defined and birationally equivalent over k, and let T be a birational trans-

† This work was supported by a research project at Harvard University, sponsored by the Office of Ordnance Research, U.S. Army, under Contract DA–020–ORD–3100.

formation of V/k into V'/k (we assume, of course, that T itself is defined over k, i.e., that the graph of T, on the direct product $V \times V'$, is a variety defined over k; this is indicated by our saying that T is a birational transformation of V/k into V'/k). The following is the

CONNECTEDNESS THEOREM FOR BIRATIONAL TRANSFORMATIONS. *If P is a point of V at which V/k is analytically irreducible, then the variety $T\{P\}$ is connected over $k(P)$.*

(By $T\{P\}$ we mean the set of points of V' which correspond to P, under T. It is known that this set of points is a variety defined over the field $k(P)$ which is generated over k by the non-homogeneous coordinates of P.)

We shall now prove this theorem in the case in which P is a simple point on V/k. In the proof we may assume that P is a fundamental point of P, for otherwise $T\{P\}$ would consist of a single point, and there would be nothing to prove. We shall now recall from our paper ([1], p. 529) the definition of *isolated* fundamental points, since in our proof we shall consider separately two cases, according as P is or is not an isolated fundamental point of T.

Let V^* be the join of V and V', i.e., let V^* be the graph of T on the direct product $V \times V'$. The projection T^* of V^* onto V is a birational transformation defined over k and semi-regular at each point of V^* (we use the terminology of our paper [1]). Let F be the fundamental locus of T. Then F is also the fundamental locus of T^{*-1}. An irreducible subvariety W of F/k is said to be an *isolated* fundamental variety of T if it corresponds to an irreducible component of $T^{*-1}\{F\}/k$, and a point P of F is said to be an *isolated fundamental point of T* if P is a general point, over k, of an isolated fundamental variety of T. It is clear that every irreducible component of F/k is an isolated fundamental variety; but the converse is not necessarily true.

It is known that if an isolated fundamental variety W of T is *simple* for V/k, then every irreducible component of $T^{*-1}\{F\}/k$ which corresponds to W has dimension $r-1$ ([1], p. 532). It follows that an irreducible simple subvariety W of V/k is an isolated fundamental variety of T if and only if the following two conditions are satisfied: (1) $\dim W < r-1$; (2) there exists an irreducible $(r-1)$-dimensional subvariety W^* of V^*/k such that W and W^* are corresponding subvarieties under T^* (i.e., such that W is the projection of W^*). Condition (2) can also be expressed as follows: if P is a general point of W/k, then there exists on V' a point P' which corresponds to P under T and which is such that $\dim (P \times P')\, k = r-1$. If we denote by s the dimension of P/k (i.e. the transcendence degree of $k(P)/k$; s is also the dimension of W), then the relation $\dim(P \times P')/k = r-1$ is equivalent to $\dim P'/k(P) = r-1-s$. Summarizing, we have the following characterization of *simple* isolated fundamental points of T, which makes no direct use of the join V^*: *A simple point P of V/k, of dimension s over k, is an isolated fundamental point of T if and only if the following two conditions are satisfied:*

$$(1)\ s < r-1; \quad (2)\ \dim T\{P\} = r-1-s.$$

(For non-isolated fundamental points P we always have

$$\dim T\{P\} < r-1-s.)$$

3. The case of a non-isolated fundamental point P

We assume now that the simple point P of V/k is a non-isolated fundamental point of T. We observe that $T\{P\}$ is the projection of $T^{*-1}\{P\}$ into V'. Since the projection is a single-valued transformation and since the V'-projection of any variety Z^* defined over some ground field K ($Z^* \subset V \times V'$) is again a variety Z' defined over K ($Z \subset V'$), it follows that $T\{P\}$ is connected over $k(P)$ if and only if $T^{*-1}\{P\}$ is connected over $k(P)$. We can therefore replace V' by V^* and T by T^{*-1}. *We may therefore assume that T^{-1} is semi-regular at each point of V'.*

We observe that the connectedness theorem is obvious in the case $r=1$ (always in the case of birational transformations), for in that case there are no fundamental points at all. We shall therefore use induction on r. We assume therefore that $r \geqq 2$ and that the connectedness theorem, for birational transformations, is true for simple points of varieties of dimension $< r$.

We assume that V lies in some affine space of dimension n and that P is at finite distance. Let $x_1, x_2, \ldots, x_n$ be non-homogeneous coordinates of a general point of V/k. We denote by $\mathfrak{o}$ the local ring of P on V/k ($\mathfrak{o}$ = set of all quotients $f(x)/g(x)$, where f and g are polynomials with coefficients in k and where $g(P) \neq 0$) and by $\mathfrak{m}$ the maximal ideal of $\mathfrak{o}$ ($f(x)g(x) \in \mathfrak{m}$ if and only if $f(P)=0$). Since the fundamental point P is not isolated, the isolated fundamental varieties of T which pass through P are all of dimension $> s$. Let $F'_1, F'_2, \ldots, F'_h$ be the $(r-1)$-dimensional subvarieties of V' which correspond to the isolated fundamental varieties passing through P and let F_i be the variety which corresponds to F'_i. Let $\mathfrak{p}_i$ be the prime ideal in $\mathfrak{o}$ which corresponds to F_i. Since $\dim F_i > s$, $\mathfrak{p}_i$ is a proper subideal of $\mathfrak{m}$ ($i = 1, 2, \ldots, h$).

We fix a polynomial $f(X)$ in $k[X]$ such that the following conditions are satisfied:

(1) $$f(x) \in \mathfrak{m}, \quad f(x) \notin \mathfrak{m}^2;$$

(2) $$f(x) \notin \mathfrak{p}_i \quad (i = 1, 2, \ldots, h).$$

We denote by H the hypersurface $f(X) = 0$. By (1), this hypersurface has a regular intersection with V at the point P. That means that the intersection $H \cap V$ has only one irreducible component passing through P (we shall call that component W), that P is a simple point of W/k and that the principal ideal generated by $f(x)$ in $\mathfrak{o}$ is the prime ideal of W. We also note that W has dimension $r-1$. Conditions (2) signify that $F_i \not\subset W (i = 1, 2, \ldots, h)$.

Since $\dim W = r-1$, W is not fundamental for T, and hence T is regular at any general point of W/k (since T^{-1} is semi-regular). Thus to W there corresponds a unique irreducible subvariety W' of V'/k; this variety W' is also of dimension $r-1$, and the restriction of T^{-1} to W' is a semi-regular birational transformation of W' into W. We denote the inverse of this induced birational transformation by T_1. Since P is a simple point of W/k, we have, by our induction hypothesis, that the variety $T_1\{P\}$ is connected over $k(P)$. We also have that $T_1\{P\} \subset T\{P\}$, since the graph of T_1 is contained in the graph of T. We shall now show that $T_1\{P\} = T\{P\}$, and this will complete the proof of the connectedness theorem in the present case.

Let P' be any point of $T\{P\}$ and let $\mathfrak{o}'$ be its local ring on V'/k. Since T^{-1} is semi-regular, we have $\mathfrak{o} \subset \mathfrak{o}'$, whence $f(x) \in \mathfrak{o}'$. Since P and P' are corresponding points under T, every non-unit of $\mathfrak{o}$ is also a non-unit of $\mathfrak{o}'$. Hence $f(x)$ is a non-unit of $\mathfrak{o}'$. We fix some isolated prime ideal $\mathfrak{p}'$ of the principal ideal generated by $f(x)$ in $\mathfrak{o}'$ and we denote by G' the irreducible subvariety of V'/k which contains P' and is defined by $\mathfrak{p}'$. Then G' is of dimension $r-1$, and the irreducible subvariety G of V/k which corresponds to G' under T passes through P and is defined in $\mathfrak{o}$ by the prime ideal $\mathfrak{p} = \mathfrak{o} \cap \mathfrak{p}'$. Since $f(x) \in \mathfrak{p}$, it follows from (2) that $\mathfrak{p} \neq \mathfrak{p}_i$ $(i = 1, 2, \ldots, h)$. Hence G is not an isolated fundamental variety of T. On the other hand, G' corresponds to G and has dimension $r-1$. Hence G is not a fundamental variety at all. It follows that G has dimension $r-1$, and $\mathfrak{p}$ is a minimal prime ideal of $\mathfrak{o}$. Since $f(x) \in \mathfrak{p}$, it follows that the variety W coincides with G. Hence $G' = W'$ and $P' \in W'$. The point of W which corresponds to P' under the *semi-regular* transformation T_1^{-1} must be the point P (since the graph of T_1 is contained in the graph of T and since P is the only point of V which corresponds to P' under T). Hence $P' \in T_1\{P\}$, as asserted.

4. The case of an isolated fundamental point P

We now consider the case in which the point P is an isolated fundamental point of T. We apply to V a locally quadratic transformation ϕ, defined over k and having center P (this is another way of saying that ϕ is a monoidal transformation of V/k whose center is the irreducible variety having P as general point over k). Let V_1 denote the ϕ-transform of V. By known properties of monoidal transformations we have that the variety $\phi\{P\}$ is defined and irreducible over $k(P)$, that it is non-singular over $k(P)$, has dimension $r-1-s$, where $s=\dim P/k$, and that all its points are simple for V_1/k. The given birational transformation T of V into V' and the birational transformation ϕ of V into V_1 define, in an obvious fashion, a birational transformation T_1 of V_1 into V'. It is clear that $T\{P\}$ and $\phi\{P\}$ are the total transforms of each other under T_1 and T_1^{-1} respectively. We now show that if for every point P_1 of $\phi\{P\}$ *it is true that* $T_1\{P_1\}$ *is connected over* $k(P_1)$, *then* $T\{P\}$ *is connected over* $k\{P\}$. Assume the contrary, and let $T\{P\}$ be the union of two varieties W_1' and W_2' having no points in common and neither of which is empty (it is assumed that both varieties are defined over $k(P)$). Let G_1 and G_2 be the total T_1^{-1}-transforms of W_1' and W_2' respectively. The two varieties G_1 and G_2 are defined over $k(P)$ and their union is $\phi\{P\}$. Since $\phi\{P\}$ is irreducible over $k(P)$, one of the varieties G_i contains the other. Let, say, $G_2 \subset G_1$. Then if P_1 is any point G_2, the variety $T_1\{P_1\}$ meets both varieties W_1' and W_2'. The variety $T_1\{P_1\}$ is defined over $k(P_1)$, and the field $k(P_1)$ contains the field $k(P)$ since ϕ^{-1} is semi-regular at any point of V_1. Hence the intersections of $T_1\{P_1\}$ with W_1' and W_2' are varieties defined over $k(P_1)$; these intersections are non-empty and have no points in common. Therefore, $T_1\{P_1\}$ is disconnected over $k(P_1)$, contrary to our assumption.

Let then P_1 be any point of $\phi\{P\}$. If P_1 is not an isolated fundamental point of T_1, then $T_1\{P_1\}$ is connected over $k(P_1)$, by the preceding case of the proof. Assume that P_1 is an isolated fundamental point of T_1. We now proceed with P_1 as we did with P, i.e., we apply to V_1 a quadratic transformation ϕ_1 with center P_1, we denote by V_2 the ϕ_1-transform of V_1 and by T_2 the birational transformation of V_2 into V' which is defined in an obvious fashion by the two birational transformations ϕ_1 and T_1. If $\phi_1\{P_1\}$ carries no isolated fundamental points of T_2, the proof is complete. In the contrary case, we consider a point P_2 on $\phi_1\{P_1\}$ which is an isolated fundamental point of T_2 and we repeat the above procedure. The proof of the theorem now depends on showing that this process must come to an end after a finite number of steps. This we now proceed to prove by an indirect argument.

Suppose that the contrary is true. We will have then an infinite sequence of varieties $V, V_1, V_2, \ldots, V_i, \ldots$ with the following properties: (a) V_{i+1} is the transform of V_i by a quadratic transformation ϕ_i whose center is a point P_i of V_i ($V_0 = V, \phi_0 = \phi, P_0 = P$); (b) the point P_i ($i > 0$) belongs to $\phi_{i-1}\{P_{i-1}\}$; (c) if T_{i+1} is the birational transformation of V_{i+1} into V' which is determined in a natural fashion by the birational transformations ϕ_i and T_i, then P_{i+1} is an isolated fundamental point of T_{i+1}. If V is a surface ($r=2$), then every fundamental point is isolated and the situation which we have just described signifies that it is not possible to eliminate the fundamental point P of the birational transformation T by applying to V successive quadratic transformations. However, we have proved in earlier papers of ours that the contrary is true, i.e., that the fundamental point P will be automatically eliminated if we apply a sufficient number of quadratic transformations ([3], p. 681). Our proof for surfaces is such that its generalization to varieties of any dimension leads to the conclusion that any *isolated* fundamental point can be eliminated by consecutive quadratic transformations. We have actually carried out this generalization in the case of three-dimensional varieties ([2], p. 535), and it will be sufficient to indicate here the main steps of the proof.

We assume then the existence of an infinite sequence of varieties V_i having the above indicated properties and we denote by s_i the dimension of the point P_i, over k. We have $s_i < r-1$ (since P_i is a fundamental point of T_i) and $s_i \leqq s_{i+1}$ (since ϕ_i^{-1} is semi-regular at P_{i+1}, whence $k(P_i)$ is a subfield of $k(P_{i+1})$). Hence for i sufficiently large we will have $s_i = s_{i+1} = \ldots$, and we may assume without loss of generality that this happens already for $i=0$, i.e., that $s = s_1 = s_2 = \ldots$. By the characterization of simple isolated fundamental points given at the end of § 2, we have $\dim T_i\{P_i\} = r-1-s$. On the other hand, we have

$$T_i\{P_i\} \subset T_{i+1}\{P_{i+1}\},$$

in view of the semi-regularity of ϕ_i^{-1}, and hence $T_i\{P_i\} = T_{i+1}\{P_{i+1}\} = \ldots$ for all sufficiently high values of i. There exists therefore a point P' of V' which belongs to all the varieties $T_i\{P_i\}$ and which has dimension $r-1-s$ over $k(P)$ (that point will then have dimension $r-1-s$ over each of the fields $k(P_i)$, since these fields have the same transcendence degree s over k). We have $\dim P'/k = r-1$, and therefore there is only a finite number of valuations whose center is P'; all these valuations are of dimension $r-1$, i.e., they are prime divisors. Since P' and P_i are corresponding points (under T_i), at least one of these valuations must have center P_i. Let L_i be the set of those valuations of center P' on V' whose center on V_i is P_i. Then each L_i is a non-empty finite set, and we have $L \supset L_1 \supset L_2 \supset \ldots \supset L_i \ldots$. It follows that there exists at least one prime divisor $\mathfrak{P}$ which is contained in all L_i. This prime divisor $\mathfrak{P}$ is of second kind with respect to all varieties V_i, since the center P_i of $\mathfrak{P}$ on V_i has dimension $s < r-1$. If then m_i denotes the maximal ideal of the local ring of P_i on V_i and if $v_{\mathfrak{P}}(m_i)$ denotes the minimum of all $v_{\mathfrak{P}}(x)$ for x in m_i, then we obtain as in [2], p. 536, or p. 493, Lemma 9.1, (see also [3], p. 681) the absurd sequence of inequalities

$$v_{\mathfrak{P}}(m) > v_{\mathfrak{P}}(m_1) > \ldots > v_{\mathfrak{P}}(m_i) > \ldots > 0,$$

where all the $v_{\mathfrak{P}}(m_i)$ are integers.

This completes the proof of the connectedness theorem for birational transformations, in the case of simple points.

HARVARD UNIVERSITY

REFERENCES

[1] O. ZARISKI, *Foundations of a general theory of birational correspondences*, Trans. Amer. Math. Soc., 53 (1943), pp. 490–542.

[2] ——, *Reduction of the singularities of algebraic three-dimensional varieties*, Ann. of Math., 45 (1944), pp. 472–542.

[3] ——, *Reduction of the singularities of an algebraic surface*, Ann. of Math., 40 (1939), pp. 639–689.

[4] ——, *Theory and applications of holomorphic functions on algebraic varieties over arbitrary ground fields*, Memoirs of Amer. Math. Soc., no. 5 (1951), pp. 1–90.

Part III

Papers in Topology

The Relations on Steenrod Powers of Cohomology Classes†

José Adem

1. Introduction

THE purpose of the present paper is to give complete proof of some of the results previously announced by the author (cf. [1, 2]). The main result states that the iterated Steenrod p-powers (denoted by Sq^i for $p=2$ and by $\mathscr{P}^i$ for $p>2$) satisfy a system of linear relations.

The p-power operations $\mathscr{P}^i$ have already been applied in diverse situations such as the computation of obstructions, the study of properties of differentiable manifolds (Borel-Serre, Hirzebruch, Thom and Wu; see bibliography), and the computation of the cohomology of Eilenberg-MacLane complexes by Cartan and Serre. The $\mathscr{P}^i$ are homomorphisms of cohomology groups of any space X with coefficient group Z_p (= the integers mod p):

$$\mathscr{P}^i\colon H^q(X; Z_p) \to H^{q+2i(p-1)}(X; Z_p) \quad (p>2),$$

and

$$\mathrm{Sq}^i\colon H^q(X; Z_2) \to H^{q+i}(X; Z_2).$$

These homomorphisms are topologically invariant, and provide additional structure in the cohomology ring of X. Any relations which are universally satisfied by the p-powers impose strong restrictions on the possible structures of the cohomology ring. All the relations we obtain have the form $\sum a_{ij}\mathscr{P}^i\mathscr{P}^j u = 0$ for all cohomology classes u. These can be used to express an arbitrary $\mathscr{P}^i$ in terms of those in which i is a power of p.

In this paper we shall not give any applications of these relations. Those applications announced in my PROCEEDINGS notes will appear in subsequent papers.

I have been informed that Cartan has obtained the same relations

† The research presented in this paper was developed in part while the author was a Guggenheim Fellow at Princeton University; and in part with support from the Instituto Nacional de la Investigacion Cientifica, Mexico City.

by his special method of calculating the cohomology structure of the Eilenberg-MacLane complexes $K(Z_p, n)$.†

The method we use is based on Steenrod's (cf. [19]) representation of a reduced power operation as an element of a homology group of a permutation group. Conversely, any homology class $c \in H_i(\pi)$ of a permutation group π of degree n defines a reduced power of a cohomology class u denoted by u^n/c. In particular, the p-powers $\mathscr{P}^i u$ are reduced powers associated with homology classes of the cyclic permutations of degree p.

If S_n is the symmetric group of degree n and $\pi \subset S_n$, there is an induced homomorphism of homology

$$h_*: H_i(\pi) \to H_i(S_n),$$

and we have the general relation

$$u^n/c = u^n/h_*c \qquad (1.1) \quad \text{(cf. [19], p. 216).}$$

We study the symmetric group of degree p^2 where p is prime. If G is a p-Sylow group of S_{p^2}, we give a system of generators of the homology of G which interpret as two-fold iterations of cyclic reduced powers. As pointed out by Steenrod in [19], p. 216, to obtain relations we consider the images of cycles of G on S_{p^2}. The map $h_*: H(G) \to H(S_{p^2})$ has a non-trivial kernel, and if $h_*c = 0$ then, by (1.1), we have $u^{p^2}/c = 0$. Since u^{p^2}/c is equal to a sum of two-fold iterated powers, this represents a relation. The elements in the kernel of h_* are constructed by considering an automorphism of a subgroup of G that extends to an inner automorphism of S_{p^2}. If c lies in the subgroup of G and if c' is its image under the automorphism, then $c - c'$ belongs to the kernel of h_*. It is important for the success of the construction to use appropriate complexes for the various groups.

Finally, the author wishes to express his gratitude to Professors S. Lefschetz and N. E. Steenrod for their interest and encouragement while engaged in this work. The present paper is an elaboration of a doctoral thesis, developed under the guidance of Professor Steenrod, and submitted to Princeton University in the spring of 1952. The author also wishes to thank Dr. J. C. Moore for many valuable discussions.

2. π-modules and tensor products‡

By a *π-module* we mean an additive Abelian group A on which a multiplicative group π operates from the left. That is, for each $\alpha \in \pi$

† (*Added in the proofs.*) Cartan's results have since appeared in Comment. Math. Helv., 29 (1955), pp. 40–58.

‡ For further information regarding this and some of the next sections, the reader is referred to [7].

and $a \in A$ the element $\alpha a \in A$ is defined satisfying the following conditions:

$$\alpha(a+b)=\alpha a+\alpha b, \quad \alpha_1(\alpha_2 a)=(\alpha_1\alpha_2)a, \quad 1a=a.$$

We say that a π-module A is *π-free* if A is a free group and if there exists a set $X \subset A$ such that the elements αx, for all $x \in X$, $\alpha \in \pi$, are pairwise distinct and form a basis for A. The set X will be called a *π-basis.*

An element of A that is a sum of elements of the form $\alpha a - a$, with $\alpha \in \pi$ and $a \in A$, is called *residual.* Let $R(A)$ denote the subgroup of residual elements. Define A_π as

$$A_\pi = A/R(A),$$

the factor group of A by $R(A)$.

Let A be a π-module, let B be a ρ-module and suppose it is given a homomorphism $\phi\colon \pi \to \rho$. Under these circumstances B can be made into a π-module by definining operations for $b \in B$, $\alpha \in \pi$ by $\alpha b = \phi(\alpha)b$. A homomorphism $f\colon A \to B$ is called a *ϕ-homomorphism*, *π-homomorphism* or (π-) *equivariant* homomorphism if the condition

$$f(\alpha a) = \phi(\alpha) f(a)$$

holds for all $\alpha \in \pi$, $a \in A$. Obviously, an equivariant homomorphism takes only residual values on residual elements, hence, it induces a natural homomorphism $f_\phi\colon A_\pi \to A_\rho$. Sometimes it will be convenient to write either f_π or f_ρ for f_ϕ.

The tensor product $A \otimes B$ of two Abelian groups A and B, briefly, can be described as follows. Each pair (a, b), $a \in A$, $b \in B$ determines an element denoted by $a \otimes b$. These elements generate over the integers the Abelian group $A \otimes B$ and the relations are

$$\text{(2.1)} \qquad \begin{cases} (a_1+a_2)\otimes b = a_1 \otimes b + a_2 \otimes b, \\ a \otimes (b_1+b_2) = a \otimes b_1 + a \otimes b_2. \end{cases}$$

If A and B are π-modules, then $A \otimes B$ will also be a π-module with the operations defined by

$$\text{(2.2)} \qquad \alpha(a \otimes b) = (\alpha a) \otimes (\alpha b).$$

Then $A \otimes_\pi B$, the tensor product over π of A and B, is defined by

$$A \otimes_\pi B = (A \otimes B)_\pi.$$

Under the natural map $A \otimes B \to A \otimes_\pi B$, the class corresponding to $a \otimes b$ is denoted by $a \otimes_\pi b$ or simply by $a \otimes b$, if no confusion arises. Besides (2.1) we have in $A \otimes_\pi B$ the relations

$$a \otimes b = (\alpha a) \otimes (\alpha b), \quad \text{all} \quad \alpha \in \pi.$$

Let A, B be π-modules, let C, D be ρ-modules and $\phi: \pi \to \rho$ a homomorphism. If $f: A \to C$, $g: B \to D$ are given homomorphisms, then $f \otimes g: A \otimes B \to C \otimes D$ is defined by

$$(f \otimes g)(a \otimes b) = f(a) \otimes g(b).$$

If f, g are π-homomorphisms, then $f \otimes g$ is a π-homomorphism and the induced

$$(f \otimes g)_\phi: A \otimes_\pi B \to C \otimes_\rho D$$

is denoted by $f \otimes_\rho g$.

3. Complexes

All our cell complexes will be closure finite and augmentable. The same symbol K will be used to denote the geometric cell complex K and the chain cell complex $K = \{C_q(K), \partial\}$, naturally associated with it. Then, if K, L are cell complexes by $K \times L$ we mean either the product complex as a cell complex or the chain complex $K \times L = \{C_q(K \times L), \partial\}$. The tensor product $K \otimes L$ of K and L is the chain complex with

$$C_q(K \otimes L) = \textstyle\sum_{q=r+s} C_r(K) \otimes C_s(L),$$

and ∂ defined by

$$\partial(\sigma \otimes \tau) = (\partial\sigma) \otimes \tau + (-1)^r \sigma \otimes \partial\tau,$$

where $\sigma \in C_r(K)$ and $\tau \in C_s(L)$. Under the correspondence $\sigma \times \tau \to \sigma \otimes \tau$ the complexes $K \times L$ and $K \otimes L$ are isomorphic and

$$H(K \times L) \approx H(K \otimes L).$$

The complex K is called a *π-complex* if each $C_q(K)$ is a π-module in such a fashion (cf. [10], p. 54) that $\alpha\partial = \partial\alpha$ and $\alpha In = In$ for all $\alpha \in \pi$, where In is the Kronecker index for the 0-chains. The π-complex K is *π-free* if each $C_q(K)$ is π-free.

For K, a π-complex, define K_π to be the chain complex with $C_q(K_\pi) = [C_q(K)]_\pi$ as chain groups. Let $\mu_\#: C_q(K) \to C_q(K_\pi)$ be the natural map; $\mu_\#$ is a homomorphism onto and the boundary operator in K_π is defined by $\partial\mu_\# c = \mu_\# \partial c$. Since $\mu_\# c = \mu_\# c'$ implies $\mu_\# \partial c = \mu_\# \partial c'$, the boundary is well defined in K_π and $\mu_\#: K \to K_\pi$ is a chain transformation. If K, L are π-complexes, then $K \otimes_\pi L$ is defined as $(K \otimes L)_\pi$.

Given K a π-complex, L a ρ-complex and $\phi: \pi \to \rho$ a homomorphism, we say that $f_\#: K \to L$ is a *ϕ-chain transformation* or an *equivariant* chain transformation if $f_\#$ is a chain transformation and each $f_\#: C_q(K) \to C_q(L)$ is a ϕ-homomorphism. The induced chain transformation $f_{\#\phi}: K_\pi \to L_\rho$ is defined by $f_{\#\phi}(\tau) = f_{\#\phi}(\mu_\# \sigma) = \nu_\# f_\#(\sigma)$, where $\mu_\#: K \to K_\pi$, $\nu_\#: L \to L_\rho$ are the natural maps, $\tau \in K_\pi$, $\sigma \in K$ and $\tau = \mu_\# \sigma$. If $f_\#, g_\#: K \to L$ are two ϕ-chain maps, then a chain homotopy $E_\#: f_\# \simeq g_\#$ is a *ϕ-chain homotopy* if each $E_\#: C_q(K) \to C_{q+1}(L)$ is a ϕ-homomorphism. In this case $E_{\#\phi}: f_{\#\phi} \simeq g_{\#\phi}$, where $E_{\#\phi}(\tau) = \nu_\# E_\#(\sigma)$.

Suppose now that K and L are cell complexes. A carrier C from K to L is a *ϕ-carrier* or an *equivariant* carrier, if for each cell σ of K and all $\alpha \in \pi$ the condition $C(\alpha\sigma) = \phi(\alpha) C(\sigma)$ holds. The following lemma is basic in this work:

(3.1) LEMMA. *Let K be a π-free complex, L a ρ-complex, $\phi: \pi \to \rho$ a homomorphism and C an acyclic ϕ-carrier from K to L. Then there exists a ϕ-chain map $f_\#: K \to L$ carried by C and, if $f_\#$, $g_\#$ are two such chain maps, then C carries a ϕ-chain homotopy $E_\#: f_\# \simeq g_\#$.*

This lemma is well known (cf. [10], p. 59; [13], p. 112) and the proof is a construction by induction on dimension. The equivariance of the chain maps is obtained by defining them, first on elements of a π-basis of K and then extending the maps equivariantly to the transforms. The proof is omitted, since it is essentially given in [18], p. 49 (cf. also [19], p. 214).

4. The derived complex and the cup-product

The derived K' of a cell complex K (cf. [14], p. 164) is the simplicial complex whose oriented q-simplexes are the collections of distinct cells of K, $(\sigma_{i_0}, \ldots, \sigma_{i_q})$ with $\sigma_{i_0} < \ldots < \sigma_{i_q}$. If K is a π-complex, then K' is made a π-complex by defining

$$(4.1) \qquad \alpha(\sigma_i, \ldots, \sigma_j) = (\alpha\sigma_i, \ldots, \alpha\sigma_j) \quad \text{for all} \quad \alpha \in \pi.$$

We say that a cell complex K is *regular* if $\bar{\sigma}$ is an acyclic subcomplex for every cell σ of K.

(4.2) LEMMA. *If K is a π-free complex then K' is also π-free and $(K')_\pi$ is regular.*

PROOF. The first part follows from (4.1). For the second part, since K is free, we have $(\bar{\tau})_\pi \simeq \bar{\tau}$, where τ is any simplex of K'. Therefore K' is regular.

Now suppose that K is a π-free regular complex, and let C_1, C_2 be, respectively, carriers from K to K' and K' to K defined by

$$C_1(\sigma) = Cl\{(\sigma_i, \ldots, \sigma_j, \sigma) \mid \sigma_i < \ldots < \sigma_j < \sigma\},$$

$$C_2[(\sigma_k, \ldots, \sigma_l, \sigma)] = \bar{\sigma}.$$

It follows that C_1, C_2 are equivariant acyclic carriers (cf. [14], p. 167).

(4.3) LEMMA. *Let K be a π-free regular complex and K' its derived. Then there exists equivariant chain maps $f_\#: K \to K'$, $f_\#$ carried by C_1, and $g_\#: K' \to K$, $g_\#$ carried by C_2, such that $g_\# f_\# = 1$ and $E_\#: f_\# g_\# \simeq 1$, where $E_\#$ is an equivariant chain homotopy of $f_\# g_\#$ into* 1.

PROOF. This is an adaptation to the case of equivariance, of the arguments given by Lefschetz in [14], pp. 165–167. We shall indicate the steps of the proof. Since K, K' are π-free complexes and C_1, C_2 equivariant acyclic carriers, it follows from (3.1) that C_1, C_2 carry respectively the equivariant chain maps $f_{\#}$, $g_{\#}$. The maps $f_{\#}g_{\#}$ and 1 are carried by the equivariant acyclic carrier C_1C_2, hence (3.1) implies the existence of $E_{\#}: f_{\#}g_{\#} \simeq 1$, an equivariant chain homotopy carried by C_1C_2. Finally, the composition $g_{\#}f_{\#}$ is carried by C_2C_1 and $C_2C_1(\sigma) = \bar{\sigma}$ for all σ, therefore $g_{\#}f_{\#} = 1$.

Let K be a π-free regular complex. We would like to compute cup-products in K_π, but K_π may not be regular, and the standard definition of the cup-product (cf. [18 (a)], p. 55) applies only to those complexes L that are regular (augmentable and closure finite). In this case, one constructs a chain transformation $d_{\#}: L \to L \otimes L$ carried by $C(\sigma) = \bar{\sigma} \otimes \bar{\sigma}$ and then the cup-product of two cochains u, v is defined by

$$(u \cup v) \cdot \sigma = (u \otimes v) \cdot d_{\#}(\sigma).$$

We will show how this definition extends to K_π. First, $K \otimes K$ is a $\pi \times \pi$-free complex with the operations naturally defined by

$$(\alpha \times \beta)(\sigma \otimes \tau) = (\alpha\sigma) \otimes (\beta\tau) \quad \text{for} \quad \alpha \times \beta \in \pi \times \pi \quad \text{and} \quad \sigma \otimes \tau \in K \otimes K.$$

Let $d: \pi \to \pi \times \pi$ be the diagonal map, i.e. $d(\alpha) = \alpha \times \alpha$, and let C be the diagonal carrier from K to $K \otimes K$, that is, $C(\sigma) = \bar{\sigma} \otimes \bar{\sigma}$. The carrier C is equivariant and acyclic and K is π-free, then by (3.1) we can construct $d_{\#}: K \to K \otimes K$, a d-chain map carried by C. Set $\tilde{d}_{\#} = (d_{\#})_\pi$, then $\tilde{d}_{\#}: K_\pi \to K_\pi \otimes K_\pi$ is the chain map induced by $d_{\#}$ and we have the following:

(4.4) THEOREM. *The cup-product can be defined in K_π by means of $\tilde{d}_{\#}: K_\pi \to K_\pi \otimes K_\pi$; that is, u, v are cohomology classes of K_π and $\tilde{d}^*: H^q(K_\pi \otimes K_\pi) \to H^q(K_\pi)$ is the homomorphism in classes induced by $\tilde{d}_{\#}$, then $u \cup v = \tilde{d}^*(u \otimes v)$.*

PROOF. Let $d'_{\#}: K' \to K' \otimes K'$ be a d-chain map and let $\tilde{d}'_{\#} = (d'_{\#})_\pi$ be its induced. Set $L = (K')_\pi$. The chain map $\tilde{d}'_{\#}: L \to L \otimes L$ is carried by the diagonal carrier and L is a regular complex. Hence, $\tilde{d}'_{\#}$ can be used to compute cup-products in L. Let $f_{\#}$, $g_{\#}$ be the maps of (4.3) and let $\tilde{f}_{\#} = (f_{\#})_\pi$, $\tilde{g}_{\#} = (g_{\#})_\pi$ be their induced. Let $\mu_{\#}: K \to K_\pi$, $\nu_{\#}: K' \to L$ be the natural chain maps, then commutativity holds in the following diagram.

$$\begin{array}{ccccccc}
K & \xrightarrow{f_{\#}} & K' & \xrightarrow{d'_{\#}} & K' \otimes K' & \xrightarrow{g_{\#} \otimes g_{\#}} & K \otimes K \\
\downarrow{\scriptstyle \mu_{\#}} & & \downarrow{\scriptstyle \nu_{\#}} & & \downarrow{\scriptstyle \nu_{\#} \otimes \nu} & & \downarrow{\scriptstyle \mu_{\#} \otimes \mu_{\#}} \\
K_\pi & \xrightarrow{\tilde{f}_{\#}} & L & \xrightarrow{\tilde{d}'_{\#}} & L \otimes L & \xrightarrow{\tilde{g}_{\#} \otimes \tilde{g}_{\#}} & K_\pi \otimes K_\pi
\end{array}$$

Now it follows from (3.1) that $\tilde{d}^*$ is independent of the particular $d_\#$ we choose. Therefore, we can take $d_\# = (g_\# \otimes g_\#) d'_\# f_\#$, since this composition is a d-chain map carried by the diagonal carrier. From the commutative relations above we have that $\tilde{d}_\# = (\tilde{g}_\# \otimes \tilde{g}_\#) \tilde{d}'_\# \tilde{f}_\#$, consequently $\tilde{d}^* = \tilde{f}^* \tilde{d}'^* (\tilde{g}^* \otimes \tilde{g}^*)$ for the induced homomorphisms in cohomology. Finally, if u, v are cohomology classes of K_π, we have

$$\begin{aligned} \tilde{d}^*(u \otimes v) &= \tilde{f}^* \tilde{d}'^* (\tilde{g}^* u \otimes \tilde{g}^* v) \\ &= \tilde{f}^* (\tilde{g}^* u \cup \tilde{g}^* v) = u \cup v. \end{aligned}$$

Thus the theorem is proved.

5. Homology groups of groups

Given π, a multiplicative group, let K be an augmentable π-free acyclic complex. Such a K always exists (e.g. cf. [13], p. 112). Let A be a π-module regarded as a π-complex where $C_0(A) = A$ is the only non-trivial group of chains. The complex $A \otimes K$ is a π-complex where the operations are given by (2.2). The homology groups of $A \otimes_\pi K$ are proved to be independent of K, and by definition they are the homology groups of π with coefficients in the π-module A. They are denoted by $H_i(\pi; A)$.

Suppose given $f: (\pi; A) \to (\rho; B)$, where A is a π-module, B is a ρ-module and $f = (f_1, f_2)$ with $f_1: \pi \to \rho$ a homomorphism and $f_2: A \to B$ a f_1-homomorphism. The map f induces homomorphisms $f_*: H(\pi; A) \to H(\rho; B)$ defined as follows. Take K, L, respectively, a π-free and a ρ-free acyclic complex, and construct $f_{3\#}: K \to L$, an f_1-chain map. Then $f_2 \otimes f_{3\#}: A \otimes K \to B \otimes L$ is an f_1-chain map and f_* is the homomorphism induced by $f_2 \otimes_\rho f_{3\#}$. Clearly, if $g: (\rho; B) \to (\sigma; C)$ is another map with $g = (g_1, g_2)$, then $gf: (\pi; A) \to (\sigma; C)$ is a map with $gf = (g_1 f_1, g_2 f_2)$ and $(gf)_* = g_* f_*$.

(5.1) LEMMA. *Let A be a π-module and let $f_1: \pi \to \pi$ be the inner automorphism given by an element $\lambda \in \pi$,* i.e. *for $\alpha \in \pi$, $f_1(\alpha) = \lambda \alpha \lambda^{-1}$. Let $f_2: A \to A$ be defined by $f_2(a) = \lambda a$ for all $a \in A$. Then $f = (f_1, f_2)$ is a map and the induced f_* is the identity.*

PROOF. Take K, a π-free acyclic complex, and define $f_{3\#}: K \to K$, an f_1-chain map, by $f_{3\#}(\sigma) = \lambda\sigma$, for $\sigma \in K$. Then $(f_2 \otimes f_{3\#})(a \otimes \sigma) = \lambda(a \otimes \sigma)$ and $f_2 \otimes_\pi f_{3\#}$ is the identity chain mapping, consequently $f_* = 1$.

The study of the homology groups of a group ρ and of a subgroup π, is chiefly done by means of a homomorphism $T: H_i(\rho, A) \to H_i(\pi; A)$, called the *transfer* homomorphism. The transfer has been studied, independently, by Artin and Tate (unpublished) and by Eckman (cf. [8]). We will not define the transfer here since we do not intend to use it explicitly. We set without proof the following (cf. [8], Theorem 5).

(5.2) THEOREM. *Let ρ be a group, π a subgroup of ρ of finite index n, A a ρ-module, $g_*: H_i(\pi; A) \to H_i(\rho; A)$ the homomorphism induced by the inclusion map. There exist homomorphisms $T: H_i(\rho; A) \to H_i(\pi; A)$, called transfer, such that $g_* T = n$ is multiplication of the elements of $H_i(\rho; A)$ by the integer n.*

In our applications the following corollary of (5.2) will be used.

(5.3) COROLLARY. *Let ρ be a finite group and π a subgroup containing a p-Sylow subgroup of ρ. Take Z_p, the integers modulo p, as coefficient group. Then the homomorphism induced by the inclusion,*

$$g_*: H_i(\pi; Z_p) \to H_i(\rho; Z_p),$$

maps $H_i(\pi; Z_p)$ onto $H_i(\rho; Z_p)$.

PROOF. We have that n, the index of π in ρ, is prime to p. Thus, $g_* T = n$ is an automorphism of $H_i(\rho; Z_p)$, and this implies that g_* is onto.

6. The cohomology ring of a cyclic group

Let π be a cyclic group of order n and x a chosen generator. For future reference some special elements of the integral group rings $Z(\pi)$ and $Z(\pi \times \pi)$ will be defined and some relations among those elements will be stated. Set

$$\Sigma = \textstyle\sum_{k=0}^{n-1} x^k \quad \text{and} \quad \Delta = x - 1. \tag{6.1}$$

Then $\Sigma\Delta = \Delta\Sigma = 0$. Set $\Sigma' = \sum_{k=1}^{n-1} kx^k$.
It follows that

$$\Sigma = -\Sigma'\Delta + n = -\Delta\Sigma' + n. \tag{6.2}$$

Let $\tilde{x} = x \times x$ and as above define

$$\tilde{\Sigma} = \textstyle\sum_{k=0}^{n-1} \tilde{x}^k \quad \text{and} \quad \tilde{\Delta} = \tilde{x} - 1.$$

Finally, set

$$\Omega = \textstyle\sum_{0 \leqq i < j < n} x^i \times x^j \quad \text{and} \quad \Omega' = \sum_{0 \leqq i < j < n} x^j \times x^i.$$

The following relations are easy to verify:

$$\tilde{\Delta} = 1 \times \Delta + \Delta \times x = \Delta \times 1 + x \times \Delta, \tag{6.3}$$

$$\begin{cases} \tilde{\Sigma} = 1 \times \Sigma + \Omega(\Delta \times 1) = \Sigma \times 1 + \Omega'(1 \times \Delta), \\ \Omega(\Delta \times 1) = -\Omega'(\Delta \times x), \quad \Omega(x \times \Delta) = -\Omega'(1 \times \Delta). \end{cases} \tag{6.4}$$

All these relations are due to Steenrod (cf. [18 (b)], p. 72).

Now let M be the canonical π-complex, π-free and acyclic constructed in [19], p. 218. We recall that for each $i \geqq 0$, $C_i(M)$ is a free

Abelian group generated by one cell e_i and its transforms $x^k e_i$; consequently the set $\{e_i\}$ constitutes a π-basis for M. The boundary operator is defined by

$$\partial e_{2i+1} = \Delta e_{2i}, \quad \partial e_{2i+2} = \Sigma e_{2i+1}.$$

Then for M_π it follows that each $C_i(M_\pi)$ is a free group on one generator e_i with the boundary relations

$$\partial e_{2i+1} = 0, \quad \partial e_{2i+2} = n e_{2i+1}. \tag{6.5}$$

Take as group of coefficients Z_n, the integers modulo n, and assume that π operates trivially. It follows from (6.5) that $H_i(\pi; Z_n) \approx Z_n$ is generated by e_i. For cohomology let e^i be the cochain *dual* to e_i; i.e. $e^i . e_i = 1$. The coboundary relations derived from (6.5) are

$$\delta e^{2i+1} = n e^{2i+2}, \quad \delta e^{2i} = 0. \tag{6.6}$$

Hence $H^i(\pi; Z_n) \approx Z_n$ is generated by e^i.

Let

$$\partial_*\colon H_i(\pi; Z_n) \to H_{i-1}(\pi; Z_n), \tag{6.7}$$

$$\delta^*\colon H^i(\pi; Z_n) \to H^{i+1}(\pi; Z_n), \tag{6.8}$$

be the boundary, coboundary operators associated with the exact coefficient sequence

$$0 \to Z_n \xrightarrow{\xi} Z_{n^2} \xrightarrow{\eta} Z_n \to 0, \tag{6.9}$$

where $\xi(a) = na$ and η is the natural factorization $Z_{n^2}/\xi Z_n$. From (6.5) and (6.6) it follows that

$$\partial_* e_{2i+2} = e_{2i+1}, \tag{6.10}$$

$$\delta^* e^{2i+1} = e^{2i+2}. \tag{6.11}$$

Now we will compute cup-products in M_π using (4.4). The complex $M \otimes M$ is a $\pi \times \pi$-free acyclic complex and for $d\colon \pi \to \pi \times \pi$, the diagonal map, we need to have $d_\#\colon M \to M \otimes M$, a d-chain mapping carried by the diagonal carrier $C(e_i) = \bar{e}_i \otimes \bar{e}_i$. For a different purpose, such a map has been constructed by Steenrod in [19], p. 219; it has the following form:

$$d_\# e_{2i} = \textstyle\sum_{j=0}^{i} e_{2j} \otimes e_{2i-2j} + \sum_{j=0}^{i-1} \Omega e_{2j+1} \otimes e_{2i-2j-1}, \tag{6.12}$$

$$d_\# e_{2i+1} = \textstyle\sum_{j=0}^{i} \{e_{2j} \otimes e_{2i-2j+1} + e_{2j+1} \otimes x e_{2i-2j}\}. \tag{6.13}$$

Then for the induced $\tilde{d}_\#\colon M_\pi \to M_\pi \otimes M_\pi$, these expressions become

$$\tilde{d}_\# e_{2i} = \textstyle\sum_{j=0}^{i} e_{2j} \otimes e_{2i-2j} + \frac{1}{2}n(n-1) \sum_{j=0}^{i-1} e_{2j+1} \otimes e_{2i-2j-1},$$

$$\tilde{d}_\# e_{2i+1} = \textstyle\sum_{j=0}^{2i+1} e_j \otimes e_{2i+1-j}.$$

And, as in §4, using $d_{\#}$ to compute the cup-products of integral cochains of M_π, we obtain

$$e^{2r} \cup e^s \quad = e^s \cup e^{2r} = e^{2r+s},$$

$$e^{2r+1} \cup e^{2s+1} = \tfrac{1}{2}n(n-1)\, e^{2r+2s+2}.$$

Consequently, we have proved the following:

(6.14) THEOREM. *Suppose π is a cyclic group of order n that operates trivially in a coefficient group Z_n. Then each $H^i(\pi; Z_n)$ is a cyclic group of order n generated by e^i and the cup-products among the various generators are as follows*

$$e^{2r} \cup e^s \quad = e^s \cup e^{2r} = e^{2r+s},$$

$$e^{2r+1} \cup e^{2s+1} = \begin{cases} \dfrac{n}{2} e^{2r+2s+2}, & \text{if } n \text{ is even,} \\ 0, & \text{if } n \text{ is odd.} \end{cases}$$

Let $(e^i)^k$ be the k-fold cup-product of e^i and $e^i e^j = e^i \cup e^j$. Then for $n=2$ we have $e^i = (e^1)^i$ and the cohomology ring $H^*(\pi; Z_2)$ is isomorphic to a polynomial ring over Z_2 in one 1-dimensional generator e^1. Now for n odd we have $e^{2i} = (e^2)^i$, $e^{2i+1} = (e^2)^i e^1$ and $e^1 e^1 = 0$, hence $H^*(\pi; Z_n)$ is isomorphic to the tensor product of a polynomial ring over Z_n in one 2-dimensional generator e^2, and a Grassman algebra over Z_n is one 1-dimensional generator e^1.

7. Steenrod reduced power operations

Steenrod introduced in [19] a general scheme for constructing cohomology operations and studying their properties. The reader is referred to these papers; they are fundamental for this work and here, for notation and reference purposes, briefly, we consider some of the definitions and theorems for the particular case of cyclic groups of coefficients. This restriction is made for simplicity, since this case is enough for our purpose. The proof will be omitted.

Let π be a subgroup of S_n, the permutation group of degree n. Let K be a regular cell complex and K^n the n-fold tensor product complex. Each $\alpha \in \pi$, as a permutation of the factors of K^n, gives rise to a chain map α: $K^n \to K^n$, so that K^n turns naturally into a π-complex. As coefficient group we take Z_m, the integers modulo m. Then for $u \in C^q(K; Z_m)$ the nth power of u is $u^n \in C^{nq}(K^n; Z_m)$, defined by

$$u^n \cdot (\sigma_1 \otimes \sigma_2 \otimes \ldots \otimes \sigma_n) = u(\sigma_1)\, u(\sigma_2) \ldots u(\sigma_n),$$

where each σ_i is a cell of K and $u(\sigma_i) \in Z_m$ is the value of u on σ_i. The value of u^n on $\sigma_1 \otimes \ldots \otimes \sigma_n$ may be different from zero only if each σ_i has dimension q. Therefore for each $\alpha \in \pi$ we have that

$$u^n \cdot \alpha(\sigma_1 \otimes \ldots \otimes \sigma_n) = (\operatorname{sign} \alpha)^q u^n \cdot (\sigma_1 \otimes \ldots \otimes \sigma_n).$$

Let $Z_m^{(q)}$ denote the group Z_m regarded as a π-module, where each $\alpha \in \pi$ operates as multiplication by $(\operatorname{sign} \alpha)^q$. Then for q even the operators are trivial, and in either case $\alpha^2 = \alpha\alpha$ operates as the identity. Now, if we regard u^n as having coefficients in $Z_m^{(q)}$, we have

$$u^n \cdot \alpha(\sigma_1 \otimes \ldots \otimes \sigma_n) = \alpha u^n \cdot (\sigma_1 \otimes \ldots \otimes \sigma_n).$$

Consequently, $u^n \in C^{nq}(K^n; Z_m^{(q)})$ is an equivariant cochain (cf. [9], p. 383).

Let W be a π-free acyclic complex (cf. § 5), then $Z_m^{(q)} \otimes W$ is a π-complex and $H(Z_m^{(q)} \otimes_\pi W) = H(\pi; Z_m^{(q)})$. Clearly $C_i(Z_m^{(q)} \otimes W) \approx C_i(W; Z_m^{(q)})$ as π-modules; for $k \in Z_m^{(q)}$ and $e \in W$, under the isomorphism to $k \otimes e$ corresponds ke and, for $\alpha \in \pi$, to $\alpha(k \otimes e) = (\alpha k) \otimes (\alpha e)$ corresponds $\alpha(ke) = (\alpha k)(\alpha e)$.

Define for each $\alpha \in \pi$ an operation on $W \otimes K$ by

$$\alpha(e \otimes \sigma) = (\alpha e) \otimes \sigma,$$

where $e \in W$, $\sigma \in K$. Then $W \otimes K$ is a π-complex, and since W is π-free it follows that $W \otimes K$ is π-free. Let C be the diagonal carrier from $W \otimes K$ to K^n defined by $C(e \otimes \sigma) = \bar{\sigma}^n$. The carrier C is equivariant $(\alpha C = C = C\alpha)$ and acyclic (since K is regular), therefore (3.1) implies that we can choose an equivariant chain mapping

$$\phi_\#\colon\ W \otimes K \to K^n, \tag{7.1}$$

carried by C. The induced cochain mapping

$$\phi^\#\colon\ C^r(K^n; Z_m^{(q)}) \to C^r(W \otimes K; Z_m^{(q)})$$

transforms equivariant cochains on equivariant cochains, hence $\phi^\# u^n \in C^{nq}(W \otimes K; Z_m^{(q)})$ is equivariant.

In order to define reduced powers, Steenrod introduces an auxiliary operation, the *slant* operation; it is a reduction of a cochain

$$v \in C^r(W \otimes K; Z_m^{(q)})$$

by a chain $c \in C_i(W; Z_m^{(q)})$ and it gives a cochain $v/c \in C^{r-i}(K; Z_m)$, defined as follows. Suppose $c = \sum n_j e_j$ with $n_j \in Z_m^{(q)}$ and each e_j an i-cell of W; if σ is an $(r-i)$-cell of K, then

$$(v/c) \cdot \sigma = v \cdot (c \otimes \sigma),$$

where

$$v \cdot (c \otimes \sigma) = \sum n_j v \cdot (e_j \otimes \sigma) = KI(v, c \otimes \sigma).$$

The operation v/c is bilinear, and it is easy to verify that

(7.2) $$\delta v/c = v/\partial c + (-1)^i \delta(v/c).$$

Assume now that v is an equivariant cochain; for $\alpha \in \pi$ we have $\alpha c = \alpha \sum n_j e_j = \sum (\alpha n_j)(\alpha e_j)$, then

$$(v/\alpha c)\cdot\sigma = \sum(\alpha n_j)\, v\cdot(\alpha e_j)\otimes\sigma = \sum(\alpha n_j)\,\alpha v\cdot(e_j\otimes\sigma)$$
$$= \alpha^2 \sum n_j v\cdot(e_j\otimes\sigma) = (v/c)\cdot\sigma.$$

Thus, in this case,

(7.3) $$v/\alpha c = v/c \quad \text{and} \quad v/(\alpha c - c) = 0.$$

Consequently, the definition of v/c extends to the case of v, an equivariant cochain, and c an element of $[C_i(W; Z_m^{(q)})]_\pi \approx C_i(Z_m^{(q)} \otimes_\pi W)$; the relation (7.2) holds for this extended operation.

Now take $v = \phi^\# u^n$ and $c \in C_i(Z_m^{(q)} \otimes_\pi W)$, then

$$\phi^\# u^n / c \in C^{nq-i}(K; Z_m)$$

is defined as the reduction by c of the nth power of u. Suppose that u is a cocycle, then $\phi^\# u^n$ is an equivariant cocycle, and if c is a cycle, it follows from (7.2) that $\phi^\# u^n/c$ is a cocycle. Moreover, if the cycle c is varied by a boundary, then (7.2) implies that $\phi^\# u^n/c$ varies by a coboundary. If u is varied by a coboundary $\phi^\# u^n/c$ also varies by a coboundary. We only remark here that the proof of this last fact requires a special argument and is not, as in the preceding case, an immediate consequence of (7.2). Thus the class $\{\phi^\# u^n/c\}$ is a function of the classes $\{u\}$, $\{c\}$, and it is independent of the particular $\phi_\#$, since by (3.1) any two choices of $\phi_\#$ are equivariantly homotopic. Then Steenrod defines $\{u\}^n/\{c\}$, the reduction by $\{c\}$ of the nth power of $\{u\}$, by

$$\{u\}^n/\{c\} = \{\phi^\# u^n/c\}.$$

This gives the Steenrod reduced power operations; they are operations defined for $u \in H^q(K; Z_m)$ and $c \in H_i(\pi; Z_m^{(q)})$, and the value is

$$u^n/c \in H^{nq-i}(K; Z_m).$$

In general, the reduced powers u^n/c are linear operations in c, but may not be linear in u. We will list some of their properties. Unless otherwise stated, we assume u and c as above.

First, we have

(7.4) $$u^n/c = 0 \quad \text{if} \quad i > nq - q.$$

Let $f\colon K \to L$ be a map and $f^*\colon H^q(L; Z_m) \to H^q(K; Z_m)$, the induced homomorphism; then

(7.5) $$f^*(u^n/c) = (f^* u)^n/c.$$

This result implies topological invariance for reduced powers.

Suppose $\pi \subset \rho$ are two subgroups of S_n, and let

$$h_*: H_i(\pi; Z_m^{(q)}) \to H_i(\rho; Z_m^{(q)})$$

be the homomorphism induced by the inclusion map. Then

(7.6) $$u^n/c = u^n/h_* c.$$

Finally, let

$$\partial_*: H_i(\pi; Z_m^{(q)}) \to H_{i-1}(\pi; Z_m^{(q)}), \quad \delta^*: H^r(K; Z_m) \to H^{r+1}(K; Z)$$

be the boundary, coboundary operators associated with the exact coefficient sequence $0 \to Z \to Z \to Z_m \to 0$. Then we have, for

$$u \in H^q(K; Z) \quad \text{and} \quad c \in H_i(\pi; Z_m^{(q)}),$$

(7.7) $$u^n/\partial_* c = (-1)^{i+1} \delta^*(u^n/c).$$

These properties were established in [19]; (7.4) follows from considerations in the dimension of the carrier; (7.5) can be obtained easily from the arguments given in [18], p. 60; the proof of (7.6) is direct from the definitions; (7.7) is an immediate consequence of (7.2).

The reduced powers extend to the relative case, that is, for $u \in H^q(K, L; Z_m)$ and $c \in H_i(\pi; Z_m^{(q)})$ we have $u^n/c \in H^{nq-i}(K, L; Z_m)$, and the above properties extend to this case.

8. Steenrod cyclic reduced powers

Suppose π is a cyclic group of order p and x is its generator. As in §7, for a given complex K, the p-fold product K^p is a π-complex, where x operates as the permutation of the factors of K^p, which increases the index of the factor by 1 mod p. Let M be the π-free acyclic complex of §6, and $\phi_\#: M \otimes K \to K^p$ a chosen equivariant map, as in (7.1). Then for $u \in H^q(K; Z_p)$ and each generator $e_i \in H_i(\pi; Z_p^{(q)})$ a cyclic reduced power $u^p/e_i \in H^{pq-i}(K; Z_p)$ is defined.

Here we will consider only cyclic reduced powers of prime period, that is, when p is a prime number. In this case the sign of x is negative for $p=2$ and positive for all $p>2$, hence π will always operate in the coefficient group Z_p as the identity. In this case, as we will show, (7.7) holds for $u \in H^q(K; Z_p)$ and $c \in H_i(\pi; Z_p)$, and ∂_*, δ^* the operators associated to (6.9). Clearly, from $\delta^*\delta^* = \partial_*\partial_* = 0$ and (6.10), this is equivalent to

(8.1) $$u^p/e_{2i+1} = -\delta^*(u^p/e_{2i+2}).$$

To prove (8.1), let u and e_j also denote representatives of their respective classes. Using (7.2) and (7.3) we have

$$\delta\phi_\# u^p/e_{2i+2} = p(\phi_\# u^p/e_{2i+1}) + \delta(\phi_\# u^p/e_{2i+2}).$$

Thus to establish (8.1) it is enough to show that

$$\frac{1}{p}(\delta\phi_{\#} u^p / e_{2i+2}) \sim 0 \quad (\bmod p).$$

We have $u \in Z^q(K; Z_p)$, then $\delta u = pv$ for some v. In a natural fashion, the operations of π on chains of K^p induce operations of π on cochains of K^p, so for $vu^{p-1} \in C^{pq+1}(K^p)$ the symbol Σvu^{p-1} is defined. Since p is prime, it is easy to verify that $\delta u^p = \Sigma(\delta u)\, u^{p-1} = p\Sigma vu^{p-1}$, and then

$$\frac{1}{p}(\delta\phi_{\#} u^p / e_{2i+2}) = \phi_{\#} \Sigma vu^{p-1} / e_{2i+2} = \phi_{\#} vu^{p-1} / \Sigma e_{2i+2}.$$

Using (6.2) and then (7.2) we obtain $(\bmod p)$

$$\begin{aligned} \frac{1}{p}(\delta\phi_{\#} u^p / e_{2i+2}) &= -\phi_{\#} vu^{p-1} / \Sigma' \Delta e_{2i+2} \\ &= -\phi_{\#} vu^{p-1} / \partial \Sigma' e_{2i+3} \\ &= -\delta(\phi_{\#} vu^{p-1} / \Sigma' e_{2i+3}). \end{aligned}$$

This completes the proof of (8.1).

The proposition (8.1) shows that the cyclic reduced powers obtained from e_{2i+1} can be expressed as the composition of two operations, the reduction by e_{2i+2} and δ^*. Then we may only consider u^n/e_{2i}. If $u \in H^q(K; Z_p)$ we have $u^p/e_{2i} \in H^{pq-2i}(K; Z_p)$, and the change in dimension is $(pq-2i)-q$. The following fundamental theorem is due to Thom (cf. [20]).

(8.2) THEOREM. *If p is an odd prime, then $u^p/e_{2i} = 0$ whenever $(pq-2i)-q$ is not an even multiple of $p-1$.*

Steenrod obtains a new proof of (8.2) as a consequence of the following theorem on homology of groups.

(8.3) THEOREM. *For p an odd prime, let S_p be the symmetric group of degree p, let π be a cyclic subgroup of order p (actually a p-Sylow subgroup of S_p), finally, let*

$$h_i\colon\ H_i(\pi; Z_p^{(q)}) \to H_i(S_p; Z_p^{(q)})$$

be the homomorphism induced by the inclusion map of π into S_p.

Then for q even, $h_{2i} = h_{2i-1} = 0$ if $2i$ is not an even multiple of $p-1$; for q odd, $h_{2i} = h_{2i-1} = 0$ if $2i$ is not an odd multiple of $p-1$.

For a proof of (8.3) the reader is referred to [19], p. 218. Theorem (8.2) follows from (7.6) and (8.3).

Let $u \in H^q(K; Z_p)$, then $\dim u^p/e_{q(p-1)} = q$ and, as shown in [19 (b)], p. 221, we have

$$u^p/e_{q(p-1)} = \begin{cases} u, & \text{when } \quad p = 2, \\ (-1)^{mq(q-1)/2}(m!)^q u, & \text{when } \quad p > 2, \end{cases}$$

where $m = (p-1)/2$.

In order to avoid awkward coefficients like the one above, and to obtain simple formulation of properties, it is convenient to change the notation for cyclic reduced powers as follows. For $p=2$ the Steenrod squares Sq^s are defined in terms of e_k by

$$\text{(8.4)} \qquad \mathrm{Sq}^s u = u^2/e_{q-s},$$

where $q=\dim u$. For $p>2$, as suggested by Serre (cf. [19], p. 22; [4], p. 421), the Steenrod p-powers $\mathscr{P}^s$ are defined by

$$\text{(8.5)} \qquad \mathscr{P}^s u = (-1)^{q+s+mq(q+1)/2}(m!)^q u^p/e_{(q-2s)(p-1)}.$$

Then Sq^s, $\mathscr{P}^s$ $(s=0,1,\dots)$ are cohomology operations

$$\mathrm{Sq}^s\colon H^q(K,L;Z_2)\to H^{q+s}(K,L;Z_2),$$

$$\mathscr{P}^s\colon H^q(K,L;Z_p)\to H^{q+2s(p-1)}(K,L;Z_p) \quad (p>2),$$

defined for any pair (K,L) of a complex and subcomplex and any prime p. They are homomorphisms having the following properties:

$$\text{(8.6)} \qquad \mathrm{Sq}^0 = \text{identity}; \quad \mathscr{P}^0 = \text{identity},$$

$$\text{(8.7)} \qquad \mathrm{Sq}^q u = u \cup u; \quad \mathscr{P}^{q/2}u = u^p \quad (p^{\text{th}} \text{ cup-product power}),$$

$$\text{(8.8)} \qquad \mathrm{Sq}^s u = 0, \quad \text{when} \quad i>q; \quad \mathscr{P}^i u = 0, \quad \text{when} \quad i>q/2,$$

where $q=\dim u$ and q is even when $p>2$.

In order to unify statements about Sq^s and $\mathscr{P}^s$ we will make the convention of writing

$$\text{(8.9)} \qquad \mathscr{P}^s = \mathrm{Sq}^s, \quad \text{when} \quad p=2.$$

The p-power operations commute with homomorphisms induced by mappings and with the invariant coboundary operator, that is,

$$\text{(8.10)} \qquad \mathscr{P}^s f^* = f^*\mathscr{P}^s; \quad \mathscr{P}^s\delta = \delta\mathscr{P}^s.$$

From (8.10) follows that the p-power operations $\mathscr{P}^s$ commute with the suspension isomorphism (cf. [19], p. 221; [17], § 11; [18 (b)], p. 55).

Finally, we have

$$\text{(8.11)} \qquad \mathscr{P}^s(u\cup v) = \textstyle\sum_{k=0}^{s} \mathscr{P}^k u \cup \mathscr{P}^{s-k} v \quad (p \geqq 2).$$

All these properties are established by Steenrod in [19]. The first proof of (8.11) for squares $(p=2)$ was given by H. Cartan in [6].

Now we will compute the p-powers on M_π (cf. § 6), where π is a cyclic group of order p. We recall that $H^i(M_\pi; Z_p) \approx Z_p$ is generated by e^i. The p-powers of e^i are as follows:

$$\mathscr{P}^k e^i = \mathrm{Sq}^k e^i = \binom{i}{k} e^{i+k} \quad (p=2), \tag{8.12}$$

$$\left.\begin{aligned} \mathscr{P}^k e^{2i} &= \binom{i}{k} e^{2i+2k(p-1)} \quad (8.13)\\ \mathscr{P}^k e^{2i+1} &= \binom{i}{k} e^{2i+1+2k(p-1)} \quad (8.14)\end{aligned}\right\} \quad (p>2),$$

where ( ) denotes the binomial coefficient.

The proof is by induction on i, using (6.14), (8.6), (8.7), (8.8) and (8.11). If $i=0$ the statements can be verified easily. Since the three cases are similar, as an illustration we write the induction step of (8.13):

$$\begin{aligned} \mathscr{P}^k e^{2i} &= \mathscr{P}^k(e^2 \cup e^{2i-2}) \\ &= \mathscr{P}^0 e^2 \cup \mathscr{P}^k e^{2i-2} + \mathscr{P}^1 e^2 \cup \mathscr{P}^{k-1} e^{2i-2} \\ &= \left\{\binom{i-1}{k} + \binom{i-1}{k-1}\right\} e^{2i+2k(p-1)} \\ &= \binom{i}{k} e^{2i+2k(p-1)}. \end{aligned}$$

9. The groups $H_r(S_p; Z_p^{(q)})$

Let p be an odd prime and as before set $m=(p-1)/2$. We have $(p-1)! \equiv (-1)^m (m!)^2 \equiv -1 \pmod{p}$ (Wilson's theorem), therefore,

$$(-1)^{mq} (m!)^{2q} \equiv (-1)^q \pmod{p}. \tag{9.1}$$

And from (8.5), using (9.1), we obtain ($p>2$)

$$u^p / e_{(q-2s)(p-1)} = (-1)^{s+mq(q-1)/2} (m!)^q \mathscr{P}^s u. \tag{9.2}$$

For future applications, it is convenient to translate to the original notation the results about p-power on M_π. So if $p=2$ with (8.4) the expression (8.12) becomes

$$(e^i)^2 / e_{i-k} = \binom{i}{k} e^{i+k}. \tag{9.3}$$

If $p>2$ with (9.2) the expressions (8.13) and (8.14) transform to

$$(e^{2i})^p / e_{(2i-2k)(p-1)} = (-1)^{k+i} \binom{i}{k} e^{2i+2k(p-1)}, \tag{9.4}$$

$$(e^{2i+1})^p / e_{(2i+1-2k)(p-1)} = (-1)^{k+i}\, m! \binom{i}{k} e^{2i+2k(p-1)+1}. \tag{9.5}$$

And, using (8.1), from (9.4) and (9.5) it follows that

(9.6) $$(e^{2i})^p/e_{(2i-2k)(p-1)-1} = 0,$$

(9.7) $$(e^{2i+1})^p/e_{(2i+1-2k)(p-1)-1} = (-1)^{k+i+1} m! \binom{i}{k} e^{2i+2k(p-1)+2}.$$

For p a prime, let S_p be the symmetric group of degree p and let $Z_p^{(q)}$ be the cyclic group of order p considered as S_p-module (cf. §7). Denote $Z_p^{(q)}$ by $\hat{Z}_p$ for q odd and by Z_p for q even. The above relations can be used to complement the proof of the following:

(9.8) THEOREM.

$$H_r(S_p; Z_p) = \begin{cases} Z_p, & \textit{if} \quad r = 2i(p-1) \quad \textit{or} \quad 2i(p-1)-1, \\ 0, & \textit{otherwise}. \end{cases}$$

$$H_r(S_p; \hat{Z}_p) = \begin{cases} Z_p, & \textit{if} \quad r = (2i+1)(p-1) \quad \textit{or} \quad (2i+1)(p-1)-1, \\ 0, & \textit{otherwise}. \end{cases}$$

PROOF. If $p = 2$, the two cases coincide and (9.8) follows trivially. Suppose $p > 2$. Let $h_* = h_i$ be the homomorphism of (8.3) where $h_*: H_i(\pi; Z_p^{(q)}) \to H_i(S_p; Z_p^{(q)})$ and π is a cyclic group of order p. Since π is a p-Sylow subgroup of S_p, it follows from (5.3) that h_* is onto. Then, by (8.3), we have $H_r(S_p; Z_p^{(q)}) = 0$ for those values of r according to (9.8).

As before, let e_r be the generator of $H_r(\pi; Z_p^{(q)}) \approx Z_p$. To complete the proof it is enough to show that $h_*(e_r) \neq 0$, when it is stated

$$H_r(S_p; Z_p^{(q)}) \approx Z_p.$$

In general, if $u^p/e_k \neq 0$ for some u, then $h_*(e_k) \neq 0$ (cf. (7.6)). Thus, in $H(S_p; \hat{Z}_p)$, that $h_*(e_{(2i+1)(p-1)}) \neq 0$, $h_*(e_{(2i+1)(p-1)-1}) \neq 0$ follows from (9.5) and (9.7). Similarly, in $H(S_p; Z_p)$, that $h_*(e_{2i(p-1)}) \neq 0$, follows from (9.4), and $h_*(e_{2i(p-1)-1}) \neq 0$ is easily obtained from the suspension of (9.5). This completes the proof of the theorem.

10. The Steenrod algebra

Unless otherwise stated, from now on δ^* will denote the coboundary operator associated with (6.9).

An *iterated* p-power is a composition of two or more of the operations δ^*, $\mathscr{P}^i$, e.g. $\mathscr{P}^i\mathscr{P}^j\delta^*\mathscr{P}^k$ (if $p = 2$ then $\delta^* = \mathrm{Sq}^1$ and $\mathscr{P}^i = \mathrm{Sq}^i$). In an obvious fashion, any iterated power can be considered as an operator on $H^*(K; Z_p) = \sum_{q=0}^{\infty} H^q(K; Z_p)$.

Let $\mathscr{R}$ be the free associative algebra over Z_p generated by Sq^0, Sq^1, ..., if $p = 2$ and by δ^*, $\mathscr{P}^0$, $\mathscr{P}^1$, ..., if $p > 2$. Since the generators operate on $H^*(K; Z_p)$ as reduced powers, it follows from the freeness of $\mathscr{R}$ that $H^*(K; Z_p)$ is an $\mathscr{R}$-module, where each monomial of $\mathscr{R}$ can be regarded as an iterated p-power.

Let $\mathscr{I}$ be the ideal of $\mathscr{R}$ defined by the following condition: $C \in \mathscr{I}$ if and only if for all K and all $u \in H^*(K; Z_p)$ we have $Cu = 0$. Each element of $\mathscr{I}$ is a *relation* on iterated p-powers and $\mathscr{I}$ is the *ideal of relations*. Then the *Steenrod algebra* is defined as

(10.1) $$\mathscr{S} = \mathscr{R}/\mathscr{I}.$$

Since the elements of $\mathscr{I}$ operate trivially, it follows that the operators of $\mathscr{R}$ are naturally induced on $\mathscr{S}$, so that $H^*(K; Z_p)$ becomes an $\mathscr{S}$-module.

Let $I = \{i_1, \ldots, i_k\}$ be a sequence of integers, each $i_j \geqq 0$. With Serre ([16, p. 200) we set
$$\mathrm{Sq}^I = \mathrm{Sq}^{i_1}\mathrm{Sq}^{i_2} \ldots \mathrm{Sq}^{i_k}.$$
The integer $n(I) = i_1 + \ldots + i_k$ is the *degree* of Sq^I. Clearly
$$\mathrm{Sq}^I\colon H^q(K; Z_2) \to H^{q+n(I)}(K; Z_2).$$
And Sq^I is called *basic* if the sequence I satisfies the following condition:
$$i_1 \geqq 2i_2, \quad i_2 \geqq 2i_3, \quad \ldots, \quad i_{k-1} \geqq 2i_k.$$

For $p > 2$, following H. Cartan (cf. [5]), the above generalizes as follows. Define $\mathscr{Q}^\epsilon$, where $\epsilon = 0, 1$, by $\mathscr{Q}^0 = \mathscr{P}^0$ and $\mathscr{Q}^1 = \delta^*$. Now, for a sequence of integers $I = \{\epsilon_1, i_1, \epsilon_2, i_2, \ldots, i_k, \epsilon_{k+1}\}$, with $\epsilon_j = 0, 1$ and $i_j > 0$, set
$$\mathscr{P}^I = \mathscr{Q}^{\epsilon_1}\mathscr{P}^{i_1}\mathscr{Q}^{\epsilon_2}\mathscr{P}^{i_2} \ldots \mathscr{P}^{i_k}\mathscr{Q}^{\epsilon_{k+1}}.$$
The integer
$$n(I) = \textstyle\sum_{j=1}^{k+1} \epsilon_j + 2(p-1)\sum_{j=1}^{k} i_j$$
is the *degree* of I. Then $\mathscr{P}^I\colon H^q(K; Z_p) \to H^{q+n(I)}(K; Z_p)$, and we say that $\mathscr{P}^I$ is a monomial of $\mathscr{S}$ written in a *standard* form (in general it is not unique because of relations). From (8.6) it follows that any element of $\mathscr{S}$ can be expressed as a sum of such elements. We say that $\mathscr{P}^I$ is *basic* if the sequence I satisfies the condition

(10.2) $$i_j \geqq pi_{j+1} + \epsilon_{j+1} \quad \text{for} \quad j = 1, \ldots, k-1.$$

Given an arbitrary $C \in \mathscr{S}$, let $C = \Sigma C_k$, where
$$C_k\colon H^q(K; Z_p) \to H^{q+n_k}(K; Z_p) \quad \text{and} \quad n_k > n_{k+1}.$$
Clearly $C \in \mathscr{I}$ if and only if each $C_k \in \mathscr{I}$. Therefore, in the study of $\mathscr{I}$, we can restrict our attention to homogeneous elements such as $\Sigma c_k P^{I_k}$, where $c_k \in Z_p$ and $n = n(I_k)$ for all k.

(10.3) Lemma. *Let $\{q_j\}$ be an unbounded sequence of positive integers. Suppose for a given $C \in \mathscr{R}$ that for any complex K and any $q_i \in \{q_j\}$, we have $C(H^{q_i}(K; Z_p)) = 0$. Then $C \in \mathscr{I}$.*

Proof. As it is mentioned above, we can assume $C = \Sigma c_k P^{I_k}$ to be homogeneous, that is, $n = n(I_k)$ for all k. For arbitrary q let $q_k \in \{q_j\}$ be such that $q_k > q$, and set $r = q_k - q$. Let $K^{(r)}$ be the r-fold suspension of K, and let s: $H^q(K) \approx H^{q+1}(K')$ be the suspension isomorphism (cf. §8). Let s^r: $H^q(K) \approx H^{q+r}(K^{(r)})$ be the isomorphism obtained by the following composition:

$$H^q(K) \overset{s}{\approx} H^{q+1}(K') \overset{s}{\approx} H^{q+2}(K'') \overset{s}{\approx} \dots \overset{s}{\approx} H^{q+r}(K^{(r)}).$$

Since the operations $\mathscr{Q}^\epsilon$, $\mathscr{P}^i$ commute with suspension, it follows that any element of $\mathscr{R}$ commutes with s^r and the following diagram is commutative:

$$\begin{array}{ccc} H^q(K; Z_p) & \xrightarrow{C} & H^{q+n}(K; Z_p) \\ \downarrow s^r & & \downarrow s^r \\ H^{q_k}(K^{(r)}; Z_p) & \xrightarrow{C} & H^{q_k+n}(K^{(r)}; Z_p) \end{array}$$

Now for $u \in H^q(K; Z_p)$ we have $s^r Cu = Cs^r u = 0$; but the kernel of s^r is zero, hence $Cu = 0$. Therefore (10.3) is proved.

(10.4) Corollary. *As above, suppose that $C(H^q(K; Z_p)) = 0$ for all even q. Then $C \in \mathscr{I}$.*

Consequently in the study of $\mathscr{I}$ it is enough to consider only the case q even. This fact enables us to avoid operators in Z_p. This is fortunate, because otherwise we would have to have theorems with double statements like that of (9.8).

11. The groups G_k ($k = 0, 1, 2, 3$)

As before, S_n will denote the symmetric group of degree n. Assuming p a prime, let G_0 be a p-Sylow subgroup of S_{p^2}. The order of G_0 is p^{p+1}—the highest power of p in $p^2!$.

In order to choose a particular p-Sylow subgroup G_0, we represent S_{p^2} as the group of all transformations of p^2 objects, O_s^r ($r, s = 1, \dots, p$), arranged in a square table. Let $\pi_s \subset S_{p^2}$ be the cyclic subgroup consisting of those transformations that effect a cyclic permutation of elements of the s-row, and leave fixed all the other elements. Let $\pi \subset S_{p^2}$ be the subgroup consisting exactly of the cyclic permutations of rows. Let x_s and x, respectively, be chosen generators of π_s and π: x_s will increase the index r of O_s^r by 1 mod p; x will increase the index s of O_s^r by 1 mod p, for all r.

Set

$$\rho = \pi_1 \pi_2 \dots \pi_p,$$

the product of the π_i's as subgroups of S_{p^2}. Since $\pi_i\pi_j = \pi_j\pi_i$, it follows that ρ is a group; its order is p^p. We have $\rho\pi = \pi\rho$, and G_0, the p-Sylow group of S_{p^2}, is defined as

$$G_0 = \rho\pi.$$

Clearly, G_0 is the group with $x, x_1, \ldots, x_p$ as generators and the relations

$$\text{(11.1)} \qquad \begin{cases} x^p = 1, & \\ x_i^p = 1 & (i = 1, \ldots, p), \\ x_i x_j = x_j x_i & (i, j = 1, \ldots, p), \\ x x_i = x_{i+1} x & (i + 1 \bmod p). \end{cases}$$

The group ρ is an invariant subgroup of G_0, and $G_0/\rho \approx \pi$. Actually, G_0 is a *split* extension† of ρ by π, where π operates non-trivially on ρ $(x x_i x^{-1} = x_{i+1})$.

Let $S_p^{(k)}$ be the group of all permutations of elements of the k-row that leave fixed all the other elements. Let S_p be the group of all permutations of rows. We have $\pi_k \subset S_p^{(k)}$ and $\pi \subset S_p$. Like above, set

$$P = S_p^{(1)} S_p^{(2)} \ldots S_p^{(k)},$$

the product of the $S_p^{(k)}$'s as subgroups of S_{p^2}. Since $S_p^{(i)} S_p^{(j)} = S_p^{(j)} S_p^{(i)}$, it follows that P is a group.

Let G_1, G_2, G_3 be the subgroups of S_{p^2} defined by the following products:

$$G_1 = \rho S_p = S_p \rho,$$
$$G_2 = P\pi = \pi P,$$
$$G_3 = P S_p = S_p P.$$

Then $G_0 \subset G_k \subset G_3$ for $k = 1, 2$. Obviously ρ is an invariant subgroup of G_1, and P is an invariant subgroup of G_2 and G_3, and we have split extensions

$$0 \to \rho \to G_1 \to S_p \to 0,$$
$$0 \to P \to G_2 \to \pi \to 0,$$
$$0 \to P \to G_3 \to S_p \to 0.$$

12. Construction of a G_k-free acyclic complex

Let M be a π-free acyclic complex and let L be a S_p-free acyclic complex. Form the p-fold tensor products $M^p = M \otimes \ldots \otimes M$, $L^p = L \otimes \ldots \otimes L$, and then form $M^p \otimes M$, $M^p \otimes L$, $L^p \otimes M$, $L^p \otimes L$.

† A group extension

$$0 \longrightarrow \rho \overset{i}{\longrightarrow} G \overset{j}{\longrightarrow} \pi \longrightarrow 0$$

is called a *split* extension if there exists a homomorphism $k: \pi \to G$ such that $jk = 1$.

We shall show how operations of G_k on these complexes can be defined so that the last four, respectively, become G_k-free for $k = 0, 1, 2, 3$.

We regard S_p as the permutations of the columns of $\| O_s^r \|$, then for fixed i there is a natural isomorphism $S_p \approx S_p^{(i)}$ obtained by assigning the kth column to the kth element of the i-row.

First L^p will be made a G_3-complex, as follows. The element $\alpha_i \in S_p^{(i)}$ operates on $a_1 \otimes \ldots \otimes a_p$ of L^p by the rule

$$\alpha_i(a_1 \otimes \ldots \otimes a_p) = a_1 \otimes \ldots \otimes (\alpha a_i) \otimes \ldots \otimes a_p,$$

where α corresponds to α_i under $S_p \approx S_p^{(i)}$. If $\alpha_i \in S_p^{(i)}$, $\alpha_j \in S_p^{(j)}$, then $\alpha_i \alpha_j = \alpha_j \alpha_i$ as operators, consequently, the operations extend to any element of P and L^p becomes a P-complex, actually P-free, since it is $S_p^{(i)}$-free. Now each $\alpha \in S_p \subset G_3$ will operate naturally by permutation of the factors of L^p, and since $S_p^{(i)}$ operates in the ith factor alone, it follows from the definition of G_3 that the operations can be extended consistently to any element of G_3 so that L^p becomes a G_3-complex. However, L^p is only P-free since S_p fails to operate freely.

In the same fashion, by replacing above π_i by $S_p^{(i)}$ and ρ by P, the complex M^p becomes a G_1-complex and ρ-free.

Now, by letting P operate as the identity on M, L, we can regard M as a G_2-complex and L as a G_3-complex. Take M^p, M as G_0-complexes, then $M^p \otimes M$ is a G_0-complex under diagonal operations (2.2) of G_0. Since M^p is ρ-free and M is π-free, it follows that $M^p \otimes M$ is G_0-free. Hence, we have

(12.1) $\quad M^p \otimes M$ is a G_0-free acyclic complex.

Equally, under diagonal operations, we obtain that

(12.2) $\quad M^p \otimes L$ is a G_1-free acyclic complex,

(12.3) $\quad L^p \otimes M$ is a G_2-free acyclic complex,

(12.4) $\quad L^p \otimes L$ is a G_3-free acyclic complex.

13. The groups $H_r(G_k; Z_p)$

Consider only trivial operators on Z_p, and for simplicity, set $G = G_0$; then (12.1) implies (cf. §5) $H(G; Z_p) \approx H(Z_p \otimes (M^p \otimes_G M))$. Let $\tilde{M} = Z_p \otimes_\pi M$, then $Z_p \otimes (M^p \otimes_G M) \approx \tilde{M}^p \otimes_\pi M$. Now let M be the π-complex of §6 so that $\partial = 0$ in $\tilde{M}$, and consequently

$$\tilde{M} = H(\tilde{M}) \approx H(\pi; Z_p).$$

We have $\tilde{M}^p \approx H^p(\pi; Z_p) = H(\pi; Z_p) \otimes \ldots \otimes H(\pi; Z_p)$, the p-fold tensor product, and the operations of π on $H^p(\pi; Z_p)$ are naturally induced as permutation of factors. Thus $\tilde{M}^p \otimes_\pi M \approx H^p(\pi; Z_p) \otimes_\pi M$, therefore

$$H(G; Z_p) \approx H(\pi; H^p(\pi; Z_p)), \tag{13.1}$$

that is, the homology of π with coefficients in the π-module $H^p(\pi; Z_p)$.†

In order to make a direct analysis of cycles of G by means of the complex $\tilde{M}^p \otimes_\pi M$, we will first study the structure of invariant cycles of $\tilde{M}^p$.

(13.2) Lemma. *Let Σ and Δ be as in* (6.1), *and let $a \in C_r(\tilde{M}^p)$. We have*

$$\Sigma a = 0 \quad \textit{is equivalent to} \quad a = \nu + \Delta\mu, \tag{13.3}$$

$$\Delta a = 0 \quad \textit{is equivalent to} \quad a = \nu + \Sigma\mu, \tag{13.4}$$

where $\nu, \mu \in C_r(\tilde{M}^p)$ and $\nu = n.(e_i)^p$, for some n, when $r = pi$, and $\nu = 0$ otherwise.

Proof. From right to left the implication is obvious. To prove the second part we first observe that the elements $b = e_{i_1} \otimes \ldots \otimes e_{i_p}$, with $r = i_1 + \ldots + i_p$, are the generators of $C_r(\tilde{M}^p)$, and if b has at least two different i_k's then, since p is prime, all the transforms $b, xb, \ldots, x^{p-1}b$ are different. On the other hand, $(e_i)^p$ is a fixed element under x. If the chain a has a term $n.(e_i)^p$ this by definition is ν. Suppose that $c = \sum_{j=0}^{p-1} n_j \otimes (x^j b)$, where $n_j \in Z_p$ is the part of $a - \nu$ formed by $b = e_{i_1} \otimes \ldots \otimes e_{i_p}$ and its different transforms. Now, assume $\Sigma a = 0$, then $\Sigma c = (\sum_{j=0}^{p-1} n_j) \otimes (\Sigma b) = 0$, consequently $\sum_{j=0}^{p-1} n_j = 0$ and we can write

$$c = \textstyle\sum_{j=0}^{p-1} n_j \otimes [(x^j - 1)b] = \Delta \sum_{j=1}^{p-1} n_j \otimes [(1 + \ldots + x^{j-1})b].$$

Clearly this establishes (13.3).

If $\Delta a = 0$ then $\Delta c = \sum_{j=0}^{p-1} n_j \otimes [(x^{j+1} - x^j)b] = 0$, and this implies $n_0 = n_1 = \ldots = n_{p-1}$, therefore $c = \Sigma(n_0 \otimes b)$. Hence (13.4) follows and so the lemma is proved.

Returning to our objective, let $c \in C_k(\tilde{M}^p \otimes_\pi M)$ be an arbitrary chain. Using the relation $a \otimes (xe) = (x^{-1}a) \otimes e$ we can write

$$c = \textstyle\sum_{j=0}^{k} a_{k-j} \otimes e_j \quad \text{with} \quad a_j \in H_i(\tilde{M}^p).$$

The boundary of c is given by

$$\partial c = (-1)^k \textstyle\sum_j a_{k-2j} \otimes (\Sigma e_{2j-1}) + (-1)^{k+1} \sum_j a_{k-2j-1} \otimes (\Delta e_{2j}),$$

and this transforms to

$$\partial c = (-1)^k \textstyle\sum_j (\Sigma a_{k-2j}) \otimes e_{2j-1} + (-1)^k \sum_j (\Delta a_{k-2j-1}) \otimes (x e_{2j}). \tag{13.5}$$

† This fact was pointed out to me by Steenrod.

Then we have $\partial c = 0$ if and only if $\Sigma a_{k-2j} = \Delta a_{k-2j-1} = 0$ $(j > 0)$. Thus if c is a cycle, a_j has the special form given either by (13.3) or (13.4). Using this and the relations of $\tilde{M}^p \otimes_\pi M$, we obtain

$$c \sim a_k \otimes e_0 + \textstyle\sum_{j=0}^{k-1} \nu_j \otimes e_{k-j},$$

where $\nu_j = n_j . (e_i)^p$ when $j = pi$ and $\nu_j = 0$ otherwise. Hence a system of generators of $H(\tilde{M}^p \otimes_\pi M)$ is formed with classes of the following types:

$$(e_j)^p \otimes e_k, \tag{13.6}$$

$$(e_{j_1} \otimes \ldots \otimes e_{j_p}) \otimes e_0. \tag{13.7}$$

Now we consider G_1, G_2 and G_3. From (12.2) it follows

$$H(G_1; Z_p) \approx H(Z_p \otimes (M^p \otimes_{G_1} L)),$$

where $Z_p \otimes (M^p \otimes_{G_1} L) \approx \tilde{M}^p \otimes_{S_p} L$ and $\tilde{M} = H(\tilde{M})$. Therefore

$$H(G_1; Z_p) \approx H(S_p; H^p(\pi; Z_p)), \tag{13.8}$$

that is, $H(G_1; Z_p)$ can be expressed as the homology of S_p with coefficients in the S_p-module $H^p(\pi; Z_p) = H(\pi; Z_p) \otimes \ldots \otimes H(\pi; Z_p)$.

From (12.3) and (12.4) we obtain

$$H(G_2; Z_p) \approx H(Z_p \otimes (L^p \otimes_{G_2} M)),$$

$$H(G_3; Z_p) \approx H(Z_p \otimes (L^p \otimes_{G_3} L)).$$

Let $\tilde{L} = Z_p \otimes_{S_p} L$, then

$$Z_p \otimes (L^p \otimes_{G_2} M) \approx \tilde{L}^p \otimes_\pi M \quad \text{and} \quad Z_p \otimes (L^p \otimes_{G_3} L) \approx \tilde{L}^p \otimes_{G_3} L,$$

hence

$$H(G_2; Z_p) \approx H(\tilde{L}^p \otimes_\pi M),$$

$$H(G_3; Z_p) \approx H(\tilde{L}^p \otimes_\pi L).$$

We would like to conclude

$$H(G_2; Z_p) \approx H(\pi; H^p(S_p; Z_p)), \tag{13.9}$$

$$H(G_3; Z_p) \approx H(S_p; H^p(S_p; Z_p)). \tag{13.10}$$

This seems likely to be true. However, in general we have $\tilde{L} \neq H(\tilde{L})$, and we cannot establish (13.9) and (13.10) with the same argument used for (13.1) and (13.8). We leave this point unsettled. The precise statements of (13.9) and (13.10) will not be used in this work.

14. A system of generators of $H(S_{p^2}; Z_p)$

Consider the following diagram of groups and inclusion maps

$$G = G_0 \xrightarrow{\theta_1} G_1 \xrightarrow{f} G_3, \quad G_0 \xrightarrow{\theta_3} G_3 \xrightarrow{\omega} S_{p^2}, \quad G_0 \xrightarrow{\theta_2} G_2 \xrightarrow{g} G_3 \tag{14.1}$$

and let $\theta = \omega\theta_3$. For the homology groups we have the commutative diagram

$$H(G; Z_p) \xrightarrow{\theta_{1*}} H(G_1; Z_p) \xrightarrow{f_*} H(G_3; Z_p), \quad H(G; Z_p) \xrightarrow{\theta_{3*}} H(G_3; Z_p) \xrightarrow{\omega_*} H(S_{p^2}; Z_p), \quad H(G; Z_p) \xrightarrow{\theta_{2*}} H(G_2; Z_p) \xrightarrow{g_*} H(G_3; Z_p) \tag{14.2}$$

where the induced homomorphisms θ_*, θ_{k*}, f_*, g_*, ω_* are all onto (cf. (5.3)). Hence

$$H(S_{p^2}; Z_p) \approx H(G; Z_p)/\text{kernel}\,\theta_* \approx H(G_3; Z_p)/\text{kernel}\,\omega_*.$$

Also, we have

$$\text{kernel}\,\theta_{k*} \subset \text{kernel}\,\theta_{3*}. \tag{14.3}$$

We will establish some results about the kernel of θ_{k*} $(k = 1, 2, 3)$.

Let $h: \pi \to S_p$ be the inclusion map and $h_{\#}: M \to L$ a chosen h-chain map. Let $\theta_{1\#}: M^p \otimes M \to M^p \otimes L$ be the θ_1-chain map defined by

$$\theta_{1\#} = 1 \otimes \ldots \otimes 1 \otimes h_{\#} = 1^p \otimes h_{\#},$$

and let
$$\bar{\theta}_{1\#}: \tilde{M}^p \otimes_\pi M \to \tilde{M}^p \otimes_{S_p} L$$

be its induced chain map. Then the induced θ_{1*} in homology can be regarded as the one induced by the map

$$(h, 1^p): (\pi; \tilde{M}^p) \to (S_p; \tilde{M}^p),$$

where $\tilde{M}^p$ is considered as the coefficient group (cf. § 5).

(14.4) Lemma.

$$\theta_{1*}[(e_r)^p \otimes e_s] = 0 \quad \textit{whenever} \quad \begin{cases} r \ \textit{is even and} \ \ s \neq 2i(p-1) \\ \qquad\qquad\quad \textit{or} \ \ 2i(p-1)-1, \\ r \ \textit{is odd and} \ \ s \neq (2i+1)(p-1) \\ \qquad\qquad\quad \textit{or} \ \ (2i+1)(p-1)-1. \end{cases}$$

PROOF. If $p=2$, then $\pi=S_2$ and (14.4) is trivially true. Assume $p>2$; in this case the proof is the same as the one given by Steenrod for (8.3) and we only need to replace the coefficient group $Z_p^{(q)}$ by $\tilde{M}^p$. We will indicate the steps. Let $k, \gamma, \psi, \lambda, \lambda_\#$ be as defined in [19], 4.8. We have the commutative diagram

$$\begin{array}{ccc} (\pi;\, \tilde{M}^p) & \xrightarrow{(h,\, 1^p)} & (S_p;\, \tilde{M}^p) \\ \downarrow{\scriptstyle (\lambda,\gamma)} & & \downarrow{\scriptstyle (\psi,\gamma)} \\ (\pi;\, \tilde{M}^p) & \xrightarrow{(h,\, 1^p)} & (S_p;\, \tilde{M}^p) \end{array}$$

and since ψ is the inner automorphism given by γ, it follows from (5.1) that $(\psi,\gamma)_*=1$, so we obtain the commutative diagram

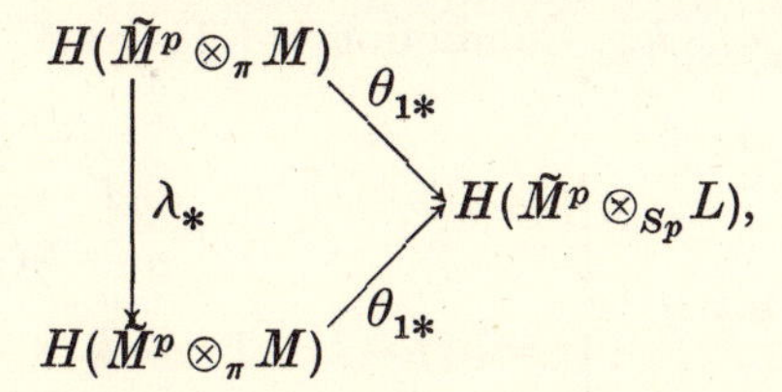

where λ_* is induced by (λ,γ). Thus $(\lambda_*-1)c$ belongs to the kernel of θ_{1*}. Now $\gamma \otimes \lambda_\#$: $\tilde{M}^p \otimes M \to \tilde{M}^p \otimes M$ is a λ-chain map, and, using the explicit expression for $\lambda_\#$ given in [19], 4.9, we obtain

$$(\lambda_*-1)\,[(e_r)^p \otimes e_{2s}] \;=[(-1)^r k^s-1]\,(e_r)^p \otimes e_{2s},$$

$$(\lambda_*-1)\,[(e_r)^p \otimes e_{2s-1}]=[(-1)^r k^s-1]\,(e^r)^p \otimes e_{2s-1}.$$

From here Lemma (14.4) follows, using the same reasoning given at the end of [19], 4.8. We leave the details to the reader.

(14.5) LEMMA.

$\theta_{2*}[(e_r)^p \otimes e_s]=0$ *whenever* $r \neq 2i(p-1)$ *or* $2i(p-1)-1$,

$\theta_{2*}[e_{r_1} \otimes \ldots \otimes e_{r_p} \otimes e_0]=0$ *whenever* $r_i \neq 2k_i(p-1)$ *or* $2k_i(p-1)-1$.

PROOF. With the above notation, let $\tilde{h}_\#$: $\tilde{M} \to \tilde{L}$ be the map induced by $h_\#$, where $\tilde{M}=Z_p \otimes_\pi M$ and $\tilde{L}=Z_p \otimes_{S_p} L$. For $p=2$ the lemma is trivial; suppose $p>2$. It follows from (8.3) and (9.8) that

(14.6) $\quad \tilde{h}_\#(e_r) \sim 0$ if and only if $r \neq 2i(p-1)$ or $2i(p-1)-1$.

The map $\theta_{2\#}$: $M^p \otimes M \to L^p \otimes M$ defined by

$$\theta_{2\#}=h_\# \otimes \ldots \otimes h_\# \otimes 1 = h_\#^p \otimes 1$$

is a θ_2-chain map. Thus, the induced map $\tilde{\theta}_{2\#}$: $\tilde{M}^p \otimes_\pi M \to \tilde{L}^p \otimes_{S_p} M$ has the form $\tilde{\theta}_{2\#}=\tilde{h}_\#^p \otimes 1$, and it can be used to compute θ_{2*}.

Let $\bar{e}_r = \tilde{h}_{\#}(e_r)$ and assume $\bar{e}_r = \partial c$, then

$$\tilde{\theta}_{\#2}[(e_r)^p \otimes e_s] = (\bar{e}_r)^p \otimes e_s = (\partial c)^p \otimes e_s.$$

Now using the relations of $\tilde{L}^p \otimes_{S_p} M$ one can prove in general that $(\partial c)^p \otimes e_s \sim 0$. The proof of this is mechanical, and it is left to the reader as an exercise. From this and (14.6) the first part of (14.5) follows.

If $\bar{e}_{r_i} = \partial c$ for some i, say $i = 1$, then

$$\partial(c \otimes \bar{e}_{r_2} \otimes \ldots \otimes \bar{e}_{r_p} \otimes e_0) = \bar{e}_{r_1} \otimes \ldots \otimes \bar{e}_{r_p} \otimes e_0.$$

Then, again using (14.6), the second part of the lemma is established.

Like in the preceding cases the map $\theta_{3\#}\colon M^p \otimes M \to L^p \otimes L$, defined by $\theta_{3\#} = h^p_{\#} \otimes h_{\#}$, is a θ_3-chain map and the induced

$$\tilde{\theta}_{3\#}\colon \tilde{M}^p \otimes_\pi M \to \tilde{L}^p \otimes_{S_p} L$$

is given by $\tilde{\theta}_{3\#} = \tilde{h}_{\#} \otimes_{S_p} h_{\#}$. Combining (14.3), (14.4) and (14.5), we obtain

(14.7) LEMMA.

$$\theta_{3*}[(e_r)^p \otimes e_s] \neq 0 \quad \textit{only if} \quad \begin{cases} r = 2i(p-1) & \textit{and} \quad s = 2j(p-1) \\ & \textit{or} \quad 2j(p-1)-1, \\ r = 2i(p-1)-1 & \textit{and} \quad s = (2j+1)(p-1) \\ & \textit{or} \quad (2j+1)(p-1)-1. \end{cases}$$

Finally, since the maps of (14.2) are all onto, using (14.5) and (14.7) we obtain the following:

(14.8) THEOREM. *A set of generators for $H(S_{p^2}; Z_p)$ can be formed with all elements of the following types:*

$$\theta_*[(e_r)^p \otimes e_s], \qquad \textit{with } r, s \quad \textit{as in } (14.7),$$
$$\theta_*[e_{r_1} \otimes \ldots \otimes e_{r_p} \otimes e_0], \qquad \textit{with } r_i \quad \textit{as in } (14.5).$$

15. Homology classes of $H(G_k; Z_p)$ and iterated reduced powers

Let $\phi_{\#}\colon M \otimes K \to K^p$ be the equivariant map considered in §8 on defining cyclic reduced powers of $u \in H^q(K; Z_p)$. We recall that K^p is a π-complex and that the generator x of π operates on $\sigma_1 \otimes \ldots \otimes \sigma_p \in K^p$ by the rule $x(\sigma_1 \otimes \ldots \otimes \sigma_p) = (-1)^\epsilon \sigma_p \otimes \sigma_1 \otimes \ldots \otimes \sigma_{p-1}$, where

$$\epsilon = \dim \sigma_p \cdot \dim(\sigma_1 \otimes \ldots \otimes \sigma_{p-1}).$$

Similarly, the complex $K^{p^2} = K^p \otimes \ldots \otimes K^p$ will be made a G-complex as follows. Let $a \in K^{p^2}$, where $a = a_1 \otimes \ldots \otimes a_p$ and each $a_i \in K^p$, then for x, x_i, generators of G, set

$$x_i a = a_1 \otimes \ldots \otimes (xa_i) \otimes \ldots \otimes a_p,$$
$$xa = (-1)^\mu a_p \otimes a_1 \otimes \ldots \otimes a_{p-1},$$

where $\mu = \dim a_p \cdot \dim (a_1 \otimes \ldots \otimes a_{p-1})$. It follows that the operations defined above are consistent with the relations (11.1); thus they extend for arbitrary elements of G and K^{p^2}, so that K^{p^2} becomes a G-complex.

Now $M^p \otimes M$ is a G-free acyclic complex (cf. (12.1)) and, like in (7.1), there exists $\psi_{\#}\colon (M^p \otimes M) \otimes K \to K^{p^2}$, a G-equivariant chain mapping carried by the diagonal carrier. Next we show how such a $\psi_{\#}$ can be constructed in terms of the $\phi_{\#}$ mentioned at the beginning of this section. Let $\kappa\colon M^p \otimes K^p \to (M \otimes K)^p$ be the map

$$(15.1) \quad \kappa(e_{r_1} \otimes \ldots \otimes e_{r_p} \otimes \sigma_1 \otimes \ldots \otimes \sigma_p) = (-1)^{\nu} (e_{r_1} \otimes \sigma_1) \otimes \ldots \otimes (e_{r_p} \otimes \sigma_p),$$

where $\nu = \sum_{i=1}^{p-1} r_{i+1} q_i$ and $q_i = \dim \sigma_1 \otimes \ldots \otimes \sigma_i$. It can be easily verified that the composition

$$M^p \otimes M \otimes K \xrightarrow{1 \otimes \ldots \otimes 1 \otimes \phi_{\#}} M^p \otimes K^p \xrightarrow{\kappa} (M \otimes K)^p \xrightarrow{\phi_{\#} \otimes \ldots \otimes \phi_{\#}} K^{p^2}$$

is G-equivariant and that it belongs to the diagonal carrier; so we set

$$(15.2) \quad \psi_{\#} = \phi_{\#}^p \kappa (1^p \otimes \phi_{\#}),$$

where $1^p = 1 \otimes \ldots \otimes 1$, $\phi_{\#}^p = \phi_{\#} \otimes \ldots \otimes \phi_{\#}$.

For arbitrary $u \in H^q(K; Z_p)$, $c \in H_i(G; Z_p^{(q)})$ we have u^{p^2}/c. The group G contains only even permutations, then it operates trivially on $Z_p^{(q)}$, and all the statements of § 13 from the beginning to (13.7) hold with $Z_p^{(q)}$ instead of Z_p. The homology classes of (13.6) and (13.7) form a set of generators of $H(G; Z_p)$; thus the study of u^{p^2}/c reduces to analyze the reductions of u^{p^2} by these elementary classes.

(15.3) Lemma.

$$u^{p^2}/[(e_r)^p \otimes e_s] = (-1)^{\mu} (u^p/e_r)^p/e_s,$$

$$u^{p^2}/(e_{r_1} \otimes \ldots \otimes e_{r_p} \otimes e_0) = (-1)^{\nu} (u^p/e_{r_1}) \cup \ldots \cup (u^p/e_{r_p}),$$

where $\mu = \frac{1}{2}p(p-1)r(q-1)$ (for ν see proof).

Proof. Let the same letter u denote the class and a representative. To establish the two parts of the lemma consider

$$\{\phi^{\#}[(\phi^{\#}u^p/e_{r_1}) \otimes \ldots \otimes (\phi^{\#}u^p/e_{r_p})]/e_s\} \cdot \sigma,$$

where σ is a cell of K of dimension $p^2 q - s - \sum r_i$. By definition of the slant operation this is equal to

$$[(\phi^{\#}u^p/e_{r_1}) \otimes \ldots \otimes (\phi^{\#}u^p/e_{r_p})] \cdot \phi_{\#}(e_s \otimes \sigma).$$

Now $\phi_{\#}(e_s \otimes \sigma) \in C(K^p)$, then assume $\phi_{\#}(e_s \otimes \sigma) = \sum c_1 \otimes \ldots \otimes c_p$. The above expression becomes

$$\sum[(\phi^{\#}u^p/e_{r_1}) \cdot c_1] \ldots [(\phi^{\#}u^p/e_{r_p}) \cdot c_p] = u^{p^2} \cdot \phi_{\#}^p \sum (e_{r_1} \otimes c_1) \otimes \ldots \otimes (e_{r_p} \otimes c_p).$$

We have that $(\phi^{\#}u^p/e_{r_i})\cdot c_i \neq 0$ only if $\dim c_i = pq - r_i$; thus we can suppose that the sum runs only over such terms. Then, using (15.1), our expression transforms to

$$\begin{aligned}(-1)^{\nu} u^{p^2}\cdot \varphi^p_{\#}\kappa\textstyle\sum e_{r_1}\otimes\ldots\otimes e_{r_p}\otimes c_1\otimes\ldots\otimes c_p\\ = (-1)^{\nu} u^{p^2}\cdot \phi^p_{\#}\kappa e_{r_1}\otimes\ldots\otimes e_{r_p}\otimes \phi_{\#}(e_s\otimes\sigma)\\ = (-1)^{\nu} u^{p^2}\cdot \psi_{\#}(e_{r_1}\otimes\ldots\otimes e_{r_p}\otimes e_s\otimes\sigma)\\ = (-1)^{\nu}[(\psi^{\#}u^{p^2}/e_{r_1}\otimes\ldots\otimes e_{r_p}\otimes e_s)]\cdot\sigma,\end{aligned}$$

where $\nu = \sum_{i=1}^{p-1} r_{i+1}(ipq - \sum_{j=1}^{i} r_j)$, and $\psi_{\#}$ is like in (15.2). Then, if we take $r = r_1 = \ldots = r_p$, we have $\nu \equiv \mu \bmod 2$, and the first part of (15.3) follows. The second part is obtained with $s = 0$, since $\phi_{\#}(e_0\otimes\ )$ is a chain approximation to the diagonal map (cf. [19], p. 217), and it can be used to define the p-fold cup-product; so (15.3) is proved.

As a corollary of (15.3) we have that for arbitrary $c \in H_i(G; Z_p)$ the reduced power u^{p^2}/c is a linear combination of iterated cyclic powers and of p-fold cup-products of cyclic powers. Similarly, extending to G_3 the operations of G on K^{p^2}, we obtain

(15.4) THEOREM. *For q even and $u \in H^q(K; Z_p)$, $c \in H_i(G_3; Z_p)$, the reduced power u^{p^2}/c is a linear combination of terms*

$$\mathscr{P}^r\mathscr{P}^s u,\quad \mathscr{P}^r\delta^*\mathscr{P}^s u,\quad \delta^*\mathscr{P}^r\mathscr{P}^s u,\quad \delta^*\mathscr{P}^r\delta^*\mathscr{P}^s u, \tag{15.5}$$

and of p-fold cup-products of terms $\mathscr{P}^r u$, $\delta^*\mathscr{P}^s u$.

PROOF. The homomorphism θ_{3*} of (14.2) is onto, hence it maps the homology classes of (13.6) and (13.7) onto a set of generators of $H(G_3; Z_p)$. From this and (7.6), (8.1), (8.5), (14.5), (14.7) and (15.3) the theorem follows.

A similar theorem holds for q odd, and it follows as above, using the corresponding statements of (14.4), (14.5) and (14.7) for the case $Z_p^{(q)}$ as coefficient group.

16. The basic scheme for relations

We will define y and z, two special elements of S_{p^2}. Using the representation of S_{p^2} as transformations of $\| O_r^s \|$ (cf. § 11), set

$$y = x_1 x_2 \ldots x_p.$$

Then $y \in G$ and it represents a cyclic permutation of columns; moreover,

$$xy = yx. \tag{16.1}$$

Let z be the reflection through the diagonal of $\| O_r^s \|$; i.e. $z(O_r^s) = O_s^r$. Clearly, z does not belong to G since the elements of G preserve rows. We have $z^2 = 1$, then $z = z^{-1}$, and

$$x = zyz, \quad y = zxz.$$

Let $\sigma\colon S_{p^2} \to S_{p^2}$ be the inner automorphism induced by z, then $\sigma(\alpha) = z\alpha z$ for $\alpha \in S_{p^2}$, and

$$\sigma(x) = y, \quad \sigma(y) = x.$$

Define G_z as the invariant set of G under σ. That is,

$$G_z = G \cap zGz.$$

It follows that G_z is the group generated by x, y. Consequently, if π' denotes the cyclic subgroup generated by y, from (16.1) we obtain that

$$G_z = \pi\pi'. \tag{16.2}$$

Set $\tau = \sigma \mid G_z$, the restriction of σ to G_z, then the map $\tau\colon G_z \to G_z$ is the one that interchanges x and y.

Let $\theta\colon G \to S_{p^2}$, $\xi\colon G_z \to G$ denote the inclusion maps and let A be an S_{p^2}-module regarded as coefficient group. We have the following commutative diagram:

$$\begin{array}{ccccc}
(G_z; A) & \xrightarrow{\xi} & (G; A) & \xrightarrow{\theta} & (S_{p^2}; A) \\
\downarrow \tau & & & & \downarrow \sigma \\
(G_z; A) & \xrightarrow{\xi} & (G; A) & \xrightarrow{\theta} & (S_{p^2}; A)
\end{array}$$

where, for simplicity, the maps of pairs are set as follows: $\xi = (\xi, z)$, $\theta = (\theta, z)$, $\tau = (\tau, z)$ and $\sigma = (\sigma; z)$, where $z\colon A \to A$ is the map given by z as element of S_{p^2}. The induced homomorphisms in homology will satisfy $\sigma_* \theta_* \xi_* = \theta_* \xi_* \tau_*$. The fact that σ is the inner autohomomorphism given by z, implies (cf. (5.1)) that $\sigma_* = 1$, hence $\theta_* \xi_* = \theta_* \xi_* \tau_*$ holds in the following diagram:

$$\begin{array}{ccccc}
H(G_z; A) & \searrow^{\xi_*} & & & \\
\downarrow \tau_* & & H(G; A) & \xrightarrow{\theta_*} & H(S_{p^2}; A). \\
H(G_z; A) & \nearrow_{\xi_*} & & &
\end{array} \tag{16.3}$$

Thus, for $c \in H(G_z; A)$, the element

$$\xi_* c - \xi_* \tau_* c \tag{16.4}$$

lies in the kernel of θ_*.

Now take $A = Z_p^{(q)}$, and let θ_3, ω be as defined in (14.1). Then $\theta = \omega\theta_3$ and $\theta_* = \omega_* \theta_{3*}$, where

$$H(G;\, Z_p^{(q)}) \xrightarrow{\theta_{3*}} H(G_3;\, Z_p^{(q)}) \xrightarrow{\omega_*} H(S_{p^2};\, Z_p^{(q)}).$$

For $u \in H^q(K;\, Z_p)$, from $\theta_* \xi_*(1-\tau_*)c = 0$ and (7.6), we obtain

$$u^{p^2}/\xi_* c = u^{p^2}/\xi_* \tau_* c, \quad u^{p^2}/\theta_{3*}\xi_* c = u^{p^2}/\theta_{3*}\xi_* c, \tag{16.5}$$

and, according to (15.3) and (15.4), these are relations on iterated reduced powers.

The expressions (16.5) constitute the pattern we shall use to obtain relations. The work of the next sections is devoted to the explicit computation of all elements of the form (16.4).

17. $H(G_z;\, Z_p)$ and τ_*

From now on we only consider the case q even (see (10.4)).

It follows from (16.2) that $G_z \approx \pi \times \pi$, where π is the cyclic group of order p and generator x. Set a fixed isomorphism λ: $G_z \approx \pi \times \pi$ by $\lambda(x) = x \times 1$, $\lambda(y) = 1 \times x$.

As before, let M be the π-free acyclic complex of § 6. In a natural form the complex $M \otimes M$ is a $\pi \times \pi$-free acyclic complex, and the set $\{e_i \otimes e_j\}$ is a $\pi \times \pi$-basis. Clearly, $M \otimes M$ also is a G_z-free acyclic complex if G_z operates through λ, i.e.

$$x(e_i \otimes e_j) = (xe_i) \otimes e_j, \quad y(e_i \otimes e_j) = e_i \otimes (xe_j).$$

The set $\{e_i \otimes e_j)$ becomes a G_z-basis. We have a natural isomorphism $Z_p \otimes_{G_z} (M \otimes M) \approx \tilde{M} \otimes \tilde{M}$, where $\tilde{M} = Z_p \otimes_\pi M$. Hence

$$H(G_z;\, Z_p) \approx H(\pi;\, Z_p) \otimes H(\pi;\, Z_p).$$

Like in § 6, let e_i represent a generator of $H(\pi;\, Z_p)$; then all $e_i \otimes e_j$ with $i+j=k$ represent a basis for $H_k(G_z;\, Z_p)$. To compute τ_*, we construct $\tau_\#$: $M \otimes M \to M \otimes M$, a τ-chain mapping, by setting

$$\tau_\#(e_i \otimes e_j) = (-1)^{ij} e_j \otimes e_i \tag{17.1}$$

for elements of the G_z-basis and then extending it equivariantly over $M \otimes M$. It is easy to verify that $\partial\tau_\# = \tau_\#\partial$ and by definition

$$\tau_\#(\alpha c) = \tau(\alpha)\, \tau_\#(c).$$

Hence, $\tau_\#$ induces (cf. § 5) τ_*. We obtain

$$\tau_*(e_i \otimes e_j) = (-1)^{ij} e_j \otimes e_i. \tag{17.2}$$

18. The map $\xi_\#$

First we consider a map $\phi_\#$ that will be used for constructing the $\xi_\#$ mentioned at the end of § 16. As in § 16, let $\xi\colon G_z \to G$ denote the inclusion homomorphism. As defined above, $M \otimes M$ is a G_z-free acyclic complex; and the group G operates on M^p in the form described in § 12, so that M^p is a G-complex. Let C be the carrier from $M \otimes M$ to M^p defined by

$$C[(x^r e_j) \otimes (x^s e_j)] = \overline{(x^s e_j)^p}, \tag{18.1}$$

where $(x^s e_j)^p$ denotes the p-fold tensor product. Clearly C is an acyclic ξ-carrier. Thus there exists (cf. (3.1))

$$\phi_\#\colon M \otimes M \to M^p, \tag{18.2}$$

a ξ-chain mapping carried by C. The definition of C implies

$$\phi_\#(e_i \otimes e_j) \neq 0 \quad \text{only if} \quad p(j-1) \geqq i \geqq 0,$$

$$\phi_\#(e_{p(j-1)} \otimes e_j) = n.(e_j)^p \quad \text{for some integer } n.$$

Let $M^p \otimes M$ be the G-free acyclic complex of (12.1); then, since $M \otimes M$ is a G_z-free acyclic complex, by (3.1) there exists

$$\xi_\#\colon M \otimes M \to M^p \otimes M,$$

a ξ-chain mapping. A map $\xi_\#$ can actually be constructed by means of the maps $d_\#$, $\tau_\#$ and $\phi_\#$, respectively defined in (6.12), (6.13), (17.1) and (18.2). Consider the compositions

$$M \otimes M \xrightarrow{d_\# \otimes 1} M \otimes M \otimes M \xrightarrow{1 \otimes \tau_\#} M \otimes M \otimes M \xrightarrow{\phi_\# \otimes 1} M^p \otimes M.$$

Then we have the following:

(18.3) THEOREM. *The map $\xi_\#$ defined by*

$$\xi_\# = (\phi_\# \otimes 1)(1 \otimes \tau_\#)(d_\# \otimes 1),$$

is a ξ-chain mapping.

The proof is only to verify equivariance and it follows automatically.

Using the expressions for $d_\#$ and $\tau_\#$ given in (6.12), (6.13) and (17.1), we obtain

$$\begin{aligned} \xi_\#(e_{2r} \otimes e_s) = {} & \textstyle\sum_{k=0}^{r} [\phi_\#(e_{2k} \otimes e_s)] \otimes e_{2r-2k} \\ & + (-1)^s \textstyle\sum_{k=0}^{r-1} \Omega[\phi_\#(e_{2k+1} \otimes e_s)] \otimes e_{2r-2k-1}, \end{aligned} \tag{18.4}$$

$$\begin{aligned} \xi_\#(e_{2r+1} \otimes e_s) = \textstyle\sum_{k=0}^{r} \{ & (-1)^s [\phi_\#(e_{2k} \otimes e_s)] \otimes e_{2r-2k+1} \\ & + [\phi_\#(e_{2k+1} \otimes e_s)] \otimes x e_{2r-2k}\}, \end{aligned} \tag{18.5}$$

where $\Omega[\phi_{\#}(e_k \otimes e_s)] \otimes e_t = \sum [x^i . \phi_{\#}(e_k \otimes e_s)] \otimes x^j e_t$, summed over the range $p-1 \geqq j > i \geqq 0$.

In order to obtain the ξ_* of (16.3), we consider the induced map $(\xi_{\#})_{G_z}\colon (M \otimes M)_{G_z} \to (M^p \otimes M)_G$. First we have

$$(M \otimes M)_{G_z} = ((M \otimes M)_{\pi'})_\pi = (M \otimes (M_\pi))_\pi = (M_\pi) \otimes (M_\pi),$$

$$(M^p \otimes M)_G = ((M^p \otimes M)_\rho)_\pi = ((M_\pi)^p \otimes M)_\pi = (M_\pi)^p \otimes_\pi M.$$

Obviously, $(\xi_{\#})_{G_z} = ((\xi_{\#})_{\pi'})_\pi$ and $(\xi_{\#})_{\pi'}\colon (M \otimes M)_{\pi'} \to (M^p \otimes M)_\rho$. Set $\tilde{\xi}_{\#} = (\xi_{\#})_{\pi'}$, then $\tilde{\xi}_{\#}\colon M \otimes (M_\pi) \to (M_\pi)^p \otimes M$, and we have the commutative diagram

$$\text{(18.6)} \qquad \begin{array}{ccc} M \otimes M & \xrightarrow{\xi_{\#}} & M^p \otimes M \\ \downarrow{\scriptstyle 1 \otimes \eta_{\#}} & & \downarrow{\scriptstyle \eta_{\#}^p \otimes 1} \\ M \otimes (M_\pi) & \xrightarrow{\tilde{\xi}_{\#}} & (M_\pi)^p \otimes M \end{array}$$

where $\eta_{\#}\colon M \to M_\pi$ denotes the natural map.

Let $(\tau_{\#})_{\pi'}\colon (M \otimes M)_{\pi'} \to (M \otimes M)_{\tau(\pi')}$, $(\phi_{\#})_{\pi'}\colon (M \otimes M)_{\pi'} \to (M^p)_\rho$, respectively, be the map induced by $\tau_{\#}$, $\phi_{\#}$. Set

$$\tilde{\tau}_{\#} = (\tau_{\#})_{\pi'}, \quad \tilde{\phi}_{\#} = (\phi_{\#})_{\pi'}, \quad \text{then} \quad \tilde{\tau}_{\#}\colon M \otimes (M_\pi) \to (M_\pi) \otimes M$$

and

$$\text{(18.7)} \qquad \tilde{\phi}_{\#}\colon M \otimes (M_\pi) \to (M_\pi)^p.$$

The following diagram is commutative:

$$\begin{array}{ccccccc} M \otimes M & \xrightarrow{d_{\#} \otimes 1} & M \otimes M \otimes M & \xrightarrow{1 \otimes \tau_{\#}} & M \otimes M \otimes M & \xrightarrow{\phi_{\#} \otimes 1} & M^p \otimes M \\ \downarrow{\scriptstyle 1 \otimes \eta_{\#}} & & \downarrow{\scriptstyle 1 \otimes 1 \otimes \eta_{\#}} & & \downarrow{\scriptstyle 1 \otimes \eta_{\#} \otimes 1} & & \downarrow{\scriptstyle \eta_{\#}^p \otimes 1} \\ M \otimes (M_\pi) & \xrightarrow{d_{\#} \otimes 1} & M \otimes M \otimes (M_\pi) & \xrightarrow{1 \otimes \tilde{\tau}_{\#}} & M \otimes (M_\pi) \otimes M & \xrightarrow{\tilde{\phi}_{\#} \otimes 1} & (M_\pi)^p \otimes M \end{array}$$

Hence from (18.3) and (18.6) it follows that

$$\tilde{\xi}_{\#} = (\tilde{\phi}_{\#} \otimes 1)(1 \otimes \tilde{\tau}_{\#})(d_{\#} \otimes 1).$$

Then $\tilde{\xi}_{\#}$ can be expressed in terms of $\tilde{\phi}_{\#}$, $\tilde{\tau}_{\#}$ and $d_{\#}$; however, in contrast with $\tilde{\tau}_{\#}$ and $d_{\#}$, the map $\tilde{\phi}_{\#}$ is not explicitly known. As we will see, enough information about $\tilde{\phi}_{\#}$ can be obtained so that ξ_* can be computed.

19. The structure of $\tilde{\phi}_\#$

All our considerations will be modulo p; up to a natural isomorphism we substitute M_π by $\tilde{M} = Z_p \otimes (M_\pi)$ and regard $\tilde{\phi}_\#\colon M \otimes \tilde{M} \to \tilde{M}^p$. As in (6.1), let Δ_x, Σ_x, Δ_y, Σ_y be the corresponding elements of $Z(G_z)$, respectively, defined with the generators x, y, e.g. $\Delta_x = x - 1$, $\Delta_y = y - 1$, etc. We have

$$\begin{aligned}\partial\phi_\#(e_{2k+2} \otimes e_{2s+1}) &= \phi_\#\partial(e_{2k+2} \otimes e_{2s+1})\\ &= \phi_\#((\Sigma e_{2k+1}) \otimes e_{2s+1} + e_{2k+2} \otimes \Delta e_{2s})\\ &= \Sigma_x\phi_\#(e_{2k+1} \otimes e_{2s+1}) + \Delta_y\phi_\#(e_{2k+2} \otimes e_{2s}),\end{aligned}$$

and, after reducing under π', we obtain

$$\partial\tilde{\phi}_\#(e_{2k+2} \otimes e_{2s+1}) = \Sigma_x\tilde{\phi}_\#(e_{2k+1} \otimes e_{2s+1}),$$

but $\partial = 0$ in $\tilde{M}^p$, thus $\Sigma_x\tilde{\phi}_\#(e_{2k+1} \otimes e_{2s+1}) = 0$. Set $\Sigma = \Sigma_x$, $\Delta = \Delta_x$. Similarly we prove (mod p)

$$\Delta\tilde{\phi}_\#(e_{2k} \otimes e_s) = 0, \quad \Sigma\tilde{\phi}_\#(e_{2k+1} \otimes e_s) = 0.$$

Therefore a direct translation of (13.2) yields

(19.1) LEMMA. *The modulo p cycle $\tilde{\phi}_\#(e_k \otimes e_s)$ has the following structure:*

$$\tilde{\phi}_\#(e_{2k} \otimes e_s) = \nu(2k, s) + \Sigma\mu(2k, s),$$

$$\tilde{\phi}_\#(e_{2k+1} \otimes e_s) = \nu(2k+1, s) + \Delta\mu(2k+1, s),$$

where $\nu(k, s)$, $\mu(k, s) \in C_{k+s}(\tilde{M}^p)$. The term $\nu(k, s)$ is different from zero only if $k + s = pi$, and in this case

$$\nu(k, s) = n_{k,s}(e_i)^p,$$

where $n_{k,s}$ is certain mod p *integer.*

20. The map ξ_*

The substitution of $\phi_\#$ by $\tilde{\phi}_\#$ in (18.4) and (18.5) gives expressions for $\tilde{\xi}_\#$, and we may further substitute $\tilde{\phi}_\#$ by the expression of (19.1). The map $\xi_*\colon H(\tilde{M} \otimes \tilde{M}) \to H(\tilde{M}^p \otimes_\pi M)$ is the one induced by $(\tilde{\xi}_\#)_\pi\colon \tilde{M} \otimes \tilde{M} \to \tilde{M}^p \otimes_\pi M$. As we will show, the terms $\Sigma\mu$, $\Delta\mu$ of $\tilde{\phi}_\#$ can be ignored under the relations of $H(\tilde{M}^p \otimes_\pi M)$. This is clear from the following:

(20.1) LEMMA. *Let $\mu \in \tilde{M}^p$, then in $\tilde{M}^p \otimes_\pi M$ we have*

$$(\Delta\mu) \otimes xe_{2r} \sim 0, \quad (\Sigma\mu) \otimes e_{2r+1} \sim 0,$$

$$\Omega[(\Delta\mu) \otimes e_{2r+1}] \sim 0, \quad (\Sigma\mu) \otimes e_{2r} \sim 0.$$

PROOF. As we have already shown (cf. (13.5)),

$$(\Delta\mu)\otimes xe_{2r} = -\mu\otimes\Delta e_{2r} = (-1)^{1+\dim\mu}\partial(\mu\otimes e_{2r+1}),$$
$$(\Sigma\mu)\otimes e_{2r+1} = \mu\otimes\Sigma e_{2r+1} = (-1)^{\dim\mu}\partial(\mu\otimes e_{2r+2}).$$

Now, using (6.4) it follows

$$\Omega[(\Delta\mu)\otimes e_{2r+1}] = -\mu\otimes\Sigma e_{2r+1} = (-1)^{1+\dim\mu}\partial(\mu\otimes e_{2r+2}).$$

Finally, from (6.2), we have

$$(\Sigma\mu)\otimes e_{2r} = -(\Delta\Sigma'\mu)\otimes e_{2r} = (\Sigma'\mu)\otimes x^{-1}\Delta e_{2r}$$
$$= (-1)^{\dim\mu}\partial[(\Sigma'\mu)\otimes x^{-1}e_{2r+1}].$$

Thus (20.1) is proved.

Obviously, $x[\nu(k,s)] = \nu(k,s)$. Hence under the relations of $\tilde{M}^p\otimes_\pi M$ we have

(20.2) $$\Omega[\nu(k,s)\otimes e_r] = \tfrac{1}{2}p(p-1)\,\nu(k,s)\otimes e_r.$$

Therefore, with the substitutions mentioned above and taking in consideration (20.1) and (20.2), we obtain ($p \geqq 2$)

(20.3) $$\xi_*(e_{2r}\otimes e_s) = \textstyle\sum_{k=0}^{r}\nu(2k,s)\otimes e_{2r-2k} + \tfrac{1}{2}p(p-1)\sum_{k=0}^{r-1}\nu(2k+1,s)\otimes e_{2r-2k-1},$$

(20.4) $$\xi_*(e_{2r+1}\otimes e_s) = \textstyle\sum_{k=0}^{r}\{(-1)^s\nu(2k,s)\otimes e_{2r-2k+1} + \nu(2k+1,s)\otimes e_{2r-2k}\}.$$

We note that the case $p=2$ is special and (20.3) and (20.4) combine into the following sole expression:

(20.5) $$\xi_*(e_r\otimes e_s) = \textstyle\sum_{k=0}^{r}\nu(k,s)\otimes e_{r-k} \quad (p=2).$$

Finally, for $p>2$, (20.3) becomes

(20.6) $$\xi_*(e_{2r}\otimes e_s) = \textstyle\sum_{k=0}^{r}\nu(2k,s)\otimes e_{2r-2k}.$$

21. The computation of $\nu(k, s)$

(21.1) LEMMA. *The map* $\tilde{\phi}_\#$ *of* (18.7) *is a π-chain map carried by the diagonal carrier, and it can be used to define cyclic reduced powers in* M_π (cf. (7.1)); *that is, if* $e^i\in C^i(\tilde{M})$ *is the dual cocycle of* $e_i\in C_i(\tilde{M})$, *then* $(e^i)^p/e_k = \{\tilde{\phi}^\#(e^i)^p/e_j\}$.

PROOF. Clearly, $\{\tilde{\phi}^\#(e^i)^p/e_j\}$ is independent of the particular chosen map $\phi_\#$, since any two such maps are equivariantly homotopic. This should be sufficient; however, the standard definition of reduced powers is stated for regular complexes and M_π is not regular. The situation is analogous to that of (4.4), and in similar form we will show that both definitions agree.

Let M' be the derived of M and assume that $f_{\#}: M \to M'$, $g_{\#}: M' \to M$ are equivariant maps as defined in (4.3). Then $g_{\#}f_{\#} = 1$ and $f_{\#}g_{\#} \simeq 1$ equivariantly. Let $\phi'_{\#}: M \otimes (M') \to (M')^p$ be a chosen equivariant map analogous to that of (18.2). The reduced map

$$\tilde{\phi}'_{\#}: M \otimes (M'_{\pi}) \to (M'_{\pi})^p$$

is a π-chain map carried by the diagonal carrier, and, since (M'_{π}) is regular (cf. (4.2)), it can be used to define reduced powers in (M'_{π}). Set $\phi_{\#} = g^p_{\#}\phi'_{\#}(1 \otimes f_{\#})$. Then it is easy to verify that $\phi_{\#}: M \otimes M \to M^p$ can be chosen as the equivariant map of (18.2). We have

$$\tilde{\phi}_{\#} = \tilde{g}^p_{\#}\tilde{\phi}'_{\#}(1 \otimes \tilde{f}_{\#}), \quad \text{where} \quad \tilde{f}_{\#}: M_{\pi} \to M'_{\pi}, \quad \tilde{g}_{\#}: M'_{\pi} \to M_{\pi}$$

denotes the respective reduced map. Hence

$$\tilde{\phi}^{\#}(e^i)^p / e_j = \tilde{f}^{\#}(\tilde{\phi}'^{\#}(\tilde{g}^{\#}e^i)^p / e_j),$$

and passing to cohomology classes the lemma follows.

Now, to compute $\nu(k, s)$, observe that

$$(21.2) \qquad (\tilde{\phi}^{\#}(e^i)^p / e_j) \cdot e_{ip-j} = (e^i)^p \cdot \tilde{\phi}_{\#}(e_j \otimes e_{ip-j}) = (e^i)^p \cdot \nu(j, ip-j).$$

On the other hand, the cyclic reduced powers are known in $\tilde{M}$ and these two facts can be combined to obtain $\nu(k, s)$ as follows. First consider the case $p = 2$. From (9.3) and (21.2) we have

$$(\tilde{\phi}^{\#}(e^{s-k})^2 / e_{s-2k}) \cdot e_s = (e^{s-k})^2 \cdot \nu(s-2k, s) = \binom{s-k}{k}.$$

Therefore

$$(21.3) \qquad \nu(s-2k, s) = \binom{s-k}{k}(e_{s-k})^2,$$

and $\nu(r, s) = 0$ otherwise (cf. (19.1)).

In similar form, for $p > 2$, using (9.4), (9.5) and (9.7), we obtain

$$(21.4) \qquad \nu((2s-2kp)(p-1), 2s) = (-1)^{s+k}\binom{s-k(p-1)}{k}(e_{2s-2k(p-1)})^p,$$

$$(21.5) \qquad \nu((2s+1-2kp)(p-1), 2s+1) = (-1)^{s+k} m! \binom{s-k(p-1)}{k}(e_{2s+1-2k(p-1)})^p,$$

$$(21.6) \qquad \nu((2s-1-2kp)(p-1)-1, 2s) = (-1)^{s+k} m! \binom{s-1-k(p-1)}{k}(e_{2s-1-2k(p-1)})^p,$$

and $\nu(r, s) = 0$ otherwise (cf. (9.6) and (19.1)).

Then since $\nu(r,s)$ is known, the expressions (20.3) and (20.4) give ξ_* explicitly and we can write all elements of the form (16.4). They are generated by the elements

$$\xi_*(e_i \otimes e_j) - (-1)^{ij} \xi_*(e_j \otimes e_i).$$

Hence in $H(S_{p^2}; Z_p)$ we have the relations

$$\theta_* \xi_*(e_i \otimes e_j) = (-1)^{ij} \theta_* \xi_*(e_j \otimes e_i). \tag{21.7}$$

Consider the various cases

$$\theta_* \xi_*(e_{2i} \otimes e_{2j}) = \theta_* \xi_*(e_{2j} \otimes e_{2i}), \tag{21.8}$$

$$\theta_* \xi_*(e_{2i} \otimes e_{2j+1}) = \theta_* \xi_*(e_{2j+1} \otimes e_{2i}), \tag{21.9}$$

$$\theta_* \xi_*(e_{2i+1} \otimes e_{2j+1}) = -\theta_* \xi_*(e_{2j+1} \otimes e_{2i+1}). \tag{21.10}$$

From the explicit expression for ξ_* and Lemma (14.7), it follows that

$$\theta_* \xi_*(e_i \otimes e_j) \neq 0 \quad \text{only if} \quad \begin{cases} i = 2r(p-1) & \text{or} \quad 2r(p-1)-1, \\ j = 2s(p-1) & \text{or} \quad 2s(p-1)-1. \end{cases} \tag{21.11}$$

Let ∂^* be the boundary operator associated to (6.9). Since (21.10) can be derived as ∂_* of (21.9), only the relations (21.8) and (21.9) will be considered. Using (21.11) they reduced to the following:

$$\theta_* \xi_*(e_{2r(p-1)} \otimes e_{2s(p-1)}) = \theta_* \xi_*(e_{2s(p-1)} \otimes e_{2r(p-1)}), \tag{21.12}$$

$$\theta_* \xi_*(e_{2r(p-1)} \otimes e_{2s(p-1)-1}) = \theta_* \xi_*(e_{2s(p-1)-1} \otimes e_{2r(p-1)}). \tag{21.13}$$

22. Reduction of the relations

The substitution of (21.3) in (20.5) gives

$$\xi_*(e_r \otimes e_s) = \sum_{k=0}^{[(r+s)/2]} \binom{k}{s-k} (e_k)^2 \otimes e_{r+s-2k} \quad (p=2). \tag{22.1}$$

Similarly for $p > 2$, replacing (21.4) and (21.6) in (20.4) and (20.6), we obtain

$$\begin{aligned} &\xi_*(e_{2r(p-1)} \otimes e_{2s(p-1)}) \\ &\quad = \sum_{k=0}^{[(r+s)/p]} (-1)^{s+k} \binom{k(p-1)}{s-k} (e_{2k(p-1)})^p \otimes e_{(2r+2s-2kp)(p-1)}, \end{aligned} \tag{22.2}$$

$$\begin{aligned} &\xi_*(e_{2r(p-1)} \otimes e_{2s(p-1)-1}) \\ &\quad = \sum_{k=0}^{[(r+s)/p]} (-1)^{s+k+1} m! \binom{k(p-1)-1}{s-k} (e_{2k(p-1)-1})^p \otimes e_{(2r+2s+1-2kp)(p-1)}. \end{aligned} \tag{22.3}$$

Finally, the expression for $\xi_*(e_{2r(p-1)-1} \otimes e_{2s(p-1)})$ can be obtained from (22.2) and (22.3) and the relation

$$\xi_*(e_{2i-1} \otimes e_{2j}) = \partial_* \xi_*(e_{2i} \otimes e_{2j}) - \xi_*(e_{2i} \otimes e_{2j-1}). \tag{22.4}$$

In order to work with these expressions we introduce the following notation:

$$(22.5)\qquad x_t^k = \begin{cases} \theta_*[(e_k)^2 \otimes e_{t-2k}] & \text{for } p=2, \\ \theta_*[(e_{2k(p-1)})^p \otimes e_{(2t-2kp)(p-1)}] & \text{for } p>2, \end{cases}$$

$$(22.6)\qquad y_t^k = -m!\,\theta_*[(e_{2k(p-1)-1}]^p \otimes e_{(2t+1-2kp)(p-1)}],$$

$$w_t^s = \begin{cases} \theta_*\,\xi_*(e_{t-s} \otimes e_s) & \text{for } p=2, \\ \theta_*\,\xi_*(e_{(2t-2s)(p-1)} \otimes e_{2s(p-1)}) & \text{for } p>2, \end{cases}$$

$$z_t^s = \theta_*\,\xi_*(e_{(2t-2s)(p-1)} \otimes e_{2s(p-1)-1}).$$

It follows easily from (22.5) that

$$(22.7)\qquad \partial_* x_t^k = \begin{cases} \theta_*[(e_k)^2 \otimes e_{t-2k-1}] & \text{for } p=2, \\ \theta_*[(e_{2k(p-1)})^p \otimes e_{(2t-2kp)(p-1)-1}] & \text{for } p>2. \end{cases}$$

For the binomial coefficients we set

$$(22.8)\qquad a_k^s = (-1)^{s+k}\binom{k(p-1)}{s-k},$$

$$(22.9)\qquad b_k^s = (-1)^{s+k}\binom{k(p-1)-1}{s-k}.$$

Then, from (22.1), (22.2) and (22.3), we obtain

$$(22.10)\qquad w_t^s = \textstyle\sum_{k=0}^{[t/p]} a_k^s x_t^k \quad (p \geqq 2),$$

$$(22.11)\qquad z_t^s = \textstyle\sum_{k=0}^{[t/p]} b_k^s y_t^k \quad (p>2).$$

The relations (21.7) for $p=2$ and (21.12) for $p>2$ are expressed by

$$(22.12)\qquad w_t^s = w_t^{t-s} \quad (p \geqq 2),$$

and the relations (21.13) become (cf. (22.4))

$$(22.13)\qquad z_t^s = \partial_* w_t^{t-s} - z_t^{t-s} \quad (p>2).$$

For *fixed* t and $0 \leqq s, k \leqq [t/p]$, set

$$a_k^{*s} = a_k^{t-s}, \quad b_k^{*s} = b_k^{t-s}.$$

From the identity $\binom{m}{n} = \binom{m}{m-n}$, it follows that

$$(22.14)\qquad a_k^{*s} = (-1)^{t+s+k}\binom{k(p-1)}{kp+s-t},$$

$$(22.15)\qquad b_k^{*s} = (-1)^{t+s+k}\binom{k(p-1)-1}{kp+s-t-1} \quad (k>0).$$

Consider the matrices $\| a_k^s \|$, $\| a_k^{*s} \|$, $\| b_k^s \|$, $\| b_k^{*s} \|$ and the vectors

$$\| x_t^s \| = (x_t^0, \ldots, x_t^{[t/p]}), \quad \| y_t^s \| = (y_t^0, \ldots, y_t^{[t/p]}),$$

where $0 \leqq s, k \leqq [t/p]$. The relations (22.12) and (22.13), for $s \leqq [t/p]$ can be expressed in matrix notation as follows (cf. (22.10) and (22.11)):

$$\| a_k^s \| \, \| x_t^s \| = \| a_k^{*s} \| \, \| x_t^s \|,$$

$$\| b_k^s \| \, \| y_t^s \| = \| a_k^{*s} \| \, \| \partial_* x_t^s \| - \| b_k^{*s} \| \, \| y_t^s \|.$$

We have $a_k^k = b_k^k = 1$ and $a_k^s = b_k^s = 0$ for $k > s$, thus $| a_k^s | = | b_k^s | = 1$. Therefore

$$\| x_t^s \| = \| a_k^s \|^{-1} \| a_k^{*s} \| \, \| x_t^s \| \qquad (p \geqq 2),$$

$$\| y_t^s \| = \| b_k^s \|^{-1} (\| a_k^{*s} \| \, \| \partial_* x_t^s \| - \| b_k^{*s} \| \, \| y_t^s \|) \qquad (p > 2).$$

Set

$$\| c_k^s \| = \| a_k^s \|^{-1} \| a_k^{*s} \|,$$

$$\| d_k^s \| = \| b_k^s \|^{-1} \| a_k^{*s} \|,$$

$$\| e_k^s \| = \| b_k^s \|^{-1} \| b_k^{*s} \|.$$

Then

(22.16) $$x_t^s = \sum_{k=0}^{[t/p]} c_k^s x_t^k,$$

(22.17) $$y_t^s = \sum_{k=0}^{[t/p]} d_k^s \partial_* x_t^k - \sum_{k=0}^{[t/p]} e_k^s y_t^k.$$

(22.18) Lemma. *The elements of the matrices* $\| c_k^s \|$, $\| d_k^s \|$, $\| e_k^s \|$ *can be described as follows:*

$$c_k^s = (-1)^{k+s+t} \binom{(k-s)(p-1)-1}{kp+s-t},$$

$$d_k^s = (-1)^{k+s+t} \binom{(k-s)(p-1)}{kp+s-t},$$

$$e_k^s = (-1)^{k+s+t} \binom{(k-s)(p-1)-1}{kp+s-t-1}.$$

Proof. We have $\| a_k^s \| \, \| c_k^s \| = \| a_k^{*s} \|$, then

$$\sum_{k=0}^{[t/p]} a_k^s c_r^k = a_r^{*s}.$$

And, by (22.8) and (22.14),

$$\sum_{k=0}^{[t/p]} (-1)^k \binom{k(p-1)}{s-k} c_r^k = (-1)^{t+r} \binom{r(p-1)}{rp+s-t}.$$

Now we use Theorem (25.3) (cf. Appendix). First, for a, b, c of (25.3) we substitute $a = s(p-1)$, $b = r(p-1) - a - 1$ and $c = rp + s - t$. We obtain

$$\sum_{i=0}^{rp+s-t} \binom{(s-i)(p-1)}{i} \binom{(r-s+i)(p-1)-1}{rp+s-t-i} = \binom{r(p-1)}{rp+s-t}.$$

Since in our case $0 \leqq r, s \leqq [t/p]$, it follows that $rp - t \leqq 0$ and $s - [t/p] \leqq 0$. Then the range $0 \leqq i \leqq rp + s - t$ can be replaced by $s - [t/p] \leqq i \leqq s$. And changing the running index i to $k = s - i$ we obtain

$$\sum_{k=0}^{[t/p]} \binom{k(p-1)}{s-k} \binom{(r-k)(p-1)-1}{rp-t+k} = \binom{r(p-1)}{rp+s-t}.$$

Then the first case of (22.18) follows from this and from one of the above expressions. The other two cases are similar and the proofs will be omitted.

Consider the generators of $H(S_{p^2}; Z_p)$ described in (14.8). Clearly, up to a numerical factor, the elements x_t^s, $\partial_* x_t^s$, y_t^s belong to this set of generators (cf. (22.5), (22.6) and (22.7)).

(22.19) THEOREM. *Among the various generators of* $H(S_{p^2}; Z_p)$ *the following relations hold:*

(22.20) *If* $sp + s > t$ *and* $p \geqq 2$, *then*

$$x_t^s = \sum_{k=[(t-s)/p]}^{[t/p]} (-1)^{k+s+t} \binom{(k-s)(p-1)-1}{kp+s-t} x_t^k.$$

(22.21) *If* $sp + s \geqq t$ *and* $p > 2$, *then*

$$y_t^s = \sum_{k=[(t-s)/p]}^{[t/p]} (-1)^{k+s+t} \binom{(k-s)(p-1)}{kp+s-t} \partial_* x_t^k$$
$$- \sum_{k=[(t-s+1)/p]}^{[t/p]} (-1)^{k+s+t} \binom{(k-s)(p-1)-1}{kp+s-t-1} y_t^k.$$

PROOF. It follows immediately from (22.16), (22.17) and (22.18). The restriction on s is only made in order to have either $s < k$ or $s \leqq k$ for all k; so that the same element does not appear in the two sides of the same equation.

23. The relations on Steenrod powers

The relations (22.19) can be translated to relations on iterated reduced powers. In order to do so we first consider some auxiliary statements. Let $p = 2$ and $u \in H^q(K; Z_2)$; combining (7.6), (8.4), (8.9), (15.3) and (22.5) we obtain

(23.1) $$u^4 / x_t^k = (u^2/e_k)^2 / e_{t-2k} = \mathrm{Sq}^{2q+k-t} \mathrm{Sq}^{q-k} u.$$

For $p > 2$ we are restricted to $Z_p = Z_p^{(2q)}$ as group of coefficients. Thus, assume $u \in H^{2q}(K; Z_p)$ for this case. In analogous fashion to the preceding case we establish

(23.2) $$u^{p^2} / x_t^k = (u^p / e_{2k(p-1)})^p / e_{(2t-2kp)(p-1)},$$

(23.3) $$u^{p^2} / y_t^k = (-1)^{m+k} m! \, [\delta^*(u^p / e_{2k(p-1)})]^p / e_{(2t+1-2kp)(p-1)}.$$

And, using (8.1) and (22.7), it follows that

$$(23.4) \qquad u^{p^2}/\partial_* x_t^k = -\delta^*(u^{p^2}/x_t^k).$$

On the other hand, we have that

$$v^p/e_{2k(p-1)} = (-1)^k \mathcal{P}^{q-k} v \quad \text{for} \quad v \in H^{2q}(K; Z_p),$$
$$v^p/e_{(2k+1)(p-1)} = (-1)^k m!\, \mathcal{P}^{q-k} v \quad \text{for} \quad v \in H^{2q+1}(K; Z_p).$$

This follows directly from (9.2) after using (9.1) to reduce the numerical factor. Thus the expressions (23.2) and (23.3) become

$$(23.5) \qquad u^{p^2}/x_t^k = (-1)^t \mathcal{P}^{pq+k-t}\mathcal{P}^{q-k}u,$$

$$(23.6) \qquad u^{p^2}/y_t^k = (-1)^{t+1} \mathcal{P}^{pq+k-t}\delta^*\mathcal{P}^{q-k}u.$$

(23.7) THEOREM. *The Steenrod p-powers satisfy the following relations:*

Case $p=2$. Let $a<2b$, then

$$(23.8) \qquad \mathrm{Sq}^a\,\mathrm{Sq}^b = \sum_{i=0}^{[a/2]} \binom{b-i-1}{a-2i} \mathrm{Sq}^{a+b-i}\,\mathrm{Sq}^i.$$

Case $p>2$. Let $a<pb$, then

$$(23.9) \qquad \mathcal{P}^a\mathcal{P}^b = \sum_{i=0}^{[a/p]} (-1)^{a+i} \binom{(b-i)(p-1)-1}{a-pi} \mathcal{P}^{a+b-i}\mathcal{P}^i.$$

Let $a \leqq pb$, then

$$(23.10) \qquad \mathcal{P}^a\delta^*\mathcal{P}^b = \sum_{i=0}^{[a/p]} (-1)^{a+i} \binom{(b-i)(p-1)}{a-pi} \delta^*\mathcal{P}^{a+b-i}\mathcal{P}^i$$
$$+ \sum_{i=0}^{[(a-1)/p]} (-1)^{a+i-1} \binom{(b-i)(p-1)-1}{a-pi-1} \mathcal{P}^{a+b-i}\delta^*\mathcal{P}^i.$$

PROOF. Assume $p=2$ and $u \in H^q(K; Z_2)$. From (22.20) and (23.1) it follows that

$$\mathrm{Sq}^{2q+s-t}\,\mathrm{Sq}^{q-s}u = \sum_{k=[(t-s)/2]}^{[t/2]} \binom{k-s-1}{2k+s-t} \mathrm{Sq}^{2q+k-t}\,\mathrm{Sq}^{q-k}u.$$

Set $i=q-k$. Then, since $t-s \leqq 2k \leqq t$, we have $2q-t \leqq 2i \leqq 2q-t+s$. On the other hand,

$$\mathrm{Sq}^{2q+k-t}\,\mathrm{Sq}^{q-k}u = \mathrm{Sq}^{3q-i-t}\,\mathrm{Sq}^i u = 0 \quad \text{if} \quad 3q-i-t > i+q$$

(cf. (8.8)), that is, if $2i < 2q-t$. Consequently the index $2i$ may be considered in the range $0 \leqq 2i \leqq 2q-t+s$. Setting $a=2q-t+s$ and $b=q-s$, the above expression yields (23.8).

The other two parts are proved in a similar fashion using (22.19), (23.4), (23.5) and (23.6). They are established first for $u \in H^{2q}(K; Z_p)$ and then in general by means of (10.4). The details are left to the reader.

The formula (23.8), in an equivalent form, was announced by the author in [1]. Originally it was conjectured by Wu and Thom ([21], p. 141). A new proof of (23.8) is due to Serre [16]; this is based on Serre's computation of the Eilenberg-MacLane groups $H^*(Z_2; q, Z_2)$. The formulae (23.9) and (23.10) were established independently and with different methods by the author [2] and H. Cartan (unpublished).† Cartan's method is based on his computation of $H^*(Z_p; q, Z_p)$ (cf. [5]).

24. Two theorems on generators for p-powers

We use the convention $\mathcal{P}^i = \mathrm{Sq}^i$ when $p=2$. Then (23.9) holds for $p \geqq 2$ and (23.8) is included in it. We will consider some particular cases. Suppose $0 \leqq a, b < p$ in (23.9), then

$$\mathcal{P}^a\mathcal{P}^b = (-1)^a \binom{b(p-1)-1}{a} \mathcal{P}^{a+b}.$$

But, since $0 \leqq a, b < p$,

$$(-1)^a \binom{b(p-1)-1}{a} \equiv (-1)^a \binom{-b-1}{a} \quad (\operatorname{mod} p).$$

Hence (cf. Appendix)

$$\mathcal{P}^a\mathcal{P}^b = \binom{a+b}{a} \mathcal{P}^{a+b}.$$

In particular, $\mathcal{P}^1\mathcal{P}^b = (1+b)\mathcal{P}^{1+b}$ and by iteration we obtain

$$(\mathcal{P}^1)^c = c!\,\mathcal{P}^c \quad (p \geqq 2), \tag{24.1}$$

where $(\mathcal{P}^j)^r$ means $\mathcal{P}^j$ iterated r times ((24.1) was conjectured by A. Borel and J. P. Serre [4], p. 424).

(24.2) Theorem. *The set* $\{\mathcal{P}^{p^k}\}$ $(k = 0, 1, \ldots)$ *constitutes a basis for p-powers in the sense that any* $\mathcal{P}^c$ *can be expressed as a sum of iterated p-powers of exponents powers of p.*

Proof. This is by induction on c. The theorem is true for $0 < c \leqq p$ (cf. (24.1)). Assume it is true for all $h < c$ and c is not a power of p. Let $c = a + b$ for some a, b such that $a < pb$. From (23.9) it follows that

$$(-1)^a \binom{b(p-1)-1}{a} \mathcal{P}^c = \mathcal{P}^a\mathcal{P}^b - \sum_{i=1}^{[a/p]} (-1)^{a+i} \binom{(b-i)(p-1)-1}{a-pi} \mathcal{P}^{a+b-i}\mathcal{P}^i.$$

Hence, if $\binom{b(p-1)-1}{a} \not\equiv 0 \ (\operatorname{mod} p)$, this expression will provide the induction step of the argument.

† See footnote ‡ on p. 192.

Let p^j be the highest power of p dividing c. Then $c=p^j(r+ps)$ where $0<r<p$ and $s\geqq 0$. We have two cases. If $s=0$ then $r>1$ and $p>2$ (since c is not a power of p). Take $a=p^j(r-1)$ and $b=p^j$. Then

$$p^j(p-1)-1=p^j(p-2)+(1+\ldots+p^{j-1})(p-1),$$

and applying (25.1) we obtain that

$$\binom{b(p-1)-1}{a}\equiv\binom{p-2}{r-1}\not\equiv 0 \pmod p.$$

If $s>0$, take $a=p^jr$ and $b=p^{j+1}s$. We have

$$b(p-1)-1=p^{j+1}(ps-s-1)+(1+\ldots+p^j)(p-1),$$

and using (25.1) it follows that

$$(-1)^a\binom{b(p-1)-1}{a}\equiv(-1)^r\binom{p-1}{r}\equiv 1 \pmod p.$$

Therefore, in both cases the induction step can be completed; so (24.2) is proved.

Clearly, the proof gives an effective procedure to write any p-power in terms of elements of the basis. As illustration, we have

$$2!\,\mathscr{P}^{2p}=(\mathscr{P}^p)^2+(\mathscr{P}^1)^{p-1}\mathscr{P}^p\mathscr{P}^1,$$

$$3!\,\mathscr{P}^{3p}=(\mathscr{P}^p)^3+\mathscr{P}^p(\mathscr{P}^1)^{p-1}\mathscr{P}^p\mathscr{P}^1+(\mathscr{P}^1)^{p-1}(\mathscr{P}^p)^2\mathscr{P}^1.$$

(24.3) THEOREM. *The basic elements* $\{\mathscr{P}^{I_c}\}$ (cf. § 10) *form an additive set of generators for iterated p-powers.*

PROOF. First we observe that the relations (23.7) can be used to express $\mathrm{Sq}^{i_1}\mathrm{Sq}^{i_2}$, and $\mathscr{Q}^{\epsilon_1}\mathscr{P}^{i_1}\mathscr{Q}^{\epsilon_2}\mathscr{P}^{i_2}$ in terms of basic elements. Now, suppose given $\mathrm{Sq}^I=\mathrm{Sq}^{i_1}\mathrm{Sq}^{i_2}\ldots\mathrm{Sq}^{i_k}$ with $i_j>0$. If Sq^I is not basic, then $i_r<2i_{r+1}$ for some r and we can expand the factor $\mathrm{Sq}^{i_r}\mathrm{Sq}^{i_{r+1}}$ by means of (23.8). And Sq^I becomes equal to a sum of iterated squares where each term has k factors of the type Sq^j. If the factor Sq^0 appears in a term we agree to move it to the k position. This procedure can then be applied to each term of the sum that is not basic. We will show that after a finite number of steps this ends with a sum of only basic elements. Set

$$\alpha_s=i_1+\ldots+i_s;$$

then $\alpha_k=n(I)$ and $\alpha_1<\alpha_2<\ldots<\alpha_k$. If $\mathrm{Sq}^{j_1}\mathrm{Sq}^{j_2}\ldots\mathrm{Sq}^{j_k}$ is a term of the sum obtained after the first operation and if $\alpha'_s=j_1+\ldots+j_s$, then $\alpha_r<\alpha'_r$. From this and the fact that the α's are always bounded by $n(I)$, it is easy to see that after a finite number of substitutions all the terms of the total sum become basic. For $p>2$ the proof is the same, using (23.9) and (23.10) instead of (23.8).

Appendix

25. On binomial coefficients

The binomial coefficient $\binom{z}{n}$ is defined for all real values of z and all integer values of n, as follows. For all z and positive integers n,

$$\binom{z}{n} = \frac{z(z-1)\ldots(z-n+1)}{n!}.$$

For all z and $n=0$, $\binom{z}{0}=1$. For all z and negative integers n, $\binom{z}{n}=0$.

They have the following properties. For any positive integer m, $\binom{m}{n}=0$ when $n>m$. For any number z and any integer n,

$$\binom{z}{n-1}+\binom{z}{n}=\binom{z+1}{n}.$$

For any $z>0$,
$$\binom{-z}{n}=(-1)^n\binom{z+n-1}{n}.$$

In dealing with binomial coefficients modulo a prime p, the following theorem of Lucas [15] proves to be very useful:

(25.1) Theorem. *Let p be a prime, and let*

$$a=a_0+a_1p+\ldots+a_rp^r \quad (0\leqq a_i<p),$$
$$b=b_0+b_1p+\ldots+b_rp^r \quad (0\leqq b_i<p).$$

That is, the p-adic expansion of a and b. Then

$$\binom{a}{b}\equiv\binom{a_0}{b_0}\binom{a_1}{b_1}\cdots\binom{a_r}{b_r} \quad (\operatorname{mod} p).$$

For a proof see [11].

I have not been able to find the next theorems in the literature, although they are of simple formulation.

(25.2) Theorem. *Let $c>1$ be an integer and p a prime. The following congruence holds:*

$$\textstyle\sum_{k=0}^{c}\binom{-k(p-1)}{k}\binom{k(p-1)}{c-k}\equiv 0 \quad (\operatorname{mod} p).$$

(25.3) Theorem. *Let a, b, c be arbitrary integers and p a prime. The following congruence holds:*

$$\textstyle\sum_{k=0}^{c}\binom{a-k(p-1)}{k}\binom{b+k(p-1)}{c-k}\equiv\binom{a+b+1}{c} \quad (\operatorname{mod} p).$$

PROOF OF (25.2). We have $\binom{-k(p-1)}{k} = (-1)^k \binom{kp-1}{k}$. Then we can consider the sum only over terms $\binom{kp-1}{k}\binom{kp-1}{c-k} \not\equiv 0 \pmod p$. To characterize those elements we shall describe the k's such that

$$\binom{kp-1}{k} \not\equiv 0 \quad \text{and} \quad \binom{k(p-1)}{c-k} \not\equiv 0 \quad (\operatorname{mod} p).$$

Assume $k > 0$ and

$$k = k_r p^r + \ldots + k_s p^s, \quad \text{where} \quad 0 \leqq k_i < p, \quad 0 < k_r.$$

Then the p-adic expansion of $kp-1$ is

$$kp - 1 = (p-1)(1 + p + \ldots + p^r) + (k_r - 1) p^{r+1} + k_{r+1} p^{r+2} + \ldots + k_s p^{s+1},$$

and, by (25.1), we have (mod p)

$$\binom{pk-1}{k} \equiv \binom{p-1}{k_r}\binom{k_r - 1}{k_{r+1}}\binom{k_{r+1}}{k_{r+2}} \cdots \binom{k_{s-1}}{k_s}. \tag{25.4}$$

Thus, if we assume $\binom{pk-1}{k} \not\equiv 0$, it follows that

$$p > k_r > k_{r+1} \geqq k_{r+2} \geqq \ldots \geqq k_s \geqq 0.$$

From this we obtain the p-adic expansion

$$k(p-1) = (p - k_r) p^r + (k_r - k_{r+1} - 1) p^{r+1} + (k_{r+1} - k_{r+2}) p^{r+2} + \ldots + (k_{s-1} - k_s) p^s + k_s p^{s+1}.$$

Let

$$c - k = a_i p^i + \ldots + a_j p^j$$

be the p-adic expansion of $c-k$. If we further assume $\binom{k(p-1)}{c-k} \not\equiv 0$ then, by (25.1), we have $a_t = 0$ for $s+1 \geqq t \geqq r$, and (mod p)

(25.5)

$$\binom{k(p-1)}{c-k} \equiv \binom{p-k_r}{a_r}\binom{k_r - k_{r+1} - 1}{a_{r+1}}\binom{k_{r+1} - k_{r+2}}{a_{r+2}} \cdots \binom{k_{s-1} - k_s}{a_s}\binom{k_s}{a_{s+1}}.$$

Therefore, from $\binom{k(p-1)}{c-k} \not\equiv 0$, it follows that

$$\begin{aligned}
0 &< a_r + k_r \leqq p, \\
0 &\leqq a_{r+1} + k_{r+1} \leqq k_r - 1 < p, \\
0 &\leqq a_{r+2} + k_{r+2} \leqq k_{r+1} < p, \\
&\cdots\cdots\cdots\cdots\cdots\cdots \\
0 &\leqq a_s + k_s \leqq k_{s-1} < p, \\
0 &\leqq a_{s+1} \leqq k_s < p.
\end{aligned}$$

In order to write the p-adic expansion of c, we have to distinguish two cases:

(1) $0 < a_r + k_r < p$, then

$$c = (a_r + k_r)\,p^r + \ldots + (a_s + k_s)\,p^s + a_{s+1}p^{s+1}.$$

(2) $a_r + k_r = p$, then, since $0 \leqq a_{r+1} + k_{r+1} + 1 < p$,

$$c = (a_{r+1} + k_{r+1} + 1)\,p^{r+1} + (a_{r+2} + k_{r+2})\,p^{r+2} + \ldots + (a_s + k_s)\,p^s + a_{s+1}p^{s+1}.$$

Now, let

$$c = c_m p^m + \ldots + c_n p^n \quad (0 \leqq c_i < p,\ 0 < c_m,\ 0 < c_n),$$

be the p-adic expansion of c. The above shows that

$$\binom{kp-1}{k}\binom{k(p-1)}{c-k} \not\equiv 0$$

only if either

(Case 1) $\qquad k = k_m p^m + \ldots + k_n p^n,$

with $0 \leqq k_i \leqq c_i$, or

(Case 2) $\qquad k = k_{m-1}p^{m-1} + \ldots + k_n p^n \quad (m > 0),$

with $0 \leqq k_{m-1} < p$, $0 \leqq k_m < c_m$ and $0 \leqq k_i \leqq c_i$ when $i > m$.

To prove (25.2) we substitute (25.4) and (25.5) in (25.2). For $m = 0$ the index of the sum will run over those k's of Case 1. For $m > 0$ the sum splits into two sums, the index running, respectively, over k's of Case 1 and Case 2.

Let $m = 0$. Clearly the total sum divides into partial sums in such a way that in each of them the following is a factor:

$$\Sigma_{k_n=0}^{c_n}(-1)^{k_n}\binom{k_{n-1}}{k_n}\binom{k_{n-1}-k_n}{c_n-k_n} \quad \text{for} \quad n \geqq 2,$$

$$\Sigma_{k_1=0}^{c_1}(-1)^{k_1}\binom{k_0-1}{k_1}\binom{k_0-k_1-1}{c_1-k_1} \quad \text{for} \quad n = 1,$$

$$\Sigma_{k_0=0}^{c_0}(-1)^{k_0}\binom{p-1}{k_0}\binom{p-k_0}{c_0-k_0} \quad \text{for} \quad n = 0.$$

But in general

$$\Sigma_{s=0}^{b}(-1)^s\binom{a}{s}\binom{a-s}{b-s} = \binom{a}{b}\Sigma_{s=0}^{b}(-1)^s\binom{b}{s} = \binom{a}{b}(1-1)^b = 0.$$

Also we have $(-1)^{k_0}\binom{p-1}{k_0} = 1$, and, since

$$c_0 - k_0 < p, \quad \binom{p-k_0}{c_0-k_0} \equiv \binom{-k_0}{c_0-k_0} (\operatorname{mod} p).$$

Thus for $n=0$ the above sum reduces (mod p) to

$$\Sigma_{k_0=0}^{c_0}\binom{-k_0}{c_0-k_0}=\Sigma_{k_0=0}^{c_0-1}(-1)^{k_0}\binom{c_0-1}{c_0-k_0}=0.$$

Hence the total sum is zero for $n \geqq 0$, and (25.2) is proved when $m=0$.

If $m>0$ then, as we mentioned before, the sum of (25.2) divides into two sums corresponding to Cases 1 and 2 of the running index. Each of these sums is analogous to the one when $m=0$, and in a similar fashion is proved to be zero. We leave the details to the reader. This ends the proof of (25.2).

PROOF OF (25.3). For $c<0$ the theorem is trivial. Let $c \geqq 0$. For $c=0, 1$ and any a and b the theorem can be verified directly. Hence, by (25.2), it holds for $a=b=0$ and all c. To establish the general case we proceed by induction; several cases will be considered. The starting step of the induction is easily verified in each case; it will be omitted.

(I) (25.3) holds with $a=b$, $b \geqq 0$. Assume it to be true for m and all c with $0 \leqq m<b$. Then

$$\begin{aligned}\Sigma_{k=0}^{c}&\binom{-k(p-1)}{k}\binom{b+k(p-1)}{c-k}\\ &=\Sigma_{k=0}^{c}\binom{-k(p-1)}{k}\left\{\binom{b-1+k(p-1)}{c-k}+\binom{b-1+k(p-1)}{c-1-k}\right\}\\ &=\binom{b}{c}+\binom{b}{c-1}=\binom{b+1}{c}.\end{aligned}$$

(II) (25.3) holds with $a=0$, $b \leqq 0$. This follows by induction on $-b+c$. Suppose it holds for all m with $0 \leqq m<-b+c$. Then

$$\begin{aligned}\Sigma_{k=0}^{c}&\binom{-k(p-1)}{k}\binom{b+k(p-1)}{c}\\ &=\Sigma_{k=0}^{c}\binom{-k(p-1)}{k}\left\{\binom{b+1+k(p-1)}{c-k}-\binom{b+k(p-1)}{c-1-k}\right\}\\ &=\binom{b+2}{c}-\binom{b+1}{c-1}=\binom{b+1}{c}.\end{aligned}$$

(III) (25.3) holds for $c \geqq 0$ and any a, b. This is proved using a composite induction on a and c as follows. Suppose that III is true for m, with $0 \leqq m<c$, and any a, b. Set

$$A=\Sigma_{k=0}^{c}\binom{a-k(p-1)}{k}\binom{b+k(p-1)}{c-k}.$$

We have

$$A = \Sigma_{k=0}^{c}\left\{\binom{a-1-k(p-1)}{k} + \binom{a-1-k(p-1)}{k-1}\right\}\binom{b+k(p-1)}{c-k},$$

$$A = \Sigma_{k=0}^{c}\left\{\binom{a+1-k(p-1)}{k} - \binom{a-k(p-1)}{k-1}\right\}\binom{b+k(p-1)}{c-k}.$$

Using the induction hypothesis we establish the following:

$$(25.6)\quad A = \Sigma_{k=0}^{c}\binom{a-1-k(p-1)}{k}\binom{b+k(p-1)}{c-k} + \binom{a+b}{c-1},$$

$$(25.7)\quad A = \Sigma_{k=0}^{c}\binom{a+1-k(p-1)}{k}\binom{b+k(p-1)}{c-k} - \binom{a+b+1}{c-1}.$$

Now we make the second induction on a. Let $a \geqq 1$ and assume III for m, with $0 \leqq m < a$ and all b, c. From (25.6) we obtain

$$A = \binom{a+b}{c} + \binom{a+b}{c-1} = \binom{a+b+1}{c}.$$

Similarly, if $a \leqq 0$ we assume III for m, with $0 \leqq m < -a$ and all b, c. Then, from (25.7), it follows that

$$A = \binom{a+b+2}{c} - \binom{a+b+1}{c-1} = \binom{a+b+1}{c}.$$

Hence III is established. This ends the proof of (25.3).

PRINCETON UNIVERSITY and
UNIVERSIDAD NACIONAL AUTÓNOMA DE MÉXICO

REFERENCES

[1] J. ADEM, *The iteration of the Steenrod squares in algebraic topology*, Proc. Nat. Acad. Sci. U.S.A., 38 (1952), pp. 720–726.

[2] ——, *Relations on iterated reduced powers*, Proc. Nat. Acad. Sci. U.S.A., 39 (1953), pp. 636–638.

[3] A. BOREL, *La cohomologie module 2 de certains espaces homogènes*, Comment. Math. Helv., 27 (1953), pp. 165–197.

[4] —— and J. P. SERRE, *Groupes de Lie et puissances réduites de Steenrod*, Amer. J. Math., 75 (1953), pp. 409–448.

[5] H. CARTAN, (a) *Sur les groupes d'Eilenberg-MacLane $H(\pi, n)$. I. Méthode des constructions*, Proc. Nat. Acad. Sci. U.S.A., 40 (1954), pp. 467–471; (b) *Sur les groupes d'Eilenberg-MacLane. II*, ibid., pp. 704–707.

[6] ——, *Une théorie axiomatique des i-carrés*, C.R. Acad. Sci. Paris, 230 (1950), pp. 425–427.

[7] —— and S. EILENBERG, Homological algebra, Princeton University Press, 1956.

[8] B. Eckmann, *Cohomology of groups and transfer*, Ann. of Math., 58 (1953), pp. 481–493.

[9] S. Eilenberg, *Homology of spaces with operators, I*, Trans. Amer. Math. Soc., 61 (1947), pp. 378–417.

[10] —— and S. MacLane, *Homology of spaces with operators. II*, Trans. Amer. Math. Soc., 65 (1949), pp. 49–99.

[11] N. J. Fine, *Binomial coefficients modulo a prime*, Amer. Math. Monthly, 54 (1947), pp. 589–592.

[12] F. Hirzebruch, *On Steenrod's reduced powers, the index of inertia, and the Todd genus*, Proc. Nat. Acad. Sci. U.S.A., 39 (1953), pp. 951–955.

[13] G. Hochschild and J. P. Serre, *Cohomology of group extensions*, Trans. Amer. Math. Soc., 74 (1953), pp. 110–134.

[14] S. Lefschetz, Algebraic Topology, Amer. Math. Soc. Colloquium Publications, 27, 1942.

[15] E. Lucas, *Théorie des fonctions numériques simplement périodiques*, Amer. J. Math., 1 (1878), p. 230.

[16] J. P. Serre, *Cohomologie modulo 2 des complexes d'Eilenberg-MacLane*, Comment. Math. Helv., 27 (1953), pp. 198–231.

[17] N. E. Steenrod, *Products of cocycles and extensions of mapping*, Ann. of Math., 48 (1947), pp. 290–319.

[18] ——, *Reduced powers of cohomology classes*, (a) Ann. of Math., 56 (1952), pp. 47–67; (b) Paris Lectures (1951) (Mimeographed).

[19] ——, (a) *Homology groups of symmetric groups and reduced power operations*, Proc. Nat. Acad. Sci. U.S.A., 39 (1953), pp. 213–217; (b) *Cyclic reduced powers of cohomology classes*, ibid., pp. 217–223.

[20] R. Thom, Une théorie axiomatique des puissances de Steenrod, Colloque de Topologie, Strasbourg, 1951 (Mimeographed).

[21] ——, *Espaces fibrés en sphères et carrés de Steenrod*, Ann. Ecole Norm. Sup., 69 (1952), pp. 109–181.

[22] ——, *Quelques propriétés globales des variétés différentiables*, Comment. Math. Helv., 28 (1954), pp. 17–85.

[23] W. T. Wu, *Classes caractéristiques et i-carrés d'une variété*, C.R. Acad. Sci. Paris, 230 (1950), pp. 508–550.

[24] ——, *Les i-carrés dans une variété grassmannienne*, C.R. Acad. Sci. Paris, 230 (1950), pp. 918–920. Also: *On squares in Grassmannian Manifolds*, Acta Scientia Sinica (1953), pp. 91–115.

[25] ——, Sur les puissances de Steenrod, Colloque de Topologie, Strasbourg, 1951 (Mimeographed).

Imbedding of Metric Complexes

C. H. Dowker

1. It is known ([2], (15.2)) that every metric complex K has the following property of Lefschetz ([5], p. 393; [6], p. 99): For every $\epsilon > 0$, the identity map of K on to itself is ϵ-homotopic to a map $\psi\phi\colon K \to K$, such that ϕ is a map of K into a Whitehead (weak) complex L and ψ is a map of L into K. It is known ([3], 28.9) that a metric space has the above Lefschetz property if and only if it is an absolute neighborhood retract (ANR) for metric spaces. And it is known ([1], Theorem 2) that a metric space is an ANR for collectionwise normal spaces if and only if it is both an ANR for metric spaces and an absolute G_δ. It follows that a metric complex is an ANR for collectionwise normal spaces if and only if it is an absolute G_δ.

It is the purpose of this paper to find under what condition a metric complex K is an absolute G_δ. It is shown in Theorem 1 that the necessary and sufficient condition on K is the ascending chain condition for its cells.

For the sake of completeness we include a proof, Theorem 2, that every metric complex is an absolute F_σ.

2. A metrizable complex is a metrizable topological space with a decomposition into mutually disjoint subsets called cells; each cell a is associated with a non-negative integer $\dim a$, called its dimension, and there is a partial order, $<$, of the cells. An exact definition is given in [2], but the properties used in this paper are the following:

(1) If $c < b$ and $b < a$ then $c < a$.

(2) If $b < a$ then $\dim b < \dim a$.

(3) For each cell a the set $\bar{a}$ defined to be $\bigcup_{b \leq a} b$, is a compact set and is the topological closure of the set a.

(4) For each cell a the set $\operatorname{St} a$, defined to be $\bigcup_{a \leq b} b$, is an open set called the star of a.

If the underlying space is provided with a metric, the complex is called a *metric complex*. If K is a metrizable complex, and $f\colon K \to L$ is a homeomorphism of K on to a metric space L, we can regard the map f either as giving a metric to the complex K, or as inducing in the space L the structure of a metric complex.

A set $\{a_i\}$ of cells, for which the given order is a simple order, is called a chain of cells. If the chain in infinite, and the cells of the chain are written in order of increasing dimension, then we have $a_1 < a_2 < \ldots$. In the theory of lattices, such a chain is called an ascending chain. A complex K is said to satisfy the *ascending chain condition* if there is no infinite chain of cells of K.

3. A subset R of a space M is called a G_δ set if R is the intersection of a sequence $\{U_n\}$ of open sets of M. R is called an F_σ set if it is the union of a sequence $\{C_n\}$ of closed sets of M. A metric space R is called an absolute G_δ [resp. absolute F_σ] if, for each metric space M with a subset S homeomorphic to R, S is a G_δ [resp. F_σ] set in M.

LEMMA 1. *If a metric complex K is a subspace of a metric space M and is not a G_δ set in M, then there is an infinite chain of cells of K.*

PROOF. For each cell a of K let

$$U_a = \{p: p \in M, \rho(p, a) < \rho(p, K - \operatorname{St} a)\},$$

where $\rho(p, a)$ is the distance from the point p to the set a. Then U_a is an open set of M. Since $\operatorname{St} a$ is open in K and $a \subset \operatorname{St} a$, each point of a has positive distance from $K - \operatorname{St} a$ but zero distance from a; hence $a \subset U_a$.

If U_a and U_b have non-empty intersection, then $a \leqq b$ or else $b \leqq a$; for otherwise $b \subset K - \operatorname{St} a$ and $a \subset K - \operatorname{St} b$ and, for each point $p \in U_a \cap U_b$, p is nearer to a than to b and also nearer to b than to a, which is absurd. Thus, for each point p, the set of all cells a, such that $p \in U_a$, is simply ordered, i.e. is a chain of cells.

For $n = 1, 2, \ldots$, let

$$V_{an} = \{p: p \in U_a, \rho(p, a) < 1/n\};$$

then V_{an} is open in M and $a \subset V_{an}$. Let $V_n = \bigcup_a V_{an}$; then V_n is open in M and $K \subset V_n$. Since K is not a G_δ set in M, there is some point $p \in M - K$ such that $p \in \bigcap_{n=1}^{\infty} V_n$. Since $\bar{a}$ is compact, $\bar{a}$ is closed in M, and since $\bar{a} \subset K$, $p \notin \bar{a}$ and therefore $\rho(p, \bar{a}) > 0$. Therefore, since $a \subset \bar{a}$, $\rho(p, a) > 0$. Hence, for sufficiently large n, $p \notin V_{an}$. But, since $p \in \bigcap_{n=1}^{\infty} V_n$, p is in some V_{an} for each n, hence is in U_a, for infinitely many cells a. Thus there is an infinite chain consisting of all cells a for which $p \in U_a$.

4. It is known ([4], p. 337) that a metric space R is an absolute G_δ if and only if there is a complete metric space homeomorphic to R.

LEMMA 2. *If a metric complex K has an infinite chain of cells, then K is not a complete metric space.*

PROOF. Let $a_1 < a_2 < \ldots$ be an infinite chain of cells of K. We construct a sequence $\{p_n\}$ of points of K such that $p_n \in a_n$ and

$$\rho(p_{n+1}, p_n) < (\tfrac{1}{3}) \rho(p_n, K - \operatorname{St} a_n).$$

The point p_1 can be chosen arbitrarily in a_1. We assume that p_n has already been chosen in a_n. Then, since p_n is in the open set $\operatorname{St} a_n$, the distance $\rho(p_n, K - \operatorname{St} a_n)$ is positive. Then, since $p_n \in a_n \subset \bar{a}_{n+1}$, we can choose a point $p_{n+1} \in a_{n+1}$ such that

$$\rho(p_{n+1}, p_n) < \tfrac{1}{3}\rho(p_n, K - \operatorname{St} a_n)$$

as required. Since $p_{n-1} \in a_{n-1} \subset K - \operatorname{St} a_n$, therefore

$$\rho(p_n, p_{n-1}) \geqq \rho(p_n, K - \operatorname{St} a_n).$$

Therefore $\rho(p_{n+1}, p_n) < (\frac{1}{3})\rho(p_n, p_{n-1})$, and hence $\{p_n\}$ is a Cauchy sequence in K.

Suppose if possible that K is a complete space. Then $\{p_n\}$ converges to a point $p \in K$; let b be the cell of K containing p. Then

$$\begin{aligned}\rho(p_n, p) &\leqq \rho(p_n, p_{n+1}) + \rho(p_{n+1}, p_{n+2}) + \cdots \\ &\leqq (\tfrac{3}{2})\rho(p_n, p_{n+1}) < (\tfrac{1}{2})\rho(p_n, K - \operatorname{St} a_n).\end{aligned}$$

Therefore $p \in \operatorname{St} a_n$ and $a_n \leqq b$. Therefore

$$\dim b \geqq \dim a_n > \dim a_{n-1} > \ldots > \dim a_1 \geqq 0;$$

hence $\dim b \geqq n - 1$ for every n, which is absurd. Therefore K is not a complete space.

Theorem. 1. *A metric complex K is an absolute G_δ if and only if it satisfies the ascending chain condition.*

Proof. *Necessity.* If K is an absolute G_δ, then there is a space K_1 homeomorphic to K which is a complete metric case. The homeomorphism induces in K_1 the structure of a metric complex isomorphic with K. By Lemma 2, since K_1 is complete, therefore K_1, and hence also K, satisfies the ascending chain condition.

Sufficiency. Let K satisfy the ascending chain condition. Let M be a metric space and let K_1 be a subspace of M homeomorphic to K. The homeomorphism induces in K_1 the structure of a metric complex isomorphic with K and hence satisfying the ascending chain condition. Hence, by Lemma 1, K_1 must be a G_δ set in M. Hence K is an absolute G_δ.

Corollary. *A metric complex K is an* ANR *for collectionwise normal spaces if and only if it satisfies the ascending chain condition.*

5. It is now to be shown that every metric complex is an absolute F_σ.

Theorem 2. *If a metric complex K is a subspace of a metric space M, then K is an F_σ set in M.*

Proof. For each cell a of K let

$$D_{an} = \{p \colon p \in a,\ \rho(p, K - \operatorname{St} a) \geqq 1/n\}.$$

Then D_{an} is a compact set and $a = \bigcup_{n=1}^{\infty} D_{an}$. Let $D_n = \bigcup_a D_{an}$; then $K = \bigcup a = \bigcup_a \bigcup_n D_{an} = \bigcup_n D_n$.

Let $B_n = \bar{D}_n - D_n$, where $\bar{D}_n$ is the closure of D_n in M. As in § 3, let

$$V_{an} = \{p \colon p \in M,\ \rho(p,a) < 1/n,\ \rho(p,a) < \rho(p, K - \mathrm{St}\, a)\}.$$

Then V_{an} is open and $D_{an} \subset a \subset V_{an}$. If $b \neq a$, then either $a < b$, in which case $a \subset K - \mathrm{St}\, b$ and $\rho(D_{bn}, a) \geqq \rho(D_{bn}, K - \mathrm{St}\, b) \geqq 1/n$ and hence $D_{bn} \cap V_{an} = 0$, or $a \nless b$, in which case $b \subset K - \mathrm{St}\, a$ and $b \cap V_{an} = 0$ and so $D_{bn} \cap V_{an} = 0$. Thus $V_{an} \cap D_n = D_{an}$ and $(V_{an} - D_{an}) \cap D_n = 0$.

Since D_{an} is compact and hence closed, $V_{an} - D_{an}$ is an open set not meeting D_n and hence not meeting $\bar{D}_n$. Therefore $(V_{an} - D_{an}) \cap B_n = 0$. But D_{an}, as a subset of D_n, does not meet $B_n = \bar{D}_n - D_n$; hence $V_{an} \cap B_n = 0$. Then, if $V_n = \bigcup_a V_{an}$, $D_n \subset V_n$ and $V_n \cap B_n = 0$. Since V_n is open and

$$V_n \cap \bar{D}_n = V_n \cap (D_n \cup B_n) = V_n \cap D_n = D_n,$$

therefore D_n is open in $\bar{D}_n$. But a set which is open in its closure is an F_σ set. Hence D_n is an F_σ set. Therefore, since $K = \bigcup_{n=1}^{\infty} D_n$, K is a countable union of F_σ sets and hence is an F_σ set.

Birkbeck College, London

References

[1] C. H. Dowker, *On a theorem of Hanner*, Arkiv. Mat., 2 (1952), pp. 307–313.

[2] ——, *Topology of metric complexes*, Amer. J. Math., 74 (1952), pp. 555–577.

[3] O. Hanner, *Retraction and extension of mappings of metric and non-metric spaces*, Arkiv. Mat., 2 (1952), pp. 315–360.

[4] C. Kuratowski, Topologie I, 2e édition, Warsaw, 1948.

[5] S. Lefschetz, *On locally connected sets and retracts*, Proc. Nat. Acad. Sci. U.S.A., 24 (1938), pp. 392–393.

[6] ——, Topics in Topology, Annals of Mathematics Studies, no. 10, Princeton, 1942.

Covering Spaces with Singularities†

Ralph H. Fox

THE familiar concept of *unbranched covering* (Unverzweigte Ueberlagerung) is a topological concept (cf. [14, 23]) abstracted from the analytical concept *Riemann surface*, or rather that part of the Riemann surface remaining after the branch points have been deleted. Hitherto the concept of *branched covering* (Verzweigte Ueberlagerung) has apparently been formulated only in combinatorial terms. For example, Heegard [12], Tietze [21], Alexander [1, 2, 3] and Reidemeister [16] considered combinatorially defined branched coverings of spherical n-dimensional space. In fact, Tietze conjectured and Alexander [1] proved that every orientable n-dimensional manifold can be represented by such a covering. Later Tucker [22] gave a combinatorial definition of a more general type of covering in which there is allowed not only 'branching' but 'folding' as well. Seifert [17] gave a combinatorial definition of a covering of a 3-dimensional manifold branched over a (single or multiple) knot, and [18, 19, 20] derived important knot-invariants therefrom.

The principal object of this note is to formulate as a topological concept the idea of a *branched covering space*. This topological concept encompasses the above-mentioned combinatorial concept used by Heegard, Tietze, Alexander, Reidemeister and Seifert. This has as a consequence that the knot-invariants defined by Seifert (the linking invariants of the cyclic coverings) are invariants of the topological type of the knot (i.e. are unaltered by an orientation-preserving autohomeomorphism of 3-space). Without the developments of this note I am unable to see any simple proof that these invariants are invariants of anything more than the combinatorial type of the knot.

It appears that the best way to look at branched covering is as a 'completion' of unbranched covering. This completion process appears

† This paper is based mainly on a course of lectures that I gave at the Instituto de Matemáticas de la Universidad Nacional Autónoma de México in the summer of 1951. I also had the opportunity of developlng my ideas by lecturing on this topic to the American Mathematical Soceity (1949), to the Summer Seminar of the Canadian Mathematical Society (1953), and at the Universities of Delft and Stockholm, while on a Fulbright grant (1952).

in its simplest form if it is applied to a somewhat wider class of objects. It is for this reason that I introduce the concept of a *spread* (a concept that encompasses, in particular, the 'branched and folded coverings' of Tucker). The basic theory of spreads is developed in §§ 1–3 for locally connected T_1-spaces. In § 4 it is shown how Freudenthal's compactification process [8] can be evolved out of the new process. In § 5 the branched covering concept is given a precise meaning. Conditions are found in § 6 ensuring that a branched covering of a complex (or manifold) be a complex (or manifold). The fundamental group of a branched covering is calculated in § 7, and a possible further line of development is indicated in § 8.

1. Spreads and their completions

A mapping g of a locally connected T_1-space Y into a locally connected T_1-space Z will be called a *spread* if the components of the inverse images of the open sets of Z form a basis of Y. The *antecedent* is Y and the *space over which the antecedent is spread* is the subset $g(Y)$ of Z. A point z of Z will be called an *ordinary point* if it has a neighborhood W in Z that is evenly covered [4] by g, i.e. if $g^{-1}(W)$ is non-vacuous and each component of $g^{-1}(W)$ is mapped topologically upon W by g. The points of Z that are not ordinary will be called *singular points*. In order that a map g: $Y \to Z$ be a spread it is necessary that $g^{-1}(z)$ be 0-dimensional for each point z of $g(Y)$. This may be expressed by saying that the antecedent of a spread must lie over the image space in thin sheets.

If g: $Y \to Z$ is a spread and Z is regular, then Y must also be regular. Let y be a point of Y and V a basic open set containing y. There is an open set W of Z containing $z = g(y)$ such that V is a component of $g^{-1}(W)$. Since Z is regular, there is an open neighborhood W_1 of z such that $\overline{W}_1 \subset W$. Let V_1 be the component of $g^{-1}(W_1)$ that contains y. Then $V_1 \subset V \cap g^{-1}(W_1)$, and hence $\overline{V}_1 \subset \overline{V} \cap g^{-1}(\overline{W}_1) \subset \overline{V} \cap g^{-1}(W) = V$.

A spread g: $Y \to Z$ will be said to be *complete* if for every point z of Z the following condition is satisfied: If to every open neighborhood W of z there is selected a component V of $g^{-1}(W)$ in such a way that $V_1 \subset V_2$ whenever $W_1 \subset W_2$, then $\bigcap_W V$ is non-vacuous (and is therefore a point).

Any locally connected subset X of the antecedent Y of a spread g is itself the antecedent of a spread; the spread with which it is associated is $f = g \mid X$, and the space over which it is spread is $f(X)$. In this circumstance the spread g is an *extension* of f. A more precise definition would be the following: an *extension* of a spread f: $X \to Z$ is a spread

g: $Y \to Z$ together with a homeomorphism i of X into Y that satisfies $gi = f$. However, I shall use the more informal definition, as this is unlikely to cause any real confusion. Two extensions g_1: $Y_1 \to Z$ and g_2: $Y_2 \to Z$ are *equivalent* if there is a homeomorphism ϕ of Y_1 upon Y_2 satisfying $g_2\phi = g_1$ and $\phi_2 | X = 1$. An extension g: $Y \to Z$ of a spread f: $X \to Z$ will be called a *completion* of f if g is complete and X is dense and locally connected† in Y.

2. The existence theorem

Existence Theorem. *Every spread has a completion.*

Given a spread f: $X \to Z$, we are going to construct a space Y in which X is contained and a mapping g of Y into Z in such a way that g is a completion of f.

(a) *The points of Y and the function g.* Let z be any point of Z. A point y of the subset $g^{-1}(z)$ of Y is a function that associates to each open neighborhood W of z a component yW of $f^{-1}(W)$ in such a way that yW_1 is contained in yW_2 whenever W_1 is contained in W_2. This defines simultaneously the set Y and the function g. (Of course there may be points z for which $g^{-1}(z)$ is vacuous.)

(b) *The topology of Y.* Given any open set W of Z and any component U of $f^{-1}(W)$ define U/W to be the set of those points of Y for which $yW = U$. For any union $\bigcup_\alpha U_\alpha$ of components U_α of $f^{-1}(W)$ define $(\bigcup_\alpha U_\alpha)/W = \bigcup_\alpha (U_\alpha / W)$. Consider components U_1 and U_2 of $f^{-1}(W_1)$ and $f^{-1}(W_2)$ respectively. It is obvious that

$$U_1/W_1 \cap U_2/W_2 \subset U_1 \cap U_2/W_1 \cap W_2.$$

If conversely, $y \in U_1 \cap U_2/W_1 \cap W_2$ then $y(W_1 \cap W_2) \subset U_j$ $(j = 1, 2)$. But yW_j is the component of $f^{-1}(W_j)$ that contains the component $y(W_1 \cap W_2)$ of $f^{-1}(W_1 \cap W_2)$. Since U_j is a component of $f^{-1}(W_j)$ that contains $y(W_1 \cap W_2)$, it follows that $U_j = yW_j$, i.e. that $y \in U_j/W_j$. Thus it has been shown that

$$U_1/W_1 \cap U_2/W_2 = U_1 \cap U_2/W_1 \cap W_2.$$

This formula justifies the use of the collection of sets U/W, W ranging over the open sets of Z and U over the components of $f^{-1}(W)$, as a basis of Y; a topology is thereby defined in Y. It is easily verified that Y is a T_1-space.

† A space X is *locally connected* in a space Y if there is a basis of Y such that $V \cap X$ is connected for every basic open set V. An example of a space X not locally connected in a space Y is the following: Y is the Cartesian plane, $Y - X$ is the origin and the positive half of the real axis. Here X fails to be locally connected at any of the points of $Y - X$ except the origin. If $Z = Y$ the identity map of Y into Z is not the completion of the identity map of X into Z; in the completion of i: $X \to Z$ each point of $Y - X$ other than the origin gets covered by two points corresponding to the two sides of the real axis.

(c) *The imbedding of X in Y.* For any point x of X and open neighborhood W of $f(x)$ define xW to be that component of $f^{-1}(W)$ in which x is contained. It is clear that $xW_1 \subset xW_2$ whenever $W_1 \subset W_2$, so that x determines a point of Y. Since X and Z are T_1-spaces, distinct points of X determine distinct points of Y. We shall identify each point of X with the point of Y that it determines; X is then a subset of Y. It is obvious that, for any basic open set U of X,

$$U/W \cap X = U,$$

so that the topology of X is identical with the relativization topology induced in X by Y. Since the intersection of X with any basic open set of Y is non-vacuous and connected it follows that X is dense and locally connected in Y. Furthermore $f = g \mid X$ (and hence

$$f(X) \subset g(Y) \subset \overline{f(X)}).$$

(d) *The continuity of g.* This is an immediate consequence of the fact that, for any open set W of Z,

$$g^{-1}(W) = f^{-1}(W)/W.$$

(e) *The spread property of g.* For any open set W of Z and component U of $f^{-1}(W)$, we have

$$U \subset U/W \subset \overline{U},$$

so that each U/W is seen to be connected. (This shows that Y is locally connected.) On the other hand,

$$g^{-1}(W) = \cup_U U/W,$$

U ranging over the components of $f^{-1}(W)$, and

$$U_1/W \cap U_2/W = 0 \quad \text{if} \quad U_1 \neq U_2,$$

so that each U/W is clopen (closed and open) in $g^{-1}(W)$. Thus the components of $g^{-1}(W)$ are the sets U/W, U ranging over the components of $f^{-1}(W)$.

(f) *The completeness of g.* It was shown in (e) that a component V of $g^{-1}(W)$ is of the form U/W, where U is a component of $f^{-1}(W)$. The condition '$U_1/W_1 \subset U_2/W_2$ whenever $W_1 \subset W_2$' is equivalent to the condition '$U_1 \subset U_2$ whenever $W_1 \subset W_2$'. Thus $\cap_W V = \cap_W U/W$ contains the point y, where $yW = U$.

LEMMA. *If X and Z are separable then Y is also separable.*

If X is separable the components U of $f^{-1}(W)$ are enumerable, and if Z is also separable a countable basis of Y is made up of the sets U/W, W ranging over a countable basis of Z, and U ranging over the components of $f^{-1}(W)$.

3. The uniqueness theorem and the extension theorem

LEMMA. *If X is dense and locally connected in Y then the intersection of X with any connected open set of Y is connected.*

Let V be a connected open subset of Y and suppose that the set $U = V \cap X$ (which is not vacuous, because X is dense in Y) is not connected. Then $U = A_1 \cup A_2$, where A_1 and A_2 are disjoint non-vacuous open subsets of X. Since X is locally connected in Y, any point y of V has an open neighborhood $N(y)$ contained in V whose intersection $M(y)$ with X is connected. Clearly either $M(y) \subset A_1$ or $M(y) \subset A_2$. Let $B_j = \{y \mid M(y) \subset A_j\}$ $(j = 1, 2)$. Then $V = B_1 \cup B_2$ and $B_1 \cap B_2 = 0$; B_j is open because $B_j = \bigcup_{y \in B_j} N(y)$; B_j is non-vacuous because $B_j \supset A_j$. Hence V cannot be connected. This contradiction shows that U must be connected.

Let f_1: $X_1 \to Z_1$ and f_2: $X_2 \to Z_2$ be spreads. A mapping a of X_1 into X_2 *covers* a mapping c of Z_1 into Z_2 if $f_2 a = c f_1$. Let g_1: $Y_1 \to Z_1$ and g_2: $Y_2 \to Z_2$ be completions of f_1 and f_2 respectively.

EXTENSION THEOREM. *The mapping a: $X_1 \to X_2$ can be extended to a mapping b: $Y_1 \to Y_2$ that also covers the mapping c: $Z_1 \to Z_2$.*

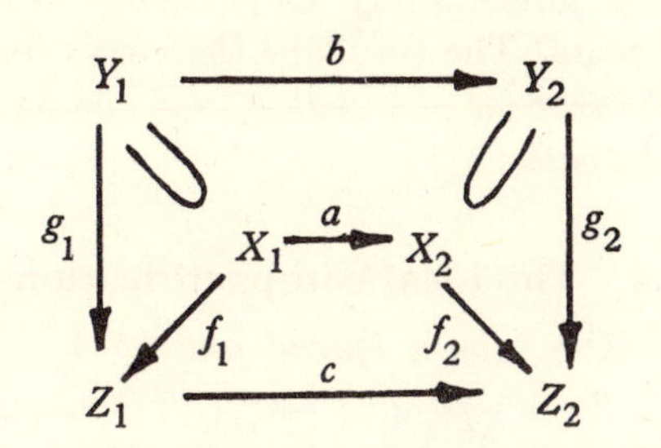

Let y_1 be any point of Y_1 and consider any open neighborhood W_2 of $z_2 = c(g_1(y_1))$. Then $W_1 = c^{-1}(W_2)$ is an open neighborhood of $z_1 = g_1(y_1)$. Let V_1 be the component of $g_1^{-1}(W_1)$ that contains y_1. By the lemma, $U_1 = V_1 \cap X_1$ is a component of $f_1^{-1}(W_1)$. Since $a(U_1)$ is a connected subset of $f_2^{-1}(W_2)$, it is contained in a component V_2 of $g_2^{-1}(W_2)$. Clearly $V_2 \subset V_2'$ whenever $W_2 \subset W_2'$. Since g_2 is complete, $\cap V_2$ is a point y_2. Define $b(y_1) = y_2$. It is obvious that $b \mid X_1 = a$ and that $g_2 b = c g_1$.

To prove that b is continuous, consider a basic open set of Y_2, i.e. a component V_2 of $g_2^{-1}(W_2)$ for some open set W_2 of Z_2. Let $U_2 = V_2 \cap X_2$ and $W_1 = c^{-1}(W_2)$. It is easily seen that $b^{-1}(V_2)$ is the union of those components of $g_1^{-1}(W_1)$ that intersect $a^{-1}(U_2)$. Hence $b^{-1}(V_2)$ is an open set.

UNIQUENESS THEOREM. *Any two completions of a spread are equivalent.*

Let $g_1: Y_1 \to Z$ and $g_2: Y_2 \to Z$ be completions of a spread $f: X \to Z$. By the extension theorem, there exist mappings $\phi: Y_1 \to Y_2$ and $\psi: Y_2 \to Y_1$ such that $\phi \mid X = \psi \mid X = 1$, $g_2\phi = g_1$, and $g_1\psi = g_2$. The map $\psi\phi: Y_1 \to Y_1$ is an extension of the identity map $1: X \to X$; since X is dense in Y_1, $\psi\phi = 1$. Similarly $\phi\psi = 1$; therefore ϕ is a homeomorphism of Y_1 upon Y_2 and $\psi = \phi^{-1}$.

By virtue of the uniqueness theorem we may speak of *the* completion of a spread.

COROLLARY OF THE EXTENSION THEOREM. *Let $f: X \to Z$ be a spread and $g: Y \to Z$ its completion. Let Q be any locally connected T_1-space, let $c: Q \times [0, 1] \to Z$ be a homotopy of Q in Z and let the 'open' homotopy $c \mid Q \times [0, 1)$ be covered by an 'open' homotopy*

$$a: Q \times [0, 1) \to X.$$

Then a can be extended to a homotopy $b: Q \times [0, 1] \to Y$ that covers c.

The identity mapping of $Q \times [0, 1)$ into $Q \times [0, 1]$ is clearly a spread, and its completion is the identity mapping of $Q \times [0, 1]$ upon itself. The corollary follows immediately. Of particular interest is the special case where Q is a point. The corollary then says that *an open path a in X that covers the interior of a path c in Z can be extended to a path b in Y that covers the path c.*

4. The ideal compactification

LEMMA. *Let $f: X \to Z$ be a spread and $g: Y \to Z$ its completion. Suppose that X and Z are separable, that $\overline{f(X)}$ is compact, and that Z has a basis such that, for each basic open set W, the number of components of $f^{-1}(W)$ is finite. Then Y is compact.*

Since Y is separable and X is dense in Y, it suffices to show that any sequence of points $x_1, x_2, \dots$ of X has a subsequence converging in Y. Let $z_j = f(x_j)$; since $\overline{f(X)}$ is compact it is no loss of generality to assume that the sequence $z_1, z_2, \dots$ converges to a point z_0. Let $W_1 \supset W_2 \supset \dots$ be a local basis of Z at z_0 such that the number of components of each set $f^{-1}(W_n)$ is finite. Let U_1 be a component of $f^{-1}(W_1)$ that contains an infinite subsequence of $\{x_j\}$. Select, inductively, for each $n > 1$, a component U_n of $f^{-1}(W_n)$ in such a way that $U_n \subset U_{n-1}$ and U_n contains an infinite subsequence of $\{x_j\}$. This can be done because $f^{-1}(W_n) \cap U_{n-1}$ contains an infinite subsequence of $\{x_j\}$ and has only a finite number of components. Define $yW_n = U_n$. Any neighborhood W of z_0 contains W_n for some index n; define yW to be that component of $f^{-1}(W)$ that contains U_n. Thus a point y of Y is defined. It is obvious from the construction that some subsequence of $\{x_j\}$ converges to y.

It is well known that any locally compact Hausdorff space X that is not already compact can be compactified by the adjunction of one point, i.e. there is a compact Hausdorff space Z containing X such that $Z-X$ is a point. Furthermore, if X is connected, locally connected, separable, and regular, then so is Z. On the other hand Freudenthal [8] has shown that any connected, locally connected, locally compact, separable, regular space X has an *ideal compactification*, i.e. X is contained in a connected, locally connected, compact, separable, regular space Y in such a way that X is dense, open and locally connected in Y, and the set $Y-X$ is 0-dimensional, hence discrete. The concept of completion of a spread allows us to establish a relation between these two kinds of compactification.

COMPACTIFICATION THEOREM. *Let X be connected, locally connected, locally compact, separable, regular, and not already compact, and let $Z=X\cup z_0$ be its one-point compactification. The identity mapping $1: X\to Z$ is a spread; let $g: Y\to Z$ be its completion. Then Y is the ideal compactification of X.*

Since the ideal compactification is determined by the properties listed above, it suffices to check that Y has them. Compactness of Y is the only one of these properties that is not clear from the preceding sections. According to the lemma it suffices to show that Z has a local basis at z_0 such that, for each open set W of this local basis, the number of components of $f^{-1}(W)$ is finite.

Consider any neighborhood W_1 of z_0. There is an open neighborhood W_2 of z_0 such that $\overline{W_2}\subset W_1$. Since the boundary B of W_2 is a compact subset of the locally connected space X, it can meet only a finite number of the components of W_1-z_0, say $U_1, \ldots, U_n$. Define

$$W=z_0\cup U_1\cup\ldots\cup U_n.$$

Since X is connected, no component of W_1-z_0 lies within W_2. Hence W is an open neighborhood of z_0. Obviously $W\subset W_1$, and

$$f^{-1}(W)=W-z_0$$

has only a finite number of components.

5. Covering spaces

A spread f: $X \to Z$ over a connected† set Z is an *unbranched* (or *non-singular*) *covering* if the antecedent space X is connected and there are no singular points; the antecedent X is an *unbranched* (or *non-singular*) *covering space*, the map f is onto, and the space $f(X) = Z$ over which the antecedent is spread is the *base space*. If z and z' are any two points of Z, the number of points in $f^{-1}(z)$ and $f^{-1}(z')$ is the same (the set of points z' for which the number of points in $f^{-1}(z')$ is the same as the number of points in $f^{-1}(z)$ for some fixed z is easily seen to be non-vacuous and clopen in Z); it is the *index* j_f (Blaetterzahl) of f.

If g: $Y \to Z$ is any spread, the set Z_o of ordinary points is obviously an open subset of Z. Hence $X = g^{-1}(Z_o)$ is an open subset of Y, and therefore locally connected. Thus $f = g \mid X$ is a spread. If Z_o is non-vacuous and connected and its inverse image X is connected, the spread f: $X \to Z_o$ is an unbranched covering; I shall call it *the unbranched covering associated with g*.

I shall call a spread g: $Y \to Z$ a *branched covering*, or simply a *covering*,‡ if (1) Z_o is connected, dense and locally connected in Z, (2) $g^{-1}(Z_o)$ is connected (so that g has an associated unbranched covering), and (3) g is the completion of its associated unbranched covering. The space Y is a *covering space* (or a *branched covering space*); Z is the *base space* of g. The set $Z_s = Z - Z_o$ is the *singular set*. An unbranched covering is a covering whose singular set is vacuous. Riemann surfaces [23] and Riemann spreads [1, 2] are covering spaces.

If y is any point of the covering space Y, W any connected open neighborhood of $z = g(y)$ such that $W_o = W \cap Z_o$ is also connected,

† Here I have adopted the customary requirement that an unbranched covering space is connected. Although this is convenient, it is not really essential. It could be weakened to the requirement that the inverse image of each component of Z be connected, without causing any other than verbal difficulties. Of course this last condition is absolutely indispensable if one has any hopes of defining a universal covering space.

‡ Condition (3) excludes 'adhesions' of all sorts, in particular the 'folded coverings' of [22] are excluded. (An example of a spread with an 'adhesion' is the projection onto the plane $z = 0$ of the double cone $x^2 + y^2 = z^2$.) Condition (1) excludes 'slits' (exemplified in footnote, p. 245) and certain undesirable pathological singularities (such as isolated points).

A puzzling kind of spread is given by Fox and Kershner [7]. Here an open 2-dimensional manifold (of infinite genus) is mapped onto the plane. The branch points lie over a dense subset of the plane, so that every point is singular. Nevertheless, the branch points are isolated and the projection is a local homeomorphism at all other points. According to the present definition this is not a covering space, although its exclusion might be debatable.

V the component of $g^{-1}(W)$ that contains y, $U = V \cap X$, $q = g \mid V$ and $p = q \mid U$, then q: $V \to W$ is a covering with p: $U \to W_0 = W \cap Z_0$ its associated unbranched covering. Denote by $j(y, W)$ the index of p (over W_0). Obviously $j(y, W) \leqq j(y, W')$ whenever $W \subset W'$. Denote by $j(y)$ the minimum of the numbers $j(y, W)$; this is the *index of branching* of the point y. The number $\mu(y) = j(y) - 1$ is the classical *order of branching* [23] of y. Clearly z is a singular point if $j(y) > 1$; the converse need not be true. It is not clear from the literature what a *branch point* is, but it seems most probable that it is a point y for which $j(y) > 1$.

I shall call a covering g: $Y \to Z$ *finitely branched* if the index of branching $j(y)$ is finite for each point y of Y. I shall call a covering *regular* if its associated unbranched covering is regular.

6. Covering complexes

It is obvious that *a simplicial mapping g of a locally finite simplicial complex Y into a locally finite simplicial complex Z is a spread if and only if no simplex is mapped degenerately.* Such a mapping may be called a *simplicial spread.* Its singular set $Z_s = Z - Z_0$ is a subcomplex of Z. Furthermore, a principal open simplex of Z (i.e. one which is not on the boundary of any other simplex of Z) belongs to Z_0 or not according as it does or does not belong to $g(Y)$. Thus Z_0 is dense in Z if and only if g maps Y onto Z. The condition that Z_0 be dense and locally connected in Z is equivalent to the condition that, for each simplex τ of the subcomplex Z_s, the intersection $S(\tau)$ of Z_0 with the open star $\mathrm{st}_Z \tau$ of τ be non-vacuous and connected. Thus we are led to the following statement:

If the locally finite simplicial complex Z is connected, a simplicial spread g: $Y \to Z$ is a covering if and only if (1) $S(\tau) = Z_0 \cap \mathrm{st}_Z \tau$ *is non-vacuous and connected for every simplex τ of Z_s,* (2) $X = g^{-1}(Z_0)$ *is connected and,* (3) $S(\sigma) = X \cap \mathrm{st}_Y \sigma$ *is non-vacuous and connected for every simplex σ of $Y - X$.* Such a mapping may be called a *simplicial covering*, and the antecedent may be called a *covering complex.*

THEOREM. *Let Z be a barycentrically subdivided, connected, locally finite simplicial complex and let g: $Y \to Z$ be any (not necessarily simplicial) covering whose singular set Z_s is a subcomplex such that $S(\tau) = Z_0 \cap \mathrm{st}_Z \tau$ is non-vacuous and connected for every simplex τ of Z_s. If the index of branching $j(y)$ is finite for each point y of Y, then Y is a locally finite simplicial complex and g is a simplicial covering.*

Let $X = g^{-1}(Z_0)$ and $f = g \mid X$, so that f: $X \to Z_0$ is an unbranched covering. We are going to define a locally finite simplicial complex Y'

containing X and a simplicial mapping $g'\colon Y' \to Z$ such that $g' \mid X = f$; it will be clear that g' is a completion of f and hence equivalent to g.

If τ is any open n-dimensional simplex of Z_o, the components of $f^{-1}(\tau) = g'^{-1}(\tau)$ are open n-dimensional simplexes σ_i, i ranging over the cosets of the subgroup H of $\pi_1(Z_o)$ to which f belongs. If τ and τ^* are simplexes of Z_o such that $\tau < \tau^*$ (i.e. τ is on the boundary of τ^*) then there is a permutation $\rho = \rho_{\tau,\tau^*}$ of the cosets of H such that $\sigma_i < \sigma_j^*$ if and only if $i = \rho(j)$.

If τ is any open n-dimensional simplex of Z_s, the components of $g'^{-1}(\tau)$ are to be open n-dimensional simplexes σ_i, i ranging over the components $S_i(\tau)$ of $f^{-1}(S(\tau)) = f^{-1}(\operatorname{st}\tau)$. Since the index of branching of g is finite at each point of $g^{-1}(\tau)$, the number of simplexes comprised in any $S_i(\tau)$ is finite.

If τ and τ^* are simplexes of Z_s such that $\tau < \tau^*$, then $S(\tau) \supset S(\tau^*)$, so that each $S_j(\tau^*)$ is contained in some $S_i(\tau)$. The incidence relations in Y' that are to cover the incidence relation $\tau < \tau^*$ are: $\sigma_i < \sigma_j^*$ if and only if $S_i(\tau) \supset S_j(\tau^*)$.

If τ is a simplex of Z_s and τ^* a simplex of Z_o such that $\tau < \tau^*$ then the incidence relations in Y' covering this are to be: $\sigma_i < \sigma_j^*$ if and only if $S_i(\tau) \supset \sigma_j^*$.

It is easy to verify that the simplexes σ of Y', with the incidence relations described above, form a locally finite simplicial complex. (In order to prove that no two simplexes of Y' have the same vertices it is necessary to use the fact that Z has been barycentrically subdivided. For example, a simplicial subdivision of the 2-sphere Z might not be covered by a simplicial subdivision of a given Riemann surface if there were branching over both end-points of some 1-dimensional simplex of Z.) It is also easy to verify that g' is a completion of f and hence equivalent to g. The homeomorphism of Y' on Y induces the triangulation of Y.

Of special interest are the finitely branched coverings of a connected n-dimensional (combinatorial) manifold Z whose singular sets are pure $(n-2)$-dimensional simplicial complexes tamely imbedded in the interior of Z. (An $(n-2)$-dimensional simplicial complex is *pure* if every principal simplex is $(n-2)$-dimensional.) If we assume, as we may, that Z is triangulated in such a way that Z_s is a subcomplex, and then barycentrically subdivided, then, by the preceding theorem, such a covering is simplicial and the antecedent Y is a locally finite simplicial complex. Under what conditions is the covering complex Y also an n-dimensional manifold?

It is well known that, for $n = 2$, Y is always a manifold. For $n > 2$

the situation is more complicated. In any particular case it can be decided (in principle) by the following method. Let z be any vertex of the singular set Z_s, assumed to be in the interior of Z; then the boundary B of the star of z is a triangulated $(n-1)$-dimensional sphere and $B \cap Z_s$ is a pure $(n-3)$-dimensional subcomplex. The components of $g^{-1}(B)$ are finitely branched covering complexes of B whose singular sets are subcomplexes of $B \cap Z_s$. One has only to examine these components and decide whether or not they are $(n-1)$-dimensional spheres. The answer is affirmative in one general case (which includes all the examples that have been considered in the literature):

THEOREM. *Let Z be a connected, barycentrically subdivided, combinatorial n-dimensional manifold and let L be a polyhedrally imbedded combinatorial $(n-2)$-dimensional manifold such that the star of any vertex in L is flat*[11] *in Z. Then any finitely branched covering complex of Z whose singular set is a subcomplex of L is a combinatorial n-dimensional manifold.*

For simplicity let it be assumed that L is in the interior of Z. If L intersects the boundary of Z the proof following has to be modified.

Let g be the finitely branched covering, Y the covering complex, $K = g^{-1}(L)$, so that $Y-K$ is an unbranched covering space, with associated covering $e = g \mid Y-K$. If τ is any q-dimensional open simplex of L and τ_i, a q-dimensional open simplex of K, one of the components of $g^{-1}(\tau)$, then the closed star $\mathrm{St}\,\tau_i$ is a covering space of the closed star $\mathrm{St}\,\tau$ (the associated covering being the restriction h of g to $\mathrm{St}\,\tau_i$) whose singular set is a subcomplex of $\mathrm{St}_L\,\tau = L \cap \mathrm{St}\,\tau$. Since there is a homeomorphism that maps $\mathrm{St}\,\tau$ onto the Cartesian product $C \times E^{n-2}$ of the plane disc C: $x_1^2 + x_2^2 \leqq 1$ and the $(n-2)$-cell E^{n-2}: $0 \leqq x_i \leqq 1$ $(i = 3, \ldots, n)$ in such a way that $\mathrm{St}_L\,\tau$ is mapped onto $p \times E^{n-2}$ (where p denotes the point $x_1 = x_2 = 0$), the covering h must be of the form $d_m \times i$, where i maps E^{n-2} identically upon itself and d_m is the cyclic covering of C with branching index m at p exemplified in the branch point that the Riemann surface of the function $w = \sqrt[m]{z}$ has at the origin. Since all of these maps are simplicial it is clear that $\mathrm{St}\,\tau_i$ is a combinatorial n-cell.

7. The fundamental group of a branched covering

If S is an open subset of a space Y, an element of $\pi_1(Y)$ will be said to be *represented in* S if it is represented by a loop of the form $\alpha\gamma\alpha^{-1}$, where γ is a loop in S and α is a path in Y from the base point of $\pi_1(Y)$ to the base point of γ. Note that, if $S_1, S_2, \ldots$ are the components of S, an element of $\pi_1(Y)$ is represented in S if and only if it is represented in some S_i.

LEMMA. *Let Y be a connected, barycentrically subdivided, locally finite complex and let K be a subcomplex such that, for each vertex u of K, the intersection $S(u)$ of $Y-K$ with the open star* st u *of u is non-vacuous and connected. Then the injection homomorphism:*

$$\phi\colon \pi_1(Y-K)\to\pi_1(Y)$$

is onto, and its kernel is the consequence† *of those elements of $\pi_1(Y-K)$ that are represented in $\bigcup_u S(u)$.*

Since Y has been barycentrically subdivided, the stars st u are the components of $\bigcup_u$ st u (and the sets $S(u)$ are the components of $\bigcup_u S(u)$). Let T be a simplicial tree in $Y-K$ rooted at the base point of $\pi_1(Y-K)$ and meeting each St u at exactly one point. The set $T\cup\bigcup_u S(u)$ is connected, and the image of the injection homomorphism $\pi_1(T\cup\bigcup_u S(u))\to\pi_1(Y-K)$ is the consequence of the elements of $\pi_1(Y-K)$ that are represented in $\bigcup_u S(u)$. The image of the injection homomorphism $\pi_1(T\cup\bigcup_u S(u))\to\pi_1(Y)$ is clearly 1. The theorem follows from an application of van Kampen's theorem [13], regarding Y as the union of $Y-K$ and $T\cup\bigcup_u$ st u.

THEOREM. *Let Z be a barycentrically subdivided, connected, locally finite complex and let L be a subcomplex such that, for each vertex v of L, the intersection $S(v)$ of $Z-L$ with the open star* st v *of v is non-vacuous and connected. Let Y be a finitely branched covering of Z whose singular set Z_s is a subcomplex of L. Let H be the subgroup of $G=\pi_1(Z-L)$ to which the associated unbranched covering of $Z-L$ belongs. Then $\pi_1(Y)\approx H/N$, where N is the consequence of those elements of H that are represented in $\bigcup_v S(v)$.*

By a preceding theorem, Y is a locally finite complex, mapped simplicially onto Z. After another barycentric subdivision, Y and K, the inverse image of L, satisfy the conditions of the lemma. The theorem follows from the observations that an element of G is covered by an element of $\pi_1(Y-K)$ if and only if it belongs to H and that an element of $\pi_1(Y-K)$ is represented in $\bigcup_u S(u)$ if and only if the element of G that it covers is represented in $\bigcup_v S(v)$.

The following application of this theorem may be of some interest. In [5], I proved that the group $F=(S_1,S_2,\ldots,S_d\colon \prod_{i=1}^{d}S_i=1, S_i^{n_i}=1$ $(i=1,\ldots,d))$, where each n_i is a positive integer greater than 1, has a normal subgroup N with finite index in F and contains no element

† By the *consequence* of a set of elements in a group is meant the smallest normal subgroup that contains all these elements.

of finite order other than the identity. Let Z be the 2-sphere and select $d>1$ points $s_1, s_2, \ldots, s_d$ of Z. The fundamental group of $Z-(s_1 \cup s_2 \cup \ldots \cup s_d)$ is $(x_1, x_2, \ldots, x_d\colon \prod_{i=1}^{d} x_i = 1)$, where x_i is represented by a small loop around s_i. Denote by ϕ the homomorphism $x_i \to S_i$ of this group upon the group F. Since N has no elements of finite order, $x_i^m \in W = \phi^{-1}(N)$ if and only if $m \equiv 0 \pmod{n_i}$. Let X be the unbranched covering space of $Z-(s_1 \cup s_2 \cup \ldots \cup s_d)$ determined by W. Thus X is a regular covering and $W \approx \pi_1(X)$. The branched covering space Y of Z to which X is associated has fundamental group N. Thus we have proved the following theorem:

If $s_1, s_2, \ldots, s_d$ $(d>1)$ are points of the 2-sphere Z and $n_1, n_2, \ldots, n_d$ any positive integers greater than 1, there exists a regular covering Y of Z of finite index for which the index of branching is equal to n_i at each point over s_i.

Naturally Y is an orientable surface of genus

$$p = 1 - n + (n/2) \sum_{i=1}^{d} (1 - (1/n_i)),$$

where n is the index of N in F.

8. Generalizations

In § 1, I defined a spread $f\colon X \to Z$ only when X and Z are locally connected. If X and Z are arbitrary T_1-spaces, which are not necessarily locally connected, a mapping $f\colon X \to Z$ may be defined to be a spread† if the clopen subsets of the sets $f^{-1}(W)$, W ranging over the open sets of Z, form a basis of X. To such a spread a 'completion' $g\colon Y \to Z$ may be constructed, by a generalization of the process of § 2. A point y of $g^{-1}(z)$ is a function that associates to each open neighborhood W of z a quasi-component yW of $f^{-1}(W)$ in such a way that $yW_1 \subset yW_2$ whenever $W_1 \subset W_2$. Basic open sets U/W are defined as in § 2 for any clopen subset U of $f^{-1}(W)$. However, there are difficulties with this generalization in connection with the uniqueness theorem. Furthermore, its relation to Freudenthal's generalized ideal

† If Z is separable, regular, and X is compact, Hausdorff, a mapping f of X into Z is a spread if and only if $f^{-1}(z)$ is totally disconnected for every z, i.e. if and only if f is a so-called *light* mapping. (Let x be any point in any open set G of X and let $W_1 \supset W_2 \supset \ldots$ be a basic sequence of neighborhoods of $z = f(x)$. Let F_n be the component of $f^{-1}(\overline{W}_n)$ that contains x. Since $f^{-1}(z)$ is totally disconnected, and $\cap_n F_n$ is connected, $\cap_n F_n = x$. Hence, for some index n, $F_n \subset G$. Thus f is a spread.) That the compactness of X is essential here is shown by the following example constructed by John Milnor: Let X be the plane set consisting of all straight lines $y = ax + b$, a and b rational; let Z be the x-axis and let f map X upon Z by orthogonal projection. This is a light mapping (and X is locally connected), but f is not a spread. In fact, for any open interval W of Z the set $f^{-1}(W)$ is connected.

compactification is unclear. For these reasons I am not certain that it is the proper generalization, and have accordingly restricted myself to the locally connected case.

It would be interesting to generalize our theory of covering spaces with singularities to a theory of fibre spaces with singularities. A satisfactory definition of 'fibre space with singularities' should encompass at least the types considered by Seifert [17] and probably also the type considered by Montgomery and Samelson [15]. In an attempt at such a generalization, I replaced the set of components of quasi-components of $f^{-1}(W)$ by a decomposition of $f^{-1}(W)$ subject to suitable conditions. However, the resulting theory turned out to be rather unsatisfactory, in that the associated non-singular fibre space has to have a 'totally disconnected group'. Such a restriction is obviously much too severe. The example of the lens spaces, which are singular fibre spaces in the sense of Seifert [17], shows that a singular fibre space cannot be uniquely recovered from its associated non-singular fibre space, at least unless some additional structure is posited. In the case of the Seifert singular fibre spaces the additional structure is roughly the type of torus knot determined by a non-singular fibre in the neighborhood of a singular fibre, and is given by the numbers α, β in the 'symbol' (cf. [17]).

PRINCETON UNIVERSITY

REFERENCES

[1] J. W. ALEXANDER, *Note on Riemann spaces*, Bull. Amer. Math. Soc., 26 (1920), pp. 370–372.

[2] ——, *Topological invariants of knots and links*, Trans. Amer. Math. Soc., 30 (1928), pp. 275–306.

[3] —— and G. B. BRIGGS, *On types of knotted curves*, Ann. of Math., 28 (1926), pp. 562–586.

[4] C. CHEVALLEY, Theory of Lie groups, Princeton, 1946.

[5] R. H. FOX, *On Fenchel's conjecture about F-Groups*, Matematisk Tidsskrift B (1952), pp. 61–65.

[6] ——, *Recent development of knot theory at Princeton*, Proc. Int. Congress of Math., 2 (1950), pp. 453–457.

[7] —— and R. B. KERSHNER, *Concerning the transitive properties of geodesics on a rational polyhedron*, Duke Math. J., 2 (1936), pp. 147–150.

[8] H. FREUDENTHAL, *Über die Enden topologischer Räume und Gruppen*, Math. Zeit., 33 (1931), pp. 692–713.

[9] ——, *Neuaufbau der Endertheorie*, Ann. of Math., 43 (1942), pp. 261–279.

[10] ——, *Enden und Primenden*, Fund. Math., 39 (1952), pp. 189–210.

[11] V. K. A. M. GUGENHEIM, *Piecewise linear isotropy and embedding of elements and spheres*, Proc. London Math. Soc., 3 (1953), pp. 29–53.

[12] P. Heegard, Forstudier til en topologisk Teori for de algebraiske Fladers Sammenhaeng, Dissertation, København (1898).

[13] E. van Kampen, *On the connection between the fundamental groups of some related spaces*, Amer. J. Math., 55 (1933), pp. 261–267.

[14] B. v. Kerékjártó, Vorlesungen über Topologie, Berlin, 1923.

[15] D. Montgomery and H. Samelson, *Fiberings with singularities*, Duke Math. J., 13 (1946), pp. 51–56.

[16] K. Reidemeister, *Knoten und Gruppen*, Hamburg Abh., 5 (1926), pp. 7–23.

[17] H. Seifert, *Topologie dreidimensionaler gefasterter Raüme*, Acta Math., 60 (1933), pp. 147–238.

[18] ——, *Über das Geschlecht von Knoten*, Math. Ann., 110 (1934), pp. 571–592.

[19] ——, *Die Verschlingungsinvarianten der Zyklischen Knotenüberlagerungen*, Hamb. Abh., 11 (1935), pp. 84–101.

[20] ——, *La théorie des nœuds*, L'enseignement Math., 35 (1936), pp. 201–212.

[21] H. Tietze, *Über die topologischen Invarianten mehrdimensionaler Mannigfaltigkeiten*, Monatshefte für Math. und Physik, 19 (1908), pp. 1–118.

[22] A. W. Tucker, *Branched and folded coverings*, Bull. Amer. Math. Soc., 42 (1936), pp. 859–862.

[23] H. Weyl, Die Idee der Riemannschen Fläche, Berlin, 1923.

A Relation Between Degree and Linking Numbers

F. B. Fuller

BY imbedding a complex in Euclidean space and selecting a neighborhood of it Lefschetz reduced the fixed-point formula for a complex to the fixed-point formula for a manifold with boundary. This note uses the imbedding method to derive the fixed-point formula from a relation involving linking numbers.

Let f be a mapping of Euclidean n-space into itself with an isolated fixed point x_0. Let S denote a sphere with center at x_0 whose radius r is small enough so that S contains no other fixed points. Then

$$g(x)=x_0+r\frac{f(x)-x}{|f(x)-x|}$$

defines a mapping g of S into itself. The degree of this mapping is called the *index* of f at x_0. Since homotopic mappings of S into itself have the same degree, the index is independent of the choice of r and also of deformations of f which keep other fixed points away from x_0. To define the index of f at x_0 it is sufficient that f be defined only on a neighborhood of x_0. A more convenient definition than the index is the *multiplicity*, equal to $(-1)^n$ times the index.

LEMMA 1. *Let ρ be a retraction of Euclidean n-space E^n onto a p-plane E^p, and let f be a mapping of E^p into itself with an isolated fixed point x_0. Then x_0 is likewise an isolated fixed point of $f\rho$ and the multiplicities of f and $f\rho$ at x_0 are equal.*

PROOF. The multiplicity of $f\rho$ at x_0 is the same for any retraction ρ. Hence ρ may be chosen to be a projection of E^n onto E^p obtained as the product of a sequence of projections $E^n\rightarrow E^{n-1}\rightarrow\ldots\rightarrow E^{p+1}\rightarrow E^p$ so that the proof reduces to proving the case $n=p+1$.

Let $n=p+1$ and let S be a sphere in E^n about x_0 containing no other fixed points. The index of $f\rho$ at x_0 is the degree of the g which maps S into itself and the index of f is the degree with which the equator (intersection of S with E^p) maps into itself. g interchanges the two hemispheres of S so that if g is followed by a reflection in the plane E^p, a mapping $\bar{g}$ is obtained which preserves the hemispheres. Hence degree $\bar{g}$ = degree of the mapping on the equator. But degree $\bar{g} = -$degree g. Hence the indices of f and $f\rho$ have opposite sign, so the multiplicities are equal.

Let κ be a mapping of a finite simplicial complex K into itself for which each fixed point is isolated and contained in the interior of a principal simplex of K. Since a principal simplex σ is defined as one which is not a face of some other simplex, a fixed point in the interior of σ must have a neighborhood which maps into σ. The multiplicity of κ at the fixed point can then be defined and can be shown [1] to be independent of the simplicial decomposition of K. The sum of the multiplicities of the fixed points is called the *algebraic number of fixed points* of κ.

The above complex K can be rectilinearly imbedded in a Euclidean n-space E^n. In E^n one can construct [3] a rectilinearly imbedded n-manifold N with boundary $\dot{N}$ which contains K in its interior in such a way that K is a deformation retract of N. The retraction ρ of N onto K, followed by κ, defines a mapping $\kappa\rho$ of N into its interior with the same fixed points as κ and, according to Lemma 1, with the same multiplicity at each fixed point. Thus κ and $\kappa\rho$ have the same algebraic number of fixed points.

Lemma 2. *Let N be a finite n-manifold rectilinearly imbedded in Euclidean n-space E^n with boundary $\dot{N}$. Let f be a mapping of N into its interior with only isolated fixed points. Then*

$$g(x) = \frac{f(x) - x}{|f(x) - x|}$$

defines a mapping from the boundary $\dot{N}$ to the unit sphere S^{n-1} with center at the origin. If $\dot{N}$ and S^{n-1} are naturally oriented by E^n, then the degree of g is equal to the sum of the indices of the fixed points of f.

Proof. Select about each fixed point P_i a sphere S_i, small enough so that it is contained in N and contains no other fixed points. Let e_i denote the open ball bounded by S_i. The function

$$g(x) = \frac{f(x) - x}{|f(x) - x|}$$

extends to $N - \bigcup e_i$, defining a mapping $\bar{g}$ from $N - \bigcup e_i$ to S^{n-1}. If $N - \bigcup e_i$, $\dot{N}$ and S^{n-1} are naturally oriented by E^n then, considering them as chains, the equation $\partial(N - \bigcup e_i) = \dot{N} - \sum S_i$ holds. Thus the degree of $\bar{g}$ on $\dot{N}$ is equal to the sum of the degrees of $\bar{g}$ on the S_i. But the degree of $\bar{g}$ on S_i is the index of f at the fixed point P_i and the degree of $\bar{g}$ on $\dot{N}$ is the degree of g, so the desired relation holds.

Lemma 2, combined with the discussion preceding it, shows at once that the algebraic number of fixed points of a mapping κ of a complex into itself is an invariant of the homotopy class of κ.

THEOREM 1. *Let f and g be two mappings from an oriented n-manifold M into Euclidean $(n+1)$-space E^{n+1} whose images fM and gM are disjoint. The vectors*

$$\frac{g(x)-f(x)}{|g(x)-f(x)|}$$

define a mapping $|f,g|$ of M into the unit sphere S^n with center at the origin. Let the cycles u_i^p form a basis for the rational p-cycle classes on M and let S^n be naturally oriented by E^{n+1}. Then the degree of $|f,g|$ is equal to

$$\sum(-1)^{p+1}\epsilon_{ij}^p \operatorname{Lk}(f_{\#}u_i^p, g_{\#}u_j^{n-p}),$$

where the matrix (ϵ_{ij}^p) of coefficients is the transposed inverse of the intersection matrix $[(u_i^p, u_j^{n-p})]$ and $\operatorname{Lk}(f_{\#}u_i^p, g_{\#}u_j^{n-p})$ is the linking number of the disjoint singular cycles $f_{\#}u_i^p$ and $g_{\#}u_j^{n-p}$.

PROOF. Let d be the diagonal mapping $x \to (x,x)$ of M into the product manifold $M \times M$. $|f,g|$ is equal to d followed by the mapping

$$\sigma(x,y) = \frac{g(y)-f(x)}{|g(y)-f(x)|}$$

from $M \times M$ to S^n. dM, the diagonal cycle in $M \times M$, is equal to the sum of product cycles

$$\sum \epsilon_{ij}^p u_i^p \times u_j^{n-p},$$

where the matrix (ϵ_{ij}^p) of coefficients is equal [2] to the transposed inverse of the intersection matrix $[(u_i^p, u_j^{n-p})]$. Furthermore, by a known theorem on linking numbers [1]

$$\sigma(u_i^p \times u_j^{n-p}) = (-1)^{p+1}\operatorname{Lk}(f_{\#}u_i^p, g_{\#}n_j^{n-p}) \cdot S^n.$$

Hence the degree of $|f,g| =$ degree of $\sigma d =$

$$\sum(-1)^{p+1}\epsilon_{ij}^p \operatorname{Lk}(f_{\#}u_i^p, g_{\#}u_j^{n-p}),$$

which is the desired relation.

THEOREM 2. (Lefschetz.) *The algebraic number of fixed points of a mapping κ of a finite complex into itself is equal to the Lefschetz number $\sum(-1)^p$ trace T_p, where T_p is the linear transformation induced by κ in the rational p-dimensional homology group of K.*

PROOF. Imbed K rectilinearly in a Euclidean space E^n of dimension $n > 1 + 2\dim K$. K is then contained as a deformation retract in the interior of an n-manifold N rectilinearly imbedded in E^n with boundary $\dot{N}$. The retraction of N onto K, followed by κ, defines a mapping

f of N into its interior with the same algebraic number of fixed points and the same Lefschetz number as κ. Let 1 denote the inclusion of $\dot{N}$ into E^n. By Lemma 2 the algebraic number of fixed points of f is equal to $(-1)^n$ times the degree of $|1, f|$ (see Theorem 1), mapping $\dot{N}$ into a unit sphere.

Let β_i^p be a basis for the rational relative p-cycle classes of the relative manifold $(N, \dot{N})$ and let α_j^q be a basis for the rational absolute p-cycle classes of N. These bases may be chosen [4] so that the intersection matrices satisfy $(\beta_i^p, \alpha_j^{n-p}) = (\delta_{ij})$. Because N is imbedded in Euclidean space, it has the special property that the inclusion map j: $H_p(N) \to H_p(N, \dot{N})$ is zero. For, since $H_p(E^n) = 0$, the further inclusion of $H_p(N)$ into $H_p(E^n, E^n - \operatorname{int} N)$ is zero, but this can only be because $j = 0$ inasmuch as the inclusion of $H_p(N, \dot{N})$ into $H_p(E^n, E^n - \operatorname{int} N)$ is an isomorphism. The case $p = 0$ is not covered by this argument, but follows from $H_0(N, \dot{N}) = 0$. Exactness of the homology sequence

$$\xrightarrow{0} H_{p+1}(N, \dot{N}) \xrightarrow{\partial} H_p(\dot{N}) \longrightarrow H_p(N) \xrightarrow{0} H_p(N, \dot{N}) \longrightarrow$$

with the condition $j = 0$, shows that the α_j^p may be chosen to lie on $\dot{N}$, whereupon the α_j^p and the $\partial\beta_i^{p+1}$ together form a basis for the rational p-cycle classes of $\dot{N}$.

The intersection matrices for these bases on $\dot{N}$ can be found as follows. Generally the relation $(\partial\beta, \alpha)$ in $\dot{N} = (\beta, \alpha)$ in N holds so that $(\partial\beta_i^p, \alpha_j^{n-p}) = (\beta_i^p, \alpha_j^{n-p}) = \delta_{ij}$. Furthermore, since $n - 1 > 2 \dim K$ and since N has the homotopy type of K, there are no basis cycles α^{p-1}, α^{n-p} of dual dimensions on N. The duality between the α's and the β's then shows that there are likewise no basis cycles $\partial\beta^p$, $\partial\beta^{n-p+1}$ of dual dimensions on N. Thus all intersection matrices are of the form $[(\partial\beta_i^p, \alpha_j^{n-p})] = [\delta_{ij}]$ or of the form $[(\alpha_j^p, \partial\beta_i^{n-p})]$.

If Theorem 1 is now applied to these bases and the mappings 1 and f, one obtains (noting that the terms $\operatorname{Lk}(\alpha_j^p, f_{\#}\partial\beta_j^{n-p})$ vanish):

$$\begin{aligned} \text{degree } |1, f| &= \sum(-1)^{p+1} \operatorname{Lk}(\partial\beta_i^{p+1}, f_{\#}\alpha_i^{n-1-p}) \\ &= \sum(-1)^{p+1} (\beta_i^{p+1}, f_{\#}\alpha_i^{n-1-p}) \\ &= \sum(-1)^{p+1} \operatorname{trace} T_{n-1-p} \\ &= (-1)^n \sum(-1)^p \operatorname{trace} T_p. \end{aligned}$$

But degree $|1, f|$ is also equal to $(-1)^n$ times the algebraic number of fixed points, so that the desired relation follows at once.

CALIFORNIA INSTITUTE OF TECHNOLOGY

REFERENCES

[1] P. ALEXANDROFF and H. HOPF, Topologie, Berlin, 1935, pp. 539 and 496.

[2] W. V. D. HODGE, The theory and applications of harmonic integrals, Cambridge, 1941, p. 66.

[3] S. LEFSCHETZ, Algebraic topology, New York, 1942, p. 292.

[4] ——, Topology, New York, 1930, p. 181.

Die Coinzidenz-Cozyklen und eine Formel aus der Fasertheorie

H. Hopf

Die Coinzidenzpunkte zweier Abbildungen F und G eines Raumes X in einen Raum Y, also die Punkte $x \in X$ mit $F(x) = G(x)$, bilden den Gegenstand klassischer Untersuchungen von Lefschetz. Ein zentraler Begriff dieser Untersuchungen, derjenige der 'algebraischen Anzahl' von Coinzidenzpunkten, läßt sich, wenn Y eine n-dimensionale Mannigfaltigkeit ist, im Rahmen der Lefschetzschen Ideen und Methoden folgendermaßen erklären: es sei P die durch $P(x) = F(x) \times G(x)$ gegebene Abbildung von X in das cartesische Quadrat $Y \times Y$; dann ist die 'algebraische Anzahl der Coinzidenzpunkte von F und G auf einem n-dimensionalen Zyklus z von X' erklärt als die Schnittzahl des Zyklus $P(z)$ mit der Diagonale D in $Y \times Y$. Diese Zahl ändert sich nicht, wenn man z durch einen homologen Zyklus ersetzt, und die so erklärte Funktion der n-dimensionalen Homologieklassen stellt eine n-dimensionale Cohomologieklasse $\bar{\Omega}\ (f, g)$ in X dar—die Klasse der 'Coinzidenz-Cozyklen' von F und G. Diese Klasse $\bar{\Omega}$ hängt nur von den Homotopieklassen, und sogar nur von den Homologietypen der Abbildungen F und G ab, und eine berühmte Formel von Lefschetz drückt $\bar{\Omega}(F, G)$ explizit durch Größen des Cohomologieringes von X aus, welche durch diese Homologietypen bestimmt sind. Wir werden später noch Gelegenheit haben, an diese Formel zu erinnern (5.3); im Augenblick ist dies nicht nötig, da unser Hauptziel andere Eigenschaften von $\bar{\Omega}$ betrifft.

Es zeigt sich nämlich, daß die Klasse $\bar{\Omega}$ nicht nur für den Zweck, für den sie ursprünglich definiert worden ist, sondern auch für andere Zwecke eine Rolle spielt, und zwar in der Erweiterungstheorie der Abbildungen, also der Theorie, die von den Aufgaben handelt, eine für einen Teil $X' \subset X$ gegebene Abbildung $X' \to Y$ zu einer Abbildung $X \to Y$ zu erweitern, und von den Hindernissen, die sich solchen Erweiterungen entgegenstellen. Nun weiß man, daß sich diese Theorie weitgehend von den Abbildungen $X \to Y$ auf die Schnittflächen ('cross-sections') oder, wie ich lieber sagen will, auf die 'Felder' in

Faserräumen mit der Basis X und der Faser Y übertragen läßt; daher ist es kein Wunder, daß auch die Klasse $\bar{\Omega}$ nicht nur für Abbildungen, sondern auch für Felder f, g in einem Faserraum erklärt werden kann; diese verallgemeinerte Klasse soll $\bar{\omega}(f,g)$ heißen. Ihre Definition wird in dem nachstehenden § 1 durchgeführt; die darin enthaltene naheliegende Definition der Klasse $\bar{\Omega}(F,G)$ für Abbildungen F, G weicht von der eingangs ausgesprochenen nur in der Form, nicht im Inhalt ab.

Unser Ziel ist eine Formel, die ein spezielles Erweiterungs- und Hindernisproblem in Faserräumen betrifft und in welcher die Klasse $\bar{\omega}(f,g)$ auftritt. Nach Vorbereitungen in den §§ 2 und 3, welche zum großen Teil bekannte Dinge rekapitulieren, erfolgen Aufstellung und Beweis der Formel im § 4. In dem kurzen § 5 wird die Formel von der Fasertheorie wieder in die Abbildungstheorie hinein spezialisiert, wobei dann auch die schon erwähnte Lefschetzsche Formel eingreift. Auf Anwendungen der Formel gehe ich in dieser Arbeit nicht ein.

Ich habe die Formel bereits vor einigen Jahren ausgesprochen [4], den Beweis bisher aber nicht veröffentlicht, da ich immer hoffte, die Formel würde als Spezialfall allgemeinerer Tatbestände in der Fasertheorie erkannt und es würde dadurch die ausführliche Darstellung eines umständlichen Beweises für einen sehr speziellen Satz überflüssig gemacht werden. Da aber einerseits bisher nichts derartiges erfolgt ist, andererseits die Formel bereits mehrfach angewendet worden ist [3, 4, 5, 2], will ich nicht noch länger warten.

Vorbemerkungen. Ich werde mich möglichst ausgiebig auf das Buch von Steenrod über Faserräume [6] stützen Durchweg sollen die folgenden Festsetzungen gelten: Es ist R ein Faserraum mit der Basis X und der Faser Y; die natürliche Projektion $R \to X$ heißt p. Die Basis X ist ein endliches Polyeder, K eine Simplizialzerlegung von X, K^s das s-dimensionale Gerüst von K; die Simplexe von K heißen σ. Die Abbildungen, welche die Teile $p^{-1}\sigma = \sigma \times Y$ von R in eine Faser retrahieren, heißen r. Ein 'Feld' f über einer Teilmenge X' von X ist eine Abbildung $X' \to R$ mit $pf(x) = x$ für jeden Punkt $x \in X'$. Die Faser Y ist eine n-dimensionale geschlossene orientierbare Mannigfaltigkeit.

Bei zwei Gelegenheiten werde ich—obwohl es sich durch Heranziehung von Koeffizientenbündeln vermeiden ließe ([6], 150 ff.)—der Einfachheit halber voraussetzen, daß in einer gewissen Dimension q die q-te Homologiegruppe $H_q(Y)$ 'stabil' in folgendem Sinne ist: wenn man einen Punkt x_1 längs einem Wege W in einen Punkt x_2 von X überführt, so ist die dadurch bewirkte isomorphe Abbildung der Gruppe H_q der Faser $p^{-1}x_1$ auf die Gruppe H_q der Faser $p^{-1}x_2$

unabhängig vom Wege W; man kann also die Gruppen H_q der verschiedenen Fasern in eindeutiger Weise miteinander identifizieren. Für $q=n$ bedeutet dies, daß die Faserung 'orientierbar' ist, d.h. daß sich die Fasern so orientieren lassen, daß diese Orientierung sich stetig und eindeutig durch den ganzen Raum fortsetzt; wir setzen voraus, daß unser Raum R in diesem Sinne orientierbar sei (es wird aber später noch eine zweite Stabilitätsvoraussetzung gemacht werden). Bekanntlich ist die Voraussetzung der Stabilität in allen Dimensionen von selbst erfüllt, wenn X einfach zusammenhängend oder wenn die Gruppe der Faserung zusammenhängend ist. Diese Gruppe wird übrigens keine Rolle spielen.

Alle vorkommenden Abbildungen sollen stetig sein.

Cozyklen werden durch griechische Buchstaben, Cohomologieklassen durch überstrichene griechische Buchstaben bezeichnet (z.B.: $\bar{\alpha}$ ist die Klasse des Cozyklus α).

∂ und δ bezeichnen Rand und Corand.

§ 1. Die Coinzidenz-Cozyklen

1.1. f und g seien Felder über K^n. Wir setzen vorläufig voraus, daß sie 'in allgemeiner Lage' sind, d.h. über K^{n-1} keinen Coinzidenzpunkt haben. Für jedes σ^n ist dann von den topologischen Bildern $f(\sigma^n)$ und $g(\sigma^n)$ jedes fremd zum Rande des andern; daher ist in der (berandeten) $2n$-dimensionalen Mannigfaltigkeit $p^{-1}\sigma^n = \sigma^n \times Y$, die durch eine Orientierung von σ^n und die feste Orientierung von Y selbst orientiert ist, in bekannter Weise die Schnittzahl $s(f(\sigma^n), g(\sigma^n))$ erklärt. Diese Funktion der Simplexe σ^n ist ein Cozyklus in K^n, der 'Coinzidenz-Cozyklus' von f und g; wir bezeichnen ihn durch $\omega(f, g)$.

Aus bekannten Eigenschaften der Schnittzahlen folgt unmittelbar: Wenn man f und g homotop so deformiert, daß die allgemeine Lage in keinem Augenblick verletzt wird, so ändert sich ω nicht; sowie: ist f' zu f und g' zu g hinreichend benachbart, so ist $\omega(f', g') = \omega(f, g)$.

Die Cohomologieklasse von $\omega(f, g)$ nennen wir $\bar{\omega}(f, g)$. Unser nächstes Ziel ist, $\bar{\omega}(f, g)$ auch für Felder f, g zu erklären, die nicht in allgemeiner Lage sind.

1.2. f, g seien beliebige Felder über K^n.

Behauptung. *Es gibt zu ihnen homotope Felder f', g', die in allgemeiner Lage sind; und zwar kann man sogar durch beliebig kleine Deformationen von f, g zu f', g' übergehen.*

Beweis. Für jedes σ^0, das Coinzidenzpunkt ist, verschiebe man den Punkt $g(\sigma^0)$ stetig in einen Punkt $g_0(\sigma^0) \neq f(\sigma^0)$ über σ^0 und erweitere diese Verschiebungen zu einer Homotopie von g über K^n ([6], 176);

man erhält so zu f, g homotope Felder $f_0=f$, g_0 ohne Coinzidenzpunkt über K^0. Man habe bereits zu f, g homotope Felder f_i, g_i ohne Coinzidenzpunkt über K^i, und es sei $i<n-1$; da für jedes σ^{i+1} die Mannigfaltigkeit $\sigma^{i+1}\times Y$ die Dimension $i+1+n$, jedes der Bilder $f(\sigma^{i+1})$ und $g(\sigma^{i+1})$ die Dimension $i+1$ hat, $(i+1)+(i+1)<i+1+n$ ist, und da von den beiden Bildern jedes zum Rande des andern fremd ist, kann man die Bilder, ohne Veränderungen an den Rändern, stetig so deformieren, daß sie zueinander fremd werden; diese Deformationen erweitert man wieder zu Homotopien der ganzen Felder. Es ist klar, daß man immer mit beliebig kleinen Deformationen auskommt. $f'=f_{n-1}$ und $g'=g_{n-1}$ erfüllen die Behauptung.

1.3. Daß zwei Felder g_0, g_1 über K^n zueinander homotop sind, bedeutet bekanntlich: bezeichnet I das Intervall $0\leqq t\leqq 1$, so existiert eine Abbildung G von $K^n\times I$ in R, sodaß $pG(X\times t)=x$, $G(x\times 0)=g_0(x)$, $G(x\times 1)=g_1(x)$ für alle $x\in K^n$ ist.—Außer den homotopen Feldern g_0, g_1 sei ein Feld f über K^n gegeben, das zu g_0 und zu g_1 in allgemeiner Lage ist.

Behauptung. *Man kann durch beliebig kleine Deformationen, welche g_0 und g_1 nicht ändern, f und G in Abbildungen f' und G' überführen, welche über K^{n-2} keinen Coinzidenzpunkt haben*

$$(\text{d.h. } f'(x)\neq G'(x\times t) \quad \textit{für} \quad x\in K^{n-2},\quad t\in I).$$

Beweis, *analog zu* 1.2. Man beseitige durch kleine Deformationen von f und G, unter Festhaltung von G für $t=0$ und $t=1$, die Coinzidenzpunkte von f und G der Reihe nach über K^i für $i=0,1,\ldots,n-2$; dies ist möglich, da die Mannigfaltigkeit $\sigma^{i+1}\times Y$ die Dimension $i+1+n$, das Bild $f(\sigma^{i+1})$ die Dimension $i+1$, das Bild $G(\sigma^{i+1}\times I)$ die Dimension $i+2$ hat, für $i<n-2$ aber noch $(i+1)+(i+2)<i+1+n$ ist.

1.4. g_0 und g_1 seien miteinander homotop und in allgemeiner Lage zu f.

Behauptung.

$$\omega(f,g_0)\sim\omega(f,g_1), \quad \textit{also} \quad \overline{\omega}(f,g_0)=\overline{\omega}(f,g_1). \tag{1.4.1}$$

Beweis. Wir wenden zunächst 1.3 an: da f' durch eine beliebig kleine Deformation aus f entsteht, dürfen wir annehmen, daß bei dieser Deformation stets die allgemeine Lage zu g_0 und g_1 gewahrt bleibt; daher ist, wie schon in 1.1 festgestellt, $\omega(f',g_0)=\omega(f,g_0)$, $\omega(f',g_1)=\omega(f,g_1)$; somit dürfen wir annehmen, daß das Feld f in (1.4.1) bereits die Eigenschaften von f' aus 1.3 hat. Die Abbildung G' aus 1.3 wollen wir jetzt G nennen. Dann haben für jedes σ^{n-1}

die Bilder $f(\sigma^{n-1})$ und $G(\sigma^{n-1}\times I)$ in $p^{-1}\sigma^{n-1}=\sigma^{n-1}\times Y$ die Eigenschaft, daß jedes zum Rande des anderen fremd ist (wegen der Eigenschaften aus 1.3 und wegen der allgemeinen Lage von f zu g_0 und g_1). Folglich ist die Schnittzahl $s'(f(\sigma^{n-1}), G(\sigma^{n-1}\times I))$ in der $(2n-1)$-dimensionalen (berandeten) Mannigfaltigkeit $\sigma^{n-1}\times Y$ definiert. Der Corand $\delta s'$ der als Cokette aufgefaßten Funktion s' der Simplexe σ^{n-1} hat auf einem σ^n den Wert

$$\delta s'(\sigma^n)=s'(f(\partial\sigma^n),\ G(\partial\sigma^n\times I)). \tag{1.4.2}$$

Wir werden (1.4.1) dadurch beweisen, daß wir zeigen: es ist

$$\omega(f,g_1)-\omega(f,g_0)=\pm\delta s',$$

mit anderen Worten:

$$s(f(\sigma^n),g_1(\sigma^n))-s(f(\sigma^n),g_0(\sigma^n))=\pm s'(f(\partial\sigma^n),G(\partial\sigma^n\times I)), \tag{1.4.3}$$

wobei hier—wie auch im Folgenden—die unbestimmt gebliebenen Vorzeichen nur von n abhängen und wobei Schnittzahlen in $\sigma^n\times Y$ mit s, solche in $Z=\partial\sigma^n\times Y$ mit s' bezeichnet sind.

Für den Beweis von (1.4.3) denken wir uns σ^n im Inneren eines größeren Simplexes $\bar\sigma^n$ gelegen (das nichts mit unserem Faserraum zu tun hat) und entsprechend $\sigma^n\times Y$ in $\bar\sigma^n\times Y$ eingebettet. Das Feld f über σ^n läßt sich zu einem Feld über $\bar\sigma^n$ fortsetzen, sodaß also $f(\bar\sigma^n)=A$ erklärt ist; Schnittzahlen in $\bar\sigma^n\times Y$ dürfen wir ebenso wie diejenigen in $\sigma^n\times Y$ mit s bezeichnen. Der $(n-1)$-dimensionale Zyklus ∂A liegt über $\partial\bar\sigma^n$ und ist daher fremd zu der Kette $G(\sigma^n\times I)$; er hat daher mit dem n-dimensionalen Zyklus $\partial G(\sigma^n\times I)$ in $\bar\sigma^n\times Y$ die Verschlingungszahl 0, und folglich ist auch die Schnittzahl $s(A,\partial G(\sigma^n\times I))=0$. Nun ist, wenn wir noch zur Abkürzung $G(\partial\sigma^n\times I)=B$ setzen,

$$s(A,\partial G(\sigma^n\times I))=s(A,B)+(-1)^n s(A,g_1(\sigma^n))-(-1)^n s(A,g_0(\sigma^n)),$$

also

$$s(f(\sigma^n),g_1(\sigma^n))-s(f(\sigma^n),g_0(\sigma^n))=\pm s(A,B). \tag{1.4.4}$$

Da B auf der Mannigfaltigkeit Z liegt, ist

$$s(A,B)=\pm s'(A\cdot Z,B),$$

wobei $A\cdot Z$ den Schnitt von A und Z in $\bar\sigma^n\times Y$ bezeichnet, und da

$$A\cdot Z=\pm\partial f(\sigma^n)=\pm f(\partial\sigma^n)$$

ist, folgt somit

$$s(A,B)=\pm s'(f(\partial\sigma^n),B).$$

Hieraus und aus (1.4.4) folgt (1.4.3).

1.5. Die Definition von $\overline{\omega}(f,g)$ für beliebige Felder f,g über K^n erfolgt jetzt durch die Festsetzung: $\overline{\omega}(f,g)=\overline{\omega}(f',g')$, wobei f', g' zu f, g homotop und zueinander in allgemeiner Lage sind. Die Möglichkeit und Eindeutigkeit dieser Definition ergibt sich unmittelbar aus 1.2 und 1.4, wobei wir 1.4 auch under Vertauschung der Rollen von f und g berücksichtigen.

Ferner folgt aus 1.4: Die Klasse $\overline{\omega}(f,g)$ ändert sich nicht bei homotoper Abänderung von f und g.

1.6. Bisher haben wir f und g nur über K^n betrachtet; daher war es trivial, daß die in 1.1 erklärte Cokette ω ein Cozyklus (in K^n) ist. Jetzt seien f und g über K^{n+1} (und vielleicht noch über einem größeren Teil von K) gegeben; wie in 1.1 sollen sie keine Coinzidenzpunkte über K^{n-1} haben.

Behauptung. *Auch dann ist $\omega(f,g)$ ein Cozyklus (in K).*

Beweis. Wir haben für ein beliebiges σ^{n+1} zu zeigen, daß

$$\omega(f,g)\,(\partial\sigma^{n+1})=0$$

ist.—Bezeichnet r eine Abbildung, die $p^{-1}\sigma^{n+1}=\sigma^{n+1}\times Y$ auf eine Faser Y retrahiert, so folgt aus der Existenz von f und g im Innern von σ^{n+1}, daß die Randsphäre $\dot{\sigma}$ von σ^{n+1} durch rf und rg 0-homotop in Y abgebildet wird. Verstehen wir unter f_0 und g_0 die Teile der Felder f und g über $\dot{\sigma}$, so folgt aus dieser 0-Homotopie: f_0 und g_0 sind homotop zu Feldern f_0' und g_0' mit der Eigenschaft: rf_0' und rg_0' bilden $\dot{\sigma}$ auf je einen Punkt von Y ab. Da wir annehmen dürfen, daß diese beiden Punkte verschieden sind, besitzen f_0' und g_0' keinen Coinzidenzpunkt; es ist also $\omega_0(f_0',g_0')=0$ und nach 1.5 auch $\omega_0(f_0,g_0)=0$ (dabei hat ω_0 für den Komplex $\dot{\sigma}$ dieselbe Bedeutung wie ω für K). Da aber

$$\omega(f,g)\,(\partial\sigma^{n+1})=\omega_0(f_0,g_0)\,(\partial\sigma^{n+1})$$

ist, ist damit die Behauptung bewiesen.

Es ist also auch für beliebige Felder über K^{n+1} (die nicht in allgemeiner Lage zu sein brauchen) gemäß 1.5 $\overline{\omega}(f,g)$ als Cohomologieklasse in K (nicht nur in K^n) erklärt.

1.7. Nach bekannten Vorzeichenregeln für Schnittzahlen gilt in 1.1: $\omega(g,f)=(-1)^n\,\omega(f,g)$, und daraus folgt für beliebige f, g:

$$\overline{\omega}(g,f)=(-1)^n\,\overline{\omega}(f,g). \tag{1.7.1}$$

Für jedes Feld f ist speziell $\overline{\omega}(f,f)$ erklärt. Aus 1.7.1. folgt: Bei ungeradem n ist $2\overline{\omega}(f,f)=0$.

1.8. Die Faserung sei trivial, d.h. $R=X\times Y$; ein Feld f stellt die Abbildung F in Y dar, die durch $f(x)=x\times F(x)$ gegeben ist. Betrachten wir den Spezialfall, in dem $F(x)$ ein konstanter Punkt y_0 ist, und sei g ein zweites Feld, G die durch g dargestellte Abbildung; nehmen wir ferner für den Augenblick an, daß G simplizial ist und y_0 im Innern eines n-dimensionalen Simplexes τ^n der benutzten Zerlegung von Y liegt. Dann sieht man leicht (man prüfe das Vorzeichen!), daß die Schnittzahl $s(f(\sigma^n), g(\sigma^n))$ gleich dem Grade ist, mit dem das Simplex σ^n von K durch g auf τ^n abgebildet wird (also 1, -1 oder 0). Daraus folgt, wenn wir den n-dimensionalen Grundcozyklus von Y mit η bezeichnen und unter G^* die zu G duale Abbildung der Cozyklen verstehen: $\omega(f,g) = G^*(\eta)$; daraus ergibt sich für beliebiges g:

(1.8.1) $\quad \bar{\omega}(f,g) = G^*(\bar{\eta})$, wenn F konstant ist.

Hieraus und aus (1.7.1) folgt

(1.8.2) $\quad \bar{\omega}(f,g) = (-1)^n F^*(\bar{\eta})$, wenn G konstant ist.

1.9. Der Koeffizientenbereich ist in diesem Paragraphen ein beliebiger Ring. Wir werden später immer den Ring der ganzen Zahlen nehmen.

§ 2. Die Differenz-Cozyklen

2.1. Unsere Klasse $\bar{\omega}(f,g)$ darf als ein erstes Hindernis angesehen werden, das sich der 'Trennung' von f und g, d.h. der Beseitigung der Coinzidenzpunkte, widersetzt: diese Trennung ist immer möglich über K^{n-1}, über K^n aber unmöglich, falls nicht $\bar{\omega} = 0$ ist. In diesem Sinne ist $\bar{\omega}$ ein Gegenstück zu der Klasse $\bar{\alpha}(f,g)$ der Differenz-Cozyklen $\alpha(f,g)$—(sie heißen in [6] nicht α und $\bar{\alpha}$, sondern d und $\bar{d}$): die Klasse $\bar{\alpha}$ ist das erste Hindernis, das sich dem Versuch widersetzt, f und g über einem möglichst hochdimensionalen Gerüst miteinander zu vereinigen, d.h. durch homotope Abänderungen miteinander zusammenfallen zu lassen.—Ich erinnere an Definitionen und bekannte Eigenschaften ([6], 181):

Es sei q die kleinste positive Zahl, für welche die Homotopiegruppe $\pi_q(Y) \neq 0$ ist; wir dürfen für unsere späteren Zwecke immer annehmen, daß $q > 1$ ist; dann ist $\pi_q(Y)$ zugleich die ganzzahlige Homologiegruppe $H_q(Y)$. Ferner wollen wir von jetzt an annehmen, daß $H_q(Y)$ 'stabil' in dem Sinne ist, den wir in den 'Vorbemerkungen' erklärt haben.

f und g seien Felder über K^q und vielleicht über einem noch größeren Teilkomplex von K. Aus $\pi_i(Y) = 0$ für $i < q$ folgt: f und g sind miteinander homotop über K^{q-1}; daher gibt es eine Abbildung F von $K^{q-1} \times I$ in R mit $pF(x \times t) = x$, $F(x \times 0) = f(x)$, $F(x \times 1) = g(x)$; ich nenne F eine 'Verbindung' von f und g über K^{q-1}. Für jedes σ^q ist $f(\sigma^q) - g(\sigma^q) + F(\partial\sigma^q \times I)$ eine q-dimensionale Sphäre; sie wird durch eine Abbildung r, die $p^{-1}\sigma^q$ auf eine Faser retrahiert, in diese

Faser abgebildet und bestimmt daher—hier wird die 'Stabilität' benutzt- ein Element $\alpha(\sigma^q) \in H_q(Y)$. Diese Funktion $\alpha = \alpha(f, g)$ der σ^q ist ein q-dimensionaler Cozyclus, ein 'Differenz-Cozyklus' von f und g. Er hängt von der Verbindung F ab, aber seine Cohomologieklasse $\bar{\alpha}(f, g)$ ist unabhängig von F. Sie ändert sich auch nicht bei homotoper Abänderung von f und g. Diese Klasse ist das Hindernis gegen die Verbindung von f und g über K^q. Der Koeffizientenbereich für die α und $\bar{\alpha}$ ist die Gruppe $H_q(Y)$.

2.2. Die folgende Interpretation ist mitunter vorteilhaft. Man zeichne in Y einen q-dimensionalen Cozyklus ζ aus, etwa—vorläufig—mit ganzen Koeffizienten; für jedes σ^q sei $\alpha_\zeta(\sigma^q)$ das skalare Produkt von ζ mit dem oben definierten Element $\alpha(\sigma^q) \in H_q(Y)$. Dann gilt analog wie oben: die α_ζ sind Cozyklen, die von F abhängen, aber die Cohomologieklasse $\bar{\alpha}_\zeta(f, g)$ ist, bei festem ζ, eine Invariante von f und g. Der Koeffizientenbereich für diese α_ζ und $\bar{\alpha}_\zeta$ ist der der ganzen Zahlen.

Es ist aber besser, statt ganzzahligen ζ solche q-dimensionale Cozyklen heranzuziehen, deren Koeffizientenbereich zu demjenigen von $H_q(Y)$ dual ist. Das System dieser Klassen $\bar{\alpha}_\zeta(f, g)$ ist äquivalent mit der ursprünglichen Klasse $\bar{\alpha}(f, g)$. Übrigens darf man statt $\bar{\alpha}_\zeta$ immer $\bar{\alpha}_{\bar{\zeta}}$ sagen, wobei $\bar{\zeta}$ die Cohomologieklasse von ζ ist.

Spezialisieren wir die Betrachtung auf den Fall, daß die Faserung trivial (die Produktfaserung) ist, daß also f, g Abbildungen F, G: $K \to Y$ darstellen. Für simpliziale F, G, bestätigt man leicht:

$$\alpha_\zeta(f, g) = F^*(\zeta) - G^*(\zeta), \tag{2.2.1}$$

wobei F^*, G^* die zu den durch F, G bewirkten Zyklen-Abbildungen dualen Cozyklen-Abbildungen sind. Hieraus folgt für die Abbildungen der Cohomologieklassen

$$\bar{\alpha}_{\bar{\zeta}}(f, g) = F^*(\bar{\zeta}) - G^*(\bar{\zeta}). \tag{2.2.2}$$

2.3. Wir werden später den untenstehenden Hilfssatz 2 benutzen.—In R sei eine Metrik eingeführt, deren Entfernungsfunktion ρ heiße. X' sei ein Teilpolyeder von X und f ein Feld über X'. Für $\epsilon > 0$ verstehen wir unter $U(f, \epsilon)$ die Menge der Punkte $u \in R$ mit $p(u) \in X'$, $\rho(fp(u), u) < \epsilon$. Wir nennen eine für $0 \leqq t \leqq 1$ erklärte stetige Abbildungsschar h_t von $U(f, \epsilon)$ in R eine 'fastertreue Retraktion von $U(f, \epsilon)$ auf f', wenn sie folgende Eigenschaften hat:

$$\left.\begin{aligned} h_0(u) = u, \quad h_1(u) = fp(u), \quad ph_t(u) = p(u) \quad & \text{für} \quad u \in U(f, \epsilon) \\ h_t f(x) = f(x) \quad & \text{für} \quad x \in X' \end{aligned}\right\} \quad (0 \leqq t \leqq 1).$$

HILFSSATZ 1. *Zu X' und f existiert ein $\epsilon=\epsilon(X',f)>0$, sodaß sich $U(f,\epsilon)$ fasertreu auf f retrahieren läßt.*

Ich deute den Beweis dieser naheliegenden Tatsache nur an: Man stelle von X' eine so feine Simplizialzerlegung K' her, daß für jedes Grundsimplex σ' von K' Folgendes gilt: für eine Abbildung r, welche $p^{-1}\sigma'$ auf eine Faser Y retrahiert, liegt das Bild $rf(\sigma')$ ganz in einem euklidschen Element der Mannigfaltigkeit Y. Man beweist dann die Behauptung des Hilfssatzes 1 induktiv, indem man K' schrittweise aus diesen σ' aufbaut.

Diese Retraktionseigenschaft bringen wir in Zusammenhang mit den Differenz-Cozyklen:

HILFSSATZ 2. *f und g seien Felder über K^q; die Zahl $\epsilon(K^q,f)$ sei wie oben erklärt; der Teilkomplex L^q von K^q habe die Eigenschaft, daß*

$$\rho(f(x),g(x))<\epsilon(K,f) \quad \text{für} \quad x\in L^q$$

ist. Dann existiert ein Differenz-Cozyklus $\alpha=\alpha(f,g)$ in K^q, sodaß $\alpha(\sigma^q)=0$ für jedes Simplex σ^q von L^q ist.

BEWEIS. Da für $x\in L^q$ die Punkte $g(x)$ zu der auf f retrahierbaren Menge $U(f,\epsilon)$ gehören, ist g über L^q mit f homotop. Diese Homotopie liefert eine 'Verbindung' (cf. 2.1) zwischen f und g nicht nur über L^{q-1}, sondern sogar über L^q. Die durch sie induzierte Verbindung F über L^{q-1} hat die Eigenschaft, daß die mit ihrer Hilfe konstruierten Sphären $f(\sigma^q)-g(\sigma^q)+F(\partial\sigma^q\times I)$ 0-homotop in die Fasern retrahiert werden. Dann sind die zugehörigen $\alpha(\sigma^q)=0$. Dieser auf L^q konstruierte Cozyklus α läßt sich, wegen $\pi_i(Y)=0$ für $i<q$, zu einem Differenz-Cozyklus $\alpha(f,g)$ in K^q erweitern.

ZUSATZ. *Wir werden den Hilfssatz 2 nicht genau in der obigen, sondern in der folgenden modifizierten Form anwenden, deren Gültigkeit evident ist: f, g, ϵ sind wie oben erklärt; K_1 sei eine Unterteilung von K und L_1^q ein Teilkomplex von K_1^q, sodaß f und g über L_1^q dieselbe ϵ-Bedingung erfüllen wie oben über L^q.*

BEHAUPTUNG. *Es gibt in K_1 einen Differenz-Cozyklus $\alpha_1=\alpha_1(f,g)$ mit $\alpha_1(\sigma_1^q)=0$ für die Simplexe σ_1^q von L_1^q.*

§3. Abbildungen $S^{N-1}\to M^n$, $N-1>n$

3.1. Wir werden auch hier hauptsächlich bekannte Dinge wiederholen. Wir betrachten Abbildungen der Sphäre S^{N-1} in eine geschlossene orientierbare Mannigfaltigkeit M^n, wobei $N-1>n$ ist; wir setzen $N-n=q$, sodaß also $q>1$ ist. Nach Gysin [1] bewirkt jede solche Abbildung F Homomorphismen der Homologiegruppen $H_k(M^n)\to H_{k+q}(M^n)$, $k=0,1,\ldots,n-q$, die in der Homotopieklasse von

F invariant sind. Für uns wird nur der Fall $k=0$ eine Rolle spielen; um ihn zu beschreiben, genügt es, für das durch einen einfachen Punkt y^0 repräsentierte Element von H_0 das Bild in H_q anzugeben; dieses Bild nennen wir $c(F)$. Seine Definition läßt sich für simpliziale F so skizzieren: Bei angemessener natürlicher Erklärung der Umkehrungsabbildung F^{-1} ist $F^{-1}(y^0)=x^{q-1}$ ein $(q-1)$-dimensionaler Zyklus in S^{N-1}; es gibt eine Kette X^q, deren Rand

(3.1.1) $$\partial X^q = x^{q-1} = F^{-1}(y^0)$$

ist; das Bild $F(X^q)$ ist ein Zyklus, da der Rand von X mit Erniedrigung der Dimension abgebildet wird; die Homologieklasse dieses Zyklus ist $c(F)$.

Da F Repräsentant eines Elements der Homotopiegruppe $\pi_{N-1}(M^n)$ ist, bewirkt c eine Abbildung $\pi_{N-1}(M^n)\to H_q(M^n)$; man sieht übrigens leicht, daß dies ein Homomorphismus ist. Jedenfalls gilt: wenn F 0-homotop ist, so ist $c(F)=0$.

3.2. In manchen Fällen wird allerdings diese Abbildung

$$\pi_{N-1}(M^n)\to H_q(M^n)$$

trivial, d.h. bei manchen S^{N-1} und M^n wird $c(F)=0$ für alle F sein. Ein solcher Fall liegt z.B. dann vor, wenn q ungerade ist und H_q kein Element der Ordnung 2 enthält ([1], 89, 99). Andererseits gibt es sicher Fälle, in denen c nicht trivial ist. Die bekanntesten Beispiele sind: $M^n=S^n$, $N=2n$, n gerade; sowie: $M^n=M^{2k}$ ist der komplexe projektive Raum mit k komplexen Dimensionen, $N=n+2$.

3.3. Bekanntlich (cf. [1, 7]) läßt sich $c(F)$ auch im Rahmen der Cohomologietheorie und mit Hilfe des Cup-Produktes ausdrücken; dies wollen wir noch skizzieren. Für eine q-dimensionale Cohomologieklasse $\bar\zeta$ von M^n betrachten wir das skalare Produkt

$$\bar\zeta\cdot c(F)=c_\zeta(F);$$

(wir wollen skalare Produkte immer durch einen Punkt andeuten). Die Kenntnis der $c_{\bar\zeta}$ für beliebige $\bar\zeta$ (in bezug auf den zu der Gruppe H_q dualen Koeffizientenbereich) ist äquivalent mit der Kenntnis von c.

F sei simplizial, F^* die duale Abbildung der Coketten usw.; das Bild $F^*\eta$ des n-dimensionalen Grundcozyklus η von M^n ist ein n-dimensionaler Cozyklus in S^{N-1}; es gibt Coketten γ in S^{N-1} mit dem Corand

(3.3.1) $$\delta\gamma = F^*\eta.$$

BEHAUPTUNG.

(3.3.2) $$c_\zeta(F)=(-1)^q\,(F^*\zeta\cup\gamma)\cdot S^{N-1},$$

wobei wir durch S^{N-1} auch den Grundzyklus der gleichnamigen Sphäre bezeichnen und unter ζ einen Cozyklus aus der Klasse $\bar{\zeta}$ verstehen.

BEWEIS. Die Dualitäts-Operatoren D und Δ in S^{N-1} und M^n, die den Coketten Ketten zuordnen, sind definiert durch

$$\xi \cdot D\gamma = (\xi \cup \gamma) \cdot S^{N-1} \tag{1}$$

(für beliebige Coketten ξ, γ), und analog für Δ in M^n; speziell ist

$$y^0 = \Delta\eta \tag{2}$$

(wie in 3.1 ist y^0 ein einfacher Punkt). Rand- und Corandbildung ∂ und δ verhalten sich bei Dualisierung so:

$$D\delta\xi^p = (-1)^{N-1-p}\,\partial D\xi^p;$$

es ist also speziell

$$D\delta\xi^{n-1} = (-1)^q\,\partial D\xi^{n-1}. \tag{3}$$

Die duale Abbildung F^* der Coketten usw. ist charakterisert durch

$$F^*\xi \cdot x = \xi \cdot Fx \tag{4}$$

(für beliebige Coketten ξ und Ketten x) und die Umkehrungsabbildung F^{-1} der Ketten usw. durch

$$F^{-1} = DF^*\Delta^{-1}. \tag{5}$$

Nun sei γ eine Cokette, die (3.3.1) erfüllt; für die Cokette

$$X = (-1)^q\,D\gamma$$

folgt dann, indem man auf ∂X nacheinander (3), (3.3.1), (5), (2) anwendet, daß sie die Eigenschaft (3.1.1) besitzt. Es ist also

$$c_{\bar{\zeta}}(F) = (-1)^q\,\zeta \cdot FD\gamma;$$

wendet man hierauf (4) und (1) an, so erhält man (3.3.2).

3.4. Unsere Sphäre S^{N-1} berande eine Zelle Z^N; es sei also, wenn wir auch die Grundkette einer simplizialen Zerlegung der Zelle mit Z^N bezeichnen sowie, wie schon früher, den Grundzyklus der Sphäre auch S^{N-1} nennen: $\partial Z^N = S^{N-1}$. Wie früher sei F eine simpliziale Abbildung von S^{N-1} in M^n, und auch ζ und η sollen dieselbe Bedeutung haben wie vorhin. Ferner seien in der simplizial zerlegten Zelle zwei Cozyklen ϕ, ψ der Dimensionen q und n gegeben, die auf S^{N-1} mit $F^*\zeta$ und $F^*\eta$ übereinstimmen.

BEHAUPTUNG.

$$c_{\bar{\zeta}}(F) = (\phi \cup \psi) \cdot Z^N. \tag{3.4.1}$$

BEWEIS. Es gibt in Z^N eine Cokette Γ mit $\delta\Gamma = \psi$; für ihre Restriktion γ auf S^{N-1} gilt dann $\delta\gamma = F^*\eta$, d.h. γ erfüllt (3.3.1) und folglich auch (3.3.2). Andererseits folgt aus $\phi \cup \psi = (-1)^q \delta(\phi \cup \Gamma)$, daß auch

$$(\phi \cup \psi) \cdot Z^N = (-1)^q (\phi \cup \Gamma) \cdot S^{N-1} = (-1)^q (F^* \zeta \cup \gamma) \cdot S^{N-1}$$

ist. Hieraus und aus (3.3.2) folgt (3.4.1).

§4. Die Formel

4.1. Unsere bisherigen Festsetzungen bleiben gültig: die Faser Y des Raumes R ist eine n-dimensionale geschlossene orientierbare Mannigfaltigkeit; q hat die Bedeutung aus 2.1; die Faserung ist stabil in den Dimensionen n und q; die Basis X ist ein Komplex K. Es sind also $\bar{\alpha}$ und $\bar{\omega}$ für Paare von Feldern über K^n erklärt. Wir setzen jetzt $n+q=N$ und betrachten Felder über K^{N-1}. Ist f ein solches Feld, so ist jedem Simplex σ^N das Element $c(F) \in H_q(Y)$ zugeordnet, das gemäß 3.1 zu der Abbildung $F=rf$ des Randes $\dot{\sigma}^N$ in Y gehört, wobei r die Retraktion von $p^{-1}\sigma^N$ in eine Faser bezeichnet. Diese Funktion $c(F)$ der Simplexe σ^N definiert eine N-dimensionale Cokette $\Gamma(f)$ in K; wir wollen übrigens von jetzt an annehmen, daß $K=K^N$ ist; dann ist $\Gamma(f)$ ein Cozyklus. Er ist invariant bei homotoper Abänderung von f. Sein Koeffizientenbereich ist $H_q(Y)$. Analog erklären wir, bei gegebenem q-dimensionalen Cozyklus ζ aus Y, den Cozyklus $\Gamma_\zeta(f)$, indem wir $c(F)$ wie in 3.3 durch $c_\zeta(F)$ ersetzen.

Die Γ sind Hindernisse, die sich der Erweiterung der Felder f zu Feldern über K^N widersetzen: das Verschwinden von Γ ist notwendig für die Erweiterbarkeit. Es besteht die allgemeine Aufgabe, eine Übersicht über alle $\Gamma(f)$ zu gewinnen, die in R auftreten; insbesondere wird man fragen: sind die Γ, die zu den verschiedenen Feldern über K^{N-1} gehören, einander cohomolog?

In gewissen Fällen ist allerdings die Antwort auf diese Frage trivial, weil in ihnen bereits alle $c(F)=0$, also auch alle $\Gamma(f)=0$ sind; solche Fälle haben wir schon in 3.2 erwähnt.

4.2. Wir betrachten zwei Felder f, g. Der Koeffizientenbereich für ω soll immer der Ring der ganzen Zahlen sein, für α ist er die Gruppe $H_q(Y)$; man kann das Produkt $\alpha \cup \omega$ bilden; sein Koeffizientenbereich ist $H_q(Y)$ und seine Dimension N; dieses Produkt stimmt also in Dimension und Koeffizentenbereich mit $\Gamma(f)$ und $\Gamma(g)$ überein. Dazu ist noch zu bemerken: der Koeffizientenbereich für die Homologiegruppe $H_q(Y)$ kann eine beliebige Gruppe J sein, sodaß wir besser $H_q(V; J)$ schreiben sollten.

Die Formel, die unser Ziel bildet, lautet:

$$\bar{\Gamma}(f) - \bar{\Gamma}(g) = \bar{\alpha}(f,g) \cup \bar{\omega}(f,g); \tag{4.2.1}$$

dabei ist für gerades n der Koeffizientenbereich J beliebig; für ungerades n aber beschränken wir die Formel auf den Fall, daß J die Gruppe der Ordnung 2 ist.

Ziehen wir wieder einen q-dimensionalen Cozyklus ζ von Y heran, so folgt aus (4.2.1)

$$\overline{\Gamma}_{\zeta}(f) - \overline{\Gamma}_{\zeta}(g) = \overline{\alpha}_{\zeta}(f, g) \cup \overline{\omega}(f, g); \tag{4.2.2}$$

dabei muß der Koeffizientenbereich J' von ζ natürlich derart sein, daß man seine Elemente mit denen von J multiplizieren kann; die Gruppe, der diese Produkte angehören, ist dann Koeffizientenbereich für α_{ζ} und Γ_{ζ}. Umgekehrt ist (4.2.1) eine Folge aus dem System aller Formeln (4.2.2) mit den soeben genannten ζ (sogar wenn man als J' nur die zu J duale Gruppe heranzieht); dieses System ist also mit der Formel (4.2.1) äquivalent.

Daher genügt es, für einen beliebigen zugelassenen Cozyklus ζ die Formel (4.2.2) zu beweisen.

4.3. Beweis. Man kann ein auf K^{N-1} gegebenes Feld in das Innere jedes Simplexes σ^N hinein so fortsetzen, daß dort eine einzige Singularität a ensteht (mit anderen Worten: man kann das Feld in das Gebiet $\sigma^N - a$ fortsetzen); dabei hat man soviel Freiheit, daß man folgende Situation herstellen kann: In jedem σ_i^N sind A_i, B_i zueinander fremde N-dimensionale Zellen, a_i, b_i Punkte in ihrem Inneren; f ist in $A_i - a_i$, g ist in $B_i - b_i$ stetig; die Abbildung $F_i = r_i f$ ist über B_i, die Abbildung $G_i = r_i g$ ist über A_i konstant; F_i ist auf dem Rande $\dot{A}_i$, G_i ist auf dem Rande $\dot{B}_i$ simplizial (r_i hat dieselbe Bedeutung wie r in unseren 'Vorbemerkungen').

X' sei das durch Herausnahme der Innengebiete der A_i und B_i aus X entstehende Polyeder. Die positive Zahl $\epsilon = \epsilon(X', f)$ sei wie in 2.3, Hilfssatz 1, erklärt. C sei die Menge aller Coinzidenzpunkte von f und g in X'; dann besitzt C eine Umgebung U, sodaß $\rho(f(x), g(x)) < \epsilon$ für $x \in U$ ist. Wir stellen eine so feine simpliziale Unterteilung von K her, daß jedes Simplex, das einen Punkt von C enthält, ganz in U liegt; überdies sollen sich die A_i und B_i in diese Unterteilung einfügen. Die so entstandene Simplizialerlegung von X' heiße K_1, den Komplex der in U gelegenen Grundsimplexe von K_1 nennen wir L_1, den der übrigen Grundsimplexe von K_1 nennen wir P_1.

Jetzt nehmen wir, was nach 1.2 möglich ist, mit f und g homotope Abänderungen vor, welche die Felder über K_1^{n-1} in allgemeine Lage bringen und welche so klein sind, daß über P_1 keine Coinzidenzpunkte entstehen und die Felder über L_1 ϵ-benachbart bleiben; ferner sollen die F_i und G_i über B_i bezw. A_i konstant und über $\dot{A}_i$ bezw. $\dot{B}_i$

simplizial bleiben; auch dies läßt sich leicht erreichen. Infolge der allgemeinen Lage existiert über K_1^n ein Coinzidenz-Cozyklus $\omega_1(f,g)$; da über P_1 keine Coinzidenzpunkte liegen, ist $\omega_1 = 0$ über P_1. Nach 2.3, Hilfssatz 2, Zusatz, existiert über K_1^q ein Differenz-Cozyklus $\alpha_1(f,g)$, der über L_1^q gleich 0 ist. Die Cozyklen $\Gamma_1(f)$, $\Gamma_1(g)$, die in K_1 analog erklärt sind wie $\Gamma(f)$, $\Gamma(g)$ in K, sind überall 0, da f und g auch im Inneren der Simplexe σ_N von K_1 erklärt und stetig sind. Aus all diesem ersieht man, daß für jedes Simplex σ^N von K_1 die Formel

$$\Gamma_1(f) - \Gamma_1(g) = \alpha_1(f,g) \cup \omega_1(f,g) \tag{4.3.1}$$

gültig ist.

Wir zeigen jetzt, daß diese Formel auch für die Zellen A_i und B_i gilt; diese Zellen sind simplizial untergeteilt (dabei sollen die a_i und b_i im Inneren N-dimensionaler Simplexe liegen), und α_1, ω_1 sind Cozyklen dieses Unterteilungskomplexes, welche die gleichnamigen Cozyklen über K_1 in die A_i und B_i hinein fortsetzen. Der Cozyklus ζ spiele dieselbe Rolle wie früher (z.B. in (4.2.2)).—Zur Entlastung der Formeln will ich den Index 1, der die Unterteilung andeutet, weglassen; wir haben also die Formel

$$(\Gamma_\zeta(f) - \Gamma_\zeta(g)) \cdot Z = (\alpha_\zeta(f,g) \cup \omega(f,g)) \cdot Z \tag{4.3.2}$$

zu beweisen, wobei Z eine beliebige der Zellen A_i, B_i ist.

Nehmen wir zunächst $Z = B_i$. Auf $B_i^\bullet$ ist nach (2.2.1)

$$\alpha_\zeta = F_i^*(\zeta) - G_i^*(\zeta),$$

also, da F_i konstant ist: $\alpha_\zeta = -G_i^*(\zeta)$. Nach (1.8.1) ist $\omega_\zeta = G_i^*(\eta)$ auf $B_i^\bullet$. Folglich können wir (3.4.1) anwenden: es zeigt sich, daß die rechte Seite von (4.3.2) gleich $-\Gamma_\zeta(g) \cdot B_i$ ist; da $\Gamma_\zeta(f) \cdot B_i = 0$ infolge der Konstanz von F_1 ist, gilt also (4.3.2).

Jetzt sei $Z = A_i$. Analog wie soeben ergibt sich nach (2.2.1)

$$\alpha_\zeta = F_i^*(\zeta) - G_i^*(\zeta) = F_i^*(\zeta)$$

und nach (1.8.2) $\omega_\zeta = (-1)^n F_i^*(\eta)$ auf $A_i^\bullet$; nach (3.4.1) hat daher die rechte Seite von (4.3.2) den Wert $(-1)^n \Gamma_\zeta(f) \cdot A_i$. Wegen der Konstanz von G_i ist $\Gamma_\zeta(g) \cdot A_i = 0$. Da wir bei ungeradem n nur mit Cohomologien modulo 2 arbeiten, ist damit auch für $Z = A_i$ die Gültigkeit von (4.3.2) bewiesen.

Da somit (4.3.2) für alle N-dimensionalen Zellen einer Unterteilung von K gilt, und da die Cohomologieklassen $\bar{\alpha}$, $\bar{\omega}$, $\bar{\Gamma}$ der hier auftretenden Cozyklen natürlich mit denen der gleichnamigen Klassen des ursprünglichen Komplexes zu identifizieren sind, ist damit der Beweis der Formel (4.2.2) erbracht.

4.4. KOROLLARE. (1) *Wenn f und g überall über K^q oder wenn sie nirgends über K^n coinzidieren, so ist* $\Gamma(f) \sim \Gamma(g)$; *denn im ersten Fall ist* $\bar{\alpha} = 0$, *im zweiten* $\bar{\omega} = 0$.

(2) *Wenn die q-te oder die n-te Cohomologiegruppe von X (bezüglich der zuständigen Koeffizientenbereiche) trivial ist, so gilt der Invarianzsatz*: $\Gamma(f) \sim \Gamma(g)$ *für beliebige f und g über* K^{N-1}.

(3) *Besonders einfach und für gewisse Anwendungen besonders interessant ist der Fall, in dem* $X = K^{n+q}$ *eine (orientierbare) Mannigfaltigkeit ist. Dann ist die Cohomologieklasse* $\bar{\alpha} \cup \bar{\omega}$ *bereits durch ihr skalares Produkt mit dem Grundzyklus* Z^{n+q} *von X bestimmt; dieser Wert ist* 0, *falls eine der Klassen* $\bar{\alpha}$, $\bar{\omega}$ *Element endlicher Ordnung in ihrer Cohomologiegruppe ist. Daher läßt sich für den Fall einer Mannigfaltigkeit X das Korollar* (2) *folgendermaßen aussprechen*: *Wenn die n-te Bettische Zahl von X verschwindet, so gilt der Invarianzsatz.* (Nach dem Poincaréschen Dualitätssatz ist die q-te Bettische Zahl gleich der n-ten.)

Wegen spezieller Anwendungen der Formel (4.2.2) verweise ich auf die Arbeiten [5] and [2] sowie auf die Skizzen [3] und [4].

§ 5. Spezialisierung auf Abbildungen

Es sei $R = X \times Y$; dann stellen die Felder f, g Abbildungen F, G von K^{N-1} in Y dar (wie in 1.8). Dann läßt sich die Formel (4.2.2) ohne Begriffe aus der Fasertheorie, insbesondere auch ohne Benutzung der Cozyklen α, ω, Γ, ganz im Rahmen der Abbildungstheorie aussprechen.—Zur Vereinfachung der Formeln will ich jetzt in den Cupprodukten das Zeichen $\cup$ weglassen und die Faktoren einfach nebeneinander schreiben.

Nach (2.2.1) geht (4.2.2) zunächst über in

$$\bar{\Gamma}_{\bar{\zeta}}(f) - \bar{\Gamma}_{\bar{\zeta}}(g) = (F^*(\bar{\zeta}) - G^*(\bar{\zeta}))\,\bar{\omega}(f,g). \tag{5.1}$$

Indem man für F eine konstante Abbildung nimmt, erhält man aus (5.1) und (1.8.1) für beliebige Abbildungen G:

$$\bar{\Gamma}_{\bar{\zeta}}(G) = G^*(\bar{\zeta})\, G^*(\bar{\eta});$$

analog, mit konstantem G aus (5.1) und (1.8.2) für beliebige f:

$$\bar{\Gamma}_{\bar{\zeta}}(F) = (-1)^n F^*(\bar{\zeta})\, F^*(\bar{\eta}),$$

worin wir aber den Faktor $(-1)^n$ weglassen dürfen, da wir bei ungeradem n ja nur modulo 2 rechnen. Einsetzen der beiden letzten Ausdrücke in die linke Seite von (5.1) ergibt für beliebige F und G

$$F^*(\bar{\zeta})\, F^*(\bar{\eta}) - G^*(\bar{\zeta})\, G^*(\bar{\eta}) = (F^*(\bar{\zeta}) - G^*(\bar{\zeta}))\,\bar{\omega}(f,g). \tag{5.2}$$

Wir wollen jetzt auch noch $\bar{\omega}(f,g)$ durch F^* und G^* ausdrücken; dies geschieht mit Hilfe der Lefschetzschen Formel, von der schon in der Einleitung die Rede war. Ich erinnere zunächst an diese Formel: In der Mannigfaltigkeit $Y = M^n$ seien $(\xi_1^r, \ldots, \xi_p^r)$ und $(\eta_1^{n-r}, \ldots, \eta_p^{n-r})$ duale Cohomologiebasen der Dimensionen r und $n-r$, d.h. es seien die Cupprodukte $\xi_i^r \eta_j^{n-r} = \delta_{ij}\eta$, wobei $\eta = \eta^n$ wieder der Grundcozyklus ist; dabei sei der Koeffizientenbereich die Gruppe der rationalen Zahlen oder auch, was damit äquivalent ist, die der ganzen Zahlen, wobei wir aber nur schwache Cohomologien, d.h. solche modulo der Torsionsgruppen zulassen; der Koeffizientenbereich darf aber z.B. auch die Gruppe der Ordnung 2 sein. Sind nun F und G zwei Abbildungen eines beliebigen Polyeders X in die Mannigfaltigkeit Y, so gilt für die Klasse $\bar{\Omega}(F,G)$ der Coinzidenz-Cozyklen, die wir in der Einleitung definiert haben, die Lefschetzsche Formel

$$\bar{\Omega}(F,G) \sim \textstyle\sum_{r=0}^{n} (-1)^r \sum_i F^*(\xi_i^r)\, G^*(\eta_i^{n-r}). \tag{5.3}$$

Da natürlich $\bar{\omega}(f,g) = \bar{\Omega}(F,G)$ ist, können wir die rechte Seite von (5.3) in (5.2) einsetzen; dabei will ich aber eine kleine Änderung der Bezeichnung vornehmen: den Grundcozyklus η will ich jetzt ζ^n nennen und unseren alten Cozyklus ζ genauer mit ζ^q bezeichnen; dann lautet die aus (5.2) und (5.3) kombinierte Formel:

$$\begin{aligned} &F^*(\zeta^q)\, F^*(\zeta^n) - G^*(\zeta^q)\, G^*(\zeta^n) \\ &\quad \sim (F^*(\zeta^q) - G^*(\zeta^q)) \textstyle\sum_{r=0}^{n} (-1)^r \sum_i F^*(\xi_i^r)\, G^*(\eta_i^{n-r}). \end{aligned} \tag{5.4}$$

Dabei sind, um daran zu erinnern, F und G Abbildungen von K^{N-1} in Y, und die Cupprodukte sind in K^N zu bilden $(N = n+q)$.

Man kann (5.4) noch etwas verkürzen, da sich die beiden Produkte auf der linken Seite gegen zwei Produkte auf der rechten Seite wegheben, die beim Ausmultiplizieren der Klammer mit der Summe entstehen (dabei hat man zu berücksichtigen, daß der Grundcozyklus ζ^n sowohl in der ξ-Basis wie in der η-Basis auftritt und daß das zu ihm duale Element das Einselement des Cohomologieringes ist); bringt man dann noch zwei Glieder von rechts nach links, so erhält man:

$$\begin{aligned} &F^*(\zeta^n)\, G^*(\zeta^q) - G^*(\zeta^n)\, F^*(\zeta^q) \\ &\quad \sim (F^*(\zeta^q) - G^*(\zeta^q)) \textstyle\sum_{r=q}^{n-q} (-1)^r \sum_i F^*(\xi_i^r)\, G^*(\eta_i^{n-r}). \end{aligned} \tag{5.5}$$

Ein Beispiel zu (5.4): Y sei der komplexe projektive Raum mit k komplexen Dimensionen; dann ist $n = 2k$, $q = 2$; man kann bekanntlich eine 2-dimensionale Cohomologieklasse ξ so wählen, daß ihre Cup-Potenzen $\xi^0, \xi^1, \xi^2, \ldots, \xi^k$ (wobei also diese oberen Indizes nicht

Dimensionszahlen sind, sondern Exponenten) je eine Basis in den Dimensionen $0, 2, 4, \ldots, 2k$ bilden, während die Cohomologiegruppen ungerader Dimension trivial sind; ξ^s ist dual zu ξ^{k-s}. Das Polyeder $X = K^{n+q}$ hat also die Dimension $2k+2$; sein Gerüst K^{2k+1} ist durch F und G in Y abgebildet; wir setzen zur Abkürzung noch $F^*(\xi) = \phi$, $G^*(\xi) = \psi$. Als Cozyklus ζ brauchen wir keinen anderen heranzuziehen als $\zeta = \xi$. Dann lautet (5.4):

$$\phi^{k+1} - \psi^{k+1} \sim (\phi - \psi)(\psi^k + \phi\psi^{k-1} + \ldots + \phi^{k-1}\psi + \phi^k).$$

Diese Identität ist trivial, aber wir dürfen sie als spezielle Bestätigung unserer allgemeinen Formel (5.4) ansehen. Es wäre interessant, auch Beispiele zu finden, in denen die Formeln (5.4) oder (5.5) etwas Neues liefern.

ZÜRICH, SWITZERLAND

LITERATUR

[1] W. GYSIN, *Zur Homologietheorie der Abbildungen und Faserungen von Mannigfaltigkeiten*, Comment. Math. Helv., 14 (1941), pp. 61–122.

[2] F. HIRZEBRUCH, *Übertragung einiger Sätze aus der Theorie der algebraischen Flächen auf komplexe Mannigfaltigkeiten von zwei komplexen Dimensionen*, Reine Angew. Math. (Crelle), 191 (1953), pp. 110–124.

[3] H. HOPF, *Sur les champs d'éléments de surface dans les variétés à 4 dimensions*, Colloque de Topologie Algébrique, Paris, 1947 (erschienen Paris 1949).

[4] ——, *Sur une formule de la théorie des espaces fibrés*, Colloque de Topologie (Espaces fibrés), Bruxelles, 1950.†

[5] E. KUNDERT, *Über Schnittflächen in speziellen Faserungen und Felder reeller und komplexer Linienelemente*, Ann. of Math., 54 (1951), pp. 215–246.

[6] N. E. STEENROD, *Cohomology invariants of mappings*, Ann. of Math., 50 (1949), pp. 954–988; angekündigt in Proc. Nat. Acad. Sci. U.S.A., 33 (1947), pp. 124–128.

[7] ——, The Topology of fibre bundles, Princeton, 1951.

† Ich benutze die Gelegenheit, um auf zwei Inkorrektheiten in dieser Note hinzuweisen: (1) Bei ungeradem n hat man für die Formel (6.5) den Koeffizientenbereich modulo 2 zu nehmen; (2) in den Zeilen 6–12 auf Seite 121 hat man vorauszusetzen, daß $n = 2$ ist.

Isotopy of Links

John Milnor

1. Introduction

LET M be a three-dimensional manifold, and let C_n be the space consisting of n disjoint circles. *It will always be assumed that M is open, orientable and triangulable.* By an *n-link* $\mathfrak{L}$ is meant a homeomorphism $\mathfrak{L}: C_n \to M$. The image $\mathfrak{L}(C_n)$ will be denoted by L. Two links $\mathfrak{L}$ and $\mathfrak{L}'$ are *isotopic* if there exists a continuous family $h_t: C_n \to M$ of homeomorphisms, for $0 \leqq t \leqq 1$, with $h_0 = \mathfrak{L}$, $h_1 = \mathfrak{L}'$.

For polygonal links in Euclidean space, K. T. Chen has proved [3] that the fundamental group $F(M-L)$ of the complement of L, modulo its qth lower central subgroup $F_q(M-L)$, is invariant under isotopy of $\mathfrak{L}$, for an arbitrary positive integer q. (The lower central subgroups $F_1 \supset F_2 \supset F_3 \supset \ldots$ are defined by $F_1 = F$, $F_{i+1} = [F, F_i]$, where $[F, F_i]$ is the subgroup generated by all $aba^{-1}b^{-1}$ with $a \in F$, $b \in F_i$.)

In § 2 of this paper, Chen's result is extended in three directions:

(1) The restriction to polygonal links is removed.

(2) A corresponding result is proved for links in arbitrary (open, orientable, triangulable) 3-manifolds. In fact let $K(M-L, M)$ denote the kernel of the inclusion homomorphism $F(M-L) \to F(M)$. Then it is proved that the group $F(M-L)/K_q(M-L, M)$ is invariant under isotopy of $\mathfrak{L}$.

(3) Certain special elements of the group $F(M-L)/K_q(M-L, M)$ are constructed: the meridians and parallels to the components of $\mathfrak{L}$. It is proved that their conjugate classes are invariant under isotopy of $\mathfrak{L}$.

In § 3 these results are applied to construct certain numerical isotopy invariants for links in Euclidean space. It is first shown that the ith parallel β_i in $F(M-L)/F_q(M-L)$ can be expressed as a word w_i in the meridians $\alpha_1, \ldots, \alpha_n$. Let $\mu(i_1 \ldots i_r)$ denote the coefficient of $\kappa_{i_1} \ldots \kappa_{i_{r-1}}$ in the Magnus expansion of w_{i_r} (where $\kappa_i = \alpha_i - 1$). Then it is shown that the residue class $\bar{\mu}$ of $\mu(i_1 \ldots i_r)$, modulo an integer $\Delta(i_1 \ldots i_r)$ determined by the μ, is an isotopy invariant of $\mathfrak{L}$.

It is shown that $\bar{\mu}(i_1 \ldots i_r)$ is a homotopy invariant of $\mathfrak{L}$ (in the sense of the author's paper [10]) whenever the indices $i_1 \ldots i_r$ are all distinct.

Furthermore, in this case the $\bar{\mu}(i_1 \ldots i_r)$ are essentially identical with the homotopy invariants defined in [10].

A number of symmetry relations between the $\bar{\mu}(i_1 \ldots i_r)$ are established; in particular cyclic symmetry. The behavior of the $\bar{\mu}$ under certain simple transformations of $\mathfrak{L}$ is described.

In §4 some specific examples of these invariants are given. (For further examples see [10], Figs. 3–9, and accompanying text.) A link is given for which all of the $\bar{\mu}$ are zero; but which is not isotopically trivial, at least under isotopies which satisfy a certain smoothness condition. Finally, a pathological example of a link is studied.

2. The theorem of Chen

The fundamental group of any topological space X will be denoted by $F(X)$. For $X \subset Y$ the kernel of the inclusion homomorphism $F(X) \to F(Y)$ will be denoted by $K(X, Y)$, and its qth lower central subgroup by $K_q(X, Y)$.

THEOREM 1. *Let $\mathfrak{L}$ be a link in the 3-manifold M. For any positive integer q there exists a neighborhood N of L, such that a homotopy of $\mathfrak{L}$ onto any second link $\mathfrak{L}'$ within N induces an isomorphism of*

$$F(M-L)/K_q(M-L, M) \quad \textit{onto} \quad F(M-L')/K_q(M-L', M).$$

In fact the neighborhood N will be constructed so that the inclusion homomorphisms

$$F(M-\bar{N}) \to F(M-L)/K_q(M-L, M)$$

and

$$F(M-\bar{N}) \to F(M-L')/K_q(M-L', M)$$

are both onto, and have the same kernel (where $\bar{N}$ denotes the closure of N). *To simplify the notation, we will first consider the case where $\mathfrak{L}$ has only one component.*

Choose a sequence $M = N_0 \supset N_1 \supset \ldots \supset N_q$ of connected open neighborhoods of L such that N_j can be deformed into L within N_{j-1}, for each $j = 1, 2, \ldots, q$. (That is, there exists a homotopy r_t: $N_j \to N_{j-1}$ such that r_0 is the inclusion map and $r_1(N_j) \subset L$.) Such a sequence of neighborhoods can certainly be constructed inductively,† since L and N_{j-1} are both absolute neighborhood retracts.

The base point for the fundamental groups under consideration is to be a point in $N_q - L$.

† Let $X \subset V_{j-1}$ be a compact ANR which contains a neighborhood of L. Then by a lemma of Borsuk ([1], p. 251) there exists a neighborhood V_j of L in X which can be deformed into L within X.

Choose fixed orientations for M and for the circle C_1. It will next be shown that every element of $K(N_j - L, N_j)$ has a well defined linking number with $\mathfrak{L}$, for $j \geqq 1$. A representative loop of an element of $K(N_j - L, N_j)$ may be considered as a singular 1-cycle in N_j. This 1-cycle bounds a singular 2-chain c in N_j. Furthermore, $\mathfrak{L}$ can be considered as a singular 1-cycle. The linking number is now defined as the intersection number of $\mathfrak{L}$ and c. This is certainly defined, since the manifold M is triangulable.

To show that this number does not depend on the choice of c, it is necessary to show that every singular 2-cycle in N_j has intersection number zero with $\mathfrak{L}$. Since N_j can be deformed into $\mathfrak{L}$ within N_{j-1}, it follows that every 2-cycle in N_j bounds in N_{j-1}. But a bounding 2-cycle certainly has intersection number zero with $\mathfrak{L}$. Therefore the linking number is well defined.

Let $Q(N_j - L, N_j)$ denote the subgroup of $K(N_j - L, N_j)$ formed by all elements which have linking number zero with $\mathfrak{L}$. (This subgroup is defined for $j \geqq 1$.)

LEMMA 1. *The inclusion homomorphism*

$$K(N_j - L, N_j) \to K(N_{j-1} - L, N_{j-1}) \quad (j \geqq 1),$$

carries the subgroup $Q(N_j - L, N_j)$ of elements with linking number zero into the commutator subgroup $K_2(N_{j-1} - L, N_{j-1})$.

Let U_j denote the universal covering space of N_j, and let V_j denote the inverse image of L in U_j. Then $U_j - V_j$ is a covering space of $N_j - L$, and its group $F(U_j - V_j)$ is naturally isomorphic to the subgroup $K(N_j - L, N_j)$ of $F(N_j - L)$. Furthermore, the inclusion homomorphism $K(N_j - L, N_j) \to K(N_{j-1} - L, N_{j-1})$ is induced by a map f: $U_j \to U_{j-1}$ (where f carries $U_j - V_j$ into $U_{j-1} - V_{j-1}$, and V_j into V_{j-1}). We may restrict our attention to the Abelianized group $K(N_j - L, N_j)/K_2(N_j - L, N_j)$, which is isomorphic to the first singular homology group $H_1(U_j - V_j)$ with integer coefficients.

It will first be shown that this group $H_1(U_j - V_j)$ is generated by certain elements α_k, there being one generator α_k corresponding to each component $V_j^{(k)}$ of V_j.

Let $U_j \cup \infty$ denote the one point compactification of U_j. By the Lefschetz duality theorem, the group

$$H_1(U_j - V_j) = H_1((U_j \cup \infty) - (V_j \cup \infty))$$

is isomorphic to the Čech cohomology group $H^2(U_j \cup \infty, V_j \cup \infty)$. In the exact sequence

$$H^1(V_j \cup \infty, \infty) \xrightarrow{\delta} H^2(U_j \cup \infty, V_j \cup \infty) \to H^2(U_j \cup \infty, \infty),$$

observe that $H^2(U_j \cup \infty, \infty) \approx H_1(U_j) = 0$, since U_j is a universal covering space; and that $H^1(V_j \cup \infty, \infty)$ is a free Abelian group having one generator γ_k corresponding to each component of V_j. This means that $H^2(U_j \cup \infty, V_j \cup \infty)$ is generated by the coboundaries $\delta(\gamma_k)$. But $\delta(\gamma_k)$ is a 2-cocycle which makes a small circuit around the kth component of V_j. Or more precisely the corresponding element α_k of $H_1(U_j - V_j)$ is a 1-cycle which makes a small loop about this component.

Clearly the image of each α_k in $K(N_j - L, N_j)/K_2(N_j - L, N_j)$ will have linking number ± 1 with $\mathfrak{L}$; and we may assume that this linking number is $+1$. This means that an arbitrary element $\sum i_k \alpha_k$ of $H_1(U_j - V_j)$ will correspond to an element of $K(N_j - L, N_j)/K_2$ which has linking number $\sum i_k$ with $\mathfrak{L}$. In particular, elements of $Q(N_j - L, N_j)/K_2$ will be represented by sums $\sum i_k \alpha_k$ in $H_1(U_j - V_j)$ with $\sum i_k = 0$.

Now observe that the map $f\colon U_j \to U_{j-1}$ (which is induced by the inclusion $N_j \subset N_{j-1}$) must carry all of the components of V_j into a single component of V_{j-1}. This follows immediately from the fact that N_j can be deformed into L within N_{j-1}. Therefore the induced homomorphism

$$f_*\colon H_1(U_j - V_j) \to H_1(U_{j-1} - V_{j-1})$$

must carry all of the generators α_k of $H_1(U_j - V_j)$ into a single element $\bar{\alpha}$ of $H_1(U_{j-1} - V_{j-1})$. This means that f_* carries an arbitrary sum $\sum i_k \alpha_k$ into $(\sum i_k)\bar{\alpha}$. In particular, a sum $\sum i_k \alpha_k$ with $\sum i_k = 0$ is carried into 0. But this means that the corresponding homomorphism

$$K(N_j - L, N_j)/K_2(N_j - L, N_j) \to K(N_{j-1} - L, N_{j-1})/K_2(N_{j-1} - L, N_{j-1})$$

carries $Q(N_j - L, N_j)/K_2$ into 1. This completes the proof of Lemma 1.

Lemma 2. *The inclusion homomorphism* $K(N_j - L, N_j) \to K(M - L, M)$ $(j \geqq 1)$ *carries* $Q(N_j - L, N_j)$ *into* $K_{j+1}(M - L, M)$.

Proof by induction on j. For $j = 1$ this is an immediate consequence of Lemma 1.

Let α be an element of $K(N_{j-1} - L, N_{j-1})$ which has linking number $+1$ with $\mathfrak{L}$. Then every element of $K(N_{j-1} - L, N_{j-1})$ can be expressed in the form $\alpha^h \lambda$, where h is an integer and λ is an element of

$$Q(N_{j-1} - L, N_{j-1}).$$

Now consider the homomorphisms

$$K(N_j - L, N_j) \to K(N_{j-1} - L, N_{j-1}) \to K(M - L, M).$$

By Lemma 1 every element σ of $Q(N_j - L, N_j)$ corresponds to a commutator in $K(N_{j-1} - L, N_{j-1})$ under the first homomorphism. This commutator can be written in the form

$$\Pi_i\, [\alpha^{h_i}\lambda_i, \alpha^{k_i}\mu_i],$$

with $\lambda_i, \mu_i \in Q(N_{j-1}-L, N_{j-1})$. Let

$$\overline{\sigma} = \prod_i [\overline{\alpha}^{h_i}\overline{\lambda}_i, \overline{\alpha}^{k_i}\overline{\mu}_i]$$

denote the image of this commutator in $K(M-L, M)$, under the second homomorphism. Then by the induction hypothesis, $\overline{\lambda}_i$ and $\overline{\mu}_i$ are elements of $K_j(M-L, M)$. Now using standard commutator formulas to expand this expression, it follows that $\overline{\sigma}$ is an element of $K_{j+1}(M-L, M)$. This completes the proof of Lemma 2.

Choose a neighborhood N of L in N_q so that $N_q - \overline{N}$ is a connected set containing

(1) the base point,

(2) a loop a, homotopic to a constant in N_q and having linking number $+1$ with $\mathfrak{L}$, and

(3) a loop b homotopic to the loop $\mathfrak{L}$ within N_q. (To prove that such a loop b exists within $N_q - L$, it is only necessary to observe that the one-dimensional set V_q cannot disconnect the universal covering space U_q of N_q.)

LEMMA 3. *The natural homomorphism*

$$F(M-\overline{N}) \to F(M-L)/K_q(M-L, M)$$

is onto, for $q \geqq 2$.

We must prove that every loop in $M-L$ can be written (up to homotopy) as a product of loops in $M-\overline{N}$ and loops in $M-L$ which are q^{th} commutators of loops contractible within M.

Since $M-\overline{N}$ and $N_q - L$ are connected open sets with connected intersection and with union $M-L$, it follows that every loop in $M-L$ can be written as a product of loops in $M-\overline{N}$ and loops in $N_q - L$ (up to a homotopy). Thus to prove Lemma 3 it is sufficient to show that every loop g in $N_q - L$ can be written as the product of a loop g' in $N_q - \overline{N} = (M-\overline{N}) \cap (N_q - L)$ and a loop g'' which represents an element of $K_q(M-L, M)$.

It will first be shown that the group $F(N_q-L)/Q(N_q-L, N_{q-1})$ is generated by the loops a and b. Since N_q can be deformed into L within N_{q-1}, it follows that the image of $F(N_q-L)$ in $F(N_{q-1})$ is generated by the loop $\mathfrak{L}$, or equivalently by the loop b. Since this image is isomorphic to $F(N_q-L)/K(N_q-L, N_{q-1})$, it follows that this last group is generated by b. On the other hand,

$$K(N_q-L, N_{q-1})/Q(N_q-L, N_{q-1})$$

is clearly generated by a. Therefore $F(N_q-L)/Q(N_q-L, N_{q-1})$ is generated by a and b.

This means that every loop g in $N_q - L$ can be written as the product of a loop $g' = a^i b^j$ in $N_q - \bar{N}$ and a loop g'' in $Q(N_q - L, N_{q-1})$. Now note that the homomorphisms

$$F(N_q - L) \to F(N_{q-1} - L) \to F(M - L)$$

carry $Q(N_q - L, N_{q-1})$ into $Q(N_{q-1} - L, N_{q-1})$ into $K_q(M - L, M)$. (The last step follows from Lemma 2.) Therefore the loop g'' represents an element of $K_q(M - L, M)$. This completes the proof of Lemma 3.

In order to prove Theorem 1, it will be necessary to obtain an exact description of the kernel of $F(M - \bar{N}) \to F(M - L)/K_q(M - L, M)$. Let Q' denote the normal subgroup of $F(M - \bar{N})$ generated by the image of $Q(N_q - \bar{N}, N_{q-1})$ under inclusion.

LEMMA 4. *The kernel of the homomorphism*

$$F(M - \bar{N}) \to F(M - L)/K_q(M - L, M)$$

is equal to $Q'K_q(M - \bar{N}, M)$, *for* $q \geqq 2$.

It is clear that $K_q(M - \bar{N}, M)$ is contained in this kernel. To show that Q' is contained in the kernel, it is sufficient (by the definition of Q') to show that every element of $Q(N_q - \bar{N}, N_{q-1})$ corresponds to the identity element of $F(M - L)/K_q(M - L, M)$. But the homomorphisms

$$F(N_q - \bar{N}) \to F(N_{q-1} - L) \to F(M - L)$$

carry $Q(N_q - \bar{N}, N_{q-1})$ into $Q(N_{q-1} - L, N_{q-1})$ into $K_q(M - L, M)$.

This proves that there is a natural homomorphism

$$F(M - \bar{N})/Q'K_q(M - \bar{N}, M) \xrightarrow{\eta} F(M - L)/K_q(M - L, M).$$

We must prove that η is an isomorphism.

An inverse homomorphism

$$F(M - L)/K_q(M - L, M) \xrightarrow{\tau} F(M - \bar{N})/Q'K_q(M - \bar{N}, M)$$

will be constructed roughly as follows. Let $F(M - \bar{N}) \xrightarrow{\sigma_1} F(M - \bar{N})/Q'$ be the quotient homomorphism, and let $F(N_q - L) \xrightarrow{\sigma_2} F(M - \bar{N})/Q'$ be a homomorphism which sends each element of $F(N_q - L)$ into an appropriate expression $a^i b^j$. Then σ_1 and σ_2 will be pieced together to construct a homomorphism $F(M - L) \xrightarrow{\sigma} F(M - \bar{N})/Q'$; and σ will induce the required homomorphism τ.

To define the homomorphism σ_2, it is first necessary to consider the following:

$$F(N_q - \bar{N})/Q(N_q - \bar{N}, N_{q-1}) \to F(N_q - L)/Q(N_q - L, N_{q-1}).$$

This homomorphism is onto, since the second group is generated by a and b. It has kernel 1, since a loop in $N_q - \bar{N}$ represents the identity element of either group if and only if it is contractible within N_{q-1} and has linking number zero with $\mathfrak{L}$. Now σ_2 can be defined as the composition of the following homomorphisms:

$$F(N_q - L) \to F(N_q - L)/Q(N_q - L, N_{q-1})$$
$$\overset{\approx}{\leftarrow} F(N_q - \bar{N})/Q(N_q - \bar{N}, N_{q-1}) \to F(M - \bar{N})/Q'.$$

In order to piece σ_1 and σ_2 together, it will be necessary to make use of the following theorem (see [8]). The fundamental group $F(X \cup Y)$ of the union of two connected open sets with connected intersection is isomorphic to the free product $F(X)*F(Y)$ modulo the normal subgroup generated by all expressions $\xi_1(\gamma)\,\xi_2(\gamma^{-1})$, where

$$F(X \cap Y) \xrightarrow{\xi_1} F(X) \quad \text{and} \quad F(X \cap Y) \xrightarrow{\xi_2} F(Y)$$

are the inclusion homomorphisms. In particular, in order to define a homomorphism $F(X \cup Y) \xrightarrow{\sigma} G$ it is sufficient to define homomorphisms $F(X) \xrightarrow{\sigma_1} G$ and $F(Y) \xrightarrow{\sigma_2} G$, and to verify that $\sigma_1 \xi_1 = \sigma_2 \xi_2$.

Applying this to the case $X = M - \bar{N}$, $Y = N_q - L$, with σ_1, σ_2 as above, the condition $\sigma_1 \xi_1 = \sigma_2 \xi_2$ is certainly satisfied. Hence we have defined a homomorphism $F(M-L) \xrightarrow{\sigma} F(M-\bar{N})/Q'$. It is now easy to verify that σ induces a homomorphism

$$F(M-L)/K_q(M-L, M) \xrightarrow{\tau} F(M-\bar{N})/Q'K_q(M-\bar{N}, M),$$

and that τ and η are inverse homomorphisms. This completes the proof of Lemma 4.

The proof of Theorem 1, for links with one component, follows. We may assume $q \geqq 2$. Let $\mathfrak{L}'$ be any link which is homotopic to $\mathfrak{L}$ within N. Then each N_j can be deformed, not only into L but also into L', within N_{j-1}. Hence Lemmas 1, 2, 3 and 4 apply also to the link $\mathfrak{L}'$. Since a 1-cycle in $N_q - \bar{N}$ has the same linking number with $\mathfrak{L}'$ as with $\mathfrak{L}$, it follows that the inclusion homomorphisms

$$F(M-\bar{N}) \to F(M-L)/K_q(M-L, M)$$

and

$$F(M-\bar{N}) \to F(M-L')/K_q(M-L', M)$$

have identical kernels $Q'K_q(M-\bar{N}, M)$. Both of these homomorphisms are onto by Lemma 3. This completes the proof.

Now let $\mathfrak{L}$ be a link with any number of components. Choose a fixed numbering for the components of C_n, and a fixed orientation for each component. Let $N_0^1, \ldots, N_0^n$ be pairwise disjoint connected neighborhoods of the components $L^1, \ldots, L^n$ of L. For each component L^i construct a sequence of neighborhoods N_j^i of L^i and a neighborhood N^i of L^i, just as before; so that $N_0^i \supset N_1^i \supset \ldots \supset N_q^i \supset \bar{N}^i$; so that each N_j^i $(j \geqq 1)$ can be deformed into L^i within N_{j-1}^i; and so that $N_q^i - \bar{N}^i$ contains a base point x_i and loops a_i, b_i. Let $N = N^1 \cup \ldots \cup N^n$. Then as before it is proved that the homomorphism

$$F(M-\bar{N}) \to F(M-L)/K_q(M-L, M)$$

is onto, and that its kernel is equal to $Q^1 \ldots Q^n K_q(M-\bar{N}, M)$, where Q^i is the normal subgroup of $F(M-\bar{N})$ generated by the image of $Q(N_q^i - \bar{N}^i, N_{q-1}^i)$ under inclusion.

There is one minor difficulty with the above procedure, since we cannot use a common base point for all of the groups under consideration. However, let x_0 be a standard base point in $M-\bar{N}$, and for each i let p_i be a path in $M-\bar{N}$ which leads from x_0 to the supplementary base point x_i in $N_q^i - \bar{N}^i$. Then, for example, the inclusion

$$N_q^i - \bar{N}^i \subset M - L,$$

together with the path p_i, induces a homomorphism

$$F(N_q^i - \bar{N}^i, x_i) \to F(M-L, x).$$

With this change in the interpretation of inclusion homomorphisms, all the details of the above proof go through without essential change. This completes the proof of Theorem 1.

THEOREM 2. *If* $\mathfrak{L}_0$ *is isotopic to* $\mathfrak{L}_1$, *then* $F(M-L_0)/K_q(M-L_0, M)$ *is isomorphic to* $F(M-L_1)/K_q(M-L_1, M)$, *where* q *denotes an arbitrary positive integer.*

Let $\mathfrak{L}_t$ $(0 \leqq t \leqq 1)$ be the isotopy. For each t there is a neighborhood N_t of L_t satisfying the conditions of Theorem 1, and a number $\epsilon(t) > 0$ such that $L_{t'} \subset N_t$ for all t' with $|t'-t| < \epsilon(t)$. Construct a sequence $0 = t_0 \leqq t_1 \leqq \ldots \leqq t_{2r} = 1$ of numbers so that $(t_{2i+1} - t_{2i}) < \epsilon(t_{2i+1})$ and $(t_{2i+2} - t_{2i+1}) < \epsilon(t_{2i+1})$ for $i = 0, 1, \ldots, r-1$. Then $L_t \subset N_{t_{2i+1}}$ for $t_{2i} \leqq t \leqq t_{2i+2}$. The required isomorphism is now obtained by combining the isomorphisms

$$F(M-L_{t_0})/K_q(M-L_{t_0}, M) \approx F(M-L_{t_1})/K_q(M-L_{t_1}, M) \approx \ldots$$
$$\approx F(M-L_{t_{2r}})/K_q(M-L_{t_{2r}}, M),$$

each of which is given by Theorem 1.

REMARK 1. If a fixed base point x_0 is given, so that $x_0 \in M - L_t$ for all t, then it is not hard to show that the isomorphism of Theorem 2 does not depend on the various choices made. Since this fact will not be needed, the proof will not be given.

REMARK 2. A link $\mathfrak{L}$ will be called *almost smooth* if L is a deformation retract of some neighborhood. An isotopy $\mathfrak{L}_t$ will be called *almost smooth* if each link $\mathfrak{L}_{t_0}$ is almost smooth. Denote the intersection of the lower central subgroups K_q by K_∞. Then the following modified versions of Theorems 1 and 2 hold:

1′. *If L is almost smooth then there exists a neighborhood N of L such that a homotopy of $\mathfrak{L}$ onto any link $\mathfrak{L}'$ within N induces an isomorphism of $F(M-L)/K_\infty(M-L, M)$ onto $F(M-L')/K_\infty(M-L', M)$.*

2′. *If $\mathfrak{L}$ is isotopic to $\mathfrak{L}'$ under an almost smooth isotopy, then $F(M-L)/K_\infty(M-L, M)$ is isomorphic to $F(M-L')/K_\infty(M-L', M)$.*

The proof of these assertions are easy modifications of the proofs of Theorems 1 and 2.

The concept of a meridian to a component of a link will next be defined. Let $p_j(t)$ $(0 \leqq t \leqq 1)$ be a path from the base point x_0 to the jth component of L, such that $p_j(t) \notin L$ for $t < 1$. Form a closed loop in $M-L$ as follows: first traverse the path p_j to a point in the neighborhood N_q^j of L^j, then traverse a closed loop in $N^j - L^j$ which has linking number $+1$ with $\mathfrak{L}^j$, and finally return to x_0 along p_j. It follows from Lemma 2 that this procedure defines a unique element α_j of $F(M-L)/K_q(M-L, M)$. This element α_j will be called the jth *meridian* of $\mathfrak{L}$ with respect to the path p_j. If p_j is replaced by some other path, then α_j is clearly replaced by some conjugate element.

By a jth *parallel* β_j of $\mathfrak{L}$ with respect to the path p_j will be meant an element of $F(M-L)/K_q(M-L, M)$ obtained as follows: traverse p_j from x_0 to a point in $N_q^j - L^j$, then traverse a closed loop in $N_q^j - L^j$ which is homotopic to $\mathfrak{L}^j$ within N_q^j, and finally return to x_0 along p_j. If β_j' is another parallel with respect to the same path, then the element $\beta_j'\beta_j^{-1}$ of $F(M-L)/K_q(M-L, M)$ is essentially represented by a closed loop in $N_q^j - L^j$ which is homotopic to a constant within N_q^j. Therefore (making use of Lemma 2) $\beta_j'\beta_j^{-1}$ is equal to some power of α_j. Thus for a given path p_j, the parallel β_j is well defined up to multiplication by a power of α_j. If p_j is replaced by some other path, then the pair (α_j, β_j) is clearly replaced by some pair $(\lambda\alpha_j\lambda^{-1}, \lambda\alpha_j^i\beta_j\lambda^{-1})$.

THEOREM 3. *The isomorphisms of Theorems 1 and 2 preserve the meridian and parallel pairs (α_j, β_j) up to multiplication of β_j by a power of α_j, and up to simultaneous conjugation of α_j and β_j.*

In order to prove this theorem it is only necessary to look again at the proof of Theorem 1. In fact, the jth meridian and parallel to $\mathfrak{L}$ are both represented by loops in $N_q^j - \bar{N}^j$, and these loops also represent the jth meridian and parallel to $\mathfrak{L}'$; which completes the proof.

In case the jth component of $\mathfrak{L}$, considered as a 1-cycle in M, is a boundary, then the linking number of $\mathfrak{L}^j$ and β_j is well defined. In this case (which certainly occurs for links in Euclidean space) it will always be assumed that β_j has linking number zero with $\mathfrak{L}^j$. It follows from this assumption that:

3′. *For links in Euclidean space, the isomorphisms of Theorems* 1 *and* 2 *preserve the pair* (α_j, β_j) *up to a simultaneous conjugation.*

3. Some numerical invariants of links in Euclidean space

In case the manifold M is Euclidean space, the group $K_q(M-L, M)$ reduces to $F_q(M-L)$.

THEOREM 4. *If $\mathfrak{L}$ is an n-link in Euclidean space, then the group* $F(M-L)/F_q(M-L)$ *has the presentation*

$$\{\alpha_1, \ldots, \alpha_n / \alpha_1^{-1} w_1^{-1} \alpha_1 w_1 = 1, \ldots, \alpha_n^{-1} w_n^{-1} \alpha_n w_n = 1, A_q = 1\},$$

where the α_i are meridians, the w_i are words in $\alpha_1, \ldots, \alpha_n$ which represent parallels, and where A_q is the qth lower central subgroup of the free group $A = \{\alpha_1, \ldots, \alpha_n\}$.

(K. T. Chen has proved in [2] that F/F_q is isomorphic to a group G/G_q, where G has n generators and n relations. Thus the only new fact in Theorem 4 is the precise form of the relations. A corresponding assertion for the 'link group' of a link was proved in [10].)

By Theorems 1 and 3 it is sufficient to consider the case of a polygonal link. Hence we may use the Wirtinger presentation of $F(M-L)$. (See, for example, [11], p. 44.) This presentation has the form $\{a_{ij}/R_{ij} = 1\}$, where the generators a_{ij} ($i = 1, \ldots, n$; $j = 1, \ldots, r_i$) correspond to the components of a plane projection of $\mathfrak{L}$; and where the relations R_{ij} correspond to the crossings of this projection. These relations have the form

$$(1) \qquad \begin{cases} R_{ij} = a_{ij+1}^{-1} u_{ij}^{-1} a_{ij} u_{ij} & \text{for} \quad 1 \leqq j < r_i, \\ R_{ir_i} = a_{i1}^{-1} u_{ir_i}^{-1} a_{ir_i} u_{ir_i}, \end{cases}$$

where the u_{ij} are certain words in the generators. Defining

$$v_{ij} = u_{i1} u_{i2} \ldots u_{ij},$$

an equivalent set of relations is given by

$$(2) \qquad \begin{cases} S_{ij} = a_{ij+1}^{-1} v_{ij}^{-1} a_{i1} v_{ij} & \text{for} \quad 1 \leqq j < r_i. \\ S_{ir_i} = a_{i1}^{-1} v_{ir_i}^{-1} a_{i1} v_{ir_i}. \end{cases}$$

In fact, the following identities in the free group show that relations (1) and (2) are equivalent:

$$(3)\qquad \begin{cases} R_{i1} = S_{i1} \\ R_{ij} = S_{ij}(u_{ij}^{-1} S_{i\,j-1}^{-1} u_{ij}) \quad \text{for} \quad 1 < j \leqq r_i. \end{cases}$$

Let $\bar{A}$ denote the free group $\{a_{ij}\}$ and let A denote the free subgroup generated by $a_{11}, a_{21}, \ldots, a_{n1}$. Define a sequence of homomorphisms $\eta_k \colon \bar{A} \to A$ as follows, by induction on k:

$$\eta_1(a_{ij}) = a_{i1},$$

$$\eta_{k+1}(a_{ij+1}) = \eta_k(v_{ij}^{-1} a_{i1} v_{ij}), \quad \eta_{k+1}(a_{i1}) = a_{i1}.$$

Let N denote the normal subgroup of $\bar{A}$ generated by the S_{ij} (or the R_{ij}), so that $\bar{A}/N \approx F(M-L)$. The following assertion will be proved by induction on k:

$$(4_k)\qquad \eta_k(a_{ij}) \equiv a_{ij} \quad (\text{mod}\, \bar{A}_k N).$$

This assertion is certainly true for $k=1$ or $j=1$. If we assume (4_k) it follows that

$$\eta_k(v_{ij}) \equiv v_{ij} \quad (\text{mod}\, \bar{A}_k N).$$

Since $\eta_k(a_{i1}) = a_{i1}$, this implies that

$$(5)\qquad \eta_k(v_{ij}^{-1} a_{i1} v_{ij}) \equiv v_{ij}^{-1} a_{i1} v_{ij} \quad (\text{mod}\, \bar{A}_{k+1} N).$$

(We are making use of the following easily proved proposition: if $x \equiv x_1$ $(\text{mod}\, G_k)$, then $x^{-1}yx \equiv x_1^{-1}yx_1$ $(\text{mod}\, G_{k+1})$.) But the left-hand side of (5) is, by definition, equal to $\eta_{k+1}(a_{i\,j+1})$, while the right-hand side is congruent to $a_{i\,j+1}$ modulo N. This proves (4_{k+1}).

The following assertion will be proved by a similar method:

$$(6_k)\qquad \eta_k(a_{ij}) \equiv \eta_{k+1}(a_{ij}) \quad (\text{mod}\, A_k).$$

Again this is trivial for $k=1$ or $j=1$. Proposition (6_k) implies that $\eta_k(v_{ij}) \equiv \eta_{k+1}(v_{ij})$ $(\text{mod}\, A_k)$. Since $\eta_k(a_{i1}) = \eta_{k+1}(a_{i1}) = a_{i1}$, this implies that

$$\eta_k(v_{ij}^{-1} a_{i1} v_{ij}) \equiv \eta_{k+1}(v_{ij}^{-1} a_{i1} v_{ij}) \quad (\text{mod}\, A_{k+1}).$$

But the left-hand side of this congruence equals $\eta_{k+1}(a_{i\,j+1})$ and the right-hand side equals $\eta_{k+2}(a_{i\,j+1})$, which proves (6_{k+1}).

Now consider the presentation

$$(7)\qquad \{a_{ij} / S_{ij} = 1, \bar{A}_q = 1\},$$

which we have obtained for the group F/F_q. According to (4_q) the relations $a_{ij}^{-1}\eta_q(a_{ij}) = 1$ are consequences of the relations in (7), hence we may introduce these new relations into this presentation. But

$a_{ij}=\eta_q(a_{ij})$ may be taken as a definition of the a_{ij} $(j>1)$ in terms of the a_{i1}. Substituting $\eta_q(a_{ij})$ for all of these a_{ij}, we are left with the presentation

$$\{a_{i1}/\eta_q(S_{ij})=1, \eta_q(\bar{A}_q)=1\}. \tag{8}$$

It is clear that $\eta_q(\bar{A}_q)=A_q$. According to (6_q) we have

$$\eta_q(a_{i\,j+1})\equiv\eta_{q+1}(a_{i\,j+1})=\eta_q(v_{ij}^{-1}a_{i1}v_{ij}) \pmod{A_q},$$

and therefore

$$\eta_q(S_{ij})=\eta_q(a_{i\,j+1}^{-1}v_{ij}^{-1}a_{i1}v_{ij})\equiv 1 \pmod{A_q} \quad \text{for} \quad 1\leqq j<r_i. \tag{9}$$

This means that the relations $\eta_q(S_{ij})=1$ $(1\leqq j<r_i)$ are consequences of the remaining relations in (8). Thus the presentation (8) has been reduced to the following:

$$\{a_{i1}/\eta_q(S_{ir_i})=1, A_q=1\}. \tag{10}$$

Now defining $\alpha_i=a_{i1}$, $w_i=\eta_q(v_{ir_i})$, we have

$$\eta_q(S_{ir_i})=\alpha_i^{-1}w_i^{-1}\alpha_i w_i, \tag{11}$$

so that the presentation (10) can be written in the desired form

$$\{\alpha_i/\alpha_i^{-1}w_i^{-1}\alpha_i w_i=1, A_q=1\} \quad (i=1, \ldots, n).$$

It is clear from the geometrical construction of the Wirtinger presentation that $\alpha_i=a_{i1}$ is an ith meridian to the link $\mathfrak{L}$, and that $v_{ir_i}=u_{i1}u_{i2}\cdots u_{ir_i}$ is an ith parallel. Since $w_i=\eta_q(v_{ir_i})$, it follows from (4_q) that w_i is also an ith parallel; which completes the proof of Theorem 4.

We will next define certain integers $\mu(i_1, \ldots, i_r)$ which are associated with a link. For each $i=1, \ldots, n$, let w_i be a fixed word in $\alpha_1, \ldots, \alpha_n$ which represents the ith parallel to $\mathfrak{L}$ in the group $F(M-L)/F_q(M-L)$, and which has linking number zero with $\mathfrak{L}^i$. The Magnus expansion (see [9]) of w_i is obtained by substituting

$$\alpha_j=1+\kappa_j, \quad \alpha_j^{-1}=1-\kappa_j+\kappa_j^2-\kappa_j^3+-\ldots$$

in the word w_i and multiplying out, so that w_i is expressed as a formal, non-commutative power series in the indeterminates $\kappa_1, \ldots, \kappa_n$. Let

$$w_i=1+\sum\mu(j_1, \ldots, j_s, i)\,\kappa_{j_1}\cdots\kappa_{j_s}$$

be the resulting expansion. Thus a coefficient $\mu(j_1\cdots j_s i)$ is defined for each sequence $j_1, \ldots, j_s, i$ $(s\geqq 1)$ of integers between 1 and n. (Only those $\mu(j_1\cdots j_s i)$ with $s<q$ will turn out to be useful.)

Alternatively, using the notation of R. H. Fox [5], the integers $\mu(j_1 \dots j_s i)$ can be defined by the formula

$$\mu(j_1 \dots j_s i) = \left(\frac{\partial^s w_i}{\partial \alpha_{j_1} \dots \partial \alpha_{j_s}} \right)^0$$

It is proved in [5] that these two definitions are equivalent.

Let $\Delta(i_1 \dots i_r)$ denote the greatest common divisor of $\mu(j_1 \dots j_s)$, where $j_1 \dots j_s$ $(2 \leqq s < r)$ is to range over all sequences obtained by cancelling at least one of the indices $i_1 \dots i_r$, and permuting the remaining indices cyclicly. (For example, $\Delta(i_1 i_2) = 0$.) Let $\bar{\mu}(i_1 \dots i_r)$ denote the residue class of $\mu(i_1 \dots i_r)$ modulo $\Delta(i_1 \dots i_r)$.

THEOREM 5. *The residue classes $\bar{\mu}(i_1 \dots i_r)$ are isotopy invariants of $\mathfrak{L}$, providing that $r \leqq q$.*

Of course the integer q can be chosen arbitrarily large; so the condition $r \leqq q$ is no real restriction.

The proof will be by induction on q. In view of Theorems 2, 3′ and 4, it is sufficient to prove the following four assertions:

(12) The residue class $\bar{\mu}(j_1 \dots j_s i)$ $(s < q)$ is not changed if w_i is replaced by a conjugate.

(13) It is not changed if some α_j is replaced by a conjugate.

(14) It is not changed if w_i is multiplied by a product of conjugates of the words $\alpha_j^{-1} w_j^{-1} \alpha_j w_j$.

(15) It is not changed if w_i is multiplied by an element of A_q.

In the ring of all Magnus power series in the indeterminates $\kappa_1, \dots, \kappa_n$, let D_i denote the set of all $\sum \nu(j_1 \dots j_s)\, \kappa_{j_1} \dots \kappa_{j_s}$ such that the coefficients satisfy

$$\nu(j_1 \dots j_s) \equiv 0 \quad (\operatorname{mod} \Delta(j_1 \dots j_s i)$$

for all $j_1 \dots j_s$ with $s < q$. (There is no restriction on $\nu(j_1 \dots j_s)$ for $s \geqq q$.) Now in order to prove that two words w_i and w_i' give rise to the same residue classes $\bar{\mu}(j_1 \dots j_s i)$ $(s < q)$, it is sufficient to prove that $w_i - w_i' \in D_i$.

Since the set D_i can be defined in terms of the $\bar{\mu}(j_1 \dots j_s i)$ $(s < q-1)$, it follows from the induction hypothesis that this set D_i is invariant under isotopy of $\mathfrak{L}$. The proofs of assertions (12)–(15) will be based on the following five assertions:

(16) *D_i is a two-sided ideal.* In fact, if $\nu(j_1 \dots j_s)\, \kappa_{j_1} \dots \kappa_{j_s}$ is a monomial in D_i, and $\lambda \kappa_{h_1} \dots \kappa_{h_t}$ is an arbitrary monomial, then either $s \geqq q$ so that the product is trivially in D_i, or

$$\nu(j_1 \dots j_s) \equiv 0 \quad (\operatorname{mod} \Delta(j_1 \dots j_s i))$$

and $\quad \Delta(j_1 \dots j_s i) \equiv 0 \quad (\operatorname{mod} \Delta(h_1 \dots h_t j_1 \dots j_s i)),$

hence $\quad \lambda \nu(j_1 \dots j_s) \equiv 0 \quad (\operatorname{mod} \Delta(h_1 \dots h_t j_1 \dots j_s i));$

which proves that $\lambda\nu(j_1\cdots j_s)\kappa_{h_1}\cdots\kappa_{h_t}\kappa_{j_1}\cdots\kappa_{j_s}\in D_i$. This proves that D_i is a left ideal, and similarly it follows that D_i is a right ideal.

(17) *Let* $1+\omega_i$ *be the Magnus expansion of* w_i. *Then* $\omega_i\kappa_j\equiv\kappa_j\omega_i\equiv 0$ $(\text{mod}\, D_i)$ *for any* κ_j. This is proved as follows. Let $\mu(j_1\cdots j_s i)\kappa_{j_1}\cdots\kappa_{j_s}$ be any term of ω_i. Then the congruence

$$\mu(j_1\cdots j_s i)\equiv 0 \quad (\text{mod}\,\Delta(j_1\cdots j_s j i))$$

implies that

$$\mu(j_1\cdots j_s i)\kappa_{j_1}\cdots\kappa_{j_s}\kappa_j\equiv 0 \quad (\text{mod}\, D_i).$$

The second congruence follows by the same method. More generally we have:

(18) *If one or more new factors* κ_j *are inserted anywhere in the term* $\mu(j_1\cdots j_s i)\kappa_{j_1}\cdots\kappa_{j_s}$, *then the resulting term is congruent to zero modulo* D_i. This follows from the congruence

$$\mu(j_1\cdots j_s i)\equiv 0 \quad (\text{mod}\,\Delta(j_1\cdots j_t j j_{t+1}\cdots j_s i)).$$

(19) *If* $1+\omega_j$ *is the Magnus expansion of* w_j, *then* $\omega_j\kappa_j\equiv\kappa_j\omega_j\equiv 0$ $(\text{mod}\, D_i)$. In fact let $\mu(j_1\cdots j_s j)\kappa_{j_1}\cdots\kappa_{j_s}$ be any term of ω_j. Then the congruences

$$\mu(j_1\cdots j_s j)\equiv 0 \quad (\text{mod}\,\Delta(j_1\cdots j_s j i))$$

and

$$\mu(j_1\cdots j_s j)\equiv 0 \quad (\text{mod}\,\Delta(j j_1\cdots j_s i))$$

show that

$$\mu(j_1\cdots j_s j)\kappa_{j_1}\cdots\kappa_{j_s}\kappa_j\equiv 0 \quad (\text{mod}\, D_i)$$

and

$$\mu(j_1\cdots j_s j)\kappa_j\kappa_{j_1}\cdots\kappa_{j_s}\equiv 0 \quad (\text{mod}\, D_i).$$

Finally we have

(20) $$A_q\equiv 1 \quad (\text{mod}\, D_i).$$

Let $1+\gamma$ be the Magnus expansion of an element of A_q. It is known (see [9] or [5]) that γ contains only terms of degree $\geqq q$. But this proves assertion (20).

Note that assertion (15) follows immediately from (16) and (20). The proofs of assertions (12)–(14) follow.

Proof of (12). Suppose that $w_i=1+\omega_i$ is replaced by $\alpha_j w_i\alpha_j^{-1}$. Then ω_i is replaced by

$$\begin{aligned}(1+\kappa_j)\,&\omega_i(1-\kappa_j+\kappa_j^2-+\cdots)\\ &=\omega_i+(\text{terms involving } \kappa_j\omega_i \text{ or } \omega_i\kappa_j)\\ &\equiv\omega_i \quad (\text{mod}\, D_i)\end{aligned}$$

(where the last step follows from (16) and (17)). Since any conjugation can be broken down into a sequence of conjugations by the α_j and α_j^{-1}, this proves (12).

PROOF OF (13). Suppose that $\alpha_j = 1 + \kappa_j$ is replaced by $\bar{\alpha}_j = \alpha_h \alpha_j \alpha_h^{-1}$. Then $\alpha_j = \alpha_h^{-1} \bar{\alpha}_j \alpha_h$, hence

$$\kappa_j = (1 - \kappa_h + \kappa_h^2 - + \ldots)\bar{\kappa}_j(1 + \kappa_h)$$
$$= \bar{\kappa}_j + (\text{terms involving } \kappa_h \bar{\kappa}_j \text{ or } \bar{\kappa}_j \kappa_h).$$

Each time the factor κ_j occurs in the Magnus expansion of w_i it is to be replaced by this last expression. It follows from (18) that the terms in brackets give rise to terms in the new Magnus expansion of w_i which are congruent to zero modulo D_i. Hence the new expansion gives rise to the same residue classes $\bar{\mu}$ as the old. Again any conjugation may be built up from a sequence of elementary conjugations.

PROOF OF (14). The identity

$$\alpha_j w_j - w_j \alpha_j = (1 + \kappa_j)(1 + \omega_j) - (1 + \omega_j)(1 + \kappa_j) = \kappa_j \omega_j - \omega_j \kappa_j$$

shows that

$$\alpha_j^{-1} w_j^{-1} \alpha_j w_j = 1 + \alpha_j^{-1} w_j^{-1}(\alpha_j w_j - w_j \alpha_j)$$
$$= 1 + \alpha_j^{-1} w_j^{-1}(\kappa_j \omega_j - \omega_j \kappa_j) \equiv 1 \pmod{D_i}$$

(where the last step follows from (19)). Hence multiplication of w_i by $\alpha_j^{-1} w_j^{-1} \alpha_j w_j$ and its conjugates does not change the residue class of w_i modulo D_i. This proves assertion (14) and completes the proof of Theorem 5.

To what extent are these invariants $\bar{\mu}(i_1 \ldots i_r)$ independent of each other? In the case $r = 2$ it is easily seen that $\bar{\mu}(ij)$ equals the linking number of the ith and j^{th} components of $\mathfrak{L}$ for $i \neq j$; and that $\bar{\mu}(ii) = 0$. But it is well known that the linking number is symmetric: $\bar{\mu}(ij) = \bar{\mu}(ji)$. It will next be shown that this is a special case of much more general symmetry relations.

By a *proper shuffle* $h_1 \ldots h_{r+s}$ of two sequences $i_1 \ldots i_r$ and $j_1 \ldots j_s$ will be meant one of the $(r+s)!/r!s!$ sequences obtained by intermeshing $i_1 \ldots i_r$ with $j_1 \ldots j_s$.

THEOREM 6. *The following symmetry relations hold between the invariants* $\bar{\mu}$:

(21) *cyclic symmetry*, $\bar{\mu}(i_1 i_2 \ldots i_r) = \bar{\mu}(i_2 i_3 \ldots i_r i_1)$;

(22) *if* $i_1 \ldots i_r$ *and* $j_1 \ldots j_s$ *are given sequences with* $r, s \geq 1$, *then*

$$\Sigma \bar{\mu}(h_1 \ldots h_{r+s} k) \equiv 0 \quad (\text{mod G.C.D. } \Delta(h_1 \ldots h_{r+s} k)),$$

where the summation extends over all proper shuffles of $i_1 \ldots i_r$ *and* $j_1 \ldots j_s$.

(The symbol G.C.D. stands for the greatest common divisor over all proper shuffles. k is to be any fixed index.) For example, for the case $s = 1$, relation (22) asserts that

$$(23) \quad \bar{\mu}(j i_1 \ldots i_r k) + \bar{\mu}(i_1 j i_2 \ldots i_r k) + \bar{\mu}(i_1 i_2 j i_3 \ldots i_r k) + \ldots$$
$$+ \bar{\mu}(i_1 i_2 \ldots i_r j k) \equiv 0 \quad (\text{mod G.C.D. } \Delta(i_1 \ldots j \ldots i_r k)).$$

The relations (22) are immediate consequences of a theorem of Chen, Fox and Lyndon [4]. By a *shuffle* of $i_1 \dots i_r$ and $j_1 \dots j_s$ will be meant a sequence $h_1 \dots h_t$ which contains $i_1 \dots i_r$ and $j_1 \dots j_s$ as subsequences and which is the union of these two subsequences; together with a specification of exactly how the two sequences $i_1 \dots i_r$ and $j_1 \dots j_s$ are to be imbedded in $h_1 \dots h_t$. A shuffle is proper if the two subsequences are disjoint. (For example, the sequences 12 and $\dot{2}$ have three proper shuffles $12\dot{2}$, $1\dot{2}2$, $\dot{2}12$, and one improper shuffle 12.) The theorem of Chen, Fox and Lyndon follows:

If $1 + \sum \nu(i_1 \dots i_r) \kappa_{i_1} \dots \kappa_{i_r}$ *is the Magnus expansion of an element of a free group, then*

$$\nu(i_1 \dots i_r) \nu(j_1 \dots j_s) = \sum \nu(h_1 \dots h_t),$$

where the summation extends over all shuffles of $i_1 \dots i_r$ *and* $j_1 \dots j_s$.

For the word $w_k(\alpha_1 \dots \alpha_n)$ this proposition asserts that

$$\mu(i_1 \dots i_r k) \mu(j_1 \dots j_s k) = \sum \mu(h_1 \dots h_t k).$$

But modulo G.C.D. $\Delta(h_1 \dots h_t k)$, the left-hand side is congruent to zero, and each improper shuffle on the right is congruent to zero. This completes the proof of the relations (22).

The proof of (21) (cyclic symmetry) will be based on the following:

Lemma 5. *If* $\{a_{ij} / R_{ij} = 1\}$ $(i = 1, \dots, n;\ j = 1, \dots, r_i)$ *is the Wirtinger presentation of the fundamental group of the complement of a link, then the elements* R_{ij} *of the first group* $\{a_{ij}\}$ *satisfy an identity of the form*

$$\prod_{i=1}^{n} \prod_{j=1}^{r_i} (\lambda_{ij}^{-1} R_{ij} \lambda_{ij}) = 1,$$

where the λ_{ij} *are also elements of the free group.*

It is well known that any one relation of the Wirtinger presentation is a consequence of the remaining relations. (See, for example, [11], p. 54.) Lemma 5 is a sharpening of this assertion. Since it is proved in exactly the same manner, no details will be given.

We will apply the homomorphism η_q to the identity of Lemma 5. (See proof of Theorem 4 for the definition of η_q.) By formulas (3), (9) and (11) it follows that

$$\eta_q(R_{ij}) \equiv 1 \pmod{A_q} \quad \text{for} \quad 1 \leq j < r_i$$

and

$$\eta_q(R_{ir_i}) \equiv \alpha_i^{-1} w_i^{-1} \alpha_i w_i \pmod{A_q}.$$

Hence, defining $\lambda_i = \eta_q(\lambda_{ir_i})$, we have

$$\begin{aligned} 1 &= \prod_{i=1}^{n} \prod_{j=1}^{r_i} \eta_q(\lambda_{ij}^{-1} R_{ij} \lambda_{ij}) \\ &\equiv \prod_{i=1}^{n} (\lambda_i^{-1} \alpha_i^{-1} w_i^{-1} \alpha_i w_i \lambda_i) \pmod{A_q}. \end{aligned}$$

Let D denote the ideal formed by all elements $\sum \nu(i_1 \ldots i_s)\kappa_{i_1} \ldots \kappa_{i_s}$ of the Magnus ring such that

$$\nu(i_1 \ldots i_s) \equiv 0 \pmod{\Delta(i_1 \ldots i_s)}$$

for all sequences $i_1 \ldots i_s$ with $s < q$. We will next study the preceding identity modulo D. This is possible since $A_q \equiv 1 \pmod D$. Let $w_i = 1 + \omega_i$. Observe that if $\kappa_i \omega_i$ or $\omega_i \kappa_i$ is multiplied on the left or right by any κ_j, then the result is congruent to zero modulo D. (The proof is an easy modification of the proofs of (17) and (19).) Therefore

$$\lambda_i^{-1}\alpha_i^{-1}w_i^{-1}\alpha_i w_i \lambda_i = 1 + \lambda_i^{-1}\alpha_i^{-1}w_i^{-1}(\alpha_i w_i - w_i \alpha_i)\lambda_i$$
$$= 1 + \lambda_i^{-1}\alpha_i^{-1}w_i^{-1}(\kappa_i\omega_i - \omega_i\kappa_i)\lambda_i \equiv 1 + (\kappa_i\omega_i - \omega_i\kappa_i) \pmod D.$$

Our identity now takes the form

$$1 \equiv \prod (\lambda_i^{-1}\alpha_i^{-1}w_i^{-1}\alpha_i w_i \lambda_i) \equiv \prod (1 + (\kappa_i\omega_i - \omega_i\kappa_i)$$
$$\equiv 1 + \sum (\kappa_i\omega_i - \omega_i\kappa_i) \pmod D.$$

Since $\omega_i = \sum_{(j)} \mu(j_1 \ldots j_s i)\kappa_{j_1} \ldots \kappa_{j_s}$, this means that

$$\sum_{i,(j)} \mu(j_1 \ldots j_s i)(\kappa_i\kappa_{j_1} \ldots \kappa_{j_s} - \kappa_{j_1} \ldots \kappa_{j_s}\kappa_i) \equiv 0 \pmod D.$$

The coefficient of $\kappa_i\kappa_{j_1} \ldots \kappa_{j_s}$ in this last expression is

$$\mu(j_1 \ldots j_s i) - \mu(i j_1 \ldots j_s);$$

which proves that

$$\mu(j_1 \ldots j_s i) \equiv \mu(i j_1 \ldots j_s) \pmod{\Delta(i j_1 \ldots j_s)}$$

for all sequences $j_1 \ldots j_s i$ with $s + 1 < q$. Since q can be chosen arbitrarily large, this completes the proof of Theorem 6.

A good way of getting some insight into the meaning of these invariants $\bar{\mu}(i_1 \ldots i_r)$ is to study their behavior under simple transformations of the link. The following two assertions are fairly easy to prove (although proofs will be omitted).

If the orientation of the jth component of $\mathfrak{L}$ is reversed, then $\bar{\mu}(i_1 \ldots i_r)$ is multiplied by $+1$ or -1 according as the sequence $i_1 \ldots i_r$ contains j an even or an odd number of times.

If the orientation of Euclidean space is reversed, then $\bar{\mu}(i_1 \ldots i_r)$ is multiplied by $(-1)^{r-1}$.

Now suppose that $\mathfrak{L}$ is a link which possesses a smooth tubular neighborhood; for example, a polygonal link. Let $\mathfrak{L}'$ be a link obtained by replacing each component of $\mathfrak{L}$ by a collection of parallel components, all lying in the tubular neighborhood of the original, and having linking numbers zero with each other. Suppose that the ith

component of $\mathfrak{L}'$ corresponds to the $h(i)^{\text{th}}$ component of $\mathfrak{L}$. Let $\bar{\mu}, \bar{\mu}'$ denote the invariants of $\mathfrak{L}$ and $\mathfrak{L}'$ respectively.

THEOREM 7. *Under these conditions*

$$\bar{\mu}'(i_1 \dots i_s) = \bar{\mu}(h(i_1), \dots, h(i_s)).$$

Thus the invariants of $\mathfrak{L}'$ are completely determined by those of $\mathfrak{L}$. It is clearly sufficient to consider the special case where $\mathfrak{L}'$ is an $(n+1)$-link, obtained by replacing the last component of $\mathfrak{L}$ by two parallel components.

Let T be the tubular neighborhood of L. It is assumed that the pair $(\bar{T}, L)$ is homeomorphic to n copies of the pair consisting of a solid torus and its central circle. It follows easily from this assumption that the homomorphism

$$F(M-T) \to F(M-L)$$

is an isomorphism. Under this isomorphism, the meridians and parallels to the boundary of the tube T may be identified with the meridians and parallels α_i, β_i of $\mathfrak{L}$.

Consider the homomorphism $F(M-T) \to F(M-L')$. The parallels $\beta_1, \dots, \beta_n$ of the tubular neighborhood map into corresponding parallels $\beta'_1, \dots, \beta'_n$ of $\mathfrak{L}'$ under this homomorphism. Note that the last parallel β'_{n+1} is equal to β'_n. Similarly, the meridians $\alpha_1, \dots, \alpha_{n-1}$ map into meridians $\alpha'_1, \dots, \alpha'_{n-1}$; however, the last meridian α_n maps into the product $\alpha'_n \alpha'_{n+1}$.

Let $w_j(\alpha_1, \dots, \alpha_n)$ be a word for the jth parallel of $\mathfrak{L}$. Then the relation $\beta_j = w_j(\alpha_1, \dots, \alpha_{n-1}, \alpha_n)$ in $F(M-T)$ goes into the relation $\beta'_j = w_j(\alpha'_1, \dots, \alpha'_{n-1}, \alpha'_n \alpha'_{n+1})$ in $F(M-L')$. Let w'_j denote this last word. The Magnus expansion of w'_j can be obtained from the expansion

$$w_j = 1 + \sum \mu(i_1 \dots i_s j) \, \kappa_{i_1} \dots \kappa_{i_s}$$

by substituting κ'_i for κ_i $(1 \leq i < n)$, and by substituting

$$\alpha'_n \alpha'_{n+1} - 1 = \kappa'_n + \kappa'_{n+1} + \kappa'_n \kappa'_{n+1} \quad \text{for} \quad \kappa_n.$$

Making use of (18), it is easily seen that the terms involving $\kappa'_n \kappa'_{n+1}$ may be ignored in computing the residue classes $\bar{\mu}'$. But if we substitute $\kappa'_n + \kappa'_{n+1}$ for κ_n in $\sum \mu(i_1 \dots i_s j) \, \kappa_{i_1} \dots \kappa_{i_s}$ and substitute κ'_i for κ_i $(i < n)$, then the coefficient $\mu'(i_1 \dots i_s j)$ of $\kappa'_{i_1} \dots \kappa'_{i_s}$ in the result is clearly equal to $\mu(h(i_1), \dots, h(i_s), j)$; where $h(i) = i$ for $i \leq n$, $h(n+1) = n$. Since the word w'_{n+1} is equal to w'_n, we also have

$$\mu'(i_1 \dots i_s n+1) = \mu(h(i_1), \dots, h(i_s), h(n+1)).$$

Thus the $\bar{\mu}'$ have the required form in all cases.

We will next study the relationship between isotopy of links and homotopy, as studied in [10].

THEOREM 8. *If the indices $i_1 \dots i_r$ are pairwise distinct, then $\bar{\mu}(i_1 \dots i_r)$ is invariant, not only under isotopy of $\mathfrak{L}$, but also under homotopy.*

It is clearly sufficient to consider polygonal links. Let $(\alpha_i)_2$ denote the commutator subgroup of the normal subgroup of $F(M-L)$ generated by the meridian α_i. It was proved in [10] that the group $\mathcal{G}(L) = F(M-L)/(\alpha_1)_2 \dots (\alpha_n)_2$ is invariant under homotopy of $\mathfrak{L}$. Furthermore, the conjugate classes $\lambda\alpha_i\lambda^{-1}$ in $\mathcal{G}(L)$ and the conjugate classes $\lambda\beta_i\lambda^{-1}$ in $\mathcal{G}(L)/(\alpha_i)$ are invariant under homotopy. (The group $\mathcal{G}$ may be considered as a factor group of F/F_{n+1}.) Hence in order to prove that $\bar{\mu}(i_1 \dots i_r)$ is a homotopy invariant, it is sufficient to prove that it is not changed when:

(12′) the word $w_{i_r}(\alpha_1 \dots \alpha_n)$ for β_{i_r} is replaced by a conjugate,

(13′) some α_j is replaced by a conjugate,

(14′) w_{i_r} is multiplied by a product of conjugates of $\alpha_j^{-1}w_j^{-1}\alpha_j w_j$,

(24) w_{i_r} is multiplied by an element of $(\alpha_j)_2$,

(25) w_{i_r} is multiplied by an element of (α_{i_r}).

The proofs of (12′), (13′) and (14′) are trivial modifications of the corresponding proofs of Theorem 5. To prove (24), let $1+\gamma$ be the Magnus expansion of an element of $(\alpha_j)_2$. Then it is easily verified that each term of γ contains the factor κ_j at least twice. But $\mu(i_1 \dots i_r)$ is the coefficient of a term in the Magnus expansion of w_{i_r} which contains the factor κ_j at most once.

Similarly, to prove (25), if $1+\gamma$ is the Magnus expansion of an element of (α_{i_r}), then each term of γ contains the factor κ_{i_r}. But $\mu(i_1 \dots i_r)$ is the coefficient of a term in w_{i_r} which does not involve κ_{i_r} as a factor. This completes the proof of Theorem 8.

In [10] certain homotopy invariants of $\mathfrak{L}$ were defined by a rather different procedure. To conclude this section we will show that these invariants are essentially identical with those of Theorem 8.

The definition of these invariants follows (with some changes in notation). For each element $\sigma = \sum_j \epsilon_j \gamma_j$ of the integral group ring $J\mathcal{G}$ (where the ϵ_j are integers and the γ_j are elements of $\mathcal{G}(L)$) define

$$\alpha_i^\sigma = \prod_j (\gamma_j \alpha_i^{\epsilon_j} \gamma_j^{-1}).$$

Every element of the normal subgroup (α_i) can be expressed in this form. Therefore every element of $\mathcal{G}$ can be expressed as $\alpha_i^\sigma u$, where u is a word in $\alpha_1 \dots \alpha_n$ which does not involve α_i. In particular, for $j \neq i$, the parallel β_j can be expressed as

$$\beta_j = \alpha_i^{\sigma_{ij}} u.$$

Let $$\sum \mu^*(h_1 \ldots h_s ij)\, \kappa_{h_1} \ldots \kappa_{h_s}$$

be the Magnus expansion of σ_{ij}. Let $\Delta^*(h_1 \ldots h_s ij)$ denote the greatest common divisor of $\mu^*(k_1 \ldots k_r)$, where $k_1 \ldots k_r$ $(2 \leqq r \leqq s+1)$ is to range over all permutations of proper subsequences of $h_1 \ldots h_s ij$. Then for distinct integers $h_1 \ldots h_s ij$ it was proved in [10] that the residue classes $\mu^*(h_1 \ldots h_s ij) \bmod \Delta^*(h_1 \ldots h_s ij)$ are invariant under homotopy of $\mathfrak{L}$.

THEOREM 9. *If $h_1 \ldots h_s ij$ are distinct integers between* 1 *and n, then*

$$\mu^*(h_1 \ldots h_s ij) \equiv \mu(h_1 \ldots h_s ij) \quad (\operatorname{mod} \Delta^*(h_1 \ldots h_s ij)).$$

Thus the invariants $\mu^* \bmod \Delta^*$ are merely a weaker form of the invariants $\bar{\mu}$. It also follows that

$$\Delta^*(h_1 \ldots h_s ij) = \text{G.C.D.}\, \Delta(k_1 \ldots k_{s+2}),$$

where $k_1 \ldots k_{s+2}$ is to range over all permutations of $h_1 \ldots h_s ij$.

The proof will make use of the free differential calculus. Let $w_j = \alpha_i^\sigma u$ be a word for the jth parallel, where neither σ nor u involves α_i. Then

$$\begin{aligned} \frac{\partial w_j}{\partial \alpha_i} &= \frac{\partial(\alpha_i^\sigma u)}{\partial \alpha_i} \\ &= \frac{\partial(\prod_j (\gamma_j \alpha_i^{\epsilon_j} \gamma_j^{-1})\, u)}{\partial \alpha_i} = \textstyle\sum_j (\gamma_1 \alpha_i^{\epsilon_1} \gamma_1^{-1}) \ldots (\gamma_{j-1} \alpha_i^{\epsilon_{j-1}} \gamma_j^{-1})\, \gamma_j \dfrac{\partial \alpha_i^{\epsilon_j}}{\partial \alpha_i} \\ &\equiv \textstyle\sum_j 1 \cdot \ldots \cdot 1 \cdot \gamma_j \cdot \epsilon_j = \sigma \quad (\operatorname{mod} (\kappa_i)), \end{aligned}$$

where (κ_i) denotes the two-sided ideal in $J\mathcal{G}$ generated by κ_i. This implies that

$$\left(\frac{\partial^{s+1} w_j}{\partial \alpha_{h_1} \ldots \partial \alpha_{h_s} \partial \alpha_i}\right)^0 = \left(\frac{\partial^s \sigma}{\partial \alpha_{h_1} \ldots \partial \alpha_{h_s}}\right)^0$$

for any sequence $h_1 \ldots h_s$ of integers distinct from i. But the left-hand side of this last equation equals $\mu(h_1 \ldots h_s ij)$, and the right-hand side equals $\mu^*(h_1 \ldots h_s ij)$; which completes the proof.

One consequence of Theorem 9 is that the following symmetry relations between the invariants $\mu^* \bmod \Delta^*$ carries over to the $\bar{\mu}$ (see [10], § 5)

$$\bar{\mu}(h_1 \ldots h_s i j_1 \ldots j_t k) \equiv (-1)^t \sum \bar{\mu}(l_1 \ldots l_{s+t} ik) \quad (\operatorname{mod} \Delta^*(h_1 \ldots h_s i j_1 \ldots j_t k)), \tag{26}$$

where the summation extends over all proper shuffles $l_1 \ldots l_{s+t}$ of $h_1 \ldots h_s$ and $j_t j_{t-1} \ldots j_1$.

For example, for the case $s=0$, this relation becomes

$$\bar{\mu}(ij_1 \dots j_t k) \equiv (-1)^t \bar{\mu}(j_t \dots j_1 ik).$$

Taking account of cyclic symmetry, this can be put in the form

$$\bar{\mu}(i_1 \dots i_r) \equiv (-1)^r \bar{\mu}(i_r \dots i_1) \pmod{\Delta^*(i_1 \dots i_r)}. \tag{27}$$

Although formula (26) has only been derived for sequences of distinct integers, it actually holds for arbitrary sequences. This can be proved in two ways: (a) Formula (26) can be deduced directly from (22) by a straightforward argument. (b) Making use of Theorem 7, it can be shown that any symmetry relation which holds for sequences of distinct integers must automatically hold for arbitrary sequences. We will not go into the details of these arguments.

4. Discussion and examples

In order to illustrate the invariants of the preceding section, and the symmetry relations between them, we will study the case of a link with two components. By an invariant $\bar{\mu}(i_1 \dots i_{r+s})$ of type $[r, s]$ will be meant one which involves the index '1' r times and the index '2' s times. To simplify the discussion it will be assumed that $\Delta^*(i_1 \dots i_{r+s}) = 0$, so that $\bar{\mu}$ is an integer.

First observe that all invariants of type $[r, 0]$ and $[r, 1]$ $(r \geqq 2)$ are zero. This can easily be proved, using formulas (23) and (21). The invariants of type $[1, 1]$ (the linking numbers) are not necessarily zero. But these are too well known to be worth much discussion.

All invariants of type $[2m+1, 2]$ are also zero. This can be proved using formulas (27) and (21). Thus the first interesting invariants are those of type $[2m, 2]$. (For example, $\bar{\mu}(1212)$, $\bar{\mu}(122111)$, etc.) For these invariants we have, by (26),

$$\bar{\mu}(1\dots1212) = -\binom{2m}{1}\bar{\mu}(1\dots1122),$$

$$\bar{\mu}(1\dots12112) = \binom{2m}{2}\bar{\mu}(1\dots1122),$$

$$\bar{\mu}(1\dots121112) = -\binom{2m}{3}\bar{\mu}(1\dots1122), \quad \text{etc.}$$

In view of cyclic symmetry this means that all of the invariants of type $[2m, 2]$ are completely determined by $\bar{\mu}(1\dots1122)$. An example of a link for which the invariant $\bar{\mu}(1\dots1122)$ of type $[2m, 2]$ is equal to $(-1)^m$ is given in Fig. 1. (The computations are much too involved to be given here.)

As a final example, for invariants of type [3, 3] it can be shown that the following identities are always satisfied:

$$(\tfrac{1}{2})\,\bar{\mu}(111222) = -(\tfrac{1}{3})\,\bar{\mu}(112122) = (\tfrac{1}{12})\,\bar{\mu}(121212).$$

A link can be constructed for which $(\frac{1}{2})\,\bar{\mu}(111222) = 1$.

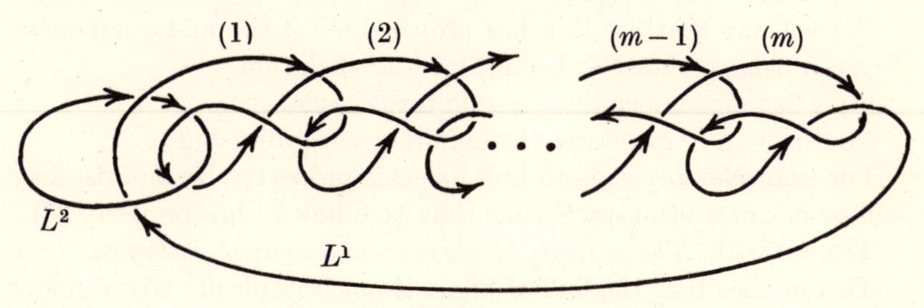

Fig. 1

Suppose that all of the invariants $\bar{\mu}$ of $\mathfrak{L}$ are equal to zero. Does it follow that $\mathfrak{L}$ is isotopically trivial?

The answer to this question is almost certainly no. For example, consider the link of Fig 2 (obtained from the clover-leaf knot by

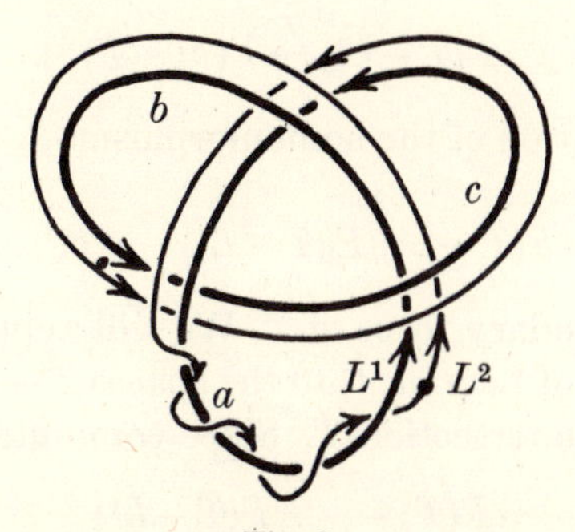

Fig. 2

replacing the one component by two parallel components). It follows from Theorem 7 that all of the invariants $\bar{\mu}$ of this link are zero. However, we will prove that this link is not isotopically trivial; at least under isotopies which satisfy a certain smoothness condition.

This is in contrast to the results of [10] where the following theorem was proved: If all the invariants $\mu^* \bmod \Delta^*$ of $\mathfrak{L}$ are zero, then $\mathfrak{L}$ is homotopically trivial.

We will say that a link $\mathfrak{L}$ has property (*P*) *if the loop $\mathfrak{L}^2$ can not be contracted within $M-L^1$.*

Observe that the link $\mathfrak{L}$ of Fig. 2 possesses this property. In fact the Wirtinger presentation of $F(M-L^1)$ is $\{a,b,c/ba=ac=cb\}$. The loop $\mathfrak{L}^2$ represents the element $caba^{-3}$ of this group. But the representation

$$a \to (1234), \quad b \to (1324), \quad c \to (1243)$$

carries $caba^{-3}$ into $(13)(24)$.

We will say that a 2-link has property (S) if the first component L^1 has a neighborhood T disjoint from L^2 such that

(a) $\bar{T}$ is homeomorphic to a solid torus, and

(b) the loop $\mathfrak{L}^1$ generates the infinite cyclic group $F(T)$.

For example, any smooth link has this property. By an S-isotopy will be meant an isotopy $\mathfrak{L}_t$ such that each link $\mathfrak{L}_t$ has property (S).

THEOREM 10. *The property* (*P*) *is invariant under S-isotopies.*

This implies that the link of Fig. 2 is not isotopically trivial under S-isotopies. The proof will be based on the following:

LEMMA 6. *Let T be a neighborhood of L^1 which satisfies conditions* (a) *and* (b). *Then the natural homomorphism $F(M-T) \xrightarrow{\eta} F(M-L^1)$ has kernel* 1.

A left inverse to η will be constructed as follows. Let

$$F(\bar{T}-L^1) \xrightarrow{\tau} F(M-T)$$

denote the composition of the homomorphisms

$$F(\bar{T}-L^1) \to F(\bar{T}-L^1)/F_2(\bar{T}-L^1) \xleftarrow{\approx} F(\dot{T}) \to F(M-T),$$

where $\dot{T}$ is the boundary torus of T. We will apply the van Kampen theorem (see proof of Lemma 4) to the spaces $\bar{T}-L^1$ and $M-T$, with union $M-L^1$ and intersection $\dot{T}$. Since commutativity holds in the following square

$$\begin{array}{ccc} F(\dot{T}) & \longrightarrow & F(\bar{T}-L^1) \\ \downarrow & & \downarrow \tau \\ F(M-T) & \xrightarrow{i} & F(M-T), \end{array}$$

it follows that the homomorphisms τ and i can be pieced together into a homomorphism $F(M-L^1) \xrightarrow{\tau'} F(M-T)$. Since the composition $F(M-T) \xrightarrow{\eta} F(M-L^1) \xrightarrow{\tau'} F(M-T)$ is the identity map i, it follows that η has kernel 1.

The proof of Theorem 10 follows. Let $\mathfrak{L}$ and T be as above and let $\mathfrak{L}_0$ be any link homotopic to $\mathfrak{L}$ within $M - \dot{T}$. Then the second components $\mathfrak{L}^2$ and $\mathfrak{L}_0^2$ are homotopic loops in $M - T$. Now by Lemma 6, if $\mathfrak{L}^2$ is contractible within $M - L^1$ then it is contractible within $M - T$, and therefore $\mathfrak{L}_0^2$ is contractible within $M - T \subset M - L_0^1$. Since the argument can be reversed, it follows that $\mathfrak{L}$ has property (P) if and only if $\mathfrak{L}_0$ has property (P). But Theorem 10 clearly follows from this last assertion (just as Theorem 2 followed from Theorem 1).

In several places in the present paper (Theorems 1′, 2′ and 10) it has been necessary to make certain smoothness assumptions. We will next give an example to show that these assumptions were necessary (at least for Theorems 1′ and 10).

Consider the link of Fig. 3. (A more precise description of this link could easily be given, using the techniques of [7].) It will be shown that this link has the following four properties:†

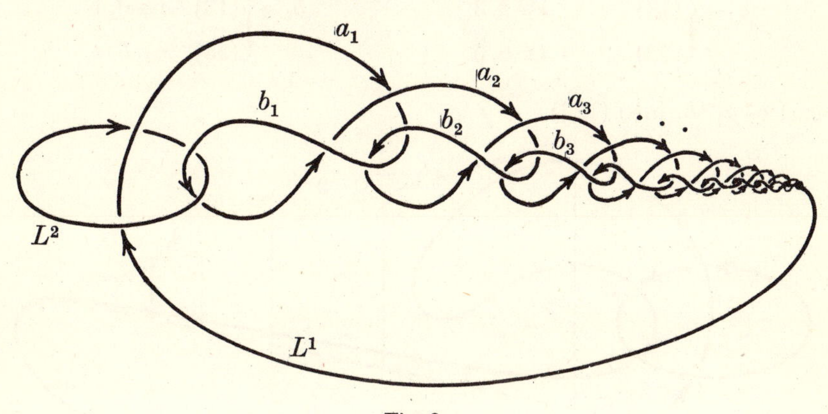

Fig. 3

(1) It is isotopically trivial. In fact for every $\epsilon > 0$ it is ϵ-isotopic to a link which is trivial in the ordinary sense of knot theory.

(2) It has property (P). ($\mathfrak{L}^2$ is not contractible within $M - L^1$.)

(3) For every ϵ there exists an $\mathfrak{L}_0$ which is ϵ-homotopic to $\mathfrak{L}$ and for which some invariant $\bar{\mu}(1\ldots122)$ is not zero.

(4) The set L is not a deformation retract of any neighborhood.

Note that property (4) is a consequence of (1) and (3). In fact, if $\mathfrak{L}$ were a deformation retract of some neighborhood, then Theorem 1′ and property (1) would imply that the group $F(M - L)/F_\infty$ is a free group on two generators. On the other hand, Theorem 1′ and property

† A closely related example has been studied by R. H. Fox [6].

(3) would imply the contrary.

To prove (3) it is only necessary to observe that the link of Fig. 3 can be approximated arbitrarily closely by links of the type illustrated in Fig. 1.

To prove (2) consider the Wirtinger presentation of $F(M-L^1)$. (See [7] for justification.) This presentation has generators $a_1, a_2, \ldots, b_1, b_2, \ldots$ and relations

$$a_2 = b_1,$$

$$a_{i+2} = b_{i+1} b_i^{-1} b_{i+1} b_i b_{i+1}^{-1}, \quad b_i = b_{i+1}^{-1} a_{i+1} a_i a_{i+1}^{-1} b_{i+1},$$

for $i \geqq 1$. The loop $\mathfrak{L}^2$ corresponds to the element $a_1^{-1} b_1$ of this group. But the representation

$$a_i \to \begin{cases} (12) & \text{for} \quad i \equiv 1, 4 \pmod 6 \\ (13) & \qquad\quad i \equiv 2, 3 \\ (23) & \qquad\quad i \equiv 0, 5 \end{cases} \qquad b_i \to \begin{cases} (12) & i \equiv 2, 5 \\ (13) & i \equiv 0, 1 \\ (23) & i \equiv 3, 4 \end{cases}$$

carries $a_1^{-1} b_1$ into (123).

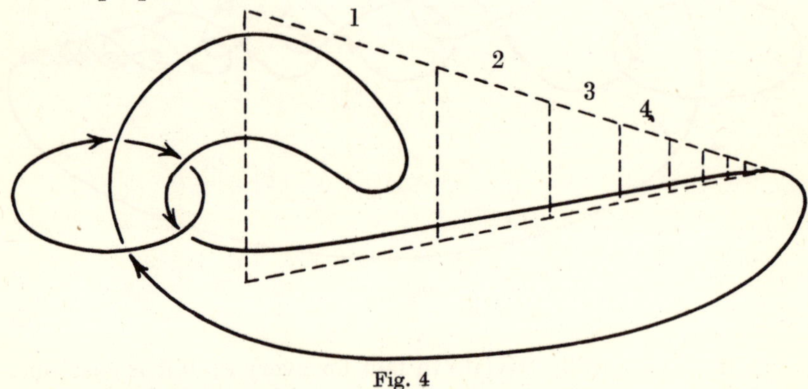

Fig. 4

To prove (1), let $\mathfrak{L}_0$ denote the trivial link illustrated in Fig. 4. During the interval $[0, \frac{1}{2}]$ perform the isotopy illustrated in Fig. 5 on the portion of $\mathfrak{L}_0$ which lies within trapezoids 1 and 2, leaving the rest of the link fixed. This defines $\mathfrak{L}_t$ for $0 \leqq t \leqq \frac{1}{2}$. Now perform this same isotopy in the trapezoids 2 and 3 during the interval $[\frac{1}{2}, (\frac{3}{4})]$; and continue by induction. (Performing the isotopy on trapezoids r and $r+1$ during the interval $[1-(\frac{1}{2})^{r-1}, 1-(\frac{1}{2})^r]$.) This defines $\mathfrak{L}_t$ for $0 \leqq t < 1$. But the limit of $\mathfrak{L}_t$ as $t \to 1$ clearly exists, and is just the link of Fig. 3. This completes the proof of properties 1, 2, 3, 4.

In conclusion, the following is a list of several related problems which the author does not know how to solve, but which seem of interest:

(a) To find other, more sensitive, isotopy invariants of links. This problem is illustrated by the link of Fig. 6. It can be shown that all of

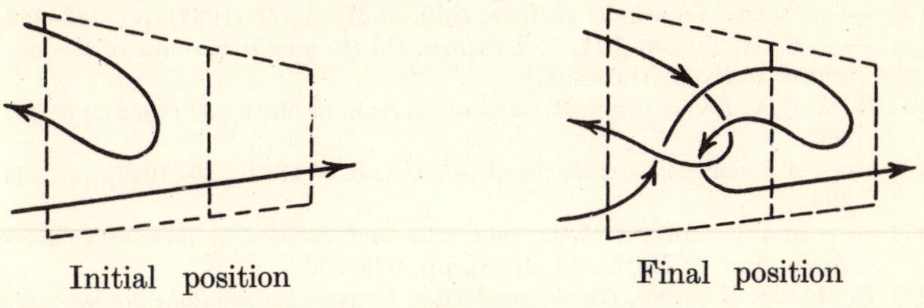

Initial position   Final position

Fig. 5

the invariants $\bar{\mu}$ of this link are zero. It is also clear that either component can be contracted within the complement of the other component. Yet intuitively it seems unlikely that this link is isotopically trivial.

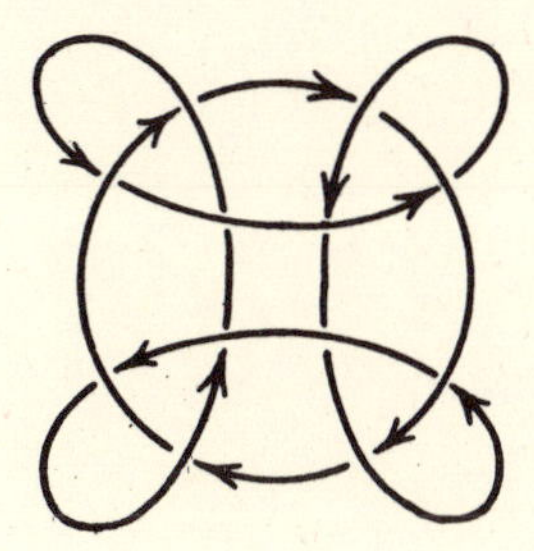

Fig. 6

(b) To find some method of attacking the group F/F_{∞} and extracting invariants from it.

(c) To determine whether Theorems 1 and 2 can be generalized to imbeddings of more general complexes in n-manifolds; or whether similar assertions about the higher homotopy groups of $M-L$ can be established.

PRINCETON UNIVERSITY

References

[1] K. Borsuk, *Quelques rétractes singuliers*, Fund. Math., 24 (1935), pp. 249–259.

[2] K. T. Chen, *Commutator calculus and link invariants*, Proc. Amer. Math. Soc., 3 (1952), pp. 44–55.

[3] ——, *Isotopy invariants of links*, Ann. of Math., 56 (1952), pp. 343–353.

[4] ——, R. H. Fox and R. C. Lyndon, *On the quotient groups of the lower central series* (forthcoming).

[5] R. H. Fox, *Free differential calculus*, 1, Ann. of Math., 57 (1953), pp. 547–560.

[6] ——, *A remarkable simple closed curve*, Ann. of Math., 50 (1949), pp. 264–265.

[7] —— and E. Artin, *Some wild cells and spheres in three-dimensional space*, Ann. of Math., 49 (1948), pp. 979–990.

[8] E. R. van Kampen, *On the connection between the fundamental groups of some related spaces*, Amer. J. Math., 55 (1933), pp. 261–267.

[9] W. Magnus, *Beziehungen zwischen Gruppen und Idealen in einem speziellen Ring*, Math. Ann., 111 (1935), pp. 259–280.

[10] J. Milnor, *Link groups*, Ann. of Math., 59 (1954), pp. 177–195.

[11] K. Reidemeister, Knotentheorie (Ergebnisse der Math. 1, 1), Julius Springer, Berlin, 1932.

Generators and Relations in a Complex

P. A. Smith

LET A be a connected cell complex. We begin by recalling the following representation of the fundamental group $\pi_1(A)$ by generators and relations. Let F_1 be the free group of loops formed with the 1-cells of A and having a fixed base point 0. Take for generators any set of free generators of F_1 and for relations those loops which are mapped onto boundaries of 2-cells by the obvious homomorphism $F_1 \to C_1$ (C_1 being the group of finite integral 1-chains). The subgroup R of F_1 generated by the relations is readily seen to be normal, and we have $\pi_1 = F_1/R$.

Suppose now that we take the relations, or rather an abstract set in one-one correspondence with the relations, as generators of a free group F_2. There is an obvious homomorphism $F_2 \to F_1$ with R for image. Hence π_1 can be regarded as the (non-Abelian) homology group of the sequence $F_2 \to F_1 \to 1$, that is, $\pi_1 = \text{kernel}/\text{image}$.

It was shown by R. Peiffer [1] and independently by the writer [2] that the sequence which yields π_1 can in a sense be imbedded in a sequence $F_3 \to F_2 \to F_1 \to 1$ which will yield π_2 as well. F_3 is generated by a set which is in one-one correspondence with those elements of kernel ($F_2 \to F_1$) which map into boundaries of 3-cells under an obvious homomorphism $F_2 \to C_2$. Actually, this description is not quite accurate since, for example, kernel ($F_2 \to F_1$)/image ($F_3 \to F_2$) would not necessarily be Abelian, hence would not be π_2. One is led, however, to consider certain equivalence relations in the F's and the required sequence is obtained by replacing the free groups by their equivalence-class groups.

We shall here consider the possibility of extending the sequence just described to higher dimensions. Specifically, we attempt to associate to an arbitrary connected simplicial complex A a homomorphism sequence

$$\text{(1)} \qquad \ldots \to F_n \to \ldots \to F_2 \to F_1 \to 1 \quad (dd = 1),$$

suggested by homotopy rather than homology concepts and described by generators and relations determined by the cellular structure. The sequence is to admit homology, that is, image d_{n+1} which is a subgroup of kernel d_n is to be normal so that we can define

$$p_n = \text{kernel}\, d_n/\text{image}\, d_{n+1}.$$

The sequence is to contain embedded in it the sequence $F_3 \rightarrow \ldots \rightarrow 1$ so that $p_2 = \pi_2$, $p_1 = \pi_1$. For $n > 2$, the groups p_n are to be Abelian (as a matter of fact, in the sequences considered, the groups F_n are Abelian for $n > 2$). Our object is to secure for the groups p_n at least some of the main properties of homotopy groups. Consider, for example, the theorem of Hurewicz that π_n is isomorphic to the homology group H_n when the lower homotopy groups are trivial. Passing to the groups p_n, we say that the sequence (1) has property h^n $(n > 1)$ if $p_n = H_n$ whenever $p_1 = \ldots = p_{n-1} = 1$. We shall give two explicit constructions Σ_{I} and Σ_{II}. In the case of Σ_{I}, h^n is true for every n. However, Σ_{I} is trivial when $n > 2$ in the sense that $p_n(A) = 1$ whenever $\dim A < n$. We regard Σ_{II} as more interesting since, although only h^2 and h^3 hold, the construction is non-trivial in the dimension 3. We shall show in fact that for a 2-sphere S, boundary of a 3-simplex, p_3 is infinite. We do not yet know whether in this case p_3 is cyclic and thus isomorphic to $\pi_3(S)$. The problem of constructing a sequence which has property h^n for every n and is non-trivial in every dimension seems to be difficult.

All sequences are constructed by induction. Having defined F_{n-1}, we use certain elements in F_{n-1} to define a free group F_n then pass to a quotient $\mathbf{F}_n$ of F_n by specifying a kernel K_n. (The constructions Σ_{I}, Σ_{II} differ only in the choice of kernels.) To make sure that the absence of n-cells will not imply $p_n = 1$, we introduce certain generators for F_n (they form the sets X_n^2, X_n^3, § 8) other than those which correspond to n-cells in the manner described above for $n > 2$, 3. These generators are degenerate in the sense that they map into zero under a natural homomorphism $F_n \rightarrow C_n$. To avoid triviality in the dimension n it is also necessary to choose K_n so that the degenerate elements will not all be liquidated in passing from F_n to $\mathbf{F}_n$.

It can be shown without difficulty that in both Σ_{I}, Σ_{II}, simplicial mappings of complexes induce homomorphism of the groups p_n and that $p_n(A) = p_n(A^*)$, where A^* is the universal covering of A. We believe that the groups p_n in the construction Σ_{I} and the group p_3 in the construction Σ_{II} are invariant with respect to simplicial subdivision, although we have not yet verified this in detail. In any case the known and conjectured properties do not permit us to identify the groups p_n. For these reasons our study is tentative, the main result being the existence of sequences with certain formal properties. On the other hand, the actual definitions which have been adopted are to a large extent forced by the requirements to be met. Thus, for example, if π_3 (like π_2, π_1) can be given by generators and relations through a sequence $(F_4 \to \dots \to 1)$, such a sequence must be essentially identical with the 4-dimensional part of the sequence defined by Σ_{II}.

The construction leading to p_2 was originally performed in [2] on a class of objects more general than complexes. This made it possible, for example, to assign a group $p_2(L)$ to every abstract local group L. The assignment is such that if L is imbedded in a group G which it generates, then $p_2(L) = p_2(A)$, where A is the complex whose n-simplexes are subsets $(x_0, \dots, x_n\}$ of G such that $x_i^{-1}x_j \in L$. We gave a precise meaning for the condition $p_2(L) = 1$ in terms of extendability. In a similar manner, $p_n(L)$ can be defined, but we leave open the question of interpretation.

Added in proof. A beautiful construction which gives the actual homotopy groups has just been obtained by D. M. Kan.

Normal sequences

1. Let X be a set in which there is defined an inversion $x \to x^{-1}$. We denote by $F[X]$ the group† of word-equivalence classes of words formed with elements of X agreeing that $F[\varnothing]$ is the trivial group (see [1]).

By a *basic sequence* we shall mean a sequence F of homomorphisms

$$\dots \to F_n[X_n] \underset{\beta_n}{\to} F_{n-1}[X_{n-1}] \underset{\beta_{n-1}}{\to} \dots \to F_2[X_2] \underset{\beta_2}{\to} F_1 \underset{\beta_1}{\to} 1,$$

where F_1 is a free group, such that

$$\beta_n \mid X_n \quad (n = 2, 3, \dots)$$

is one-one. Let $Y_n = \beta_{n+1} X_{n+1}$ $(n \geqq 1)$. We call Y_n the set of *n-relations* of F; X_n is the set of *n-generators*. Call $F_n = F[X_n]$ $(n > 1)$ the *n*-dimensional *component* of F. For convenience define $X_1 = \varnothing$. For each n, X_n and Y_n are symmetric subsets of F_n.

The basic sequence obtained from F by replacing the components of dimension $> k$ by trivial groups and the set Y_k by the empty set, is the *k-dimensional part* F^k of F.

F being a basic sequence, consider the operation of F_1 on itself defined by

$$f \cdot h = fhf^{-1}.$$

† If X contains no element x such that $x^{-1} = x$, $F[X]$ is a free group.

If Y_1 is admissible under F_1, i.e. if $f \cdot Y_1 = Y_1$ for every f in F_1, then F_1 also operates on F_2. Namely, let $x \in X_2$. Then $\beta x \in Y_1$, $f \cdot \beta x \in Y_1$. Define $f \cdot x$ to be the unique element of X_2 such that $\beta_2(f \cdot x) = f \cdot \beta x$. Evidently $f \cdot x^{-1} = (f \cdot x)^{-1}$. Having now defined $f \cdot h$ over $F_1 \times X_2$, we extend multiplicatively over $F_1 \times F_2$. Evidently $f_1 f_2 \cdot h = f_1 \cdot (f_2 \cdot h)$ and $f \cdot \beta h = \beta(f \cdot h)$. If now Y_2 is admissible under F_1, we can extend the function $f \cdot h$ over $F_1 \times F_3$ in precisely the same manner. Suppose that $Y_1, Y_2, \ldots$ are such that $f \cdot h$ can be extended over $F_1 \times (F_1 \cup F_2 \cup \ldots)$. We call F a *normal* sequence. In a normal sequence F_1 operates on every F_n and operation by F_1 permutes with β. In a normal sequence we have

$$fh \cdot k = (f \cdot h) \cdot (f \cdot k) \tag{1.1}$$

when $f, h \in F_1$. This is obvious when $k \in F_1$ and is proved by a simple induction when $f, h \in F_n$.

In a normal sequence F_2 also operates on every F_n by the rule

$$f \cdot h = \beta f \cdot h \quad (f \in F_2).$$

Operation by F_2 evidently permutes with β and satisfies (1.1). A subset of F_n which is admissible under F_1 is admissible under F_2, but the converse is not generally true. We shall agree that the word admissible without qualification will mean admissible in the stronger sense, that is, F_1-admissible.

μ-SEQUENCES

2. Let F be a normal sequence. We define a mapping $\mu_2: F_2 \times F_2 \to F_2$ by

$$\mu_2(f, h) = fhf^{-1}(f \cdot h^{-1}).$$

We establish certain properties† of μ_2. First μ_2 is F_1- and F_2-compatible, i.e.

$$\mu_2(f \cdot h_1, f \cdot h_2) = f \cdot \mu_2(h_1, h_2) \quad (f \in F_1 \text{ or } F_2). \tag{2.1}$$

This follows immediately from (1.1).

Let $Q_2 = \text{kernel } \beta_1$. Then

$$\mu_2(F_2, F_2) \subset Q_2. \tag{2.2}$$

This follows from the fact that

$$\beta(f \cdot h^{-1}) = \beta(\beta f \cdot h^{-1}) = \beta f \cdot \beta h^{-1} = \beta(fh^{-1}f).$$

If $q \in Q_2$ then

$$\mu_2(q, h) = [q, h]. \tag{2.3}$$

For, we have $q \cdot h^{-1} = (\beta q) \cdot h^{-1} = h^{-1}$.

† The function $\mu_2(x, y)$ has appeared more or less explicitly in a number of papers. See [1], [2], [3].

The subgroup Γ *of* F_2 *generated by* $\mu_2(F_2, F_2)$ *is normal.*

For, let $r \in \mu_2(F_2, F_2)$. Then $r \in Q_2$ and $r^{-1}frf^{-1} = \mu_2(r^{-1}, f) \in \Gamma$. Hence $frf^{-1} \in r\Gamma$. But $r\Gamma = \Gamma$ since $r \in \Gamma$.

Let L *be a subset of* Q_2. *The group generated by* $\mu_2(\Gamma L, F_2)$ *is normal.*

We first observe that $fLf^{-1} \subset \Gamma L$. For if $l \in L$, then

$$l^{-1}flf^{-1} = \mu_2(l^{-1}, f) \subset \Gamma \quad \text{so that} \quad flf^{-1} \in l\Gamma \in L\Gamma = \Gamma L.$$

From this it follows that $fL\Gamma f^{-1} \subset L\Gamma$. Since $L\Gamma \subset Q_2$, an element t in $\mu_2(L\Gamma, F_2)$ is of the form $[r, h]$, $r \in L\Gamma$. Hence

$$ftf^{-1} = [frf^{-1}, h] \in \mu_2(\Gamma L, F_2)$$

which proves the assertion.

We shall say that a function η on $F_2 \times F_n$ to F_m *has property M* if

$$\eta(fh, t) = \eta(f, h \cdot t)\, \eta(h, t), \tag{2.4}$$

$$\eta(f, ts) = \eta(f, t)\, \eta(f, s). \tag{2.5}$$

We shall say that η *has property M modulo a normal subgroup K of F* if the (2.4) and (2.5) hold modulo K.

The function μ_2 *has property M modulo the group generated by* $\mu_2(\Gamma, F_2)$.

One verifies in fact that

$$\mu(f_1 f_2, h) = \mu(f_1, f_2 h)\, \mu(f_2, h)\, [(ab)^{-1}, c^{-1}],$$
$$\mu(f, h_1 h_2) = \mu(f, h_1)\, \mu(f, h_2)\, [d, e],$$

where $\quad a = \mu(f_2, h), \quad b = \mu(\mu(f_2, h)^{-1}, h), \quad c = \mu(f_1, f_2 h),$

$$d = \mu(f, h_2)^{-1}, \quad e = f \cdot h_1.$$

Referring to (2.2) and (2.3) we see that the commutators belong to $\mu_2(\Gamma, F_2)$.

3. We use μ_2 as the basis for defining functions

$$\mu_n\colon\ F_2 \times F_n \to F_n,$$
$$\nu_n\colon\ F_2 \times F_n \to F_{n+1}$$

for $n \geqq 2$. Suppose that $\mu_2, \ldots, \mu_n$ and $\nu_2, \ldots, \nu_{n-1}$ have been defined and are F_1-compatible. Suppose further that

$$\mu_m(X_2, X_m) \subset Y_m \cup \{1\} \tag{3.1}$$

for $m = 2, \ldots, n$. We define ν_n. First, on $X_2 \times X_n$ take $\nu_n(x, y)$ to be the unique element of X_{n+1} such that

$$\beta\nu_n(x, y) = \mu_n(x, y), \tag{3.2}$$

assuming the right member $\neq 1$; take $\nu_n(x, y) = 1$ if the right member equals 1. We extend ν_n over $F_2 \times F_n$ as follows. Let $w = x_1 \ldots x_k$,

$v = y_1, \ldots, y_m$ be *words* formed in X_2, X_n respectively. Define ν_n for word-pairs as follows:

$$\nu_n(\omega, v) = \nu_n(w, \omega) = 1 \quad (\omega = \text{empty word}),$$
$$\nu_n(w, v) = \nu_n(w, y_1) \ldots \nu_n(w, y_n),$$

where

$$\nu_n(x, y_1) = \nu_n(x_1, x_2 \ldots x_k \cdot y_i)\, \nu_n(x_2, x_3 \ldots x_k \cdot y_i) \ldots \nu_n(x_k, y_i).$$

Now let $(f, h) \in F_2 \times F_n$ and let w, v be the reduced words of f, h. Define $\nu_n(f, h) = \nu_n(w, v)$. Define μ_{n+1} by

$$\mu_{n+1}(f, h) = h(f \cdot h^{-1})\, \nu_n(f, \beta h).$$

We verify readily that ν_n and μ_{n+1} are F_1-compatible. If now the relation (3.1) holds in the dimension $n+1$, we can define ν_{n+1}, μ_{n+2} in the same manner. A normal sequence F in which μ_m is defined for $m = 2, \ldots, n$ will be called a μ_n-sequence. If F is a μ_n-sequence for every n, or if F is a μ_n-sequence for some n and F_m is trivial for $m > n$, we call F a μ-sequence. The n-dimensional part of a μ-sequence is a μ-sequence and a μ_n-sequence.

In a μ_{n+1}-sequence F, consider the words xx^{-1}, yy^{-1} formed in X_2 and X_n. We have, by definition,

$$\nu_n(xx^{-1}, y) = \nu_n(x, x^{-1}y)\, \nu_n(x^{-1}, y), \tag{3.3}$$

$$\nu_n(x, yy^{-1}) = \nu_n(x, y)\, \nu_n(x, y^{-1}). \tag{3.4}$$

Let I_{n+1} denote the smallest normal subgroup of F_{n+1} containing all elements (3.3) and (3.4). It is easy to see that $\nu_n(f, h) = \nu_n(u, v)$ modulo $I_{n+1}[F_{n+1}, F_{n+1}]$, where u, v are any word-representations of f, g. From this and the definition of ν_n for word-pairs, we see that ν_n has property M modulo $I_{n+1}[F_{n+1}, F_{n+1}]$. It is then a matter of straightforward verification to see that for $n > 2$, μ_n has property $M \bmod I_n[F_n, F_n]$. Moreover, for $n > 2$

$$\nu_{n-1}(1, h) = \nu_{n-1}(f, 1) = \mu_n(1, s) = \mu_n(t, 1) \bmod I_n[F_n, F_n]. \tag{3.5}$$

For μ_2 these formulas hold without modulus and this is the basis for the induction by which (3.5) is proved. Finally

$$\beta\nu_n = \mu_n \quad \bmod I_n[F_n, F_n] \quad (n > 2),$$
$$\beta\nu_2 = \mu_2 \quad \bmod \mu_2(\Gamma, F_2),$$
$$\beta I_{n+1} \subset I_n[F_n, F_n] \quad (n \geqq 2),$$
$$\beta I_3 \subset \mu_2(\Gamma, F_2).$$

Kernels

4. Let F be a normal sequence and let $K=(K_n)$, $K_n \subset F_n$ be a sequence of normal subgroups. We introduce the following notation:

$$G_n = \operatorname{im} \beta_n, \quad Q_n = \beta^{-1} K_{n-1},$$
$$\mathbf{F}_n = F_n / K_n,$$
$$\kappa_n : F_n \to \mathbf{F}_n \quad \text{(canonical)}.$$

We denote the image of $X, \ldots$ under by $\mathbf{X}, \ldots$. Evidently G_n is generated by Y_n, $\mathbf{G}_n$ by $\mathbf{Y}_n$, $\mathbf{F}_n$ by $\mathbf{X}_n$ and

$$\mathbf{G}_n = \operatorname{im} d_n, \quad \mathbf{Q}_n = \ker d_{n-1}.$$

Let F be a normal sequence and K a sequence of normal subgroups. If, in the passage from F_n to $\mathbf{F}_n$, F goes over into a sequence $\mathbf{F}$

$$\ldots \mathbf{F}_n \underset{d_n}{\longrightarrow} \mathbf{F}_{n-1} \underset{d_{n-1}}{\longrightarrow} \ldots \underset{d_2}{\longrightarrow} \mathbf{F}_1 \underset{d_1}{\longrightarrow} 1,$$

and the function $f \cdot h$ is carried over to $\mathbf{f} \cdot \mathbf{h}$ in $\mathbf{F}$, we call K a *kernel* for F, and (F, K) a *kernel sequence*. If, in addition, we have $d_n d_{n+1} = 1$ $(n \geqq 1)$, and if image d_{n+1}, which is then a subgroup of kernel d_n, is normal in $\ker d_n$, $\mathbf{F}$ will have homology, that is, we can form the groups

$$p_n = \ker d_n / \operatorname{im} d_{n+1}.$$

We then call K a *homology kernel* and (F, K) a *homology kernel sequence*. If F is a μ_n-sequence and if K is a kernel such that the functions $\nu_2 \ldots \nu_{n-1}$, $\mu_2 \ldots \mu_n$ are carried over into $\mathbf{F}$ and in $\mathbf{F}$ satisfy property M (absolute) and the relations

$$d\nu_m = \mu_m \quad (m = 2, \ldots, n-1),$$

K is called a *μ_n-kernel*. K is a *μ-kernel* if it is a μ_n-kernel $(n = 2, 3, \ldots)$.

If K is a *μ_n-kernel* then in $\mathbf{F}$ we shall have

$$\nu(1, \mathbf{h}) = \nu(\mathbf{h}, 1) = 1,$$
$$\nu(\mathbf{f}, \mathbf{h})^{-1} = \nu(\mathbf{f}, \mathbf{h}^{-1}),$$

in dimension $2, \ldots, n-1$ and similar formulas for μ in dimension $2, \ldots, n$.

Let F be a normal sequence and let $K = (K_n)$ be a sequence of normal subgroups. Consider the following conditions:

(a) K_n is admissible, all n;

(b) $K_1 = 1$;

(c) $\beta K_n \subset K_{n-1}$, $\beta Y_n \subset K_{n-1}$, all $n \geqq 2$;

(d) $\mu_2(\Gamma G_2, F_2) \subset K_2$;

(e) $I_m[F_m, F_m] \subset K_m$ $(3 \leqq m \leqq n)$;

(f) $\nu_n(F_2, K_m) \subset K_{m+1}$, $\nu_m(K_2, F_m) \subset K_m$ $(2 \leqq m \leqq n-1)$.

We assert: (a, b) *imply that* K *is a kernel*; (a, b, c) *imply that* K *is a homology kernel. If* F *is a* μ_n*-sequence,* (a, b, d, e, f) *imply that* K *is a* μ_n*-kernel.*

PROOF. (a, b) imply the existence of homomorphisms d_n: $\mathbf{F}_n \to \mathbf{F}_{n-1}$ such that $d_n \kappa_{n-1} = \kappa_n \beta_n$. F then goes over into $\mathbf{F}$:

$$\ldots \underset{d_{n+1}}{\longrightarrow} \mathbf{F}_n \longrightarrow \ldots \underset{d_2}{\longrightarrow} \mathbf{F}_1 \underset{d_1}{\longrightarrow} 1.$$

If k is an element of K_n and $f \in F_1$, we have

$$f \cdot hk = (f \cdot h)(f \cdot k) = f \cdot h \bmod K_n.$$

Hence we carry the function $f \cdot h$ into $\mathbf{F}$ by the rule

$$\mathbf{f} \cdot \mathbf{h} = \kappa(f \cdot h), \quad \mathbf{f} = \kappa f, \quad \mathbf{h} = \kappa h$$

(using, of course, the fact that $K_1 = 1$). If $\mathbf{F}$ we define $\mathbf{f} \cdot \mathbf{h}$ for $\mathbf{f} \in \mathbf{F}_2$ by $\mathbf{f} \cdot \mathbf{h} = d\mathbf{f} \cdot \mathbf{h}$. Evidently, in this case also, $\mathbf{f} \cdot \mathbf{h} = \kappa(f \cdot h)$. Thus $f \cdot h$ is carried over into functions $\mathbf{f} \cdot \mathbf{h}$ which define operations on $\mathbf{F}$ by $\mathbf{F}_1$ and $\mathbf{F}_2$.—Suppose (a, b, c) hold. Since Y_n generates G_n, we have $\beta G_n \subset K_{n-1}$. Hence if $\mathbf{f} \in \mathbf{F}_n$, $dd\mathbf{f} = dd\kappa f = \kappa\beta\beta f \in \kappa\beta G = 1$. Hence $dd = 1$. —Suppose finally that F is a μ_n-sequence and that (a, b, d, e, f) hold. We note first that

$$\mu_m(K_2, F_n) \subset K_m, \quad \mu_m(F_2, K_m) \subset K_m \quad (m = 2, \ldots, n).$$

This follows immediately from (f) and the definition of μ (for $m = 2$ one uses also the fact that $\mu_2(k, f) = [k, f]$, $k \in K_2$). Now let k, k' be elements of K_2, K_m. We find

$$\begin{aligned} \nu_m(fk, hk') &= \nu_m(f, h)\, \nu_m(f, k')\, \nu_m(k, h)\, \nu_m(k, k') \bmod K_{m+1} \\ &= \nu_m(f, h) \bmod K_{m+1} \quad (2 \leqq m \leqq n-1), \end{aligned}$$

and similarly $\mu_m(fk, hk') = \mu_m(f, h) \bmod K_m$ $(2 \leqq m \leqq n)$. Hence we may define $\nu_m(\mathbf{f}, \mathbf{h}) = \nu_m(f, h)$ and similarly for μ. It follows from § 3 that in F, ν_m, μ_m have property M and that $d_m \nu_m = \mu_m$.

5. Let (F, K) be a kernel sequence. If Y_n is replaced by the empty set and if for dimensions $> n$, F_n and K_n are replaced by trivial groups, we obtain a kernel sequence (F^n, K^n), the *n-dimensional part* of (F, K). If (F, K) is a homology kernel sequence, so is its n-dimensional part.

Let (F, K), (F', K') be kernel sequences. The first is a *subsequence* of the second, $(F, K) \subset (F', K')$, if $F_n \subset F'_n$, $Y_n \subset Y'_n$, $K_n \subset K'_n$. When these conditions hold, we shall have $X_n \subset X'_n$ $(n \geqq 2)$, and the functions $f \cdot h$, β, ν, μ in (F, K) will be the appropriate restrictions.

An *isomorphism* ϕ: $(F, K) \to (F', K')$ consists of a sequence of isomorphisms ϕ_n: $F_n \to F'_n$ which permute with β such that $\phi_n Y_n \subset Y'_n$, $\phi_n K_n \subset K'_n$. If ϕ is an isomorphism then ϕ preserves the function $f \cdot h$. If F, F' are μ_n-sequences, ϕ preserves $\mu_2, \ldots, \mu_n$, $\nu_2, \ldots, \nu_{n-1}$.

SYSTEMS FOR SIMPLICIAL COMPLEXES

6. *Notation.* Let A be a simplicial complex. Let $C_n(A)$ denote the group of finite integral n-chains of A, ∂ the boundary operator, $\{\sigma_n, -\sigma_n\}$ the oriented simplexes. Each subcomplex $|\sigma_n|$ will be denoted simply by σ_n and its $(n-1)$-dimensional part by $\dot{\sigma}_n$. $\Lambda(A)$ will denote the groupoid of simplicial paths of A. To indicate that a path λ begins at P and ends at Q, write $\lambda \to PQ$.

7. Our object now is to construct a system

$$\Sigma = (F(A,O), K(A,O), \xi, \lambda *),$$

where A is an arbitrary simplicial complex, O a vertex of A, (F,K) a homology kernel μ-sequence associated with (A,O), ξ a system of homomorphisms

$$\xi_n: F_n(A,O) \to C_n(A),$$

and $\{\lambda *, \lambda \in \Lambda(A)\}$ an associative system of isomorphisms

$$\lambda * (F(A,P), K(A,P)) = (F(A,O), K(A,O)), \quad \lambda \to OP,$$

such that

(7.1) $F_1(A,O) = \Lambda(A,O)$, $f * h = f \cdot h$ when $f \in F_1$;

(7.2) $\xi\beta = \partial\xi$, $\xi(f \cdot h) = \xi h$, $\xi(\lambda * h) = \xi h$;

(7.3) ξ maps each n-generator $(n > 1)$ into O or a generator of C_n;

(7.4) $\xi(K_n) = \{O\}$;

(7.5) if $O \in B \subset A$, then

$$(F(B,O), K(B,O)) \subset (F(A,O), K(A,O))$$

and ξ, $\lambda *$ in (B,O) are the appropriate restrictions.

Let Σ be a system. If we replace each (F,K) by its n-dimensional part and for dimensions $>n$ replace ξ, $\lambda *$ by trivial mappings, we obtain a system Σ^n which we call the *n-dimensional part* of Σ.

In a system Σ, *the following relations hold:*

$$(7.6) \qquad \begin{cases} \xi_n \mu_n = 0, \quad \xi_n \nu_{n-1} = 0, \\ \mu_n(\lambda * f, \lambda * h) = \lambda * \mu_n(f,h), \\ \nu_n(\lambda * f, \lambda * h) = \lambda * \nu_n(f,h). \end{cases}$$

The last two formulas are proved by an easy induction starting with μ_2 then $\nu_3, \mu_3, \nu_4, \ldots$. To prove (7.6) we first note that

$$\xi\mu_2(f,h) = \xi(fhf^{-1}(f \cdot h^{-1})) = 0.$$

Let $x, y \in X_2$. Then $\nu_2(x,y)$ is either 1 or an element of X_3; hence $\xi\nu_2(x,y) = \epsilon\sigma_3$ where $\epsilon = \pm 1$ or 0. We have

$$\epsilon\partial\sigma = \xi\beta\nu_2(x,y) = \xi\mu_2(x,y) = 0,$$

hence $\epsilon = 0$. Using property M we get $\xi_3\nu_2 = 0$. From this $\xi_3\mu_3 = 0$, then by the same argument, $\xi_4\nu_3 = 0$ and so on.

In a system Σ the mappings ξ, $\lambda *$ like the functions μ, ν, $f \cdot h$, are transferable to $\mathbf{F}(A,O)$; namely, in each (A,O) we use the rules

$$\xi f \doteq \xi f, \quad \lambda * f = \kappa(\lambda * f) \quad (f = \kappa f);$$

this is made possible by (7.7) and the relations $\lambda * K_n(A,P) = K_n(A,O)$.

Mappings into homologies

8. For a given Σ let $\kappa_n(A,O)$ be the canonical homomorphism $F_n \to \mathbf{F}_n$ for (A,O) and let $\mathbf{X}_n(A,O), \ldots$ denote the images of $\mathbf{X}_n(A,O), \ldots$ under $\kappa_n(A,O)$. Let $p_n(A,O) = \mathbf{Q}_n(A,O)/\mathbf{G}_n(A,O)$. The isomorphisms $\lambda *$ induce isomorphisms $\lambda * p_n(A,P) = p_n(A,O)$. If A is connected, $p_n(A,O)$ is independent, up to isomorphism, of O and we may write $p_n(A,O) = p_n(A)$. The ξ's induce homomorphisms $\xi_n \colon \mathbf{F}_n(A,O) \to C_n(A)$ such that $\xi\beta = d\xi$. Hence they induce homomorphisms

$$\xi_{*n} \colon p_n(A,O) \to H_n(A),$$

where the homology group H is based on finite integral chains.

We shall say that Σ *has property* h^1 if $p_1(A) = \pi_1(A)$ for every finite connected A. We shall say that Σ *has property* h^n $(n > 1)$ if ξ_{*n} is an isomorphism onto, whenever A is connected and $p_1 A = \ldots p_{n-1} A = 1$.

We shall say that Σ is *n-trivial* or trivial for the dimension n if $p_n(A,O)$ is trivial whenever $\dim A < n$.

Theorem I. *There exists a Σ which has property h^n for every n and is trivial for every $n > 2$.*

Theorem II. *There exists a Σ which has property h^1, h^2, h^3 and is non-trivial for the dimension* 3.

We shall give a procedure for the construction of a class of Σ's which differ from each other only in the definition of K. The systems required in Theorems I and II will belong to this class.

We define Σ by defining successively its n-dimensional parts. Assume that Σ^n has been defined, F^n being given by

$$\ldots \to 1 \to F^n \to \ldots \to F_1 \to 1.$$

We define Σ^{n+1} by the following steps. First we define symmetric admissible subsets $Y_n(A,O) \subset F_n(A,O)$ which are to be the n-relations for F^{n+1}; those for F^n are of course empty. Taking for $X_{n+1}(A,O)$ a set with inversion, in one-one inversion-preserving correspondence with $Y_n(A,O)$, we form $F^{n+1}(A,O)$

$$\ldots \to 1 \to F_{n+1} \to F_n \to \ldots \to F_1 \to 1,$$

where $F_{n+1}(A,O)=F[X_{n+1}(A,O)]$. We then define mappings ξ, $\lambda *$ for the dimension $n+1$ and a kernel K_{n+1}. For lower dimensions these are already defined; for higher dimensions everything is trivial. Assuming that the definitions are properly made, we obtain in this way a system Σ^{n+1} of which Σ^n is the n-dimensional part. The obvious limit of Σ^n gives a Σ.

Definition of Σ^1. Take

$$F_1(A,O)=\Lambda(A,O), \quad K_1(A,O)=1.$$

Take ξ_1 to be the obvious mapping $F_1 \to C_1(A)$ (paths $\to$ 1-chains). If $f \in F_1(A,P)$, define $\lambda * f = \lambda f \lambda^{-1} \quad (\lambda \to OP).$

Take the 1-relation sets to be empty and everything trivial in dimensions >1. All properties in § 7 are satisfied and Σ^1 is defined.

Assume Σ^n is defined. We define $Y_n = Y_n(A,O)$. To simplify notation, assume that all symbols refer to a fixed (A,O). Then

$$Y_n = Y_n^1 \cup Y_n^2 \cup Y_n^3,$$

where Y_n^1 is the totality of elements of F_n of the form

$$\lambda * q, \quad q \in Q_n(\dot{\sigma}_{n+1}, P), \quad \lambda \to OP, \quad \xi_n q = \pm \partial\sigma_{n+1};$$

Y_n^2 is empty when $n=1,2$; for $n>2$, Y_n^2 is the totality of elements y of F_n such that y or y^{-1} is of the form

$$\lambda * u_1 u_2^{-1} r^{-1}, \quad \textit{where} \quad u_1 u_2^{-1} r^{-1} \in Q_n(\sigma_n, P),$$
$$\{u_1, u_2\} \subset X_n^1(\sigma_n, P), \quad \xi u_1 = \xi u_2, \quad r \in F_n(\dot{\sigma}_n, P)$$

(X_n^i means of course the set of generators which correspond to elements of Y_{n-1}^i);

Y_n^3 is empty for $n=1$ and for $n>1$ is the smallest symmetric set containing the non-trivial elements of $\mu_n(X_2, X_n)$.

Note that for each n, Y_n does not contain the element 1.

The set X_{n+1}^3 of $(n+1)$-generators which correspond to elements of Y_n^3 is the union of X^+ and $(X^+)^{-1}$, where X^+ consists of the non-trivial elements $\nu_{n-1}(x,y)$, $x, y \in X_{n-1}$.

Consider $Y_1 = Y_1^1$. Each element is of the form $\lambda s_1 s_2 s_3 \lambda^{-1}$, where $s_1 s_2 s_3$ is the path boundary of a 2-simplex. Hence Y_1 is symmetric and admissible. We verify that Y_n is symmetric, admissible; in fact each set Y_n^i $(i=1,2,3)$ is symmetric and admissible. In verifying this, use the fact that these properties hold for X_n by the induction hypothesis. To define $\lambda * x$ for $x \in X_n(A,P)$ we first verify that

$$\lambda * Y_n^i(A,P) = Y_n^i(A,O),$$

then define $\lambda * x$ to be the unique element of $X_{n+1}(A,O)$ such that $\beta(\lambda * x) = \lambda * \beta x$. Having thus defined $\lambda *$ on $X_{n+1}(A,P)$ we extend multiplicatively over $F_{n+1}(A,P)$.

To define ξ_{n+1} it is sufficient to define it on X_{n+1}. Say $x \in X_{n+1}$. Then $\beta x \in Y_n^1$, so $\xi \beta x = \epsilon \partial \sigma_{n+1}$, $|\epsilon| = 1$. Take $\xi x = \epsilon \sigma_{n+1}$. For

$$x \in X_{n+1}^2 \cup X_{n+1}^3, \quad \text{take} \quad \xi x = 0.$$

We must prove that $\xi_n \beta_{n+1} = \partial \xi_{n+1}$. This is obvious on X_{n+1}^1. For X_{n+1}^3 it is sufficient to show that $\xi Y_n^3 = \{0\}$, hence that $\xi \mu_n(x,y) = 0$ $(x, y \in X_n)$. This, however, follows from (7.6) applied in Σ^n. For X_{n+1}^2 it is sufficient to show that $\xi Y_n^2 = \{0\}$. Let $\lambda * u_1 u_2^{-1} r^{-1} \in Y_n^2$. We have $\xi u_1 = \xi u_2$ so $\xi u_1 u_2^{-1} = 0$. It remains to be shown that $\xi r = 0$. Now $r \in F_n(\dot{\sigma}_n, P)$. Since $\dim \dot{\sigma}_n < n$, there are no generators of the first kind in $F_n(\dot{\sigma}_n, P)$. Hence r is in the group generated by $X_n^2 \cup X_n^3$, hence $\xi r = 0$.—The remaining properties (§ 7) of the mappings ξ, β, $\lambda *$ are readily verified in the dimension $n+1$ except as they concern K^{n+1}. To conclude the definition of Σ^{n+1} therefore, it remains only to define K_{n+1} so that

$$K_{n+1}(B,O) \subset K_{n+1}(A,O), \quad O \in B \subset A, \tag{8.1}$$

$$\lambda * K_{n+1}(A,P) = K_{n+1}(A,O), \quad \lambda \to OP. \tag{8.2}$$

and such that $K^{n+1} = (\ldots 1, K_{n+1}, K_n, \ldots K_1)$ is a homology kernel.

SYSTEMS Σ_{I} AND Σ_{II}

9. From now on we shall consider *only systems obtained by the procedure given in the preceding section.* We define systems Σ_{I}, Σ_{II} by giving the kernels K_{I}, K_{II} for arbitrary (A,O).

K_{I}: $K_1 = 1$, $K_2 = \Gamma$,

K_n $(n > 2)$ is the smallest normal subgroup of F_n containing X_n^2, X_n^3, Y_n^2, Y_n^3, $[F_n, F_n]$.

K_{II}: $K_1 = 1$, $K_2 = \mu_2(G_2\Gamma, F_2)$,

K_n $(n > 2)$ is the smallest normal subgroup of F_n containing $I_n[F_n, F_n]$, $\nu_{n-1}(K_2, F_{n-1})$, $\nu_{n-1}(F_2, K_{n-1})$.

The systems Σ_{I} and Σ_{II} satisfy the properties stated in Theorems I and II respectively. We shall treat system Σ_{II} in detail; we consider this the more interesting case.

Let F_n^{23} be the subgroup of F_n generated by $X_n^{23} = X_n^2 \cup X_n^3$. We observe that for K_{II} we have $K_n \subset [F_n, F_n] F_n^{23}$, from which it follows that $\xi K_n = \{0\}$ as required. To verify this for K_{I} we need the relations $\xi Y_n^2 = \xi Y_n^3$. They are established in § 7.

It is readily verified that (8.1) and (8.2) hold for K_{n+1} in both K_{I} and K_{II}. To show that K_{I} and K_{II} are homology μ-kernels, we show that $(a, \ldots, f,$ § 4) are satisfied. We do this for K_{II}. The only relations which do not follow immediately from the definition of K_{II} are

$$\beta Y_m \subset K_{m-1}, \quad \beta K_m \subset K_{m-1}. \tag{9.1}$$

In any case we have $\beta Y_m^1 \subset K_{m-1}$ and $\beta Y_m^2 \subset K_{m-1}$. For example

$$\begin{aligned} \beta Y_3^1 &\subset \mathsf{U}_{\lambda,\sigma} \lambda * \beta Q_2(\dot{\sigma}_3, P) \\ &\subset \mathsf{U}\, \lambda * K_2(\dot{\sigma}_3, P) \\ &\subset \mathsf{U}\, \lambda * K_2(A, P) = K_2(A, O) = K_2. \end{aligned}$$

We show that $\beta Y_3^3 \subset K_2$. Say $y = \mu_3(x, t) \in Y_3^3$, $x, t \in X_2$. Then

$$\begin{aligned} \beta y &= \beta(t(x \cdot t^{-1})\, \nu_2(x, \beta t)^{-1}) \\ &= \beta t(x \cdot \beta t^{-1})\, \mu_2(x, \beta t)^{-1} \mod K_2 \quad (\S\, 3). \end{aligned}$$

From the definition of μ_2 we find that $\beta y = [\beta t, x] \bmod K_2$. Since $\beta t \in G_2$, and since G_2 is generated by $Y_2 = Y_2^1$, we have $\beta t \in Q_2$. Hence $[\beta t, x] = \mu_2(\beta t, x) \in \mu_2(G_2, F_2) \subset K_2$. Hence $\beta y \in K_2$.—Let us next prove that $\beta K_3 \subset K_2$. We have $\beta I_3 \subset \mu_2(\Gamma, F_2) \subset K_2$ (§4). Next,

$$\beta[F_3, F_3] \subset [G_3, G_3].$$

Let $g, h \in G_3$. Then $[g, h] = \mu_2(g, h) \in \mu_2(G_2 \Gamma, F_2) \subset K_2$, so that

$$\beta[F_3, F_3] \subset K_2.$$

Next let $k \in K_2$. Then

$$\beta \nu_2(k, f) = \nu_2(k, f) = [k, f] \in K_2,$$

and

$$\beta \nu_2(f, k) = \mu_2(f, k) = fkf^{-1}(f \cdot k^{-1}) \in K_2,$$

since K_2 is admissible.

We have now shown that (9.1) holds for $m = 3$. A straightforward induction will establish these relations for $m > 3$.

DECOMPOSITION OF F_n

10. We shall prove that *in the system* Σ_{II}, $\mathbf{F}_n = \mathbf{F}_n^1 \times \mathbf{F}_n^{23}$, *where* $\mathbf{F}_n^1$ *is the subgroup of* $\mathbf{F}_n$ *generated by* X_n^1. *Moreover,* F_n^1 *is free with generators* X_n^1, $\mathbf{F}_n^1$ *is free Abelian with free generators* $\mathbf{X}_n^1$; κ *maps* X_n^1 *into* $\mathbf{X}_n^1$ *in a one-one manner.*

We observe first that if $x \in X^1$, then $x \neq x^{-1}$, otherwise $\xi x = -\xi x$ which implies $\xi x = 0$ which is impossible. Hence F_n^1 is free and is freely generated by X_n^1. We observe that $X^1 \cap X^{23} = \varnothing$, since ξx is 0 or $\neq 0$ according as $x \in X^1$ or $x \in X^{23}$. Hence $F^1 \cap F^{23} = \varnothing$. Our assertions now follow readily from the fact that $K_n \subset F_n^{23}[F_n, F_n]$.

LOCAL ELEMENTS

11. Let σ be a simplex of A and let

$$F_n(\sigma) = \bigcup_{\lambda, P} \lambda * F_n(\sigma, P), \quad \lambda \to OP,$$

and let $X_n(\sigma), \ldots,$ be similarly defined. The sets $F_n(\sigma)$ are not disjoint and their union is not F_n. The situation is different, however, for the sets $X_n^1(\sigma_n)$. Since each element of $X_n^1(\sigma_n)$ mapped by ξ into $\pm \sigma_n$, we see that these sets, for the different σ_n's, are disjoint and form a decomposition of $X_n(A, O)$. The same is true of the sets $Y_n^1(\sigma_n)$—they form a decomposition for Y_n^1. The sets $X_n^1(\sigma_n)$, $Y_n^1(\sigma_n)$ are admissible and symmetric and $\beta Y_n^1(\sigma_n) = X_n^1(\sigma_n)$. It will be convenient to decompose X_n still further:

$$X_n^1(\sigma_n) = X_n^+(\sigma_n) \cup (X_n^+(\sigma_n))^{-1},$$

where X^+ consists of those generators x such that $\xi x = +\sigma_n$.

PROPERTY l^n

12. In order to prove that a given Σ possesses property h^n, we must of course show the existence of suitable mappings from $C_n(A)$ back into $\mathbf{F}_n(A, O)$. Such mappings will be defined by means of mappings ζ, α whose properties we now postulate.

We shall say that Σ *possesses property* l^n $(n > 1)$ if for every (A, O) such that A is connected and $p_1 A = \ldots = p_{n-1} A = 1$ there exist homomorphisms

$$\zeta_m\colon \mathbf{F}_m \to \mathbf{Q}_m \quad (m = 2, \ldots, n),$$

$$\alpha_m\colon \mathbf{F}_m \to \mathbf{F}_{m+1} \quad (m = 1, \ldots, n-1),$$

such that

$$\zeta_m \mathbf{f} = (\alpha_m d\mathbf{f}^{-1})\mathbf{f}, \quad \mathbf{f} \in \mathbf{F}_m, \tag{12.1}$$

$$\xi\alpha_m \mathbf{F}_m^{23} = \{0\}, \quad \zeta\mathbf{F}_m^{23} = \{0\} \quad \text{mod}\, \mathbf{G}_m, \tag{12.2}$$

$$\xi\alpha_m \text{ is constant on every } \mathbf{X}_m^+(\sigma_m), \tag{12.3}$$

$$\xi\zeta_m \text{ is constant mod}\, \mathbf{G}_m \text{ on every } X_m^+(\sigma_m). \tag{12.4}$$

(12.5) THEOREM. *Every Σ possesses property h^1. If Σ possesses property l^n, $n > 1$, it also possesses properties $h^2, h^3, \ldots, h^n$.*

PROOF. The proof that Σ possesses h^1 involves no difficulty and will be omitted.

Assume that Σ has property l^n. We may then assume that $h^1, \ldots, h^{n-1}$ are true and prove h^n.

13. We consider a fixed (A, O). Immediate consequences of l^n are:

$$\zeta_m \mathbf{q} = \mathbf{q}, \quad \mathbf{q} \in \mathbf{Q}_n \tag{13.1}$$

$$\xi\zeta_m = \xi - \xi\alpha_m d_m \quad (m = 2, \ldots, n). \tag{13.2}$$

We next show: if ξ_m is the homomorphism $\mathbf{F}_m \to C_m$, then

$$\ker \xi_m \subset \ker(\xi_{m+1}\alpha_m) \quad (m = 1, \ldots, n-1), \tag{13.3}$$

$$\zeta_m(\ker \xi_m) \subset \mathbf{G}_m \quad (m = 2, \ldots, n). \tag{13.4}$$

PROOF. Let $\mathbf{f} \in \mathbf{F}_m$. We may write

$$\mathbf{f} = (\textstyle\prod_i (\prod_{j=1}^{M_i} \mathbf{x}^{ij} \prod_{k=1}^{N_i} \mathbf{y}^{ik}))\mathbf{hc}, \tag{13.5}$$

where $\mathbf{x}^{ij} \in \mathbf{X}_m^+(\sigma_m^i)$, $\mathbf{y}^{ij} \in (\mathbf{X}_m^+(\sigma_m^i))^{-1}$, $\mathbf{h} \in \mathbf{F}^{23}$, and $\mathbf{c} \in [\mathbf{F}_m, \mathbf{F}_m]$. ($\mathbf{c}$ is of course trivial except when $m = 2$.) Suppose $\xi\mathbf{f} = 0$. Then evidently $M_i = N_i$. Let $\check{\mathbf{f}}_i$ be the element obtained by replacing $\mathbf{x}^{ij}$ by $\mathbf{x}^{i1}$ and $\mathbf{y}^{ik}$ by $(\mathbf{x}^{i1})^{-1}$ in the right member of (13.5). Then $\xi\alpha_m\check{\mathbf{f}} = \xi\alpha_m\mathbf{h} = 0$ (12.2). But by (12.3), $\xi\alpha_m\check{\mathbf{f}} = \xi\alpha_m\mathbf{f}$. Hence $\xi\alpha_m\mathbf{f} = 0$, proving (12.3).—We also have $\zeta\check{\mathbf{f}} = \zeta\mathbf{h} \text{ mod}\, \mathbf{G}_m$ (12.2). But by (12.4), $\zeta\check{\mathbf{f}} = \zeta\mathbf{f} \text{ mod}\, \mathbf{G}_m$. Hence $\zeta\mathbf{f} \in \mathbf{G}_m$, which proves (13.4).

We show next that $\mathbf{X}_n^+(\sigma_n) \neq \varnothing$. For, by h^{n-1}, ξ_{*n-1} maps $p_{n-1}(\dot{\sigma}_{n-1}, P)$ isomorphically onto $H_{n-1}(\dot{\sigma}_n)$. Hence ξ_{n-1} maps $\mathbf{Q}_{n-1}(\dot{\sigma}_n, P)$ onto $Z_{n-1}(\dot{\sigma}_n)$ ($(n-1)$-cycles), and so $\partial\sigma_n$ has a pre-image under ξ_{n-1} in $\mathbf{Q}_{n-1}(\dot{\sigma}_n, P)$. It follows that there exists an element $\mathbf{y}$ of $\mathbf{Y}_{n-1}^1$ such that $\xi\mathbf{y} = \partial\sigma_n$, hence an element $\mathbf{x}$ of $\mathbf{X}_n^+$ such that $\xi\mathbf{x} = \sigma_n$; this element of course is in $\mathbf{X}_n^1(\sigma_n)$.

Using the result in the preceding paragraph, we define a homomorphism

$$\phi\colon C_n \to \mathbf{F}_n$$

by extending over C_n the mapping which assigns to σ_n a definite element in $X_n^+(\sigma_n)$. Evidently

$$\xi_n \phi_n = 1. \tag{13.6}$$

Let
$$\omega_n \colon C_n \to \mathbf{Q}_n$$
be the homomorphism defined by
$$\omega_n = \zeta_n \phi_n.$$

We shall show eventually that

$$\omega B_n \subset \mathbf{G}_n, \tag{13.7}$$

where B_n is the group of n-boundaries. For the moment let this be assumed. As a consequence, ω (restricted to Z) induces a homomorphism
$$\omega_* n \colon Z_n/B_n \to \mathbf{Q}_n/\mathbf{G}_n.$$
We shall show that

$$\omega_* \xi_* = 1, \quad \xi_* \omega_* = 1, \tag{13.8}$$

from which h^n follows. To show the first of these, it is sufficient to show that

$$\omega \xi \mathbf{q} = \mathbf{q} \mod \mathbf{G}, \quad \mathbf{q} \in \mathbf{Q}. \tag{13.9}$$

We have $\omega\xi\mathbf{q} = \zeta\phi\xi\mathbf{q}$ and by (13.6), $\xi\mathbf{q} = \xi(\phi\xi\mathbf{q})$. From this last we have, by (13.4), $\zeta(\phi\xi\mathbf{q}) = \zeta\mathbf{q} \bmod \mathbf{G}$. Hence $\omega\xi\mathbf{q} = \zeta\mathbf{q} \bmod \mathbf{G}$. But $\zeta\mathbf{q} = \mathbf{q}$, hence (13.9) is true.

To prove $(13.8)_2$ it is sufficient to show that $\xi\omega z = z \bmod B$ when $z \in Z_n$. As a matter of fact, we shall show that $\xi\omega z = z$. The relations (13.3) make it possible to define a homomorphism
$$\theta_m \colon C_m \to C_{m+1} \quad (m = 1, \ldots, n-1)$$
by the formula

$$\theta_m = \xi_{m+1} \alpha_m \xi_m^{-1}. \tag{13.10}$$

We then have
$$\begin{aligned} \xi_n \omega = \xi_n \zeta \phi &= \xi_n \phi - \xi_n \alpha_{n-1} d\phi && (13.2) \\ &= 1 - \theta_{n-1} \xi_{n-1} d\phi && (13.10) \\ &= 1 - \theta_{n-1} \partial \xi_n \phi && \\ &= 1 - \theta_{n-1} \partial && (13.6). \end{aligned}$$

Then if $z \in Z_n$, we have $\xi\omega z = (1 - \theta\partial) z = z$.

We have now proved h^n assuming that $\omega B \subset \mathbf{G}_n$. This certainly holds if $\dim A \leqq n$, for then $B_n = \{0\}$. Consequently h^n holds for connected complexes A of dimension $\leqq n$. We can then prove for general A:

Let σ_{n+1} be a simplex of A. The set $\mathbf{Y}_n^1(\dot\sigma_{n+1})$ is not empty.

The proof is the same as the proof (above) that $\mathbf{X}_n^+(\sigma_n) \neq \varnothing$, this time applying h^n to the n-complex $\dot{\sigma}_{n+1}$.

We can now establish the relation $\omega B_n \subset \mathbf{G}_n$ for the general case. It is sufficient to show that $\omega d\sigma_{n+1} \in \mathbf{G}_n$ for each σ_n. Let $\mathbf{y} \in \mathbf{Y}_n^1(\sigma_{n+1})$. We may choose $\mathbf{y}$ such that $\xi\mathbf{y} = \partial\sigma_{n+1}$. We may write $\xi\mathbf{y} = \xi\phi\,\partial\sigma_{n+1}$ (13.6). It follows from (13.4) that $\zeta\mathbf{y} = \zeta\phi\,\partial\sigma_{n+1} \mod \mathbf{G}_n$. Since $\mathbf{y} \in \mathbf{G}_n \subset \mathbf{Q}_n$ we have $\zeta\mathbf{y} = \mathbf{y} = 1 \mod \mathbf{G}_n$. Hence $\zeta\phi\,\partial\sigma_{n+1} = \omega\,\partial\sigma_{n+1} \in \mathbf{G}_n$.

Some lemmas

14. In this section we consider a fixed $F(A,O)$ in a given Σ. We establish some simple lemmas needed to prove (§ 16) that Σ_{II} possesses property l^3.

(14.1) *Let* $fyf^{-1} = y, f \in F_1, y \in Y_1$. *Then* f *is a power of* y.

Proof. f and y generate an Abelian subgroup of the free Abelian subgroup of the free group F_1. This subgroup must be free, hence it is cyclic. Hence f and y are both powers of, say, h. Say $y = h^p$. Then ξy is divisible by p, hence $p = 1$. Hence $y = h$ and f is a power of y.

(14.2) *Let* $f \cdot x = x, x \in X_2, f \in F_2$. *Then* $f = x^k q, q \in Q_2$.

Proof. We have $f \cdot y = y, y = \beta x$. Hence $\beta f y (\beta f)^{-1} = y$, which implies (14.1) $\beta f = y^k$. Then $\beta(x^{-k} f) = 1$ so $x^{-k} f = q \in Q_2$ and $f = x^k q$.

(14.3) *Let* $x, t \in X_2$. *Then the word* $xtx^{-1}(x.t^{-1})$ *is irreducible unless* $x = t^{\pm 1}$, *in which case it is of course reducible to the empty word.*

Proof. Suppose, for example, that x^{-1} cancels with $x \cdot t^{-1}$. Then $x = x \cdot t^{-1}$, so $x^{-1} \cdot x = x^{-1} \cdot (x \cdot t^{-1}) = t^{-1}$ which gives $x = t^{-1}$.

(14.4) *Let* $\Sigma = \Sigma_{\text{II}}$, *and let* $p_1(A,O) = 1$. *Let* $f_1 \cdot x = f_2 \cdot x$, $x \in X_2(A,O)$, $n \geqq 2, f \in F_2$. *Then* $\nu_2(f_1, x) = \nu_2(f_2, x) \mod K_{n+1}$.

Proof. We have $f_1 = f_2 x_q^k$, $q \in Q_2$ (14.2). Since $p_1 = 1$, $q \in G_2$. Then

$$\begin{aligned} \nu(f_1, x) &= \nu(f_2, x^k q \cdot x)\, \nu(x^k, q \cdot x)\, \nu(q, x) \quad \mod K_2. \\ &= \nu(f_2, x)\, \nu(x^k, x) \quad \mod K_2. \end{aligned}$$

Using property M and the relation $\nu_2(x,x) = 1$, we find $\nu_2(x^k, x) = 1 \mod K_2$ from which (14.8) follows.

(14.5) *Let* U *be a complete set of representatives of the orbits in* $X_n^1(\sigma_n, P)$ *under operation by* $F_2(\sigma_n, P)$. *Assume* $p_2(A,O) = 1$. *Then* $\lambda * U$ *contains a complete set of representatives of the orbits in* $X_n^1(\sigma_n)$ *under* $F_2(A,P)$.

Proof. Let P, P' be vertices of $\sigma, \lambda' \to OP'$, $\mu \to P'P$, $\mu \in \Lambda(\sigma)$. Then $\lambda'\mu * X^1(\sigma, P) = \lambda' * X^1(\sigma, P')$. Since $p_1 = 1$, there exists a relation $\beta f = \lambda'\mu\lambda^{-1}$, $f \in F_2(A,O)$. Then $f \cdot (\lambda * X^1(\sigma, P)) = \lambda' * X^1(\sigma, P')$. Now $X^1(\sigma)$ is the union of sets $\lambda' * X^1(\sigma, P')$. We have shown that each of these is the image under operation by an element in $F_2(A, O)$ of $\lambda * X^1(\sigma, P)$, and the elements of this set are clearly images of elements in $\lambda * U$ under operations by elements in $F_2(A, O)$.

(14.6) *Let* $f_i \in F_n(\sigma, P_i)$ $(i=1,2)$, *and* $f_i \neq 1$. *Let* $\lambda_1 * f_1 = \lambda_2 * f_2$, $\lambda_i \to OP_i$. *Then* $\lambda_2\lambda_1^{-1} \in \Lambda(\sigma)$.

Proof. Let $n=1$. Then $\lambda_2^{-1}\lambda_1 f_1 \lambda_1^{-1}\lambda_2 = f_2$. Since f_1, f_2 are non-trivial elements of $\Lambda(\sigma)$, it is easy to see that $\lambda_2^{-1}\lambda_1 \in \Lambda(\sigma)$. Let $n>1$. It is readily seen that the assumed relation implies at least one relation $\lambda_1 * x_1 = \lambda_2 * x_2$, where x_1, x_2 are generators. We then have

$$\lambda_1 * \beta x_1 = \lambda_2 * \beta x_2$$

and so by the induction hypothesis, $\lambda_2^{-1}\lambda_1 \in \Lambda(\sigma)$.

Let σ be a simplex in A and P a vertex of σ. A path $\lambda \to OP$ is *irreducible* rel σ if it is not expressible as $\lambda\mu$, where μ is a non-trivial element of $\Lambda(\sigma)$.

(14.7) *For each non-trivial element* f *in* $F_n(\sigma)$ *there is a unique* λ, *irreducible* rel σ, *such that* $f \in \lambda * F_n(\sigma, P)$.

Call λ the σ-*stem* of f.

Proof. Say $f \in \lambda' * F(\sigma, P')$, $\lambda' \to OP'$. Evidently we can write $\lambda' = \lambda\mu$, where λ is irreducible rel σ and $\mu \in \Lambda(\sigma)$. Then

$$f \in \lambda\mu * F(\sigma, P') = \lambda * F(\sigma, P),$$

where P is the terminal point of λ. Suppose also $f \in \lambda_1 * F(\sigma, P_1)$, where λ_1 is irreducible. Then (14.10) $\lambda\lambda_1^{-1} \in \Lambda(\sigma)$, say $\lambda\lambda_1^{-1} = \mu$. Then $\lambda = \lambda_1\mu$, hence μ is trivial and $\lambda = \lambda_1$.

(14.8) *Let* $f \in F_n(\sigma)$, *then* $\text{stem}_\sigma f = \text{stem}_\sigma \beta f$. (Obvious.)

(14.9) *Let* $f \in \lambda^1 * F_n(\sigma, P')$, *and let* $\lambda = \text{stem}_\sigma f$. *Then*

$$\lambda^1 * F_n(\sigma, P^1) = \lambda * F_n(\sigma, P).$$

For, $\lambda^1 = \lambda\mu$, $\mu \in \Lambda(\sigma)$, so

$$\lambda^1 * F_n(\sigma, P^1) = \lambda * (\mu * (\mu * F_n(\sigma, P^1) = \lambda * F_n(\sigma, P).$$

15. Let F be a normal sequence and D a symmetric admissible subset of F_n. D falls into orbits under operation by F_2; choose a representative in each orbit. The function $u(h)$ which assigns to every $h \in D$ the representative of the orbit $\omega(h)$ which contains h will be called a *selector*. Let u be a selector for D. To each h assign an element $t(h) \in F_2$ such that

$$h = t(h) \cdot u(h).$$

Then $b = (t, u)$ is called a *base* for D. Call b *normalized if* $t(u(h)) = 1$. There obviously exists a normalized base for D such that the values of $u(h)$ fall in any pre-assigned complete set of representatives of the orbits of D. For any base $b=(t,u)$, we have $u(f\cdot h)=u(h)$. If $f\cdot h\neq h^{-1}$ for every h in D we may and shall assume that in every base $b=(t,u)$ for D we have the further normalization

$$u(h^{-1})=u(h)^{-1}.$$

This condition, therefore, will hold for every base for $X_n^1(\sigma_n)$.

(15.1) *Let λ be irreducible* rel σ, $\sigma=\sigma_n$. *There exists a base $b=(t,u)$ for $X_n^1(\sigma)$ such that* $\text{stem}_\sigma u(x)=\lambda$ *for $x\in X_n^1(\sigma)$.*

This follows readily from (14.5) and (14.9).

(15.2) *Let $p_1(A,O)=1$. If x, y are elements of $X_2^1(\sigma_2)$, then $f\cdot x_1=x_2^{\pm 1}$ for some $f\in F_2$.*

This follows from (14.5) and the obvious fact that $X_2^1(\sigma_2,P)$ consists of a single element and its inverse.

Proof of l^3 for Σ_{II}

16. *We shall show that Σ_{II} has property l^3*, hence has properties h^1, h^2, h^3. Let A be a connected complex such that $p_1=p_2=1$. We are to show the existence of α_2, ζ_2, ζ_3.

Definition of α_1. Let S be the totality of oriented 1-simplexes of A. Associate to each vertex P a path $\alpha_0(P)\to OP$. If $s\to PQ$ is an element of S, define

$$\zeta(s)=\alpha_0(P)\,s\,\alpha_0(Q)^{-1}.$$

ζ maps S into $\mathbf{F}_1$ (which can be identified with F_1). We extend over the free group generated by S. It is easy to see that the resulting homomorphism ζ leaves the elements of $\mathbf{F}_1$ fixed. Since $p_1=1$, $\mathbf{F}_1=d\mathbf{F}_2$. Choose for every s an element $\alpha_1(s)$ of $\mathbf{F}_2$ such that $d\alpha_1 s=\zeta s$. This can be done so that $\alpha_1(s^{-1})=(\alpha_1 s)^{-1}$. Now extend α_1 over the free group generated by S and take the restriction to $\mathbf{F}_1$. In this way we get a homomorphism

$$\alpha_1\colon\ \mathbf{F}_1\to\mathbf{F}_2.$$

Let

$$\zeta_2\mathbf{f}=(\alpha_1 d\mathbf{f}^{-1})\,\mathbf{f}_1.$$

We have $d\zeta_2\mathbf{f}=(d\mathbf{f}^{-1})\,d\mathbf{f}=1$, hence $\zeta_2\mathbf{F}_2\subset\mathbf{Q}_2$. Evidently ζ_2 leaves the elements of $\mathbf{Q}_2$ fixed. We show that ζ_2 is a homorphism. We have

$$\begin{aligned}\zeta_2\mathbf{f}_1\mathbf{f}_2&=\alpha d(\mathbf{f}_1\mathbf{f}_2)^{-1}\mathbf{f}_1\mathbf{f}_2\\&=\alpha d\mathbf{f}_2^{-1}(\alpha d\mathbf{f}_1^{-1})\,\mathbf{f}_1\mathbf{f}_2\\&=(\alpha d\mathbf{f}_2^{-1})\,(\zeta_2\mathbf{f}_1)\,\mathbf{f}_2.\end{aligned}\tag{16.1}$$

Since $p_2=1$, $\mathbf{Q}_2=\mathbf{G}_2$ so $\mathbf{f}_1\in\mathbf{G}_2$. Hence $[\zeta\mathbf{f}_1,\mathbf{f}_2]\in\mu_2[\mathbf{G}_2,\mathbf{F}_2]=\{1\}$. Hence $\zeta_2\mathbf{f}_1$ permutes with $\mathbf{f}_2$ and (16.1) equals $(\zeta_2\mathbf{f}_1)\,(\zeta_2\mathbf{f}_2)$.

We define α_2. Choose a base $b_2 = (t, u)$ for X_2^1. For $\mathbf{x} \in \mathbf{X}_2^1$, define

$$\mathbf{u}(\mathbf{x}) = \kappa u(x), \quad \mathbf{t}(\mathbf{x}) = \kappa t(x), \quad (\mathbf{x} = \kappa(x)).$$

This is possible since κ maps X_2^1 into $\mathbf{X}_2^1$ in a one-one manner (§ 10). For each $x \in X_2^1$ we have $\mathbf{x} = \mathbf{t}(\mathbf{x}) \cdot \mathbf{u}(\mathbf{x})$.

Since $\mathbf{F}_2^1$ is free Abelian with generators X_2^1 it is sufficient to define α_2 in $\mathbf{X}_2 = \mathbf{X}_2^1$. For $\mathbf{x} \in \mathbf{X}_2^1$ take $\alpha\mathbf{u}(\mathbf{x})$ to be any element of $\mathbf{F}_3$ such that

$$d\alpha\mathbf{u}(\mathbf{x}) = \zeta_2 \mathbf{u}(\mathbf{x}); \tag{16.2}$$

such an element exists since $p_2 = 1$. Since $\zeta\mathbf{u}(\mathbf{x}^{-1}) = (\zeta\mathbf{u}(\mathbf{x}))^{-1}$, we may define $\alpha\mathbf{u}(\mathbf{x})$ so that $\alpha\mathbf{u}(\mathbf{x}^{-1}) = (\alpha\mathbf{u}\mathbf{x})^{-1}$. Now define

$$\alpha(\mathbf{x}) = \alpha\mathbf{u}(\mathbf{x})\, \nu_2^{-1}(\mathbf{t}(\mathbf{x}), \mathbf{u}(\mathbf{x})), \quad \mathbf{x} \in \mathbf{X}_2^1.$$

This is consistent with (16.2) since $\mathbf{u}(\mathbf{u}(\mathbf{x})) = \mathbf{u}(\mathbf{x})$ and $\mathbf{t}(\mathbf{u}(\mathbf{x})) = 1$, $\nu(1, \text{—}) = 1$. We see that $\alpha_2 \mathbf{x}^{-1} = (\alpha_2 \mathbf{x})^{-1}$, hence α_2 can be extended.

We show that $d\alpha_2 = \zeta_2$. *It is sufficient to do this on* $\mathbf{X}_2^1 = \mathbf{X}_2$. We have

$$\begin{aligned} d\alpha\mathbf{x} &= (d\alpha\mathbf{u})\, d\nu(\mathbf{t}, \mathbf{u})^{-1}, \quad \mathbf{u} = \mathbf{u}(\mathbf{x}), \quad \mathbf{t} = \mathbf{t}(\mathbf{x}), \\ &= (\zeta_2 \mathbf{u})\, \mu_2(\mathbf{t}, \mathbf{u}^{-1}) \\ &= \zeta_2(\mathbf{u}\mu_2(\mathbf{t}, \mathbf{u}^{-1}), \quad \text{since } \mu_2(\ ) \in Q_2, \\ &= \zeta_2(\mathbf{u}\mathbf{t}\mathbf{u}^{-1}\mathbf{t}^{-1}(\mathbf{t} \cdot \mathbf{u})) \\ &= [\zeta_2 \mathbf{u}, \zeta\mathbf{t}]\, \zeta_2(\mathbf{t} \cdot \mathbf{u}). \end{aligned}$$

As noted above, $\zeta\mathbf{u}$ permutes with $\mathbf{t}$ hence $d\alpha\mathbf{x} = \zeta_2(\mathbf{t} \cdot \mathbf{u}) = \zeta_2 \mathbf{x}$.

We shall need to show that

$$\alpha_2(\mathbf{f} \cdot \mathbf{x}) = \alpha(\mathbf{x})\, \nu_2(\mathbf{f}, \mathbf{x})^{-1}. \tag{16.3}$$

Using $\mathbf{u}(\mathbf{f} \cdot \mathbf{x}) = \mathbf{u}(\mathbf{x})$ we have

$$\alpha(\mathbf{f} \cdot \mathbf{x}) = \alpha\mathbf{u}(\mathbf{f} \cdot \mathbf{x})\, \nu(\mathbf{t}(\mathbf{f}, \mathbf{x}), \mathbf{u}(\mathbf{x}))^{-1}. \tag{16.4}$$

Now
$$\mathbf{t}(\mathbf{f} \cdot \mathbf{x}) \cdot \mathbf{u}(\mathbf{x}) = \mathbf{f} \cdot \mathbf{x} = \mathbf{f}\mathbf{t}(\mathbf{x}) \cdot \mathbf{u}(\mathbf{x}).$$
Hence by (14.8)

$$\begin{aligned} \nu(\mathbf{t}(\mathbf{f} \cdot \mathbf{x}), \mathbf{u}(\mathbf{x})) &= \nu(\mathbf{f}\mathbf{t}(\mathbf{x}), \mathbf{u}(\mathbf{x})) \\ &= \nu(\mathbf{f}, \mathbf{t}(\mathbf{x}) \cdot \mathbf{u}(\mathbf{x}))\, \nu(\mathbf{t}(\mathbf{x}), \mathbf{u}(\mathbf{x})) \\ &= \nu(\mathbf{f}, \mathbf{x})\, \nu(\mathbf{t}(\mathbf{x}), \mathbf{u}(\mathbf{x})). \end{aligned}$$

On substituting into (16.4) we obtain (16.3).

We now define
$$\zeta_3 \mathbf{f} = (\alpha_2 d\mathbf{f})^{-1} \mathbf{f},$$

and show, as for ζ_2, that $\zeta\mathbf{F}_3 \subset \mathbf{Q}_3$, that ζ_3 is the identity over $\mathbf{Q}_3$ and that ζ_3 is a homomorphism, using this time the fact that $\mathbf{F}_3$ is Abelian.

We show that $\zeta_3 \mathbf{F}_3^{23} = \{1\}$. Since $X_3^2 = \varnothing$ it is sufficient to show that $\zeta_3 \nu_2(x, y) = 1$ $(x, y \in X_2)$. We have

$$\begin{aligned}\zeta_3 \nu_2(\mathbf{x}, \mathbf{y}) &= \alpha\mu_2(\mathbf{x}, \mathbf{y})^{-1} \nu(\mathbf{x}, \mathbf{y}) \\ &= \alpha(\mathbf{x}\mathbf{y}\mathbf{x}^{-1}(\mathbf{x}\cdot\mathbf{y})^{-1})^{-1} \nu(\mathbf{x}, \mathbf{y}) \\ &= 1 \quad \text{by (16.3).}\end{aligned}$$

We show that ζ_2 *is constant* mod $\mathbf{G}_2$ *on* $\mathbf{X}_2^+(\sigma_2)$. Let $\mathbf{x}_1$, $\mathbf{x}_2$ be elements in this set. Since $p_1 = 1$ there is a relation $\mathbf{f}\cdot\mathbf{x}_1 = \mathbf{x}_2$ (15.2). Now $\mu_2(\mathbf{f}, \mathbf{x}_1) \in \mathbf{G}_2$, that is, $\mathbf{f}\mathbf{x}_1\mathbf{f}^{-1}(\mathbf{f}\cdot\mathbf{x}_1^{-1}) \in \mathbf{G}_2$, so $\zeta_2(\mathbf{f}\mathbf{x}_1\mathbf{f}^{-1}\mathbf{f}\cdot\mathbf{x}_1^{-1}) \in \mathbf{G}_2$. Since $\zeta_2\mathbf{f}$ permutes with $\zeta_2\mathbf{x}$, we have $\zeta\mathbf{x}_1 = \zeta\mathbf{f}\cdot\mathbf{x}$ mod $\mathbf{G}_2$, i.e. $\zeta\mathbf{x}_1 = \zeta\mathbf{x}_2$ mod $\mathbf{G}_2$.

We shall show that $\xi_3\alpha_2$ *is constant on* $\mathbf{X}_1^+(\sigma_2)$. Let $\mathbf{x}_1$, $\mathbf{x}_2$ be as in the preceding paragraph. Then

$$\begin{aligned}\xi\alpha\mathbf{x}_2 = \xi\alpha(\mathbf{f}\cdot\mathbf{x}_1) &= \xi\alpha\mathbf{x}_1 \xi\nu_2(\mathbf{f}, \mathbf{x}_1^{-1}) \quad (16.3) \\ &= \xi\alpha\mathbf{x}_1, \quad \text{since} \quad \xi\nu = 0.\end{aligned}$$

We now show that ζ_3 *is constant on* $\mathbf{X}_3^+(\sigma_3)$. For each 3-simplex σ choose a base vertex P_σ and a path $\lambda_\sigma \to OP$ which is irreducible rel σ. Let $b_3 = (t, u)$ be a base for X_3^1 such that $\operatorname{stem}_\sigma u(x) = \lambda_\sigma$ when $x \in X_3^+(\sigma)$ (15.1).

We show first that $\zeta_3\mathbf{u}(\mathbf{x})$ is constant on $\mathbf{X}_3^+(\sigma)$, $\sigma = \sigma_3$. Let $\mathbf{x}_1$, $\mathbf{x}_2$ be elements in this set. Say $\mathbf{x}_i = \kappa x_i$, $x_i \in X_3^+(\sigma)$. Now $\xi u_i = \xi x_i$. Hence $u_i \in X_3^+(\sigma)$. By definition of X^1, $u_i \in \lambda_i * X_3^1(\sigma, P_i)$, $\beta u_i \in \lambda_i * Q_3(\dot\sigma, P_i)$ with $\xi q_i = \partial\sigma$. Since stem $u_i = \lambda_\sigma$ it follows (14.13) that $u_i \in \lambda_\sigma * X_3^1(\sigma, P_\sigma)$ and $\beta u_i \in \lambda_\sigma * Q_2(\dot\sigma, P_\sigma)$. Say $u_i = \lambda_\sigma * v_i$. There exists an element $r \in F_3(\dot\sigma, P_\sigma)$ such that $\beta v_1 v_2^{-1} r^{-1} \in Q_3(\sigma, P_\sigma)$. We show this as follows. Since h^2 is true, and $p_1(\dot\sigma_3, P_\sigma) = 1$, ξ induces an isomorphism $\xi_*\colon p(\dot\sigma_3, P_\sigma) \to H_2(\dot\sigma)$. Since $\xi\beta v_1 v_2^{-1} = 0$, $\beta v_1 v_2^{-1}$, which is an element of $Q_2(\dot\sigma, P_\sigma)$, represents the trivial element of $p_2(\sigma_3, P_\sigma)$ and therefore there exists an $r \in F_2(\dot\sigma, P_\sigma)$ such that $\beta r = \beta v_1 v_2^{-1}$ mod $K_2(\dot\sigma, P_\sigma)$. This relation holds also mod $(K_2(\dot\sigma, P_\sigma)$, hence $v_1 v_2^{-1} r^{-1} \in Q_2(\sigma, P_\sigma)$ as stated. It now follows that if $s = \lambda_\sigma * r$ so that $u_1 u_2^{-1} s^{-1} = \lambda_\sigma * v_1 v_2^{-1} r^{-1}$, then $u_1 u_2^{-1} s^{-1} \in Y_3^2$. Hence $\mathbf{u}_1\mathbf{u}_2^{-1}\mathbf{s}^{-1} \in \mathbf{Y}_3^2 \subset \mathbf{G}_3 \subset \mathbf{Q}_3$. Hence

$$\mathbf{u}_1\mathbf{u}_2^{-1}\mathbf{s}^{-1} = \zeta(\mathbf{u}_1\mathbf{u}_2^{-1}\mathbf{s}^{-1}) = \zeta\mathbf{u}_1 \zeta\mathbf{u}_2^{-1} \zeta\mathbf{s}^{-1}.$$

Now $\mathbf{s} \in F_3^{23}$, since $\mathbf{s} \in F_3(\dot\sigma)$ and $\dim \dot\sigma = 2$. Hence $\zeta\mathbf{s} = 1$ and $\zeta\mathbf{u}_1 = \zeta\mathbf{u}_2$ mod $\mathbf{G}_3$.

Now let $\mathbf{x} \in \mathbf{X}_3^+(\sigma_3)$. We will show that

$$\zeta_3\mathbf{x} = \zeta_3\mathbf{u}(\mathbf{x}) \bmod \mathbf{G}_3 \cdot \zeta_3\mu_3(\mathbf{t}, \mathbf{u}) = \mu_3(\mathbf{t}, \mathbf{u}) \in \mathbf{G}_3.$$

But $\zeta_3\mu_3(\mathbf{t},\mathbf{u}) = \zeta_3(\mathbf{u}(\mathbf{t}\cdot\mathbf{u}^{-1})\,\nu_2(\ldots))$. Since $\xi\nu_2 = 0$, we conclude that $\zeta_3\mathbf{u} = \zeta_3\mathbf{t}\cdot\mathbf{u} = \zeta_3\mathbf{x} \bmod \mathbf{G}_3$.—This completes the proof that ζ_3 is constant on $\mathbf{X}_3^+(\sigma) \bmod \mathbf{G}_3$. We have now shown that α_1, α_2, ζ_2, ζ_3 have the required properties, therefore l^3 and h^3 hold in Σ_{II}.

Proof that Σ_{II} is non-trivial

17. Let A be the 2-complex $\dot{\sigma}$ where σ is a 3-simplex. *We shall show that $p_3 A$ is infinite.*

It is easy to describe explicitly a set U of representatives of the orbits (§15) of X_2 such that

$$U = \{u_1, u_2, u_3, u_4\},$$

$$\xi u_i = \sigma_2^i, \quad \Sigma\sigma_2^i \text{ generates } Z_2(A),$$

$$u_1 u_2 u_3 u_4 \in Q_2.$$

Since $\dim A = 2$, $X_3^1 = X_3^2 = \varnothing$. We have then a decomposition $X_3 = X_3^3 = X^+ \cup (X^+)^{-1}$, where X^+ is the totality of elements $\nu_2(x,y)$, $x, y \in X_2$, $x \neq y^{\pm 1}$. By (14.3), $\nu_2(x,y) = \nu_2(x',y')$ if and only if $x = x'$, $y = y'$. Hence we can define a mapping

$$\chi\colon\ X^+ \to C_2 \otimes C_2$$

by

$$\chi\nu_2(x,y) = \xi u(x) \otimes \xi u(y),$$

where $u(x)$ is the selector with values in U. We extend χ over X^3 by $\chi t^{-1} = -\chi t$, then extend χ to a homomorphism $F_3 \to C_2 \otimes C_2$.

To show that $p_3 \neq 1$, it is sufficient to produce an element q of Q_3 such that q is not a member of $G_3 K_3$.

Let $t = u_1 u_2 u_3 u_4$ and let $q = \nu_2(t,t)$. Then $\beta q = \mu_2(t,t) \bmod K_2$, and since $t \in Q_2$, $\mu_2(t,t) = 1$. Hence $q \in Q_3$. We have

$$\chi q = \Sigma_{i,j=1}^4 \sigma_2^i \otimes \sigma_2^j.$$

Hence $\chi q \neq 0$; in fact $\chi q^k = k\chi q$ so $\chi q^k \neq 0$, when $k \neq 0$.

On the other hand, one verifies easily that $\chi G_3 K_3 = \{0\}$. For example, to show that $\chi G_3 = \{0\}$, it is sufficient to show that $\chi\mu_3(x,y) = 0$ when $x, y \in X_3$, since Y_3, which generates G_3, contains only relations of the third kind. We have $\chi\mu_3(x,y) = \chi\nu_2(x,\beta y)$. Say $y = \nu_2(a,b)$, $a, b \in X_2$. Then

$$\begin{aligned}\chi\nu_2(x,\beta y) &= \chi\nu_2(x,\mu_2(a,b))\\ &= \chi\nu_2(x, aba^{-1}(a\cdot b^{-1}))\\ &= \chi(\nu_2(x,a)\,\nu_2(x,b)\,\nu_2(x,a^{-1})\,\nu_2(x,a\cdot b^{-1})).\end{aligned}$$

Now $u(a^{-1}) = u(a)^{-1}$, $u(a.b^{-1}) = u(b^{-1}) = u(b)^{-1}$ (see §15). Hence the right-hand member of the last equation is 0.—The verification of $K_3 = \{0\}$ is slightly more elaborate but quite straightforward.

The element q^k $(k > 0)$ cannot be in $G_3 K_3$, otherwise $\chi q^k = 0$. Hence p_3 is infinite.

COLUMBIA UNIVERSITY

REFERENCES

[1] R. PEIFFER, *Über Identitäten zwischen Relationen*, Math. Ann., 121 (1949), pp. 67–99.

[2] P. A. SMITH, *The complex of a group relative to a set of generators, Part I*, Ann. of Math., 54 (1951), pp. 371–402.

[3] J. H. C. WHITEHEAD, *On adding relations to homotopy groups*, Ann. of Math., 42 (1941), pp. 409–428.

The Theory of Carriers and S-Theory

E. H. Spanier and J. H. C. Whitehead

Introduction

In the present work we present details of a note published earlier [7] discussing a new category. This new category is called the *suspension category* and is obtained by passing to the direct limit with respect to suspension. It has the property that suspension is always an isomorphism, which makes it simpler and easier to handle than the category of homotopy classes. Also the set of mappings from one object in the suspension category to another is always an Abelian group. We describe the theory of this category as *S-theory*.

The paper is divided into two parts. The first deals with the general theory of carriers, which provides a method of treating simultaneously the various relativizations which occur in homotopy theory. The definition of a carrier is not the same as that used in our previous note [7] because base points have been omitted. In S-theory the base point is replaced by S^{n-1}, the n-fold suspension of the empty set ($n \geqq 3$).

The second part deals with S-theory. Among other things it is shown how to obtain an exact couple from an ordered sequence of carriers and how to use this to obtain obstructions for extension and compression problems in the sense of S-theory. There are two obstruction theories, one leading to obstructions involving cohomology of the space being mapped, with coefficients in S-homotopy groups of the image space, and the other involving homology of the image space with coefficients in S-cohomotopy groups of the antecedent space. Because of the relativizations involved in the use of carriers, it is necessary to consider homology and cohomology in generalized local systems, and this is done by introducing stacks (faisceaux) similar to those defined by Leray [4].

A kind of duality between homotopy groups and cohomotopy groups has been noticed by various authors [5, 6]. This duality is present in strengthened form in S-theory and we hope to publish a precise formulation of it at a later date.

I. The theory of carriers

1. Preliminaries

We recall some of the definitions given in [7], with minor modifications. A *carrier* is an order-preserving map ϕ: $\mathfrak{a} \to \mathfrak{b}$ (i.e. $A \subset A'$ implies $\phi A \subset \phi A'$ if $A, A' \in \mathfrak{a}$), where $\mathfrak{a}$, $\mathfrak{b}$ are collections of subsets of spaces X, Y. A map f: $X \to Y$ is called a *ϕ-map*, if, and only if, $fA \subset \phi A$ for every $A \in \mathfrak{a}$. A *ϕ-homotopy*, f_t: $X \to Y$ or f: $X \times I \to Y$, is one such that f_t, or $f(.,t)$ $(f(.,t)\,x = f(x,t))$, is a ϕ-map for each $t \in I$. The corresponding equivalence classes of ϕ-maps are called *ϕ-homotopy classes*. We use $\{f, \phi\}$, or simply $\{f\}$, to denote the ϕ-homotopy class of f. The specification of $\mathfrak{a}$ includes the specification of X and we define $\mathfrak{C}$ as the category in which the objects are all collections such as $\mathfrak{a}$ and the mappings are the ϕ-homotopy classes, indexed by ϕ, for every carrier ϕ. We denote the totality of ϕ-homotopy classes by $\Pi(\phi)$. Notice that $\Pi(\phi)$ may be empty, as it is, for example, if $\phi A = \varnothing$, the empty set, for some $A \neq \varnothing$, or if ϕA, $\phi A'$ are in different components of Y and A, A' are in the same component of X. We shall invariably denote maps by Roman and carriers by Greek letters.

Let $\mathfrak{a}'$, $\mathfrak{b}'$ also be collections of subsets of X, Y and let ϕ: $\mathfrak{a} \to \mathfrak{b}$, ϕ': $\mathfrak{a}' \to \mathfrak{b}'$ be given carriers. If every ϕ-map is a ϕ'-map and conversely, then $\Pi(\phi) = \Pi(\phi')$ and we describe ϕ, ϕ' as *equivalent*. This will occur, for example, if $\mathfrak{a}'$, $\mathfrak{b}'$ consist of all the subsets of X, Y and if $\phi' A' = \bigcap \phi A$, for every $A \in \mathfrak{a}$ which contains A', or $\phi' A' = Y$ if there is no such $A \in \mathfrak{a}$. In this case we write ϕ': $X \to Y$ and say that ϕ is *generated* by $\mathfrak{a}$. When we define a carrier ϕ: $X \to Y$ by assigning values to ϕA, for certain sets A, it is to be understood that ϕ is generated by the totality of these sets.

When considering the structure of $\mathfrak{C}$ it is necessary to refer to the objects $\mathfrak{a}$, $\mathfrak{b}$, etc. In particular, they are needed to determine the equivalences. For example, if $\mathfrak{a}'$, $\mathfrak{b}'$ contain all the subsets of X, Y, then the identity $\mathfrak{a}' \to \mathfrak{a}'$ consists only of the identical map, and an equivalence $\mathfrak{a}' \to \mathfrak{b}'$ can only consist of a single homeomorphism. However, when primarily interested in the sets $\Pi(\phi)$, rather than the structure of $\mathfrak{C}$, we may confine ourselves to carriers of the form $X \to Y$.

We write $\phi \leqq \psi$: $X \to Y$ if, and only if, $\phi A \subset \psi A$ for every $A \subset X$. In this case a map γ: $\Pi(\phi) \to \Pi(\psi)$ is defined by $\gamma\{f, \phi\} = \{f, \psi\}$. We describe γ as an *injection*. If ϕ_j: $X \to Y$, for each j in a set J, we define carriers

$$\bigcup \phi_j,\ \bigcap \phi_j\colon\ X \to Y$$

by

$$(\bigcup \phi_j) A = \bigcup (\phi_j A),\ (\bigcap \phi_j) A = \bigcap (\phi_j A).$$

If $\phi: X \to Y$, $\psi: Y \to Z$, then a ϕ-map f and a ψ-map g (alternatively $\{f, \phi\}$, $\{g, \psi\}$) induce maps

$$\Pi(\psi) \xrightarrow{f^*} \Pi(\psi\phi) \xleftarrow{g_*} \Pi(\phi), \tag{1.1}$$

where $(\psi\phi)A = \psi(\phi A)$. In particular if $X \subset Y$ and $\phi A = A$ for every $A \subset X$ we describe $\psi\phi$ as the *restriction* of ψ to X and denote it by $\psi \mid X$. If f is the inclusion map, f: $X \subset Y$, we describe f^* as the *restriction map*.

If X is a CW-complex ([10]) we describe $\phi: X \to Y$ as *cellular*, if, and only if, it is generated by a collection of sub-complexes of X. If Y is a CW-complex we describe ϕ as *cocellular* if, and only if, ϕA is a subcomplex of Y for every $A \subset X$. By a *closed cell*, σ, of a CW-complex X we mean the smallest subcomplex which contains some open cell e. We write $\sigma - e = \dot{\sigma}$. Then $\dim \dot{\sigma} < \dim \sigma = \dim e$. If $\dim \sigma = n$ we write $\sigma = \sigma^n$ and describe σ as an *n-cell*. We define $\sigma^{-1} = \varnothing$ and σ^n may also be defined inductively as the union of an open n-cell, e^n, and all the closed cells σ^q $(q < n)$ such that $\sigma^q - \dot{\sigma}^q$ meets $\bar{e}^n$. By a cell, without qualification, we shall mean a closed cell. We shall write $\sigma \in X$ to indicate that σ is a cell of X. Let Y be a CW-complex and let one, at least, of X, Y be locally finite. Then $X \times Y$ is a CW-complex ([10], p. 227), and it follows without difficulty from the inductive definition of σ and induction on $\dim(\sigma \times \tau)$ that the cells of $X \times Y$ are the products $\sigma \times \tau$, for every $\sigma \in X$, $\tau \in Y$.

Let χ: $X \times I \to X \times I$ be defined by $\chi(\sigma \times I) = \sigma \times I$ for every $\sigma \in X$. Let $P = (X \times 1) \cup (A \times I)$, where A is a subcomplex of X.

LEMMA (1.2). *There is a χ-homotopy h_t,* rel P, *such that $h_0 = 1$, the identity, and $h_1(X \times I) = P$.*

This is well known if $(X, A) = (\sigma, \dot{\sigma})$ and therefore follows from the standard inductive argument.

COROLLARY (1.3). *If ϕ: $X \to Y$ is a cellular carrier and f a ϕ-map, then any $(\phi \mid A)$-homotopy of $f \mid A$ can be extended to a ϕ-homotopy of f.*

Let $\phi \leqq \psi$: $X \to Y$ be cellular carriers and let θ: $X \to Y$ be defined by $\theta A' = \phi A'$ or $\psi A'$ according as $A' \subset A$ or $A' \not\subset A$. Let f: $X \times I \to Y$ be a θ-homotopy such that $f(\cdot, 1)$ is a ϕ-map and let h_t be as in (1.2). Then $fh_t(\cdot, 0)$ is a θ-homotopy, rel A, between $f(\cdot, 0)$ and $fh_1(\cdot, 0)$. The latter is a ϕ-map and we have proved:

LEMMA (1.4). *If a θ-map is θ-homotopic to a ϕ-map, then it is so* rel A.

Let X, Y be CW-complexes and let ϕ: $X \to Y$ be both cellular and cocellular.

LEMMA (1.5). *Each ϕ-homotopy class contains a cellular map.*

This, like (1.2), follows from the standard (inductive) argument when due respect is paid to ϕ.

We recall that the join, $X * Y$, of non-vacuous spaces X, Y is the space obtained from $X \times Y \times I$ by taking each of the sets $x \times Y \times 0$, $X \times y \times 1$ to be a single point. The subspaces corresponding to $X \times Y \times 0$, $X \times Y \times 1$ are copies of X, Y, which we also denote by X, Y, and we write $x \times Y \times 0 = x$, $X \times y \times 1 = y$. We define $X * \varnothing = \varnothing * X = X$. If f: $X \to X'$, g: $Y \to Y'$, then $f * g$: $X * Y \to X' * Y'$ is defined by $(f*g)(x, y, t) = (fx, gy, t)$. If $\mathfrak{a}$, $\mathfrak{b}$ are collections of subsets in X, Y, then $\mathfrak{a} * \mathfrak{b}$ will denote the collection of sets $(i * j)(A * B)$, which we also denote by $A * B$, where i: $A \subset X$, j: $B \subset Y$ for every $A \in \mathfrak{a}$, $B \in \mathfrak{b}$. If ϕ: $\mathfrak{a} \to \mathfrak{a}'$, ψ: $\mathfrak{b} \to \mathfrak{b}'$ are carriers we define $\phi * \psi$: $\mathfrak{a} * \mathfrak{b} \to \mathfrak{a}' * \mathfrak{b}'$ by $(\phi * \psi)(A * B) = (\phi A) * (\psi B)$. Obviously $f * g$ is a $(\phi * \psi)$-map if f is a ϕ-map and g a ψ-map. If $\mathfrak{b}$ is fixed, then a covariant functor $\mathfrak{C} \to \mathfrak{C}$ is defined by $\mathfrak{a} \to \mathfrak{a} * \mathfrak{b}$, $\{f, \phi\} \to \{f * 1, \phi * \iota\}$, where $\{1, \iota\}$: $\mathfrak{b} \to \mathfrak{b}$ is the identity. After replacing Y by a homeomorph if necessary, so as to avoid 'accidental intersections', we imbed X, Y in $X * Y$ in the manner indicated above.

Let X, Y, Z be spaces such that either (a) X, Z are locally compact and regular, or (b) either X and Y or Y and Z are compact and regular. (Cf. § 5 in [9]. The possibility that conditions such as these are necessary for what follows was pointed out to us by D. E. Cohen.) Then there is a natural homeomorphism of $(X * Y) * Z$ onto $X * (Y * Z)$ by means of which we identify these two spaces and omit the parentheses. If f, g are as before and if h: $Z \to Z'$, then, for every $x \in X$, $y \in Y$, $z \in Z$, the maps $(f * g) * h$ and $f * (g * h)$ determine the same barycentric map of the 2-simplex $x * y * z$ onto $fx * gy * gz$. Therefore

$$(f * g) * h = f * (g * h) = f * g * h,$$

say, under the above identification. Similarly, we identify $X * Y$ and $f * g$ with $Y * X$ and $g * f$.

If Y is a single point we write $X * Y = TX$ and we write $X * Y = SX$ if Y is an ordered pair of points. We describe TX and SX as a *cone on* and *a suspension of* X. We do not fix the 'vertex', $T(\varnothing)$, of TX or the pair of 'poles' $S(\varnothing) \subset SX$. Therefore the same symbol, TX or SX, may denote different spaces; but if $X * y$, $X * y'$ occur in the same context we write $X * y = TX$, $X * y' = T'X$. Thus $SX = TX \cup T'X$, $TX \cap T'X = X$ if $SX = X * (y \cup y')$. If f: $X \to X'$ and if g maps the vertex of TX on the vertex of TX' or the poles of SX on the poles of SX', preserving order, we write $f * g$ as Tf: $TX \to TX'$ or Sf: $SX \to SX'$. We denote points of SX by (x, s), where $-1 \leqq s \leqq 1$ and $(x, -1) = y$,

$(x,1) = y'$. Thus $(x,t) \in T'X$, where $x \in I$, $(x,0) = x$, $(x,1) = T'(\varnothing)$. This will be the usual notation for points in a cone. We define $S^n X$ inductively as $S(S^{n-1}X)$, where $n \geqq 1$ and $S^0 X = X$. We have $S^n(\varnothing) = S^{n-1}$, where S^{n-1} is an $(n-1)$-sphere, and we write $TS^{n-1} = E^n$. Under the above identifications,

$$S^n X = X * S^{n-1}, \quad S^n TX = TS^n X = X * E^n.$$

Also $T_0 SX = X * (y \cup y') * y_0$, where $y_0 = T_0(\varnothing)$, and we may take y_0 to be the mid-point of a 1-simplex whose vertices are y, y'. Then $(y \cup y') * y_0 = y * y'$ and $T_0 SX = T'TX$. If $f\colon X \to X'$, then $T_0 Sf$, $T'Tf$ both determine the same barycentric map $x * y * y' \to fx * y * y'$, for every $x \in X$. Therefore $T_0 Sf = T'Tf$ under the identification

$$T_0 SX = T'TX.$$

Define $u\colon TX \to SX$ by $u(x,t) = (x, 2t-1)$. Then $u\colon T'TX \to T'TX$ under the identifications $T'T = TS = ST$ $(T'(\varnothing) = y' = (x',1),\ x' \in TX)$. More precisely, let $T'TX = T_0 SX = ST_0 X$, where y_0 is the mid-point of $y * y'$. Let $f\colon y * y' \to y_0 * y'$ be the linear map such that $fy = y_0$, $fy' = y'$. Then u is the composite of

$$T'TX \xrightarrow{1*f} T'T_0 X \xrightarrow{u_0} T'TX,$$

where $u_0 = u\colon T'T_0 X \to ST_0 X$ is defined as above $(u_0 y' = y')$. Define χ, $\chi_0\colon T'TX \to T'TX$ by

$$\chi TA = TA, \quad \chi SA = SA, \quad \chi T'TA = T'TA,$$

$$\chi_0 TA = T_0 TA, \quad \chi_0 SA = T_0 TA \cup T'A, \quad \chi_0 T'TA = T'TA,$$

for every $A \subset X$ $(SA = TA \cup T'A)$.

Lemma (1.6). *The map $u\colon T'TX \to T'TX$ is χ-homotopic to the identity.*

Proof. Let $g\colon y * y' \to y * y'$ be a map such that $g(y * y_0) = y$, $gy' = y'$ and let $h = 1 * g$. Obviously u and h are χ_0- and χ-homotopic to the identity and $hT_0 TA = TA$. Therefore there are χ-homotopies $u \simeq hu \simeq h \simeq 1$ and (1.6) is proved.

If $\phi\colon X \to Y$ we define $S\phi\colon SX \to SY$ by $(S\phi)(SA) = S(\phi A)$, for every $A \subset X$, and $S_*\colon \Pi(\phi) \to \Pi(S\phi)$ by $S_* \{f, \phi\} = \{Sf, S\phi\}$. We describe S_* as the *suspension* map. Evidently $S(\psi\phi) = (S\psi)(S\phi)$, $S(gf) = (Sg)(Sf)$ if ψ, $g\colon Y \to Z$ and it follows that S_* is natural for $\mathfrak{C}$. That is to say, $(Sf)^* S_* = S_* f^*$, $(Sg)_* S_* = S_* g_*$, where f^*, g_* are as in (1.1).

Let X, Y be CW-complexes, at least one of which is locally finite, and treat I as a 1-simplex. Then $X \times Y \times I$ is a CW-complex whose cells are the products $\sigma \times \tau \times \tau'$, for every $\sigma \in X$, $\tau \in Y$, $\tau' \in I$. Hence it follows ([10], p. 237) that $X * Y$ is a CW-complex whose cells are the joins $\sigma * \tau$, for every $\sigma \in X$, $\tau \in Y$ (including σ^{-1}, τ^{-1}). In particular, the cells of TX are of the form σ and $T\sigma$ and the cells of

$$SX = TX \cup T'X$$

are of the form σ, $T\sigma$, $T'\sigma$.

2. Extension and compression

The extension problem is to determine the image of the restriction map $\Pi(\phi) \to \Pi(\phi \mid A)$ for a given carrier $\phi: X \to Y$ and subset $A \subset X$. The compression problem is to determine the image of the injection $\Pi(\phi) \to \Pi(\psi)$ for given carriers $\phi \leqq \psi: X \to Y$. These two problems occupy a central position in homotopy theory, and we prove below that for cellular carriers they are equivalent. Having done this we shall restrict our attention to the compression problem.

THEOREM (2.1). *Every compression problem is equivalent to an extension problem.*

PROOF. Let $\phi \leqq \psi: X \to Y$ and let $\theta: X \times I \to Y$ be defined by $\theta(A \times I) = \psi A$, $\theta(A \times 1) = \phi A$ for every $A \subset X$. Then the compression problem for ϕ, ψ is obviously equivalent to the extension problem for θ, $X \times 0$.

THEOREM (2.2). *The extension problem for a cellular carrier $\phi: X \to Y$ and subcomplex, A, of a* CW*-complex, X, is equivalent to a compression problem.*

PROOF. Let us regard ϕ as a carrier $\phi: X \to TY$, with values in Y, and let $\psi: X \to TY$ be defined by $\psi A' = \phi A'$ or TY according as $A' \subset A$ or $A' \not\subset A$. Since TY is contractible every $(\phi \mid A)$-map $A \to Y$ has an extension $X \to TY$, which is necessarily a ψ-map. Moreover, any two such extensions are ψ-homotopic, and it follows that the extension problem for ϕ, A is equivalent to the compression problem for ϕ, ψ.

3. The exact sequence

Let $\phi \leqq \psi: X \to Y$, where X, Y are any spaces. We define

$$\phi^n_m: S^m X \to S^n Y \quad \text{by} \quad \phi^n_m S^m A = S^n \phi A \quad (m, n \geqq 0)$$

and $(\psi, \phi)_1: TX \to Y$ by $(\psi, \phi)_1 A = \phi A$, $(\psi, \phi)_1 TA = \psi A$.

Let
$$(\psi, \phi)^n_{m+1} = ((\psi, \phi)_1)^n_m = (\psi^n_m, \phi^n_m)_1,$$

under the identification $S^m T = TS^m$, and let

$$\Pi^n_{m+1}(\psi, \phi) = \Pi((\psi, \phi)^n_{m+1}), \quad \Pi^n_m(\phi) = \Pi(\phi^n_m).$$

We shall omit the superscript 0 if $n = 0$.

Let F be any set. By a *quasi-topology* in F we mean a rule which, for every topological space P, selects a class of functions $P \to F$, to be described as *quasi-continuous*, subject to the following conditions (the necessity of Condition (b) for what follows was pointed out to us by W. D. Barcus):

(a) *if P, P' are topological spaces, then the composite of a* (*continuous*) *map $P \to P'$ and a quasi-continuous function $P' \to F$ is quasi-continuous,*

(b) *if $P = P_1 \cup P_2$, where P_1, P_2 are closed or open subspaces of P, and if g_i: $P_i \to F$ $(i = 1, 2)$ are quasi-continuous functions such that $g_1 p = g_2 p$ if $p \in P_1 \cap P_2$, then the function g: $P \to F$, defined by $g \mid P_i = g_i$, is quasi-continuous,*

(c) *every constant function $P \to F$ is quasi-continuous.*

Terms such as 'map', 'homotopy class', 'path-component' will have their obvious meaning when used with reference to a quasi-topology. Evidently the homotopy extension theorem is valid for maps g: $E^n \to F$ and homotopies of $g \mid S^{n-1}$. Therefore a large part of homotopy theory is valid for quasi-topological spaces and we shall take the elements of this theory for granted.

Let F_θ be the set of θ-maps $X \to Y$, where $\theta = \phi$ or ψ. A quasi-topology in F_θ is defined by taking $P \to F_\theta$ to be quasi-continuous if, and only if, the corresponding function $X \times P \to Y$ is continuous. Let O_θ be the set of constant maps in $\theta(\varnothing)$ and let the pair (O_ψ, O_ϕ) be 1-connected, and hence 0-connected.† Then we define a sequence of maps

$$(3.1) \quad \Pi_1(F_\phi, O_\phi) \xrightarrow{\gamma'} \Pi_1(F_\psi, O_\psi) \xrightarrow{\alpha} \Pi_1(F_\psi, F_\phi, O_\psi) \xrightarrow{\beta} \Pi_0(F_\phi, O_\phi) \xrightarrow{\gamma} \Pi_0(F_\psi, O_\psi)$$

as follows. If $F_0, F_1 \subset F \subset F_\psi$ then $\Pi_1(F, F_0)$ and $\Pi_1(F, F_0, F_1)$ consist of the homotopy classes of maps from $(I, 0, 1)$ to (F, F_0, F_0) and (F, F_0, F_1). A zero in either of these sets will mean the class of a constant map. The set $\Pi_0(F, F_0)$ is the set of path components of F, indexed by F_0. A zero in $\Pi_0(F, F_0)$ is a path component which contains at least one point of F_0. We use $\{0\}$ to denote the set of zeros in any of the sets in (3.1). The maps γ, γ' in (3.1) are injections and β is defined by restriction to 0. Let

$$\Pi_1(F_\psi, O_\psi) \xleftarrow{\lambda} \Pi_1(F_\psi, O_\phi, O_\psi) \xrightarrow{\mu} \Pi_1(F_\psi, F_\phi, O_\psi)$$

be the injections. Since (O_ψ, O_ϕ) is 1-connected λ is obviously a 1-1 correspondence and $\alpha = \mu\lambda^{-1}$. We have, obviously:

LEMMA (3.2). *The sequence* (3.1) *is exact, in the sense that each of*

† We describe a pair of spaces (Z, C) $(C \subset Z)$ as 0-connected if, and only if, $Z \neq \varnothing$ and each path component of Z contains at least one point of C.

$\alpha^{-1}\{0\}$, $\beta^{-1}\{0\}$, $\gamma^{-1}\{0\}$ *is the image set of the preceding map. Moreover,* β *is* 1-1 *into if* $\Pi_1(F_\psi, O_\psi) = \{0\}$ *and* α *is* 1-1 *into if* $\Pi_1(F_\phi, O_\phi) = \{0\}$.

Let C_θ be the set of constant θ-maps. The image of such a map is in the intersection $\bigcap \theta A$, for every non-vacuous $A \subset X$, and $O_\theta \subset C_\theta \subset F_\theta$. Let the pair (C_θ, O_θ) be 0-connected. Then $\Pi_0(F_\theta, O_\theta)$ may be identified with $\Pi_0(F_\theta, C_\theta)$ and the latter with $\Pi(\theta)$, in which a 0 is the class of a constant map. Also $\Pi_0(F_{\theta_1}, O_{\theta_1})$ may be identified with $\Pi_1(F_\theta, O_\theta)$ and $\Pi_1(F_\psi, F_\phi, O_\psi)$ with $\Pi_1(\psi, \phi)$. Therefore the sequences (3.1) for ϕ_m, ψ_m $(m = 0, 1, \ldots)$ combine to form the sequence

$$(3.3) \quad \ldots \xrightarrow{\alpha} \Pi_{m+1}(\psi, \phi) \xrightarrow{\beta} \Pi_m(\phi) \xrightarrow{\gamma} \Pi_m(\psi) \xrightarrow{\alpha} \ldots \to \Pi(\phi) \to \Pi(\psi).$$

Here each γ is an injection, β is defined by restriction and $\alpha\{f\} = \{fu\}$, where $f: S^m X \to Y$ is a ψ_m-map such that

$$f(x', -1) \in \phi(\varnothing) \quad (x' \in S^{m-1}X, m \geqq 1)$$

and $u: TS^{m-1}X \to S^m X$ is defined by $u(x', t) = (x', 2t - 1)$. In consequence of (3.2) we have:

THEOREM (3.4). *The sequence* (3.3) *is exact. Moreover,*

$$\beta: \Pi_m(\psi, \phi) \to \Pi_{m-1}(\phi) \quad \textit{is 1-1 into if} \quad \Pi_m(\psi) = \{0\}$$

and $$\alpha: \Pi_m(\psi) \to \Pi_m(\psi, \phi) \quad \textit{is 1-1 into if} \quad \Pi_m(\phi) = \{0\}.$$

The sequence (3.3) is evidently natural for $\mathfrak{C}$. That is to say, if ϕ, ψ: $\mathfrak{a} \to \mathfrak{b}$, if χ: $\mathfrak{b} \to \mathfrak{c}$, where $\mathfrak{c}$ is a collection of sets in a space Z, and if g is a χ-map, then $g_* \alpha = \alpha g_*$, $g_* \beta = \beta g_*$, $g_* \gamma = \gamma g_*$, where α, β, γ are as in (3.3) and the corresponding sequence for $(\chi\psi, \chi\phi)$. A similar remark applies to a carrier θ: $\mathfrak{a}' \to \mathfrak{a}$ and a θ-map f, except that f^*: $\Pi_m(\psi, \phi) \to \Pi_m(\psi\theta, \phi\theta)$ is the map induced by the $(\hat{\theta}, \theta)_m$-map Tf, where $\hat{\theta}$ is defined by $\hat{\theta}A' = T\theta A'$ $(A' \in \mathfrak{a}')$.

Let $(3.3)^n$ denote the sequence (3.3) for the carriers ϕ_0^n, ψ_0^n $(n \geqq 0)$. Then
$$S_*: \Pi_m^n(\xi) \to \Pi_{m+1}^{n+1}(\xi) \quad (\xi = \phi, \psi, (\psi, \phi)_1),$$
and $S_* \alpha = \alpha S_*$, $S_* \beta = \beta S_*$, $S_* \gamma = \gamma S_*$. If $\psi(\varnothing) \neq \varnothing$, then $S^2\psi(\varnothing)$ is 1-connected, and $\psi(\varnothing) = \phi(\varnothing)$ if $\psi(\varnothing) = \varnothing$. Also $S^2\phi(\varnothing)$ is 0-connected. Therefore $(S^n\psi(\varnothing), S^n\phi(\varnothing))$ is 1-connected if $n \geqq 2$. Similarly $(S^n \bigcap \theta A, S^n\theta(\varnothing))$ is 0-connected if $n \geqq 1$, where the intersection $\bigcap \theta A$ is taken over every non-vacuous $A \subset X$. Therefore $(C_{\theta_0^n}, O_{\theta_0^n})$ and $(O_{\psi_0^n}, O_{\phi_0^n})$ are certainly 0-connected and 1-connected if $n \geqq 2$.

Let ϕ, ϕ', ψ: $X \to Y$ be carriers such that $\phi, \phi' \leqq \psi$, let $\theta = \phi \cap \phi'$ and let
$$(\psi; \phi, \phi')_2 = ((\psi, \phi')_1, (\phi, \theta)_1)_1: \ T'TX \to Y.$$
Thus $(\psi; \phi, \phi')_2$: $(T'TA, TA, T'A) \to (\psi A, \phi A, \phi' A) \quad (A \subset X)$.

Let $$(\psi;\, \phi, \phi')^n_{m+2} = ((\psi;\, \phi, \phi')_2)^n_m = (\psi^n_m;\, \phi^n_m, \phi'^n_m)_2$$
and let
$$\Pi^n_{m+2}(\psi;\, \phi, \phi') = \Pi((\psi;\, \phi, \phi')^n_{m+2}) = \Pi^n_{m+1}((\psi, \phi')_1, (\phi, \theta)_1).$$
Let the pair $(S^n\phi'(\varnothing), S^n\theta(\varnothing))$ be 1-connected and let $(S^n \cap \chi A, S^n\chi(\varnothing))$ be 0-connected for $\chi = (\phi, \theta)_1, (\psi, \phi')_1$. Then we have an exact sequence

$$(3.5)^n \quad \ldots \xrightarrow{\alpha} \Pi^n_{m+1}(\psi;\, \phi, \phi') \xrightarrow{\beta} \Pi^n_m(\phi, \theta) \xrightarrow{\gamma} \Pi^n_m(\psi, \phi')$$
$$\xrightarrow{\alpha} \Pi^n_m(\psi;\, \phi, \phi') \to \ldots \to \Pi^n_1(\phi, \theta) \to \Pi^n_1(\psi, \phi').$$

If $\psi = \phi \cup \phi'$ we describe each γ in $(3.5)^n$ as an *excision map.*

Let $\xi, \xi' \leqq \eta\colon X \to Y$ be carriers such that $\phi \leqq \xi$, $\phi' \leqq \xi'$, $\psi \leqq \eta$ and let
$$\iota\colon\ \Pi^n_m(\psi;\, \phi, \phi') \to \Pi^n_m(\eta;\, \xi, \xi') \quad (m \geqq 2)$$
be the injection. Evidently $\Pi^n_m(\psi, \phi) = \Pi^n_m(\psi;\, \phi, \phi)$ under the identification $TSX' = TT'X' = T'TX'$, where $X' = S^{m-2}X$. Define
$$\alpha'\colon\ \Pi^n_m(\psi, \phi) \to \Pi^n_m(\psi;\, \phi, \phi')$$
by composition of $(\psi, \phi)^n_m$-maps with
$$u\colon\ TT'X' \to TT'X' \quad (uT(\varnothing) = T(\varnothing),\ uT'X' = T'(\varnothing)).$$
Then we have, in consequence of (1.6),

LEMMA (3.6). *If also $\phi \leqq \xi'$, then $\iota\alpha'$ is the injection*
$$\Pi^n_m(\psi;\, \phi, \phi) \to \Pi^n_m(\eta;\, \xi, \xi').$$

A similar result holds for α in $(3.5)^n$.

Obviously $\Pi^n_q(\theta, \theta) = \{0\}$ for every $q \geqq 1$. Therefore it follows from (3.4) that for $\theta \leqq \phi$
$$\alpha'\colon\ \Pi^n_m(\psi, \phi) \to \Pi^n_m(\psi;\, \phi, \theta)$$
is a 1-1 correspondence, and from (3.6), with $\phi' = \theta$, $\xi = \xi' = \phi$, $\eta = \psi$, that $\iota = \alpha'^{-1}$. Since u pinches $T'X'$ to a point and β in $(3.5)^n$ is defined by restriction to TX' it follows that $\beta\alpha' = \alpha\beta = \delta$, say, where
$$\Pi^n_m(\psi, \phi) \xrightarrow{\beta} \Pi^n_{m-1}(\phi) \xrightarrow{\alpha} \Pi^n_{m-1}(\phi, \theta)$$
are as in (3.3) for the pairs (ψ, ϕ), (ϕ, θ). Therefore, on taking $\phi' = \theta \leqq \phi$ in $(3.5)^n$, we derive the exact sequence

$$(3.7)^n \qquad \ldots \xrightarrow{\alpha} \Pi^n_m(\psi, \phi) \xrightarrow{\delta} \Pi^n_{m-1}(\phi, \theta) \xrightarrow{\gamma} \Pi^n_{m-1}(\psi, \theta) \xrightarrow{\alpha} \ldots,$$

in which α, γ are injections.

Evidently $(3.5)^n$, $(3.7)^n$ are natural for the category $\mathfrak{C}$ and their maps commute with S_*. We describe (3.3), $(3.7)^0$ *as the homotopy sequences of the pair (ψ, ϕ) and of the triple (ψ, ϕ, θ).* We describe $(3.5)^0$

as the *upper homotopy sequence of the triad* $(\psi; \phi, \phi')$. The analogous sequence in which the parts played by ϕ, ϕ' are interchanged will be called the *lower homotopy sequence* of $(\psi; \phi, \phi')$. Thus α' in (3.6) belongs to the lower homotopy sequence of $(\psi; \phi, \phi')$.

Consider the diagram

$$
\begin{array}{ccc}
\Pi^n_{m+2}(\psi; \phi, \phi') & \xrightarrow{\beta} & \Pi^n_{m+1}(\phi, \theta) \\
\downarrow \beta' & & \downarrow \beta \\
\Pi^n_{m+1}(\phi', \theta) & \xrightarrow{\beta} & \Pi^n_m(\theta)
\end{array}
\tag{3.8}
$$

in which β' is taken from the lower homotopy sequence of $(\psi; \phi, \phi')$. Let $f\colon T'TS^mX \to S^nY$ be a $(\psi; \phi, \phi')^n_{m+2}$-map. Then $\beta\beta\{f\}$, likewise $\beta\beta'\{f\}$, is the θ^n_m-homotopy class of $f \mid S^mX$. Therefore (3.8) is commutative (it is assumed that the upper sequence of $(\psi; \phi, \phi')$ is defined in terms of $T'T$ and the lower sequence in terms of TT').

4. The local-global theorem

Let X, K be CW-complexes, at least one of which is locally finite and either of which may be empty. Let Y be an arbitrary space. For any carrier $\chi\colon X * K \to Y$ and each $\sigma \in X$ we define $\chi_\sigma\colon K \to Y$ by $\chi_\sigma C = \chi(\sigma * C)$, for every $C \subset K$.

Let $\phi \leqq \psi\colon X * K \to Y$ be cellular carriers and define

$$\phi' \leqq \psi'\colon \; K \times I \to Y$$

by $$\phi'(C \times I) = \psi'(C \times (0 \cup 1)) = \phi C, \quad \psi'(C \times I) = \psi C,$$

for every $C \subset K$. As in §3 we denote injections by γ, and if $K = \varnothing$ we agree that $\gamma\Pi(\phi') = \Pi(\psi')$, $\gamma\Pi(\phi \mid K) = \Pi(\psi \mid K)$. If $K = \varnothing$ and $\sigma = \sigma^0$ we take $\Pi_1(\psi_\sigma, \phi_\sigma)$ to be the set of path components of $\psi_\sigma(\varnothing) = \psi\sigma$, indexed by ϕ_σ, and a $0 \in \Pi_1(\psi_\sigma, \phi_\sigma)$ will mean a path component which contains at least one point of $\phi\sigma$.

THEOREM (4.1). *If $\gamma\Pi(\phi \mid K) = \Pi(\psi \mid K)$ and $\Pi_{p+1}(\psi_\sigma, \phi_\sigma) = \{0\}$, for every $\sigma = \sigma^p \in X$ and every $p \geqq 0$, then $\gamma\Pi(\phi) = \Pi(\psi)$. If $\gamma\Pi(\phi') = \Pi(\psi')$ and $\Pi_{p+2}(\psi_\sigma, \phi_\sigma) = \{0\}$ for every $\sigma = \sigma^p$, then $\gamma\colon \Pi(\phi) \to \Pi(\psi)$ is* 1-1 *into.*

PROOF. Let $\gamma\Pi(\phi \mid K) = \Pi(\psi \mid K)$, $\Pi_{p+1}(\psi_\sigma, \phi_\sigma) = \{0\}$, for every $\sigma = \sigma^p$, and let $f\colon X * K \to Y$ be a ψ-map. Since $\gamma\Pi(\phi \mid K) = \Pi(\psi \mid K)$ it follows from (1.3) that we may take $f \mid K$ to be a $(\phi \mid K)$-map. Let $p \geqq 0$ and assume that $f \mid X^{p-1} * K$ is a $(\phi \mid X^{p-1} * K)$-map. If $\sigma = \sigma^p \in X$ let $v_\sigma\colon E^p \to X$ be a characteristic map for σ (i.e. $v_\sigma \mid E^p - S^{p-1}$ is a homeomorphism onto $\sigma^p - \dot{\sigma}^p$) and let

$$w_\sigma = v_\sigma * 1\colon \; E^p * K \to X * K.$$

Then $fw_\sigma(E^p * C) \subset \psi(\sigma^p * C) = \psi_\sigma C$ and $fw_\sigma(S^{p-1} * C) \subset \phi_\sigma C$ since $f \mid X^{p-1} * K$ is a $(\phi \mid X^{p-1} * K)$-map. Therefore fw_σ is a $(\psi_\sigma, \phi_\sigma)_{p+1}$-map, under the identifications $E^p * K = K * E^p = TS^pK$. Since

$$\Pi_{p+1}(\psi_\sigma, \phi_\sigma) = \{0\}$$

it follows from (1.4) that there is a $(\psi_\sigma, \phi_\sigma)_{p+1}$-homotopy, g_t, rel $S^{p-1} * K$ such that $g_0 = fw_\sigma$ and g_1 is a $(\phi_\sigma, \phi_\sigma)_{p+1}$-map. Obviously

$$w_\sigma \mid (E^p - S^{p-1}) * K$$

is a homeomorphism onto $(\sigma^p - \dot{\sigma}^p) * K$ and it follows ([9], p. 1131) that $g_t w_\sigma^{-1}$ is a homotopy $f_{\sigma,t}$: $\sigma^p * K \to Y$, rel $\dot{\sigma}^p * K$. Since g_t is a $(\psi_\sigma, \phi_\sigma)_{p+1}$-homotopy, and g_1 a $(\phi_\sigma, \phi_\sigma)_{p+1}$-map, $f_{\sigma,t}$ is a $(\psi \mid \sigma^p * K)$-homotopy and $f_{\sigma,1}$ a $(\phi \mid \sigma^p * K)$-map. Since $f_{\sigma,t}$ is rel $\dot{\sigma}^p * K$ a $(\psi \mid X^p * K)$-homotopy f_t', rel $X^{p-1} * K$, is defined by $f_t' \mid \sigma^p * K = f_{\sigma,t}$ for each $\sigma^p \in X$. By (1.3) this can be extended to a ψ-homotopy f_t^p.

It follows from the preceding paragraph and induction on p that there is a sequence of ψ-homotopies, $f_t^0, f_t^1, \ldots$, such that $f_1^p \mid X^p * K$ is a $(\phi \mid X^p * K)$-map, f_t^p is rel $X^{p-1} * K$ and $f_0^p = f_1^{p-1}$ $(f_1^{-1} = f)$. Let $a_p = p/p+1$, $(t,p) = (t - a_p)/(a_{p+1} - a_p)$ and define g_t: $X * K \to Y$ by

$$g_t x = f_{(t,p)}^p x \quad \textit{if} \quad a_p \leq t \leq a_{p+1}$$
$$g_1 x = f_1^p x \quad \textit{if} \quad x \in X^p * K.$$

Clearly g_t is single-valued and it is continuous since $g_t x = f_1^p x$ if $x \in X^p * K$ and $t \geq a_{p+1}$. Obviously g_t is a ψ-homotopy and

$$g_1(\sigma^p * C) \subset \phi(\sigma^p * C)$$

for every $\sigma^p \in X$, $C \subset K$. Therefore g_1 is a ϕ-map and the first part of (4.1) is proved.

Let $\gamma\Pi(\phi') = \Pi(\psi')$, $\Pi_{p+2}(\psi_\sigma, \phi_\sigma) = \{0\}$, for every $\sigma = \sigma^p$, and let χ: $(X * K) \times I \to Y$ be defined by

$$\chi(C' \times (0 \cup 1)) = \phi C', \quad \chi(C' \times I) = \psi C',$$

for every $C' \subset X * K$. Then a χ-map, f, is a ψ-homotopy between ϕ-maps and $f \mid K \times I$ is a ψ'-map. Since $\Pi(\psi') = \gamma\Pi(\phi')$ we may assume, in consequence of (1.3), that $f \mid K \times I$ is a ϕ'-map. Then $f \mid K \times I$ is ϕ'-homotopic to the map $(k, t) \to f(k, 0)$ $(k \in K)$, and it follows, again from (1.3), that we need only consider ϕ-maps, f_0, f_1, which are related by a ψ-homotopy f_t, rel K.

Let $P = (X \times I) * K$ and let $\xi \leq \eta$: $P \to Y$ be defined by

$$\xi((\sigma \times I) * C) = \eta((\sigma \times (0 \cup 1)) * C) = \phi_\sigma C,$$

and $\eta((\sigma \times I) * C) = \psi_\sigma C$, for every $\sigma \in X$, $C \subset K$. Since f_t is rel K a map $g: P \to Y$ is defined by

$$(4.2) \qquad g((x, t), k, s) = f_t(x, k, s).$$

Since f_t is a ψ-homotopy it follows that g is an η-map. Conversely, if g is a given η-map, then a ψ-homotopy f_t, rel K, is defined by (4.2) and f_t is a ϕ-homotopy if g is a ξ-map. The second part of (4.1) now follows from the proof of the first, with ϕ, ψ, f replaced by ξ, η, g.

Let $K = \varnothing$. Then $\Pi_{q+1}(\psi_\sigma, \phi_\sigma)$ consists of homotopy classes of the form $(E^q, S^{q-1}) \to (\psi\sigma, \phi\sigma)$, and we have:

Corollary (4.3). *If the pair $(\psi\sigma^p, \phi\sigma^p)$ is p-connected for every $\sigma^p \in K$ and every $p \geqq 0$, then $\gamma\Pi(\phi) = \Pi(\psi)$. If $(\psi\sigma^p, \phi\sigma^p)$ is $(p+1)$-connected for every $\sigma^p \in X$, then $\gamma: \Pi(\phi) \to \Pi(\psi)$ is a 1-1 correspondence.*

5. Equivalences

We now consider carriers $\phi: \mathfrak{a} \to \mathfrak{b}$, where $\mathfrak{a}, \mathfrak{b}$ are collections of subsets of X, Y. We confine ourselves to collections which are closed under the (infinite) intersection operator $\bigcap$. By an *isomorphism*, $\phi: \mathfrak{a} \approx \mathfrak{b}$, we mean a carrier which is 1-1 onto and commutes with $\bigcap$. If $\phi: \mathfrak{a} \approx \mathfrak{b}$, then $\phi^{-1} \bigcap \phi A = \bigcap A$ and it follows that $\phi^{-1}: \mathfrak{b} \approx \mathfrak{a}$. It may be verified that, if a carrier $\phi: \mathfrak{a} \to \mathfrak{b}$ is 1-1 onto and if the map ϕ^{-1} is also a carrier (i.e. is order preserving) then $\phi: \mathfrak{a} \approx \mathfrak{b}$.

Let X, Y be (non-vacuous) CW-complexes, let $\phi: \mathfrak{a} \approx \mathfrak{b}$ be both cellular and cocellular and let X, Y be the unions of the subcomplexes in $\mathfrak{a}$, $\mathfrak{b}$, respectively.

Theorem (5.1). *An element $\{f\} \in \Pi(\phi)$ is an equivalence in $\mathfrak{C}$ if, and only if, the map $f_A: A \to \phi A$, determined by $f \in \{f\}$, is a homotopy equivalence for every $A \in \mathfrak{a}$.*

Proof. Let $\{f\}$ be an equivalence and let $\{g\} = \{f\}^{-1} \in \Pi(\phi^{-1})$. Then f_A is obviously a homotopy equivalence having $g_{\phi A}: \phi A \to A$ as its homotopy inverse.

Conversely, let each f_A be a homotopy equivalence. Since ϕ is both cellular and cocellular we may assume that f is a cellular map, in consequence of (1.5). Let this be so and let Z be the mapping cylinder of f. We assume that X, Y are imbedded in Z in the usual way and denote points in Z by (x, t), y $(x \in X, y \in Y)$, with $(x, 0) = x$, $(x, 1) = fx$. The cells of Z are σ, τ, for all $\sigma \in X$, $\tau \in Y$, together with the cells containing points $(x, \frac{1}{2})$. If $x \in \sigma - \dot{\sigma}$ we denote the cell containing $(x, \frac{1}{2})$ by $\hat{\sigma}$. Then $\hat{\sigma}$ consists of the image of $\sigma \times I$ in the natural map $X \times I \to Z$, together with the smallest subcomplex of Y which contains $f\sigma$. Let $Z_A \subset Z$ be the mapping cylinder of f_A and let $\mathfrak{c}$ be the collection of sets

Z_A, for every $A \in \mathfrak{a}$. Notice that, if $\sigma \subset A$, then $f\sigma \subset fA \subset \phi A$, whence $\hat{\sigma} \subset Z_A$. Define κ: $\mathfrak{a} \to \mathfrak{c}$ by $\kappa A = Z_A$ and let $\iota_\mathfrak{c}$: $\mathfrak{c} \to \mathfrak{c}$ be the identity. Let ψ, χ: $Z \to Z$ be the extensions of κ^{-1}, $\iota_\mathfrak{c}$, which are defined as in § 1. Evidently each cell $\sigma \in X$ is contained in at least one $A \in \mathfrak{a}$ and each cell $\tau \in Y$ is contained in some $B \in \mathfrak{b}$. Let $A(\sigma)$, $B(\tau)$ be the smallest subcomplexes in $\mathfrak{a}$, $\mathfrak{b}$ which contain σ, τ and let $A(\tau) = \phi^{-1}B(\tau)$. Then $\psi\sigma = \psi\hat{\sigma} = A(\sigma)$, $\psi\tau = A(\tau)$, $\chi\sigma = \chi\hat{\sigma} = Z_{A(\sigma)}$, $\chi\tau = Z_{A(\tau)}$. Since f_A is a homotopy equivalence for each $A \in \mathfrak{a}$ it follows that the pair (Z_A, A) is q-connected for every $q \geqq 0$. Therefore γ: $\Pi(\psi) \to \Pi(\chi)$ is a 1-1 correspondence, by (4.3). Since the identical map $X \to X$ is a $(\psi \mid X)$-map it follows from (1.4) that the identical map $Z \to Z$ is $\iota_\mathfrak{c}$-homotopic, rel X, to a κ^{-1}-map g: $Z \to X$. Let k: $X \subset Z$, $\iota_\mathfrak{a}$: $\mathfrak{a} \subset \mathfrak{a}$. Then $\{gk, \iota_\mathfrak{a}\}$, $\{kg, \iota_\mathfrak{c}\}$ are identities in $\mathfrak{C}$ and it follows that $\{g, \kappa^{-1}\}$ is a two-sided inverse of $\{k, \kappa\}$. Therefore the latter is an equivalence in $\mathfrak{C}$ and so obviously is $\{l, \lambda\}$, where l: $Z \to Y$, λ: $\mathfrak{c} \to \mathfrak{b}$ are defined by $l(x,t) = fx$, $ly = y$, $\lambda Z_A = \phi A$. But $lk = f$, $\lambda\kappa = \phi$ and (5.1) is proved.

If $\{f, \phi\}$: $\mathfrak{a} \to b$ is an equivalence and if θ: $\mathfrak{a}' \to \mathfrak{a}$, ψ: $\mathfrak{b} \to \mathfrak{b}'$ are any carriers, then

$$f_*: \Pi(\theta) \to \Pi(\phi\theta), \quad f^*: \Pi(\psi) \to \Pi(\psi\phi)$$

are 1-1 correspondences, as is the case with an equivalence in any category. There are mappings in $\mathfrak{C}$ which, though not equivalences, yet have this property of inducing 1-1 correspondences by composition. We give an example of such a mapping to which we shall refer later on.

Let X be a CW-complex and X_0 a subcomplex of X. Let $\mathfrak{a}$ be a collection of the subcomplexes of TX such that $A \cap TX_0 = T(A \cap X_0)$ for every $A \in \mathfrak{a}$. Let $\mathfrak{a}_1$ consist of the subcomplexes $A \cup TX_0$ $(A \in \mathfrak{a})$, let ϕ_1: $\mathfrak{a} \to \mathfrak{a}_1$ be defined by $\phi_1 A = A \cup TX_0$ and let f_1: $TX \subset TX$. Then f_1 is a ϕ_1-map and f_{1*}: $\Pi(\theta) \to \Pi(\phi_1\theta)$ is the injection if θ: $\mathfrak{a}' \to \mathfrak{a}$. Since $A \cap TX_0 = T(A \cap X_0)$ if $A \in \mathfrak{a}$, it follows from (1.3) that the 'radial' contraction of TX_0 to $T\varnothing$ can be extended to an $\iota_\mathfrak{a}$-homotopy, r_t, such that $r_0 = 1$, $r_t TX_0 \subset TX_0$, $r_1 TX_0 = T\varnothing$. Therefore each $A \in \mathfrak{a}$ is a deformation retract of $A \cup TX_0$, and it follows from (4.3) that f_{1*} is a 1-1 correspondence.

If ψ: $\mathfrak{a}_1 \to \mathfrak{b}'$ and if g is a $\psi\phi_1$-map, then

$$gr_1(A \cup TX_0) \subset gA \subset \psi\phi_1 A = \psi(A \cup TX_0).$$

Therefore gr_1 is a ψ-map. Since r_t is an $\iota_\mathfrak{a}$-homotopy we have $\{gr_1, \psi\phi_1\} = \{g, \psi\phi_1\}$. Similarly $\{hr_1, \psi\} = \{h, \psi\}$ if h is a ψ-map. Hence it follows that f_1^*: $\Pi(\psi) \to \Pi(\psi\phi_1)$ is a 1-1 correspondence, whose inverse is the map $\{g, \psi\phi_1\} \to \{gr_1, \psi\}$.

Let Y be the CW-complex which is obtained from TX by shrinking TX_0 to a point and let $f\colon TX \to Y$ be the identification map. Let $\mathfrak{b}$ be the collection of subcomplexes $f(A \cup TX_0)$, for every $A \in \mathfrak{a}$, and let $\chi\colon \mathfrak{a}_1 \to b$ be the carrier defined by $\chi(A \cup TX_0) = f(A \cup TX_0)$. Then f is a χ-map and χ is evidently a 1-1 map† of $\mathfrak{a}_1$ onto $\mathfrak{b}$. Since TX_0 is contractible each map $A \cup TX_0 \to \chi(A \cup TX_0)$, which is determined by f, is a homotopy equivalence. Let $\mathfrak{a}_1$, $\mathfrak{a}$ and hence obviously $\mathfrak{b}$, be closed under the intersection operator. Then $\chi\colon \mathfrak{a}_1 \approx \mathfrak{b}$ and it follows from (5.1) that $\{f, \chi\}$ is an equivalence in $\mathfrak{C}$. Let $\phi = \chi\phi_1\colon \mathfrak{a} \to \mathfrak{b}$. Then we have proved:

LEMMA (5.2). *If* $\theta\colon \mathfrak{a}' \to \mathfrak{a}$, $\psi\colon \mathfrak{b} \to \mathfrak{b}'$, *are given carriers, then* $\{f, \phi\}$ *induces* 1-1 *correspondences* $\Pi(\theta) \to \Pi(\phi\theta)$, $\Pi(\psi) \to \Pi(\psi\phi)$.

6. Excision

Let X_1, $X_2 \subset X$, where X is any space, and define carriers $\phi, \phi', \psi\colon \varnothing \to X$ by $\phi(\varnothing) = X_1$, $\phi'(\varnothing) = X_2$, $\psi(\varnothing) = X$. We write

$$\Pi_{m+1}(\psi) = \Pi_m(X), \quad \Pi_{m+1}(\psi, \phi) = \Pi_m(X, X_1)$$

$$\Pi_{m+2}(\psi; \phi, \phi') = \Pi_{m+1}(X; X_1, X_2) \quad (m \geqq 0).$$

We describe X, or (X, X_1), as *q-connected* $(q \geqq 0)$, if, and only if,

$$\Pi_m(X) = \{0\}, \quad \text{or} \quad \Pi_m(X, X_1) = \{0\},$$

for every $m \leqq q$. We describe $(X; X_1, X_2)$ as *r-connected* $(r \geqq 1)$ if, and only if, $\Pi_n(X; X_1, X_2) = \{0\}$ for every $n \leqq r$. When we say that X is (-1)-connected, or that $(X; X_1, X_2)$ is 0-connected, we shall mean that $X \neq \varnothing$ or that each pair $(X_i, X_1 \cap X_2)$ is 0-connected.

We describe the triad $(X; X_1, X_2)$ as *excisive* if, and only if, *either* $X = \operatorname{Int} X_1 \cup \operatorname{Int} X_2$ *or* X_1, X_2 are closed, $X = X_1 \cup X_2$ and $X_1 \cap X_2$ is a strong deformation retract of some relatively open subset‡ $N \subset X_1$ (or $N \subset X_2$). An argument similar to the proof of (M) on p. 230 of [10] shows that the second condition is satisfied if X is a CW-complex and X_1, X_2 are subcomplexes of X.

LEMMA (6.1). *Let* $(X; X_1, X_2)$ *be an excisive triad, let* $A = X_1 \cap X_2$ *and let the pair* $(S^n X_i, S^n A)$ *be* q_i*-connected* $(i = 1, 2;\ n, q_i \geqq 0)$. *Then the triad* $(S^n X; S^n X_1, S^n X_2)$ *is* $(q_1 + q_2)$*-connected.*

Let $X_2' = X_2$ or $N \cup X_2$ according as $(X; X_1, X_2)$ satisfies the first

† χ is a map of the set $\mathfrak{a}_1$, not of $\mathfrak{a}_1$ indexed to $\mathfrak{a}$ by ϕ_1. In general $\chi\phi_1\colon \mathfrak{a} \to \mathfrak{b}$ is not 1-1.

‡ We consider this condition to be satisfied if $X_1 \cap X_2 = N = \varnothing$. In this case $(X_i, X_1 \cap X_2)$ is not 0-connected.

or only the second of the above conditions for an excisive triad. In the second case the injections

$$\Pi_q(S^n X_i, S^n A) \to \Pi_q(S^n X'_i, S^n N) \quad (X'_1 = X_1)$$

and
$$\Pi_q(S^n X; S^n X_1, S^n X_2) \to \Pi_q(S^n X; S^n X_1, S^n X'_2)$$

are 1-1 onto (cf. § 6 in [2]). Therefore we may replace X_2 by X'_2, and it is sufficient to prove (6.1) on the assumption that

$$X = \operatorname{Int} X_1 \cup \operatorname{Int} X_2.$$

We represent the points of $S^n X = X * S^{n-1}$ by (x, p, r), where $x \in X$, $p \in S^{n-1}$, $r \in I$ and $(x, p, 0) = x$, $(x, p, 1) = p$. Let $U \subset S^n X$ be the open subset consisting of the points (x, p, r) such that $\frac{1}{2} < r \leqq 1$ and let $Y_1 = S^n X_1$, $Y_2 = U \cup S^n X_2$. A homotopy f_t: $S^n X \to S^n X$ such that $f_0 = 1$ and

$$f_t S^n X_i = S^n X_i, \quad f_t Y_i \subset Y_i, \quad f_1 Y_2 = S^n X_2$$

is defined by $f_t(x, p, r) = (x, p, g_t r)$, where $g_t r = (1+t)r$ or $r + t - rt$ according as $r \leqq \frac{1}{2}$ or $r \geqq \frac{1}{2}$. Hence it follows that the injections

$$\Pi_q(S^n X_i, S^n A) \to \Pi_q(Y_i, Y_1 \cap Y_2)$$

and
$$\Pi_q(S^n X; S^n X_1, S^n X_2) \to \Pi_q(S^n X; Y_1, Y_2)$$

are 1-1 correspondences. Obviously $S^n X = \operatorname{Int} Y_1 \cup \operatorname{Int} Y_2$ and (6.1) follows from Theorem I in [2] (see (1.23) in [8] for the case $q_1 = 1$ or $q_2 = 1$. The case $q_1 = 0$ or $q_2 = 0$ presents no difficulty.)

Let X be a CW-complex, let ϕ, $\phi' \leqq \psi$: $X \to Y$ be cellular carriers and let $\theta = \phi \cap \phi'$. Let $\chi = (\phi, \theta)_1$, $\chi' = (\psi, \phi')_1$, and in (4.1) let K, ϕ, ψ be v, χ, χ', where $v = T \varnothing \in TX$. If $\sigma \in X$ we have

$$\Pi_q(\chi'_\sigma, \chi_\sigma) = \Pi_q(\psi\sigma; \phi\sigma, \phi'\sigma) \quad (q \geqq 1)$$

and $\chi v = \phi(\varnothing)$, $\chi' v = \psi(\varnothing)$. Therefore $\gamma\Pi(\chi \mid K) = \Pi(\chi' \mid K)$ if the pair $(\psi(\varnothing), \phi(\varnothing))$ is 0-connected and the condition $\gamma\Pi(\phi') = \Pi(\psi')$ in the second part of (4.1) is satisfied if $(\psi(\varnothing), \phi(\varnothing))$ is 1-connected. Therefore we have, in consequence of (4.1):

THEOREM (6.2). *The injection $\Pi_1(\phi, \theta) \to \Pi_1(\psi, \phi')$ is onto if the pair $(\psi(\varnothing), \phi(\varnothing))$ is 0-connected and $\Pi_{p+1}(\psi\sigma^p; \phi\sigma^p, \phi'\sigma^p) = \{0\}$, for every $\sigma^p \in X$ and every $p \geqq 0$. It is 1-1 into if $(\psi(\varnothing), \phi(\varnothing))$ is 1-connected and $\Pi_{p+2}(\psi\sigma^p; \phi\sigma^p, \phi'\sigma^p) = \{0\}$ for every $\sigma^p \in X$.*

If Z is any space and $n \geqq 1$, then $S^n Z$ is $(n-2)$-connected and is $(n-1)$-connected if $Z \neq \varnothing$. Let $C \subset Z$. Then $C = Z$ if $Z = \varnothing$, and it follows that the pair $(S^n Z, S^n C)$ is $(n-1)$-connected. Let $(\psi\sigma; \phi\sigma, \phi'\sigma)$ be an excisive triad for each $\sigma \in X$. Then it follows from (6.1) that $(S^n\psi\sigma; S^n\phi\sigma, S^n\phi'\sigma)$ is $(2n-2)$-connected. If $\sigma^p \in S^n X$, then $p+2 \leqq \dim X + n + 2$. Therefore we have, in consequence of (6.2):

COROLLARY (6.3). *If* $\dim X \leqq n-4$ $(3 \leqq n < \infty)$ *and* $(\psi\sigma; \phi\sigma, \phi'\sigma)$ *is an excisive triad for each* $\sigma \in X$, *then the injection*

$$\Pi^n_{n+1}(\phi, \theta) \to \Pi^n_{n+1}(\psi, \phi')$$

is a 1-1 *correspondence.*

7. Suspension

Let X be a CW-complex and ϕ: $X \to Y$ a cellular carrier. Let $SY = TY \cup T'Y$ and let $\hat{\phi}, \hat{\phi}'$: $X \to SY$ be defined by $\hat{\phi}A = T\phi A$, $\hat{\phi}'A = T'\phi A$, for each $A \subset X$. Then $\hat{\phi} \cap \hat{\phi}' = \phi$ and $\hat{\phi} \cup \hat{\phi}' = \phi^1_0$. We have $\Pi_1(\phi^1_0) = \Pi(S\phi)$ and

$$\beta\colon \Pi_1(\hat{\phi}, \phi) \to \Pi(\phi), \quad \alpha\colon \Pi(S\phi) \to \Pi_1(\phi^1_0, \hat{\phi}'),$$

as in (3.3). Obviously $\Pi_m(\hat{\phi}) = \{0\}$, $\Pi_m(\hat{\phi}') = \{0\}$, for every $m \geqq 0$, and it follows from (3.4) that α, β are 1-1 correspondences. The triad $(S\phi\sigma; T\phi\sigma, T'\phi\sigma)$ is obviously excisive for every $\sigma \in X$ and the pair $(S\phi(\varnothing), T\phi(\varnothing))$ is $(q+1)$-connected if $\phi(\varnothing)$ is q-connected $(q \geqq -1)$. Clearly $\alpha^{-1}\gamma\beta^{-1} = S_*$, where γ: $\Pi_1(\hat{\phi}, \phi) \to \Pi_1(\phi^1_0, \hat{\phi}')$ is the injection. Therefore we have, in consequence of (6.2):

THEOREM (7.1). S_*: $\Pi(\phi) \to \Pi(S\phi)$ *is onto if* $\phi(\varnothing) \neq \varnothing$ *and*

$$\Pi_{p+1}(S\phi\sigma^p; T\phi\sigma^p, T'\phi\sigma^p) = \{0\},$$

for every $\sigma^p \in X$ *and every* $p \geqq 0$. *It is* 1-1 *if* $\phi(\varnothing)$ *is* 0-*connected and* $\Pi_{p+2}(S\phi\sigma^p; T\phi\sigma^p, T'\phi\sigma^p) = \{0\}$.

Let $n_p = [\frac{1}{2}(p+1)]$ $(p \geqq -1)$. Then $2n_p \geqq p$ and $n_p + 1 \geqq n_{p+1}$. If $\phi\sigma^p$ is q-connected then each of the pairs $(T\phi\sigma^p, \phi\sigma^p)$, $(T'\phi\sigma^p, \phi\sigma^p)$ is $(q+1)$-connected and $(S\phi\sigma^p; T\phi\sigma^p, T'\phi\sigma^p)$ is $(2q+2)$-connected. Therefore we have:

COROLLARY (7.2). *If* $\phi\sigma^p$ *is* $(n_{p+1}-1)$-*connected for every* $\sigma^p \in X$ *and every* $p \geqq -1$, *then* S_*: $\Pi(\phi) \to \Pi(S\phi)$ *is onto. If each* $\phi\sigma^p$ *is* n_p-*connected, then* S_* *is a* 1-1 *correspondence.*

From (6.3) we have:

COROLLARY (7.3). *If* $\dim X \leqq n-4$, *then* S_*: $\Pi(S^n\phi) \to \Pi(S^{n+1}\phi)$ *is a* 1-1 *correspondence.*

Let $\phi(\varnothing) = \phi x_0 = y_0$, where x_0, y_0 are points in X, Y. Then the above theorems are obviously valid if S is replaced by the 'reduced suspension' of [7].

8. The track addition

Let X, Y be arbitrary spaces and $\phi: X \to Y$ a given carrier such that $\phi(\varnothing)$ is 1-connected. Let F_ϕ, O_ϕ be as in § 3. We may regard

$$S^m X = X * S^{m-1} \quad (m \geqq 1)$$

as the space obtained from $X \times E^m$ by the identifications $X \times p = p$ if $p \in S^{m-1}$. We thus have $\Pi_m(\phi) = \Pi_m(F_\phi, O_\phi)$, where the latter is defined in terms of the quasi-topology of F_ϕ. Since $\phi(\varnothing)$, and hence O_ϕ, are 1-connected $\Pi_m(F_\phi, O_\phi)$ may be given the structure of a homotopy group, absolute if $m = 1$, relative, and Abelian, if $m > 1$. This group, written additively, will be called a *track-group* [1] and the addition will be called the *track addition*. Evidently S_* maps track groups homomorphically and the part of (3.3) which terminates with $\Pi_1(\phi) \to \Pi_1(\psi)$ is a sequence of homomorphisms with respect to the track addition, assuming $\phi(\varnothing)$, $\psi(\varnothing)$ to be 1-connected. Notice that, if ϕ is arbitrary and $n \geqq 3$, or $n \geqq 2$ if $\phi(\varnothing) \neq \varnothing$, then $S^n\phi(\varnothing)$ is 1-connected and $\Pi(S^n\phi) = \Pi_n^n(\phi)$ is an Abelian track-group.

II. The suspension category

9. Basic definitions

Let $\phi: \mathfrak{a} \to \mathfrak{b}$ be any carrier, where $\mathfrak{a}$, $\mathfrak{b}$ are collections of subsets in X, Y. Then $S^n\phi: S^n\mathfrak{a} \to S^n\mathfrak{b}$, where $S^n\mathfrak{a}$, $S^n\mathfrak{b}$ consists of the sets S^nA, S^nB $(A \in \mathfrak{a}, B \in \mathfrak{b})$, and we form the sequence

$$\Sigma: \ \Pi(\phi) \xrightarrow{S_*} \dots \xrightarrow{S_*} \Pi(S^n\phi) \xrightarrow{S_*} \Pi(S^{n+1}\phi) \xrightarrow{S_*} \dots.$$

We define $\Sigma(\phi) = \varinjlim \Sigma$ and describe the elements of $\Sigma(\phi)$ as *$\phi - S$-maps*. Thus a $\phi - S$-map is an equivalence class of elements in the sets $\Pi(S^n\phi)$, elements $a \in \Pi(S^n\phi)$ and $b \in \Pi(S^p\phi)$ being equivalent if, and only if, $S_*^{q-n}a = S_*^{q-p}b$ for some $q \geqq n, p$. The injection $\Sigma(S^n\phi) \to \Sigma(\phi)$ is a 1-1 correspondence and $\Sigma(S^3\phi)$ is the direct limit of a sequence of homomorphisms of Abelian track-groups. We give $\Sigma(\phi)$ the corresponding group structure.

Let $\psi: \mathfrak{b} \to \mathfrak{c}$ be a carrier, where $\mathfrak{c}$ is a collection of subsets of a space Z. Let f be an $(S^m\phi)$-map and g an $(S^n\psi)$-map which represent given elements $a \in \Sigma(\phi)$, $b \in \Sigma(\psi)$. Then $(S^m g)(S^n f)$ is an $S^{m+n}(\psi\phi)$ map and therefore represents an element $ba \in \Sigma(\psi\phi)$. It is easily verified that ba does not depend on the choice of representative maps $f \in a$, $g \in b$; also that the map $(b, a) \to ba$ is a pairing (i.e. is bilinear) of $\Sigma(\psi)$, $\Sigma(\phi)$

to $\Sigma(\psi\phi)$. In particular, if $\mathfrak{a}=\mathfrak{b}$ and $\phi\phi=\phi$ then $\Sigma(\phi)$ is a ring which operates on $\Sigma(\psi\phi)$ from the right. Similarly, $\Sigma(\psi)$ is a ring of left operators on $\Sigma(\psi\phi)$ if $\mathfrak{b}=\mathfrak{c}$ and $\psi\psi=\psi$.

The totality of pairs (a,ϕ), where $a\in\Sigma(\phi)$, may evidently be taken as the mappings (a,ϕ): $\mathfrak{a}\to\mathfrak{b}$, in a category, $\mathfrak{C}_s$, whose objects are the collections $\mathfrak{a}$. The injection ι: $\Pi(\phi)\to\Sigma(\phi)$ defines a functor $\mathfrak{C}\to\mathfrak{C}_s$. Let X be a CW-complex and ϕ a cellular carrier such that $\phi\sigma^p$ is n_p-connected for every $\sigma^p\in X$. Then the same condition is obviously satisfied by $S^n\phi$ and it follows from (7.2) that ι is a 1-1 correspondence which may be used to define $\Pi(\phi)$ as a group. If $\dim X\leqq n-4$, then ι: $\Pi(S^n\phi)\approx\Sigma(\phi)$ in consequence of (7.3). If ϕ': $X\to Y$ is defined, as in § 1, in terms of ϕ: $\mathfrak{a}\to\mathfrak{b}$, then $S^n\phi'$ is similarly defined in terms of $S^n\phi$, and $\Sigma(\phi')=\Sigma(\phi)$.

If ϕ: $X\to Y$ is any carrier, X, Y being arbitrary spaces, and if m is any integer we define $\Sigma_m(\phi)=\Sigma(\phi^0_m)$ or $\Sigma(\phi^{-m}_0)$ according as $m\geqq 0$ or $m<0$. If ϕ, $\phi'\leqq\psi$: $X\to Y$ we define

$$\Sigma_m(\psi,\phi)=\Sigma_{m-1}((\psi,\phi)_1),\quad \Sigma_m(\psi;\phi,\phi')=\Sigma_{m-2}((\psi;\phi,\phi')_2).$$

On passing to the direct limit of the sequences $(3.3)^n$ under the maps S_*: $(3.3)^n\to(3.3)^{n+1}$ we have an exact sequence of homomorphisms

$$\text{(9.1)}\qquad \cdots\xrightarrow{\alpha}\Sigma_{m+1}(\psi,\phi)\xrightarrow{\beta}\Sigma_m(\phi)\xrightarrow{\gamma}\Sigma_m(\psi)\xrightarrow{\alpha}\cdots$$

which extends to infinity in both directions. We describe (9.1) as the *S-homotopy sequence* of the pair (ψ,ϕ). Similarly (3.7), (3.5) lead to sequences which we describe as the *S-homotopy sequence of the triple* $(\psi;\phi,\theta)$ and the *upper S-homotopy sequence* of the triad $(\psi;\phi,\phi')$. Similarly, we define the *lower S-homotopy sequence of* $(\psi;\phi,\phi')$. Since (3.3), (3.5), (3.7) and S_* are natural for $\mathfrak{C}$, it follows that the corresponding S-homotopy sequences are natural for $\mathfrak{C}_s$.

Let ψ: $X\to Y$ be a given carrier and let $\psi_\varnothing$: $X\to Y$ be the carrier $X\to\varnothing$. Obviously $\Sigma_p(\psi)=0$ if Y is a single point and $\psi A=Y$ for every $A\subset X$. Therefore in this case β: $\Sigma_p(\psi,\psi_\varnothing)\approx\Sigma_{p-1}(\psi_\varnothing)$. An element of $\Sigma_{-q}(\psi_\varnothing)$ is represented by a map of the form $S^{n-q}X\to S^{n-1}$. We write $\Sigma_{-q}(\psi_\varnothing)=\Sigma^{q-1}(X)$ and describe it as the $(q-1)^{\text{st}}$ *S-cohomotopy group* of X. Now let ψ be arbitrary. Since ψ may be replaced by $S\psi$ we assume, without loss of generality, that $\psi(\varnothing)\neq\varnothing$, and we define ψ_y: $X\to Y$ by $\psi_yX=y$, where $y\in\psi(\varnothing)$. On considering the injection $\Sigma_{p+1}(\psi_y,\psi_\varnothing)\to\Sigma_{p+1}(\psi,\psi_\varnothing)$ it follows from a standard argument that β: $\Sigma_{p+1}(\psi,\psi_\varnothing)\to\Sigma_p(\psi_\varnothing)$ has a right inverse and hence that

$$\text{(9.2)}\qquad \Sigma_{p+1}(\psi,\psi_\varnothing)\approx\Sigma_{p+1}(\psi)+\Sigma^{-p-1}(X),$$

where + indicates direct summation. Moreover, (9.2) is such that, if $\Sigma_{p+1}(\psi)$, $\Sigma_p(\psi\phi)$ are imbedded in $\Sigma_{p+1}(\psi, \psi\phi)$ by means of (9.2), then the S-homotopy sequence of a triple $(\psi, \phi, \psi\phi)$ $(\phi \leqq \psi)$ is obtained from that of the pair (ψ, ϕ) by adding the direct summand $\Sigma^{-p}(X)$ to each of $\Sigma_p(\phi)$, $\Sigma_p(\psi)$ and letting γ map $\Sigma^{-p}(X)$ identically.

Consider the category in which the objects are all pairs of carriers $\phi \leqq \psi: X \to Y$, for a given pair of spaces, X, Y, and in which the mappings, $(\phi, \psi) \to (\phi', \psi')$, are all pairs of relations $\phi \leqq \phi'$, $\psi \leqq \psi'$. This category, and the groups $\Sigma_q(\psi, \phi)$, evidently satisfy the first five axioms for homology theory, with appropriate modifications, which are given on pp. 10, 11 of [3]. Moreover, it follows from (6.3) that, if X, Y are CW-complexes, if $\dim X < \infty$ and if we confine ourselves to carriers which are both cellular and cocellular, then the excision axiom, in the form

$$\gamma\colon \Sigma_q(\phi, \phi \cap \phi') \approx \Sigma_q(\phi \cup \phi', \phi'),$$

is satisfied for any pair of carriers $\phi, \phi'\colon X \to Y$. Notice that the axioms refer to pairs of spaces, or carriers in our case. Therefore it is the S-homotopy sequence of a triple $(\psi, \phi, \psi\phi)$, not a pair† (ψ, ϕ), which corresponds to the homology sequence of a pair of spaces.

Let X, Y be CW-complexes, let $\dim X < \infty$ and let $\phi, \phi'\colon X \to Y$ be carriers which are both cellular and cocellular. Let $\theta = \phi \cap \phi'$, $\psi = \phi \cup \phi'$ and let

$$\Sigma_p(\phi, \theta) \xrightarrow{\gamma} \Sigma_p(\psi, \theta) \xleftarrow{\gamma'} \Sigma_p(\phi', \theta)$$

be the injections. Then it follows from (14.2) in [3] and the preceding paragraph that

$$\gamma''\colon \Sigma_p(\phi, \theta) + \Sigma_p(\phi', \theta) \approx \Sigma_p(\psi, \theta), \tag{9.3}$$

where $\gamma''(a, a') = \gamma a + \gamma' a'$.

The axiomatic formulation of homology theory can be similarly applied to theorems concerning groups of the form $\Sigma_p(\psi, \phi)$, where $\phi \leqq \psi\colon X \to Y$ for a fixed X and variable Y. If Y is fixed and X variable, then it is the axioms for cohomology which are relevant.

10. Pairing by composition

Let $\phi \leqq \psi\colon X \to Y$, $\mu \leqq \nu\colon Y \to Z$. We have observed that $\Sigma(\phi)$, $\Sigma(\mu)$ are paired by composition to $\Sigma(\mu\phi)$. We proceed to define a pairing

$$P\colon \Sigma_q(\nu, \mu) \times \Sigma_p(\psi, \phi) \to \Sigma_{p+q}(\nu\psi; \mu\psi, \nu\phi). \tag{10.1}$$

† Let X be a CW-complex and $\phi \leqq \psi\colon X \to TY$ cellular carriers, with values in Y. Then the S-homotopy sequence of (ψ, ϕ) may be recovered as the sequence of the triple (χ, ψ, ϕ), where $\chi A = TY$ for every $A \subset X$.

We identify $\chi: X \to Y$ with the carrier $\chi: X \to TY$, which has the same values, and we define $\hat{\chi}: X \to TY$ by $\hat{\chi}A = T\chi A$, as in § 7. Elements of $\Sigma_{p+1}(\hat{\psi}; \psi, \hat{\phi})$ and $\Sigma_q(\nu, \mu)$ are represented by maps of the form

$$T'TS^{n+p+q-2}X \to TS^{n+q-1}Y \to S^nZ$$

for sufficiently large values of n. Also

$$(\nu, \mu)_1 (\hat{\psi}; \psi, \hat{\phi})_2 = (\nu\hat{\psi}; \mu\psi, \nu\hat{\phi})_2.$$

Therefore a pairing

$$\Sigma_q(\nu, \mu) \times \Sigma_{p+1}(\hat{\psi}; \psi, \hat{\phi}) \to \Sigma_{p+q}(\nu\hat{\psi}; \mu\psi, \nu\hat{\phi}) \tag{10.2}$$

is defined by composition. Obviously $\Sigma_{p+1}(\hat{\psi}, \hat{\phi}) = 0$, and it follows that, in the upper S-homotopy sequence of $(\hat{\psi}; \psi, \hat{\phi})$,

$$\beta: \Sigma_{p+1}(\hat{\psi}; \psi, \hat{\phi}) \approx \Sigma_p(\psi, \phi).$$

We define P, in (10.1), by $P(b, a) = b \circ \rho a$, where $\rho = \beta^{-1}$ and $\circ$ indicates composition.

Let $\theta \leqq \phi: X \to Y$, $\lambda \leqq \mu: Y \to Z$ and let δ denote the 'boundary' operator in the S-homotopy sequence of each triple (ψ, ϕ, θ), (ν, μ, λ) (cf. (3.7)). Then

$$P(b, \delta a) \in \Sigma_{p+q-1}(\nu\phi; \mu\phi, \nu\theta)$$
$$P(\delta b, a) \in \Sigma_{p+q-1}(\mu\psi; \lambda\psi, \mu\phi).$$

Let $\chi_1 = \mu\phi \cup \lambda\psi$, $\chi_2 = \nu\theta \cup \mu\phi$ and let

$$\iota: \Sigma_{p+q-1}(\nu\phi; \mu\phi, \nu\theta) \to \Sigma_{p+q-1}(\nu\psi; \chi_1, \chi_2),$$
$$\kappa: \Sigma_{p+q-1}(\mu\psi; \lambda\psi, \mu\phi) \to \Sigma_{p+q-1}(\nu\psi; \chi_1, \chi_2)$$

be the injections. Then we say that

$$\iota P(b, \delta a) = \kappa P(\delta b, a). \tag{10.3}$$

For consider the diagram

$$\begin{array}{ccccc} \Sigma_{p+1}(\hat{\psi}; \psi, \hat{\phi}) & \xrightarrow{\beta'} & \Sigma_p(\hat{\phi}, \phi) & \xrightarrow{\alpha'} & \Sigma_p(\hat{\phi}; \phi, \hat{\theta}) \\ \downarrow \beta & & \downarrow \beta & & \downarrow \beta \\ \Sigma_p(\psi, \phi) & \xrightarrow{\beta} & \Sigma_{p-1}(\phi) & \xrightarrow{\alpha} & \Sigma_{p-1}(\phi, \theta), \end{array} \tag{10.4}$$

in which β', α' are taken from the lower S-homotopy sequences of $(\hat{\psi}: \psi, \hat{\phi})$, $(\hat{\phi}; \phi, \hat{\theta})$. Since (3.8) is commutative so is the left-hand square of (10.4). Under the identification $T'T = TS$ a given element

$$c \in \Sigma_p(\hat{\phi}, \phi) = \Sigma_{p-1}((\hat{\phi}, \theta)_1, (\phi, \theta)_1)$$

is represented by a map $f: T'TS^{n+p-2}X \to S^nY$ and $\alpha\beta c$, $\beta\alpha' c$ are both represented by $f \mid TS^{n+p-2}X$. Therefore (10.4) is commutative and (10.3) is equivalent to

$$(10.5) \qquad \iota(b \circ \delta' \rho a) = \kappa(\delta b \circ \rho a) \quad (\delta' = \alpha'\beta').$$

Let $a' \in \Sigma_p(\hat{\phi}, \phi)$ be given. Evidently $(\nu, \mu)_1 (\hat{\phi}, \phi)_1 = (\nu\phi, \mu\phi)_1$ and an element $b \circ a' \in \Sigma_{p+q-1}(\nu\phi, \mu\phi)$ is defined by composition. Let

$$\alpha'': \Sigma_{p+q-1}(\nu\phi, \mu\phi) \to \Sigma_{p+q-1}(\nu\phi; \mu\phi, \nu\theta)$$

be the homomorphism which appears in the lower S-homotopy sequence of $(\nu\phi; \mu\phi, \nu\theta)$. Both α'' and α' are defined by composing maps of STS^nX, for sufficiently large values of n, with

$$u: TX' \to SX' \quad (u(x', t) = (x', 2t-1),\ x' \in X' = TS^nX).$$

Since the composition of maps is associative it follows that

$$(10.6) \qquad b \circ \alpha' a' = \alpha''(b \circ a').$$

Therefore it follows from (3.6) that

$$(10.7) \qquad \iota(b \circ \delta' \rho a) = \iota\alpha''(b \circ \beta' \rho a) = \gamma(b \circ \beta' \rho a),$$

where $\gamma: \Sigma_{p+q-1}(\nu\phi, \mu\phi) \to \Sigma_{p+q-1}(\nu\psi; \chi_1, \chi_2)$ is the injection.

Let $\qquad X' = S^{m+p-1}X, \quad Y' = S^mY, \quad Z' = S^{m-q+1}Z,$

where m is so large that a is represented by a map $f: TX' \to Y'$. We take $TX' = X' * 0 \subset X' * I$ and write $X' * I = T_1TX'$ $(T_1X' = X' * 1)$. Then we have $T_1f: T_1TX' \to TY'$ and we define a $(\hat{\psi}, \phi)^m_{m+p}$-homotopy $f_s: TX' \to TY'$ by $f_s(x', t) = (T_1f)(x', s, t)$, where $(x', s, t) \in x' * s$ $(x' \in X',\ s \in I)$. The map f_0 has the same values as f and $f_1 = T_1(f \mid X')$, which represents $\beta' \rho a$. Since $\lambda \leq \mu$ we have

$$\alpha: \Sigma_q(\nu, \mu) \approx \Sigma_q(\nu; \lambda, \mu),$$

and, if m is large enough,

$$T'TX' \xrightarrow{T'f_s} T'TY' \xrightarrow{g} SZ',$$

where g represents αb and also b, in consequence of (3.6), for upper sequences. Then $g \mid T'Y'$ represents δb. Since f_s is a $(\hat{\psi}, \phi)^m_{m+p}$-homotopy we have, for every $A' \subset X'$,

$$(gT'f_s)\,TA' = gf_sTA' \subset S^{m-q+2}\lambda\psi A',$$

$$(gT'f_s)\,T'A' = gT'f_sA' \subset S^{m-q+2}\mu\phi A',$$

and $(gT'f_s)\,T'TA' \subset S^{m-q+2}\nu\psi A'$. Therefore $gT'f_s$ is a $(\nu\psi; \chi_1, \chi_2)^{m-q+2}_{m+p+1}$-homotopy. The map $T'f_0$ represents ρa and has its values in $T'Y'$. Therefore $gT'f_0 = (g \mid T'Y')\,T'f_0$, which represents $\delta b \circ \rho a$. Therefore $\kappa(\delta b \circ \rho a)$ is also represented by $gT'f_1 = gT'T_1(f \mid X')$. Under the identification $T'T_1 = ST_0$, say, the map $T'T_1(f \mid X')$ represents $\beta'\rho a$. Therefore $gT'f_1$ represents $b \circ \beta'\rho a$ and hence also $\iota(b \circ \delta'\rho a)$, by (10.7). This proves (10.5) and (10.3).

Let $\xi: X' \to X$, $\eta: Y' \to Y$, $\zeta: Z \to Z'$ be given carriers and left f, g, h be ξ-, η-, ζ-maps. Since the pairing (10.2) is by composition it follows that

$$(10.8)\qquad \begin{cases} \text{(a)} & P(b, f^*a) = f^*P(b,a), \\ \text{(b)} & P(b, g_*a') = P(g^*b, a'), \\ \text{(c)} & P(h_*b, a) = h_*P(b,a), \end{cases}$$

where each f^*, g_*, g^*, h_* denotes the appropriate homomorphism induced by f, g, h.

Let $\chi \leqq \chi': X \to Z$ be such that $\nu\psi \leqq \chi'$, $\chi_1, \chi_2 \leqq \chi$ and let

$$\gamma: \Sigma_r(\nu\psi; \chi_1, \chi_2) \to \Sigma_r(\chi'; \chi, \chi) = \Sigma_r(\chi', \chi)$$

be the injection. Then the pairing

$$(10.9)\qquad \gamma P: \Sigma_q(\nu, \mu) \times \Sigma_p(\psi, \phi) \to \Sigma_{p+q}(\chi', \chi)$$

may be defined in terms of the composition pairing

$$(10.10)\qquad \Sigma_q(\nu, \mu) \times \Sigma_{p+1}(\hat{\psi}, \psi \cup \hat{\phi}) \to \Sigma_{p+q}(\chi', \chi),$$

and the composite of ρ and the injection

$$\gamma': \Sigma_{p+1}(\hat{\psi}; \psi, \hat{\phi}) \to \Sigma_{p+1}(\hat{\psi}, \psi \cup \hat{\phi})$$

instead of by (10.2). We have a commutative diagram

$$\begin{array}{ccc} \Sigma_{p+1}(\hat{\psi}; \psi, \hat{\phi}) & \xrightarrow{\gamma'} & \Sigma_{p+1}(\hat{\psi}, \psi \cup \hat{\phi}) \\ \downarrow \beta & & \downarrow \beta' \\ \Sigma_p(\psi, \phi) & \xrightarrow{\gamma''} & \Sigma_p(\psi \cup \hat{\phi}, \hat{\phi}) \end{array}$$

where γ'' is an excision since $\psi \cap \hat{\phi} = \phi$. Since $\Sigma_r(\hat{\psi}, \hat{\phi}) = 0$ it follows that β' is an isomorphism onto. Hence, if γ'' is an isomorphism, so is γ'.

11. Exact couples of carriers

Let $\theta_i: X \to Y$ be a sequence of carriers, indexed by the integers, such that $\theta_i \leqq \theta_{i+1}$ for every i. Let

$$A_{p,q} = \Sigma_{p+q+1}(\theta_p), \quad C_{p,q} = \Sigma_{p+q+1}(\theta_p, \theta_{p-1}),$$

and let A, C be the weak direct sums of the sets of groups $A_{p,q}$, $C_{p,q}$ respectively. Let

$$A_{p,q} \xrightarrow{\alpha} C_{p,q} \xrightarrow{\beta} A_{p-1,q} \xrightarrow{\gamma} A_{p,q-1}$$

be as in (9.1). Then we have the exact couples [5]

(11.1)
$$\begin{array}{ccc} A^{(n)} & \xleftarrow{\gamma^{(n)}} & A^{(n)} \\ & \alpha^{(n)} \searrow \quad \nearrow \beta^{(n)} & \\ & C^{(n)} & \end{array} \quad (n=0,1,\ldots),$$

where $A^{(0)}=A$, $C^{(0)}=C$, $\alpha^{(0)}, \beta^{(0)}, \gamma^{(0)}$ are the homomorphisms determined by α, β, γ and $\langle A^{(n+1)}, C^{(n+1)}\rangle$ is the derived couple of $\langle A^{(n)}, B^{(n)}\rangle$. We have

$$A^{(n)}_{p,q}=\gamma^{(n-1)}A^{(n-1)}_{p-1,q+1}=\gamma^n A_{p-n,q+n} \quad (n\geqq 1)$$

and $A_{r,-r-1}=\Sigma_0(\theta_r)=\Sigma(\theta_r)$. Therefore a given element $a\in\Sigma(\theta_p)$ is 'compressible' to $\Sigma(\theta_{p-n})$ if, and only if, $a\in A^{(n)}$. This being so, $a\in A^{(n+1)}$ if, and only if, $\alpha^{(n)}a=0$. Therefore $\alpha^{(n)}a$ $(n\geqq 0)$ may be defined as the $(n+1)^{\text{st}}$ *obstruction* to compressing a into $\Sigma(\theta_{p-n-1})$, assuming that the preceding obstructions all vanish.

We shall also consider decreasing sequences of carriers $\theta^i\colon X\to Y$ (i.e. $\theta^{i+1}\leqq\theta^i$). The exact couple of such a sequence will consist of the groups

$$C^p_q=\Sigma_{q-p}(\theta^{p-1},\theta^p), \quad A^p_q=\Sigma_{q-p-1}(\theta^p)$$

and the homomorphisms

$$A^{p-1}_q \xrightarrow{\alpha} C^p_q \xrightarrow{\beta} A^p_q \xrightarrow{\gamma} A^{p-1}_{q-1}.$$

We denote the corresponding groups in the nth derived couple by ${}^{(n)}C^p_q$, ${}^{(n)}A^p_q$.

By a *filtration* of a space X we mean a collection of subsets $X_k\subset X$, indexed by the integers, such that $X_k\subset X_{k+1}$, $\bigcup X_k=X$. Let X, Y be filtered spaces and define $\xi_p, \xi^p\colon X\to Y$ by $\xi_p X=Y_p$, $\xi^p X_p=\varnothing$. If $\phi\leqq\psi\colon X\to Y$, we define

(11.2)
$$\theta_p=\phi\cup(\psi\cap\xi_p), \quad \theta^p=\phi\cup(\psi\cap\xi^p).$$

Thus $\theta_p A=\phi A\cup(\psi A\cap Y_p)$, $\theta^p A=\phi A$ or ψA according as $A\subset X_p$ or $A\not\subset X_p$. Notice that θ_p is defined by (11.2) for an unfiltered X and θ^p for an unfiltered Y. If Y is a CW-complex, filtered by its skeletons, we shall use $C_{p,q}(\psi,\phi)$, $C^{(1)}_{p,q}(\psi,\phi)$, etc., to denote the groups in the exact couple of $\{\theta_p\}$ and its derived couple. If X is a CW-complex, filtered by its skeletons, then $C^p_q(\psi,\phi)$, ${}^{(1)}C^p_q(\psi,\phi)$ will have analogous meanings with respect to $\{\theta^p\}$.

12. S-homotopy and S-cohomotopy couples

If X is any space and $A\subset X$ we define $\chi_A\colon\varnothing\to X$ by $\chi_A\varnothing=A$. Then $\chi_A\leqq\chi_X$ and we write

$$\Sigma_{n+1}(\chi_A)=\Sigma_n(A), \quad \Sigma_{n+1}(\chi_X,\chi_A)=\Sigma_n(X,A).$$

We describe $\Sigma_n(A)$, $\Sigma_n(X,A)$ as the nth *S-homotopy groups* of A and of the pair (X,A). We write $\Sigma_{n-1}(\varnothing)=\Lambda_n \approx \Pi_{n+k}(S^k)$ if $k \geqq n+2$.

We define $\chi^A\colon X \to T\varnothing$ by $\chi^A A=\varnothing$. Then $\chi^X \leqq \chi^A$, and we define

$$\Sigma^n(X,A)=\Sigma_{-n}(\chi^A,\chi^X).$$

Thus an element of $\Sigma^n(X,A)$ is represented by a map of the form

$$(TS^{m-n-1}X, S^{m-n-1}(X \cup TA)) \to (E^m, S^{m-1}).$$

It may be verified that, if (X,A) has the homotopy extension property with respect to maps $X \to I$ and homotopies $A \to I$, then $\Sigma^n(X,A)$ may be identified with $\Sigma^n(X \cup TA)$, as defined in §9; likewise $\Sigma_{-n-1}(\chi^A)$ with $\Sigma^n(A)$. We describe $\Sigma^n(X,A)$ as the nth *S-cohomotopy group* of the pair (X,A). By the *S-homotopy sequence* and the *S-cohomotopy sequence* of the pair (X,A) we shall mean the corresponding sequences for (χ_X,χ_A) and (χ^A,χ^X). A similar remark applies to triples.

Let Y be CW-complex, filtered by its skeletons, and let θ_p be as in (11.2) with $X=\varnothing$, $\phi=\chi_\Phi$, $\psi=\chi_Y$. Then we describe the exact couple of $\{\theta_p\}$ as the *S-homotopy couple* of Y.

Let X be a CW-complex, filtered by its skeletons, and let θ^p be as in (11.2) with $Y=T\varnothing$, $\phi=\chi^X$, $\psi=\chi^\Phi$. Then we describe the exact couple of $\{\theta^p\}$ as the *S-cohomotopy couple* of X. We use $C_{p,q}(Y)$, $C^{(1)}_{p,q}(Y)$, $C^p_q(X)$, etc., to denote $C_{p,q}(\chi_Y,\chi_\Phi)$, $C^{(1)}_{p,q}(\chi_Y,\chi_\Phi)$, $C^p_q(\chi^\Phi,\chi^X)$, etc. If $\dim X<\infty$ there are natural isomorphisms ([5], II, §14; III, §3)

$$(12.1) \qquad C^{(1)}_{p,q}(Y)\approx H_p(Y;\Lambda_q), \quad {}^{(1)}C^p_q(X)\approx H^p(X;\Lambda_q).$$

The purpose of the two final sections is to generalize (12.1).

13. Generalized homology and cohomology

By a *covariant stack* (faisceau, [4]) in a space X we mean a covariant functor, G, from the category, $\mathfrak{X}$, of subsets of X and inclusion maps to the category, $\mathfrak{A}$, of Abelian groups. Thus G assigns a group, $G(A)$, to each $A \subset X$ and a homomorphism $G(i)\colon G(A) \to G(A')$ to every $i\colon A \subset A'$ $(A \subset A' \subset X)$, subject to the functorial conditions on identities and composites. Similarly a *contravariant stack* on X is a contravariant functor between the same categories. Thus

$$G(i)\colon G(A') \to G(A)$$

if G is contravariant and $i\colon A \subset A'$. Let $\mathfrak{X}_\mathfrak{a} \subset X$ be the subcategory consisting of the sets in a collection $\mathfrak{a}$, which contains X, and inclusion maps between them. Let $G_\mathfrak{a}\colon \mathfrak{X}_\mathfrak{a} \to \mathfrak{A}$ be a covariant functor and, for every $A \subset X$, let $G(A)$ be the inverse limit of the system $\{G_{\mathfrak{a}}(A'), G_{\mathfrak{a}}(i)\}$, for every $A' \in \mathfrak{a}$ and $i: A' \subset A''$ such that $A \subset A' \subset A''$ $(A'' \in \mathfrak{a})$. If $i: A_1 \subset A_2$ let $G(i)$ be the projection $G(A_1) \to G(A_2)$. Then G is a covariant stack which coincides with $G_{\mathfrak{a}}$, up to natural equivalence, in $\mathfrak{X}_{\mathfrak{a}}$. Similarly a contravariant functor $G_{\mathfrak{a}}: \mathfrak{X}_{\mathfrak{a}} \to \mathfrak{A}$ can be extended to a contravariant stack, G. In this case $G(A)$ is the direct limit of the system $\{G_{\mathfrak{a}}(A'), G_{\mathfrak{a}}(i)\}$, for every $A' \in \mathfrak{a}$ such that $A \subset A'$. In either case we say that G is *generated* by $\mathfrak{a}$. In particular a group G determines the stack, also denoted by G, which is generated by $\mathfrak{a} = (X)$, with $G(X) = G$. We have $G(A) = G$, $G(i) = 1$, for every $A \subset X$ and every $i: A \subset A'$, and G may be regarded as both covariant and contravariant.

By a homomorphism $\eta: G_1 \to G_2$, where G_1, G_2 are both covariant or both contravariant stacks, we mean a natural transformation. Thus if $i: A \subset A'$ and if G_1, G_2 are covariant we have a commutative diagram

$$\begin{array}{ccc} G_1(A) & \xrightarrow{\eta(A)} & G_2(A) \\ {\scriptstyle G_1(i)}\downarrow & & \downarrow{\scriptstyle G_2(i)} \\ G_1(A') & \xrightarrow{\eta(A')} & G_2(A') \end{array}$$

If G_1, G_2 are contravariant we have a similar diagram with the vertical arrows pointing upwards. The totality of homomorphisms $G_1 \to G_2$, with addition defined in the usual way, is obviously a group. We denote it by $\operatorname{Hom}(G_1, G_2)$.

Let X be a CW-complex. Then the groups $C_p(A) = \Sigma_p(A^p, A^{p-1})$, for every subcomplex $A \subset X$, together with the injections

$$C_p(A) \to C_p(A') \quad (A \subset A'),$$

are the values of a covariant stack, C_p, which is generated by the subcomplexes of X. If G is any covariant stack we write

$$\operatorname{Hom}(C_p, G) = C^p(X; G)$$

and describe it as the group of *$p - G$-cochains* in X. A homomorphism $d: C_{p+1} \to C_p$ is defined by taking $d(A)$ to be the boundary homomorphism in the S-homotopy sequence of the triple (A^{p+1}, A^p, A^{p-1}). We define $\delta: C^p(X; G) \to C^{p+1}(X; G)$ by $(\delta\eta)(A) = \eta(A)d(A)$ if $\eta \in C^p(X; G)$. Obviously $\delta\delta = 0$, and we define the *$p - G$-cohomology group* as

$$H^p(X; G) = \delta^{-1}(0)/\delta C^{p-1}(X; G).$$

The injections $C_p(A) \to C_p(A')$ $(A \subset A')$ are isomorphisms into and $C_p(A)$ may be identified with the corresponding subgroup of $C_p(X)$.

The latter is free Abelian and is freely generated by generators of the (cyclic infinite) groups $C_p(\sigma^p)$ for every $\sigma^p \in X$. Hence it follows that an element $\eta \in C^p(X; G)$ is determined by its values on the p-cells of X. If G is a group (i.e. if the stack G is generated by X and $G(X) = G$) then $H^p(X; G)$ is the ordinary $p - G$-cohomology group of X.

If A is any subcomplex of X we write $\bar{A}^p = A \cup X^p$. Then a contravariant stack, C^p, is defined by $C^p(A) = \Sigma^p(\bar{A}^p, \bar{A}^{p-1})$, the homomorphism $C^p(i)$: $C^p(A_2) \to C^p(A_1)$ being the one induced by

$$i'\colon (\bar{A}_1^p, \bar{A}_1^{p-1}) \subset (\bar{A}_2^p, \bar{A}_2^{p-1}) \quad \text{if} \quad i\colon A_1 \subset A_2.$$

If G is any contravariant stack we write

$$\operatorname{Hom}(C^p, G) = C_p(X; G).$$

A homomorphism d: $C^p \to C^{p+1}$ is defined by taking $d(A)$ to be coboundary homomorphism in the S-cohomotopy sequence of the triple $(\bar{A}^{p+1}, \bar{A}^p, \bar{A}^{p-1})$. We define ∂: $C_{p+1}(X; G) \to C_p(X; G)$ by

$$(\partial\eta)(A) = \eta(A)\, d(A) \quad \text{if} \quad \eta \in C_{p+1}(X; G).$$

We have $\partial\partial = 0$ and we define

$$\bar{H}_p(X; G) = \partial^{-1}(0)/\partial C_{p+1}(X; G).$$

If i: $A \subset A'$, then $C^p(i)$: $C^p(A') \to C^p(A)$ is 1-1 into. Thus $C^p(A)$ may be imbedded in $C^p(\varnothing) = \Sigma^p(X^p, X^{p-1})$. The latter is the strong direct sum of the (cyclic infinite) groups $\Sigma^p(\sigma^p, \dot{\sigma}^p)$ for every $\sigma^p \in X$. If X^p has but a finite number of p-cells, then $C^p(\varnothing)$ is free Abelian and the value of an element $\eta \in C_p(X; G)$ is determined by its values on the generators of the groups $\Sigma^p(\sigma^p, \dot{\sigma}^p)$. In this case we write

$$\bar{H}_p(X; G) = H_p(X; G)$$

and describe it as the $p - G$-*homology group* of X. If G is a group, then $H_p(X; G)$ is the ordinary homology group of X with coefficients in G.

Notice that the relative homology or cohomology theory of (X, A), where A is a sub-complex of X, is provided for by the condition $G(A') = 0$ if $A' \subset A$.

Let X^p contain but a finite number of p-cells and for each $\sigma = \sigma^p \in X$ let A_σ be the largest subcomplex of X which does not contain σ. Then $C^p(A_\sigma) \approx \Sigma^p(\sigma^p, \dot{\sigma}^p)$. Let c_σ be a generator of $C^p(A_\sigma)$. We say that

$$k\colon C_p(X; G) \approx \sum G(A_\sigma), \tag{13.1}$$

where $\sum$ indicates direct summation over the range of p-cells of X and k is defined by $k\eta = \{\eta(A_\sigma)\, c_\sigma\}$ if $\eta \in C_p(X; G)$. For $k^{-1}(0) = 0$, since η is determined by the set of elements $\eta(A_\sigma)c_\sigma$. Let $g_\sigma \in G(A_\sigma)$ be given, for each $\sigma = \sigma^p \in X$. Then an element $\eta \in C_p(X; G)$ is defined by

$$\eta(A)\,C^p(i_\sigma)\,c_\sigma = G(i_\sigma)\,g_\sigma,$$

where i_σ: $A \subset A_\sigma$, for every subcomplex $A \subset X$ and every $\sigma^p \not\subset A$. Obviously $k\eta = \{g_\sigma\}$ and (13.1) is proved. A similar result holds for $C^p(X; G)$.

14. The group ${}^{(1)}C_q^p(\psi, \phi)$

Let X be a CW-complex, let $\phi \leqq \psi$: $X \to Y$ be cellular carriers and let θ^p be defined by (11.2), with $X_p = X^p$. For every $A \subset X$ let

$$G_q(A) = \Sigma_{q+1}(\chi_{\psi A}, \chi_{\phi A}) = \Sigma_q(\psi A, \phi A)$$

and let $G_q(i)$: $G_q(A) \to G_q(A')$ be the injection if i: $A \subset A'$. Then G_q is a covariant stack. Let A be a subcomplex. Then $\theta^n\chi_{A^m} \leqq \chi_{\psi A}$ and $\theta^n\chi_{A^m} \leqq \chi_{\phi A}$ if $m \leqq n$. Therefore a pairing

$$C_q^p(\psi, \phi) \times C_p(A) \to G_q(A) \tag{14.1}$$

is defined as in (10.9). This pairing determines a homomorphism

$$h(A)\colon\ C_q^p(\psi, \phi) \to \operatorname{Hom}(C_p(A), G_q(A))$$

and it follows from (10.8a) that, if $c \in C_q^p(\psi, \phi)$, then

$$h(A)\,c\colon\ C_p(A) \to G_q(A)$$

is natural with respect to the injections. Therefore (14.1) determines a homomorphism h: $C_q^p(\psi, \phi) \to C^p(X; G_q)$. It follows from (10.3) that

$$h\delta = \delta h\colon\ C_q^p(\psi, \phi) \to C^{p+1}(X; G_q), \tag{14.2}$$

and we say that

$$h\colon\ C_q^p(\psi, \phi) \approx C^p(X; G_q). \tag{14.3}$$

For, if $X' = S^nX$, $X_0' = S^nX^p$ it follows from (1.2), with $(X, A) = (X', X_0')$ and $X' \times 1$ pinched to a point, that there is a homotopy h_t': $TX' \to TX'$, rel TX_0', such that $h_0' = 1$, $h_1'TX' = TX_0'$ and $h_t'T\sigma' \subset T\sigma'$ for every $\sigma' \in S^nX$. Since $\theta^pA = \theta^{p-1}A$ if $A \not\subset X^p$ it follows that

$$i^*\colon\ C_q^p(\psi, \phi) \approx C_q^p(\psi \mid X^p, \phi \mid X^p),$$

where i: $X^p \subset X$, and we assume for simplicity that $X = X^p$.

Let $\eta \in C^p(X; G_q)$ be given and let w_σ: $E^p \to \sigma^p$ be a characteristic map for $\sigma = \sigma^p \in X$. Let $b_\sigma \in C_p(\sigma^p)$ be the generator represented by w_σ and let

$$g_\sigma\colon\ (TE^m, E^m, TS^{m-1}) \to (S^n\psi\sigma^p, S^n\phi\sigma^p, y)$$

be a map which represents $\eta(\sigma)\,b_\sigma$, where $m = n + q - 1$ and $y \in S^n\phi(\varnothing)$.

Since $\dim X < \infty$ we may assume that n has the same value for every $\sigma^p \in X$ and a $(\theta^{p-1}, \theta^p)^n_{m-p+1}$-map, $f\colon TS^{m-p}X \to S^n Y$, is defined by

$$fTS^{m-p}X^{p-1} = y, \quad f(TS^{m-p}w_\sigma)\,s = q_\sigma s (s \in TE^m).$$

Obviously $(h\{f\})\,b_\sigma = \eta(\sigma)\,b_\sigma$, where $\{f\} \in C^p_q(\psi, \phi)$ is the element represented by f. Therefore $h\{f\} = \eta$ and h is onto.

Let $f\colon TS^{m-p}X \to S^n Y$ be a $(\theta^{p-1}, \theta^p)^n_{m-p+1}$-map such that $h\{f\} = 0$. Then it follows from the first part of the proof of (4.1) that f is $(\theta^{p-1}, \theta^p)^n_{m-p+1}$-homotopic to a $(\theta^p, \theta^p)^n_{m-p+1}$-map. Therefore $\{f\} = 0$ and (14.3) is proved.

It follows from (14.3), (14.2) that h induces an isomorphism

$$h_*\colon {}^{(1)}C^p_q(\psi, \phi) \approx H^p(X; G_q). \tag{14.4}$$

Therefore the second obstruction to the 'S-compression' of a θ^{p-1}-map to a θ^p-map may be regarded as a cohomology class, which is the usual algebraic expression for an obstruction.

15. The group $C^{(1)}_{p,q}(\psi, \phi)$

Let X, Y be arbitrary spaces and $\theta\colon X \to Y$ a given carrier. We define a carrier $\theta^{-1}\colon Y \to X$ by taking $\theta^{-1}B$ ($B \subset Y$) to be the union of all the sets $A \subset X$ such that $\theta A \subset B$, with $\theta^{-1}B = \varnothing$ if there is no such $A \subset X$. If $\phi \leqq \psi\colon X \to Y$, then $\psi^{-1} \leqq \phi^{-1}$. If $\theta \bigcup A = \bigcup \theta A$ for every union, $\bigcup A$, of sets $A \subset X$, then $\theta\theta^{-1}B \subset B$ provided $\theta(\varnothing) \subset B$. If Y is a CW-complex and if θ is cocellular, then $\theta^{-1}B = \theta^{-1}B'$, where B' is the largest subcomplex of Y which is contained in B. If X is a CW-complex and if θ is cellular, then θ^{-1} is cocellular. In this case θ is equivalent to $\theta'\colon X \to Y$, where θ' is defined by $\theta' A = \bigcup_{\sigma \subset A} \theta\sigma$ for every subcomplex $A \subset X$. Evidently $\theta' \bigcup = \bigcup \theta'$.

Let $\phi \leqq \psi\colon X \to Y$, and, for every $B \subset Y$, define $\xi^\theta_B\colon X \to T\varnothing$ ($\theta = \phi, \psi$) by $\xi^\theta_B \varnothing = T\varnothing$ if $\theta(\varnothing) \not\subset B$, $\xi^\theta_B \theta^{-1}B = \varnothing$ if $\theta(\varnothing) \subset B$. Then $\xi^\phi_B \leqq \xi^\psi_B$ and $\xi^\theta_{B'} \leqq \xi^\theta_B$ if $B \subset B'$. Therefore a contravariant stack G^n is defined by

$$G^n(B) = \Sigma_{-n}(\xi^\psi_B, \xi^\phi_B),$$

$G^n(i)\colon G^n(B') \to G^n(B)$ being the injection if $i\colon B \subset B'$. Clearly $G^n(B) = 0$ if $\phi(\varnothing) \not\subset B$. If $\phi(\varnothing) \subset B$, if X is a CW-complex and if $\phi^{-1}B$, $\psi^{-1}B$ are subcomplexes of X, then $G^n(B)$ may be identified with $\Sigma^n(\phi^{-1}B)$ or $\Sigma^n(\phi^{-1}B, \psi^{-1}B)$ according as $\psi(\varnothing) \not\subset B$ or $\psi(\varnothing) \subset B$.

Let X, Y be CW-complexes, let ϕ, ψ be both cellular and cocellular and let $\phi \bigcup = \bigcup \phi$, $\psi \bigcup = \bigcup \psi$. Let $\theta_p\colon X \to Y$ be defined by (11.2), with $Y_p = Y^p$, and for every subcomplex $B \subset Y$ let

$$\chi^p_B = \chi^{\bar{B}^p}\colon\ Y \to T\varnothing.$$

Then $\chi_B^n \theta_m A = \chi_B^n(\phi A \cup (\psi A)^m)$ and $\phi A \cup (\psi A)^m$ is in B if $\psi(\varnothing) \subset B$, $A \subset \psi^{-1}B$, and in $\bar{B}^m$ if $\phi(\varnothing) \subset B$, $A \subset \phi^{-1}B$. Therefore $\chi_B^n \theta_m \leqq \xi_B^\psi$ and $\chi_B^n \theta_m \leqq \xi_B^\phi$ if $m \leqq n$. Since

$$C^p(B) = \Sigma^p(\bar{B}^p, \bar{B}^{p-1}) = \Sigma_{-p}(\chi_B^{p-1}, \chi_B^p),$$

$$C_{p,q}(\psi, \phi) = \Sigma_{p+q+1}(\theta_p, \theta_{p-1}),$$

a pairing

(15.1) $$C^p(B) \times C_{p,q}(\psi, \phi) \to G^{-q-1}(B)$$

is defined as in (10.9). If $\phi(\varnothing) \not\subset B$, then $G^{-q-1}(B) = 0$. If $\phi(\varnothing) \subset B$, then (15.1) is defined by composing maps of the form

(15.2) $$TS^m \phi^{-1}B \to TS^n \bar{B}^p \to TS^{n+p} \quad (m = n+p+q+1)$$

with carriers $(\hat{\theta}_p, \theta_p \cup \hat{\theta}_{p-1})_{m+1}^n$, $(\chi_B^{p-1}, \chi_B^p)_{n+1}^{n+p+1}$, restricted to $TS^m\phi^{-1}B$, $TS^n\bar{B}^p$. This pairing determines a homomorphism

(15.3) $$h\colon C_{p,q}(\psi, \phi) \to C_p(Y; G^{-q-1}),$$

and it follows from (10.3) that

(15.4) $$h\partial = \partial h\colon C_{p,q}(\psi, \phi) \to C_{p-1}(Y; G^{-q-1}).$$

Now let X be finite-dimensional and let Y contain but a finite number of p-cells. Then we say that

(15.5) $$h\colon C_{p,q}(\psi, \phi) \approx C_p(Y; G^{-q-1}).$$

Proof. For each $\sigma = \sigma^p \in Y$ define $\theta_\sigma\colon X \to Y$ by $\theta_\sigma A = \theta_{p-1}A \cup \sigma$ or $\theta_{p-1}A$ according as $\sigma \subset \psi A$ or $\sigma \not\subset \psi A$. Then

$$(\textstyle\bigcup \theta_\sigma) A = \theta_{p-1}A \cup (\psi A)^p = \theta_p A$$

and $\theta_\sigma \cap \theta_\tau = \theta_{p-1}$ if $\sigma \neq \tau$ ($\tau = \tau^p \in Y$). If $\sigma \subset \phi(\varnothing)$, then $\sigma \subset \theta_{p-1}A$, whence $\theta_\sigma = \theta_{p-1}$. Since $\dim X < \infty$ it follows from (9.3) and induction on the number of p-cells in Y that $C_{p,q}(\psi, \phi)$ may be identified injectively with the direct sum of the groups $\Sigma_{p+q+1}(\theta_\sigma, \theta_{p-1})$, for every $\sigma = \sigma^p \in Y$. Let B_σ be the largest subcomplex of Y which does not contain σ. We choose a generator $c_\sigma \in C^p(B_\sigma)$ and identify $C_p(Y; G^{-q-1})$ with the direct sum of the groups $G^{-q-1}(B_\sigma)$ in the manner indicated by (13.1). Clearly $Y^{p-1} \subset B_\sigma$ and σ is the only p-cell not in B_σ. Therefore $\bar{B}_\sigma^p = B_\sigma \cup \sigma$, and on taking $B = B_\sigma$ in (15.2) we see that

$$h\Sigma_{p+q+1}(\theta_\sigma, \theta_{p-1}) \subset G^{-q-1}(B_\sigma),$$

both groups being 0 if $\phi(\varnothing) \not\subset B_\sigma$ (i.e. $\sigma \subset \phi(\varnothing)$). So we have only to prove that

(15.6) $$h_\sigma\colon \Sigma_{p+q+1}(\theta_\sigma, \theta_{p-1}) \approx G^{-q-1}(B_\sigma),$$

where h_σ is the homomorphism determined by h. This is trivially true if $\phi(\varnothing) \not\subset B_\sigma$, so we assume that $\phi(\varnothing) \subset B_\sigma$.

Because of our identifications we have $h_\sigma a = P(c_\sigma, a)$, where $a \in \Sigma_{p+q+1}(\theta_\sigma, \theta_{p-1})$ and P is the pairing (15.1). Since $\bar{B}^{p-1}_\sigma = B_\sigma$ it follows that, for $B = B_\sigma$, $\chi^{p-1}_B \theta_{p-1} = \xi^\phi_B$, $\chi^{p-1}_B \theta_\sigma = \xi^\psi_B$, and hence that

$$(\chi^{p-1}_B, \chi^p_B)_1 (\hat{\theta}_\sigma, \theta_\sigma \cup \hat{\theta}_{p-1})_1 = (\xi^\psi_B, \xi^\phi_B)_1.$$

Since $\dim X < \infty$ we may replace $\Sigma_{p+q+1}(\theta_\sigma, \theta_{p-1})$ by $\Pi^n_m(\theta_\sigma, \theta_{p-1})$ $(m = n+p+q+1)$, for a sufficiently large value of n, and what we have to prove is that

$$(15.7) \qquad f_*\colon \Pi^n_{m+1}(\hat{\theta}_\sigma, \theta_\sigma \cup \hat{\theta}_{p-1}) \approx \Pi^{m-q}_{m+1}(\xi^\psi_B, \xi^\phi_B),$$

where $f\colon TS^n\bar{B}^p_\sigma \to TS^{n+p}$ represents c_σ. Let $z_0 \in S^{n+p}$ and define $\lambda\colon TS^n\bar{B}^p_\sigma \to TS^{n+p}$ by

$$\lambda(TS^n\bar{B}^p_\sigma, TS^nB_\sigma \cup S^n\sigma, TS^nB_\sigma) = (TS^{n+p}, S^{n+p}, z_0).$$

Then $\lambda(\hat{\theta}_\sigma, \theta_\sigma \cup \hat{\theta}_{p-1})^n_{m+1} = \mu$, say, is defined by $\mu S^m\phi^{-1}B_\sigma = S^{n+p}$ and $\mu TS^n\psi^{-1}B_\sigma = z_0$ if $\psi(\varnothing) \subset B_\sigma$. Thus $\mu \leqq (\xi^\psi_B, \xi^\phi_B)^{m-q}_{m+1}$ and, since $TS^n\psi^{-1}B_\sigma$ is contractible,

$$(15.8) \qquad \gamma\colon \Pi(\mu) \approx \Pi^{m-q}_{m+1}(\xi^\psi_B, \xi^\phi_B),$$

where γ is the injection. We may regard TS^{n+p} as the space obtained from $TS^n\bar{B}^p_\sigma$ by shrinking TS^nB_σ to the point z_0 and we may take f to be the identification map. Then (15.7) follows from (5.2), (15.8). This proves (15.5).

It follows from (15.5) and (15.4) that, if $\dim X < \infty$ and if Y^p is a finite complex for every $p \geqq 0$, then h induces an isomorphism

$$(15.9) \qquad h_*\colon C^{(1)}_{p,q}(\psi, \phi) \approx H_p(Y; G^{-q-1}).$$

Application

Let X, Y be as above, let Y be $(n-1)$-connected and let

$$\dim X \leqq 2n-2 \quad (n \geqq 2).$$

Let ϕ, ψ be defined by $\phi X = Y^{n-1}$, $\psi X = Y$ and let $f\colon X \to Y$ be a map with values in Y^p $(p \geqq n)$. Then the first obstruction to compressing f to a map in Y^{p-1}, by means of a homotopy in Y^p, is a cycle $z_f \in Z_p(Y; G)$, where G is the cohomotopy group $\Pi^p(X)$. If σ is an oriented p-cell in Y the coefficient $z_f(\sigma)$ is obtained by composing f with a map $g\colon Y^p \to S^p$, such that $g(Y^p - (\sigma - \dot{\sigma}))$ is a single point and $g \mid \sigma$ is of degree $+1$. Let $\{z_f\} \in H_p(Y; G)$ be the homology class of z_f. Then the conditions (a) $z_f = 0$, (b) $\{z_f\} = 0$ are, respectively, necessary and sufficient for f to be homotopic in (a) Y^p, (b) Y^{p+1} to a map in Y^{p-1}.

University of Chicago
Magdalen College, Oxford

REFERENCES

[1] M. G. BARRATT, *Track Groups. I and II*, Proc. London Math. Soc., 3rd ser. vol. 5 (1955), pp. 71–106; and 285–329.

[2] A. L. BLAKERS and W. S. MASSEY, *The Homotopy Groups of a Triad. II*, Ann. of Math., 55 (1952), pp. 192–201.

[3] S. EILENBERG and N. E. STEENROD, Foundations of Algebraic Topology, Princeton University Press, 1952.

[4] J. LERAY, *L'anneau spectral et l'anneau filtré d'homologie d'un espace localement compact et d'une application continue*, J. Math., Pures Appl., 29 (1950), pp. 1–139.

[5] W. S. MASSEY, *Extract Couples in Algebraic Topology* (*I, II*), Ann. of Math., 56 (1952), pp. 364–396; (*III, IV, V*), ibid., 57 (1953), pp. 248–286.

[6] E. SPANIER, *Borsuk's Cohomotopy Groups*, Ann. of Math., 50 (1949), pp. 203–245.

[7] —— and J. H. C. WHITEHEAD, *A First Approximation to Homotopy Theory*, Proc. Nat. Acad. Sci. U.S.A., 39 (1953), pp. 655–660.

[8] H. TODA, *Generalized Whitehead products and homotopy groups of spheres*, J. Inst. Polytech., Osaka City University, Series A, Mathematics, 3 (1952), pp. 43–82.

[9] J. H. C. WHITEHEAD, *Note on a Theorem due to Borsuk*, Bull. Amer. Math. Soc., 54 (1948), pp. 1125–1132.

[10] ——, *Combinatorial Homotopy. I*, Bull, Amer. Math. Soc., 55 (1949), pp. 213–245.

The Jacobi Identity for Whitehead Products

Hiroshi Uehara and W. S. Massey*

Introduction

LET X be a topological space and x a point of X. Let $\alpha \in \pi_p(X, x)$, $\beta \in \pi_q(X, x)$ and $\gamma \in \pi_r(X, x)$, where p, q and r are integers > 2. The main purpose of this note is to prove that the following identity involving Whitehead products holds:

$$(1) \qquad (-1)^{p(r+1)}[\alpha, [\beta, \gamma]] + (-1)^{q(p+1)}[\beta, [\gamma, \alpha]] + (-1)^{r(q+1)}[\gamma, [\alpha, \beta]] = 0.$$

It is probable that this identity also holds in case p, q and r are $\geqq 2$, but this case is not entirely covered by the following proof.†

First of all, we give an outline of our method of proof.

It is readily seen that it is sufficient to prove this theorem for the following special case: $X = S_1 \vee S_2 \vee S_3$ is the union‡ of three oriented spheres S_1, S_2 and S_3 with a single point in common, dimension $S_1 = p$, dimension $S_2 = q$, dimension $S_3 = r$, and α, β and γ are represented by the inclusion maps $S_1 \to X$, $S_2 \to X$ and $S_3 \to X$ respectively (cf. the discussion in § 2 of [4]).

In order to prove the theorem in this special case, we make a more detailed study of the homotopy group $\pi_{\rho-2}(X)$, where X is the space described in the preceding paragraph, and $\rho = p + q + r$. Without loss of generality, we may obviously assume $p \leqq q \leqq r$. Under these assumptions, the structure of the group $\pi_{\rho-2}(X)$ is given by the following theorem:

* This paper was written while one of the authors was supported by the Fulbright-Smith-Mundt Exchange Person Program.

† The authors have been informed that independent proofs of this Jacobi identity have recently been found by H. Toda and M. Nakaoka in Japan, by Serre, Hilton, and Green in Europe, and by G. W. Whitehead in this country.

‡ If A and B are topological spaces, we use the notation '$A \vee B$' to denote the union of A and B with a single point in common (cf. [6], §4). When convenient, we will consider $A \vee B$ as a subset of the product $A \times B$.

THEOREM I. *The homotopy group $\pi_{p-2}(X)$ is the direct sum of seven subgroups, as follows:*

(a) *Three subgroups isomorphic to $\pi_{p-2}(S_i)$ $(i=1,2,3)$. These subgroups are the image of the injections*† $\pi_{p-2}(S_i) \to \pi_{p-2}(X)$ $(i=1,2,3)$.

(b) *Three subgroups isomorphic to $\pi_{p-1}(S_i \times S_j, S_i \vee S_j)$, where $(i,j)=(1,2)$, $(2,3)$ and $(3,1)$. The subgroups are the image under the composition of the homotopy boundary operator*

$$\partial\colon \pi_{p-1}(S_i \times S_j, S_i \vee S_j) \to \pi_{p-2}(S_i \vee S_j)$$

and the injection $\pi_{p-2}(S_i \vee S_j) \to \pi_{p-2}(X)$.

(c) *A free subgroup of rank 2; this subgroup is the intersection of the kernels of the three retractions $\pi_{p-2}(X) \to \pi_{p-2}(S_i \vee S_j)$, where $(i,j)=(1,2)$, $(2,3)$ and $(3,1)$ as before.*

It is clear that each of the triple Whitehead products $[\alpha,[\beta,\gamma]]$, $[\beta,[\gamma,\alpha]]$ and $[\gamma,[\alpha,\beta]]$ is contained in each of the kernels of the retractions $\pi_{p-2}(X) \to \pi_{p-2}(S_i \vee S_j)$; hence these three elements belong to the subgroup mentioned in part (c) of Theorem I. Therefore they are linearly dependent.

The proof is then completed by means of the following theorem:

THEOREM II. *If there exists a linear relation of the form*

$$l[\alpha,[\beta,\gamma]]+m[\beta,[\gamma,\alpha]]+n[\gamma,[\alpha,\beta]]=0,$$

where l, m and n are integers, then

$$(-1)^{p(r+1)}l=(-1)^{q(p+1)}m=(-1)^{r(q+1)}n.$$

It is thus clear that Theorems I and II together imply the Jacobi identity (1). It should be pointed out that the proof of Theorem II is completely independent of the proof of Theorem I. To carry out the proof of Theorem II, a new cohomology operation is introduced. This new operation should be of independent interest; for example, it seems likely that it should prove useful in the classification of simply connected 5-dimensional polyhedra according to homotopy type.

1. Proof of Theorem I

Let $Y=S_1 \vee S_2$; then $X=Y \vee S_3$. Let $g_1\colon S_1 \to Y$ and $g_2\colon S_2 \to Y$ denote inclusion maps, and $r_1\colon Y \to S_1$ and $r_2\colon Y \to S_2$ retractions. Consider the diagram in Fig. 1. The arrows numbered 1, 2 and 3 are homotopy boundary operators, and the arrow numbered 4 is an injection. The homomorphisms g_i' and g_i'' are induced by g_i, and r_i', r_i''

† We use the term 'injection' to denote a homomorphism induced by an inclusion map.

are induced by r_i $(i=1, 2)$. We can apply a theorem of G. W. Whitehead ([7], Theorem 4.8) to the horizontal lines of this diagram. Thus we see that the homomorphisms numbered 1, 2 and 3 are isomorphisms into, and the image of each is a direct summand; also, 4 is a homomorphism onto. On account of the fact that r_1 and r_2 are retractions, $r_1' \circ g_1'$ and $r_2' \circ g_2'$ are identity maps. One also sees that $r_2' \circ g_1'$ and $r_1' \circ g_2'$ are homomorphisms which map everything into zero. It follows that the group $\pi_{p-1}(Y \times S_3, Y \vee S_3)$ is the direct sum of the following subgroups: $g_1'[\pi_{p-1}(S_1 \times S_3, S_1 \vee S_3)]$, $g_2'[\pi_{p-1}(S_2 \times S_3, S_2 \vee S_3)]$, and the intersection of the kernels of r_1' and r_2'.

$$\begin{array}{ccccc}
\pi_{p-1}(S_1 \times S_3, S_1 \vee S_3) & \xrightarrow{1} & \pi_{p-2}(S_1 \vee S_3) & & \\
g_1' \downarrow \quad \uparrow r_1' & & g_1'' \downarrow \quad \uparrow r_1'' & & \\
\pi_{p-1}(Y \times S_3, Y \vee S_3) & \xrightarrow{2} & \pi_{p-2}(Y \vee S_3) & \xrightarrow{4} & \pi_{p-2}(Y \times S_3) \\
g_2' \uparrow \quad \downarrow r_2' & & g_2'' \uparrow \quad \downarrow r_2'' & & \\
\pi_{p-1}(S_2 \times S_3, S_2 \vee S_3) & \xrightarrow{3} & \pi_{p-2}(S_2 \vee S_3) & &
\end{array}$$

Fig. 1

By putting together all these facts about the groups and homomorphisms in Fig. 1, it is readily seen that in order to complete the proof of Theorem I, it suffices to prove the following lemma:

LEMMA 1. *The intersection of the kernels of r_1' and r_2' is a free group of rank* 2.

PROOF OF LEMMA 1. Let

$$T_1 = S_1 \times S_3,\ T_2 = S_2 \times S_3 \quad \text{and} \quad T = Y \times S_3 = T_1 \cup T_2.$$

We may regard T_1 as obtained from $S_1 \vee S_3$ by adjunction of a $(p+r)$-cell, and T_2 as obtained from $S_2 \vee S_3$ by the adjunction of a $(q+r)$-cell. By applying Lemma 2 of [3], we may choose closed subsets E_1 and B_1 of T_1 such that E_1 is a closed $(p+r)$-cell, $E_1 \cap B_1 = \dot{E}_1$, the boundary of E_1, $T_1 = E_1 \cup B_1$, and $S_1 \vee S_3$ is a deformation retract of B_1. Similarly, we may choose subsets E_2 and B_2 of T_2 such that E_2 is a closed $(q+r)$-cell, etc. Let $A = E_1 \cup E_2$, $B = B_1 \cup B_2$. Now consider the diagram in Fig. 2. All homomorphisms in this diagram are induced by inclusion maps. The three vertical columns are portions of the homotopy sequences of certain triads. Note that the middle line of Fig. 2 is equivalent to the left-hand column of Fig. 1. The first part of the proof consists in drawing certain conclusions about this diagram.

First of all, it is readily proved that there exist retractions

$$R_1\colon (T; A, B) \to (T_1; E_1, B_1) \quad \text{and} \quad R_2\colon (T; A, B) \to (T_2; E_2, B_2).$$

Hence the homomorphisms represented by horizontal arrows in the diagram below are all isomorphisms into, and the image subgroups are direct summands; moreover, the two summands thus obtained in any of the groups in the middle column are supplementary, in the sense that they have only the zero element in common.

$$\begin{array}{ccccc}
\pi_{p-1}(E_1, \dot{E}_1) & \xrightarrow{7} & \pi_{p-1}(A, A \cap B) & \xleftarrow{8} & \pi_{p-1}(E_2, \dot{E}_2) \\
\downarrow 1 & & \downarrow 2 & & \downarrow 3 \\
\pi_{p-1}(T_1, B_1) & \xrightarrow{9} & \pi_{p-1}(T, B) & \xleftarrow{10} & \pi_{p-1}(T_2, B_2) \\
\downarrow 4 & & \downarrow 5 & & \downarrow 6 \\
\pi_{p-1}(T_1; E_1, B_1) & \xrightarrow{11} & \pi_{p-1}(T; A, B) & \xleftarrow{12} & \pi_{p-1}(T_2; E_2, B_2)
\end{array}$$

Fig. 2

Secondly, by direct application of Theorem III of [2], one shows that the homomorphisms numbered 1, 2 and 3 in Fig. 2 are isomorphisms into, and each image subgroup is a direct summand (use the fact that $2 \leqq p \leqq q \leqq r$ in the proof). By another application of the same theorem, together with the exactness of the homotopy sequence of a triad, one proves that the homomorphisms numbered 4, 5 and 6 are onto.

Thirdly, observe that $\pi_{p-1}(A, A \cap B)$ is the direct sum of the images of the homomorphisms 7 and 8. This follows from the fact that $A = E_1 \vee E_2$ and Theorem 5.3.3. of [1].

We will now complete the proof of Lemma 1 for the case where $2 \leqq p = q \leqq r$. In this case Theorem I of [3] can be applied directly to determine the structure of the 'critical dimension' triad homotopy groups in the bottom line of Fig. 2. The results are as follows:

$$\pi_{p-1}(T_1; E_1, B_1) \cong \pi_{p+r}(E_1, \dot{E}_1) \otimes \pi_p(B_1, \dot{E}_1),$$

$$\pi_{p-1}(T_2; E_2, B_2) \cong \pi_{p+r}(E_2, \dot{E}_2) \otimes \pi_p(B_2, \dot{E}_2),$$

$$\pi_{p-1}(T; A, B) \cong \pi_{p+r}(A, A \cap B) \otimes \pi_p(B, A \cap B).$$

Each of these isomorphisms is natural, and is induced by the generalized Whitehead product. From this one concludes that $\pi_{p-1}(T; A, B)$ is the direct sum of the images of the homomorphisms numbered 11 and 12 in Fig. 2, and a free group of rank 2.

This completes the proof of Lemma 1 for the case where $2 \leqq p = q \leqq r$. For the rest of this proof, we assume that $2 < p < q \leqq r$. Note that under this assumption, $\pi_{\rho-1}(T_2; E_2, B_2) = 0$, since the triad $(T_2; E_2, B_2)$ is $(2q+r-2)$-connected by Theorem I of [2], and $\rho - 1 \leqq 2q + r - 2$. It follows by exactness that the homomorphism numbered 3 in Fig. 1 is an isomorphism onto.

LEMMA 2. *The group* $\pi_{\rho-1}(T; A, B)$ *is the direct sum of the image of the homomorphism numbered* 11 *in Fig.* 1 *and a free subgroup of rank* 2.

It is readily seen that Lemma 1 for the case $p < q$ follows from the facts previously stated together with Lemma 2. Therefore we will now give the proof of Lemma 2.

Let $(T^*; A^*, B^*)$ be the triad obtained from $(T; A, B)$ by identifying all of $A \cap B$ to a single point, and let $(T_1^*; E_1^*, B_1^*)$ be the image of $(T_1; E_1, B_1)$ under this identification. Then we have the following commutative diagram:

$$\begin{array}{ccc} \pi_{\rho-1}(T_1; E_1, B_1) & \xrightarrow{11} & \pi_{\rho-1}(T; A, B) \\ \downarrow & & \downarrow \\ \pi_{\rho-1}(T_1^*; E_1^*, B_1^*) & \xrightarrow{13} & \pi_{\rho-1}(T^*; A^*, B^*). \end{array}$$

The vertical arrows represent homomorphisms induced by the identification map, and the horizontal arrows are injections. It can be proved by exactly the same method as used to prove Corollary 3.5 of [5] that both these vertical arrows represent isomorphisms onto (it is at this point that one must use the assumption $p > 2$).

Next, note that $T^* = A^* \vee B^*$ and $T_1^* = E_1^* \vee B_1^*$. Hence we have the following commutative diagram:

$$\begin{array}{ccc} \pi_{\rho-1}(T_1^*; E_1^*, B_1^*) & \xrightarrow{13} & \pi_{\rho-1}(T^*; A^*, B^*) \\ \uparrow & & \uparrow \\ \pi_\rho(E_1^* \times B_1^*, T_1^*) & \xrightarrow{14} & \pi_\rho(A^* \times B^*, T^*). \end{array}$$

Here both the vertical arrows represent homotopy boundary operators, as described in § 8 of [4]. It follows from Corollary 8.3 of [4] that both vertical arrows represent isomorphisms onto. Since

$$T^* \cap (E_1^* \times B_1^*) = T_1^*,$$

the homomorphism labelled 14 in the diagram may be fitted into the exact sequence of the triad $(A^* \times B^*; T^*, E_1^* \times B_1^*)$, as follows:

$$(2) \qquad \ldots \to \pi_\rho(E_1^* \times B_1^*, T_1^*) \xrightarrow{14} \pi_\rho(A^* \times B^*, T^*) \xrightarrow{15} \pi_\rho(A^* \times B^*; T^*, E_1^* \times B_1^*).$$

We will now study the homotopy groups of this triad. For this purpose, consider the following exact sequence (cf. § 8 of [4]):

$$\ldots \xrightarrow{\partial} \pi_\rho(T^* \cup (E_1^* \times B_1^*); T^*, E_1^* \times B_1^*) \xrightarrow{i}$$

$$\pi_\rho(A^* \times B^*; T^*, E_1^* \times B_1^*) \xrightarrow{j} \pi_\rho(A^* \times B^*, T^* \cup (E_1^* \times B_1^*)) \xrightarrow{\partial} \ldots.$$

Now one can prove that the triad $(T^* \cup (E_1^* \times B_1^*); T^*, E_1^* \times B_1^*)$ is $(2p+q+r-2)$-connected by means of Theorem I of [2]. Since $2p+q+r-2 \geqq \rho$, it follows that the homomorphism

$$j\colon \pi_\rho(A^* \times B^*; T^*, E_1^* \times B_1^*) \to \pi_\rho(A^* \times B^*, T^* \cup (E_1^* \times B_1^*))$$

is an isomorphism onto. One can now compute the integral homology groups of the pair $(A^* \times B^*, T^* \cup (E_1^* \times B_1^*))$, and then apply the Hurewicz equivalence theorem to conclude that this pair is $(\rho-1)$-connected, and $\pi_\rho(A^* \times B^*, T^* \cup (E_1^* \times B_1^*))$ is a free Abelian group of rank 2. Combining this result with the exact sequence (2), we obtain the following sequence, which is exact:

$$0 \to \pi_\rho(E_1^* \times B_1^*, T_1^*) \xrightarrow{14} \pi_\rho(A^* \times B^*, T^*) \to$$

$$\pi_\rho(A^* \times B^*, T^* \cup (E_1^* \times B_1^*)) \to 0.$$

By putting all these facts together, one can now complete the proof of Lemma 2.

REMARK. It is possible to prove by the above method that the subgroup mentioned in part (c) of Theorem I is generated by the triple Whitehead products $[\alpha, [\beta, \gamma]]$ and $[\beta, [\gamma, \alpha]]$. We will not make use of this fact, however.

2. A new cohomology operation

In this section we will consider various associative rings of cochains with a coboundary operator. If A is such a ring with coboundary operator $\delta\colon A \to A$, then we will assume that:

(a) δ is linear and $\delta^2 = 0$.

(b) A is a graded ring, with degrees $\geqq 0$, i.e. the additive group A is the direct sum of a sequence of subgroups A_n $(n \geqq 0)$, and

$$A_m \cdot A_n \subset A_{m+n}.$$

(c) $\delta(A_n) \subset A_{n+1}$ for all n.

(d) If $x \in A_n$ and $y \in A$, then $\delta(xy) = (\delta x)y + (-1)^n x(\delta y)$.

It should be emphasized that all the rings of cochains which we consider will be assumed to be associative, to have a coboundary operator, and to satisfy conditions (a)–(d) above. As usual we will denote by $H(A)$ the derived ring, or cohomology ring, and $H^n(A)$ will denote the subgroup of elements of degree n.

For any elements $w \in H^r(A)$ and $u \in H^p(A)$, let

$$J^n(u, w) = u \cdot H^{n-p}(A) + H^{n-r}(A) \cdot w.$$

Then $J^n(u, w)$ is a subgroup of $H^n(A)$.

Let $u \in H^p(A)$, $v \in H^q(A)$ and $w \in H^r(A)$ be any three cohomology classes which satisfy the following condition:

$$uv = vw = 0.$$

Then the *triple product*, $\langle u, v, w \rangle$, is a certain element of the factor group $H^{p+q+r-1}(A)/J^{p+q+r-1}(u, w)$, defined as follows. Choose representative cocycles $u' \in A_p$, $v' \in A_q$ and $w' \in A_r$ for the cohomology classes u, v and w respectively. Then there exist cochains a and b of degrees $p+q-1$ and $q+r-1$ respectively such that

$$u'v' = \delta a,$$

$$v'w' = \delta b.$$

Define $$z' = aw' - (-1)^p u'b.$$

It is readily verified that $\delta(z') = 0$, i.e. z' is a cocycle. Let z denote its cohomology class. Then $\langle u, v, w \rangle$ is defined to be the coset of z modulo $J^{p+q+r-1}(u, w)$. To justify this definition, one must verify that the coset so obtained is independent of the choices made for u', v', w', a and b. However, this verification is purely mechanical, and is left to the reader.†

We will now list some properties of this triple product.

Let A and B be two associative rings of cochains, and let $f: A \to B$ be a homomorphism which preserves all the structures involved (i.e. products, degrees and commutes with the coboundary operators). Let $f^*: H(A) \to H(B)$ denote the induced homomorphism. For any homogeneous elements $u, w \in H(A)$, it is readily seen that

$$f^* J^n(u, w) \subset J^n(f^*u, f^*w).$$

Hence f^* maps cosets mod $J^n(u, w)$ into cosets mod $J^n(f^*u, f^*w)$. If $v \in H^p(A)$ is an element such that $uv = vw = 0$, then we assert that

$$f^* \langle u, v, w \rangle \subset \langle f^*u, f^*v, f^*w \rangle.$$

† The authors wish to acknowledge that the original idea for this operation arose in a discussion with A. Shapiro at the Topology Conference held at the University of Chicago in May 1950. An abstract describing this operation appeared in the Bull. Amer. Math. Soc. 57 (1951), p. 74.

The verification, which is easy, is left to the reader. This property may be expressed by saying that the triple product operation, $(u, v, w) \to \langle u, v, w \rangle$, is *natural*. By making use of this naturality, one can readily prove that if one computes triple products using any of the standard kinds of cochains on a simplicial polyhedron, a cell complex, or an arbitrary topological space, the result is an invariant of the homotopy type of the space.

For $u \in H^p(A)$ and $w \in H^r(A)$, let $I^n(u, w)$ denote the set of all elements $v \in H^n(A)$ such that $uv = vw = 0$. Then for fixed u and w, the mapping $v \to \langle u, v, w \rangle$ is a linear map of $I^q(u, w)$ into

$$H^{p+q+r-1}(A)/J^{p+q+r-1}(u, w).$$

Again the verification is trivial.

The next property makes use of the Steenrod functional cup product (cf. [6]). For our purposes, we will reformulate Steenrod's definitions as follows. Let A and B be rings of cochains and let $f\colon A \to B$ be a homomorphism preserving degrees, products, etc., as above. If $u \in H^p(A)$ and $v \in H^q(A)$ are cohomology classes such that $uv = 0$ and $f^*u = 0$, then the *left functional cup product*, $L_f(u, v)$, is a coset of $H^{p+q-1}(B)$ modulo $\{f^*H^{p+q-1}(A) + H^{p-1}(B) \cdot (f^*v)\}$ defined as follows: choose cocycles $u' \in A_p$ and $v' \in A_q$ which belong to the cohomology classes u and v respectively. Then there exists a cochain $a \in A_{p+q-1}$ such that $u'v' = \delta a$, and a cochain $b \in B_{p-1}$ such that $f(u') = \delta b$. Let

$$z' = f(a) - b \cdot f(v').$$

Then z' is a cocycle. Let z denote its cohomology class. We define $L_f(u, v)$ to be the coset of z modulo $\{f^*H^{p+q-1}(A) + H^{p-1}(B) \cdot (f^*v)\}$.

In an analogous fashion, if $u \in H^p(A)$, $v \in H^q(A)$, $uv = 0$ and $f^*(v) = 0$, the *right functional cup product*, $R_f(u, v)$, is a coset of $H^{p+q-1}(B)$ modulo $\{f^*H^{p+q-1}(A) + (f^*u) \cdot H^{q-1}(B)\}$ defined as follows. Choose representative cocycles u' and v' for u and v as before. There exist cochains $a \in A_{p+q-1}$ and $b \in B_{q-1}$ such that

$$u'v' = \delta a, \quad f(v') = \delta b.$$

Let $z' = f(a) - (-1)^p (fu') \cdot b$. Then z' is a cocycle. Let z denote its cohomology class. Define $R_f(u, v)$ to be the coset of z modulo

$$\{f^*H^{p+q-1}(A) + (f^*u) \cdot H^{q-1}(B)\}.$$

LEMMA 3. *Suppose that $u \in H^p(A)$, $v \in H^q(A)$, $uv = 0$, and $f^*u = f^*v = 0$ in the above discussion. Then $L_f(u, v)$ and $R_f(u, v)$ are both defined, and*

$$L_f(u, v) = R_f(u, v).$$

PROOF. Choose representative cocycles u' and v' for u and v, and cochains a, b and c such that

$$u'v' = \delta a,$$

$$f(u') = \delta b,$$

$$f(v') = \delta c.$$

Then the cohomology class of the cocycle

$$y' = f(a) - b \cdot f(v') = f(a) - b(\delta c)$$

belongs to the coset $L_f(u, v)$, and the cohomology class of the cocycle

$$z' = f(a) - (-1)^p f(u') \cdot c = f(a) - (-1)^p (\delta b) \cdot c$$

belongs to the coset $R_f(u, v)$. Then

$$y' - z' = (-1)^p \delta(bc),$$

hence y' and z' are cohomologous. To complete the proof, observe that under the hypotheses of this lemma, both $L_f(u, v)$ and $R_f(u, v)$ are cosets of the subgroup $f^* H^{p+q-1}(A)$.

For the statement of the next lemma, let $f: A \to B$ be a homomorphism as described above, and let $u \in H^p(A)$, $v \in H^q(A)$ and $w \in H^r(A)$ be elements such that

$$uv = vw = 0, \quad f^* v = 0.$$

Then $R_f(u, v) \cdot (f^* w)$ is defined, and is a coset of

$$\{f^*[H^{p+q-1}(A) \cdot w] + (f^* u) \cdot H^{q-1}(B) \cdot (f^* w)\};$$

likewise, $(f^* u) \cdot L_f(v, w)$ is defined and is a coset of

$$\{f^*[u \cdot H^{q+r-1}(A)] + (f^* u)\, H^{q-1}(B) \cdot (f^* w)\}.$$

Finally, $f^* \langle u, v, w \rangle$ is a coset of

$$f^*[H^{p+q-1}(A) \cdot w + u \cdot H^{q+r-1}(A)].$$

LEMMA 4. *Assuming the above hypotheses,*

$$f^* \langle u, v, w \rangle \equiv R_f(u, v) \cdot (f^* w) - (-1)^p (f^* u)\, L_f(v, w)$$

modulo the subgroup

$$f^*[H^{p+q-1}(A) \cdot w + u \cdot H^{q+r-1}(A)] + (f^* u) \cdot H^{q-1}(B) \cdot (f^* w).$$

The verification of this lemma is a straightforward application of the definitions.

LEMMA 5. *Assume that* $f: A \to B$, $u \in H^p(A)$, $v \in H^q(A)$, $w \in H^r(A)$, $uv = vw = 0$, *exactly as in the previous lemma. However, assume that*

$f^*(u)=0$ and $f^*(w)=0$, instead of $f^*(v)=0$. Then the coset $\langle u,v,w\rangle$ is contained in the kernel of the homomorphism

$$f^*\colon H^{p+q+r-1}(A)\to H^{p+q+r-1}(B).$$

PROOF. Let z be a cohomology class belonging to the coset $\langle u,v,w\rangle$. Then there exist representative cocycles u', v' and w' for u, v and w respectively, and cochains a and b such that $\delta a=u'v'$, $\delta b=v'w'$ and $z'=aw'-(-1)^p u'b$ is a cocycle belonging to the cohomology class z. Let $c,d\in B$ be cochains such that $f(u')=\delta(c), f(w')=\delta(d)$. Let

$$e=c(fv')d-(fa)d-(-1)^q c(fb).$$

Then a direct computation shows that

$$\delta(e)=(-1)^{p+q}f(z'),$$

which proves the lemma.

Lemmas 4 and 5 will be used later actually to compute the operation $\langle u,v,w\rangle$ in certain polyhedra.

3. Some special properties of the functional cup product

The main purpose of this section is to prove a theorem involving the functional cup product which will be used in the next section. This theorem, and one occurring in the course of the proof, are perhaps of independent interest.

For the statement of the first lemma, assume that we have the following commutative diagram:

$$\begin{array}{ccccccccc} 0 & \to & A' & \xrightarrow{i} & A & \xrightarrow{j} & A'' & \to & 0 \\ & & \downarrow g & & \downarrow f & & \downarrow h & & \\ 0 & \to & B' & \xrightarrow{k} & B & \xrightarrow{l} & B'' & \to & 0 \end{array}$$

Here A, A', A'', B, B' and B'' are graded rings of cochains with coboundary operators, and the homomorphisms represented by the arrows preserve products and degrees, commute with the coboundary operator, etc. Moreover, it is assumed that the two horizontal lines of this diagram are exact. Passing to the derived rings, one obtains the following commutative diagram, where the horizontal lines are again exact sequences:

$$\begin{array}{ccccccccc} \ldots\to & H^*(A) & \xrightarrow{j^*} & H^*(A'') & \xrightarrow{\Delta_1} & H^*(A') & \xrightarrow{i^*} & H^*(A) & \to\ldots \\ & \downarrow f^* & & \downarrow h^* & & \downarrow g^* & & \downarrow f^* & \\ \ldots\to & H^*(B) & \xrightarrow{l^*} & H^*(B'') & \xrightarrow{\Delta_2} & H^*(B') & \xrightarrow{k^*} & H^*(B) & \to\ldots \end{array}$$

Lemma 6. *With the above notation, assume that* $u \in H^p(A)$ *and* $v \in H^q(A)$ *satisfy the conditions* $f^*(u)=f^*(v)=0$ *and* $j^*(uv)=0$. *Then the cosets* $g^*i^{*-1}(-uv)$ *and* $\Delta_2 L_h(j^*u, j^*v)$ *are the same.*

Proof. Choose representative cocycles $u' \in A_p$ and $v' \in A_q$ for u and v respectively. Choose $a \in A''_{p+q-1}$ such that $j(u'v') = \delta a$, and $b \in B_{p-1}$ such that $f(u') = \delta b$. Then

$$z' = h(a) - l(b)\cdot hj(v') = h(a) - l[b\cdot f(v')]$$

is a representative cocycle of $L_h(j^*u, j^*v)$. Choose an element $a' \in A$ such that $j(a') = a$. Then it is readily verified that $l\delta[f(a') - b\cdot f(v')] = 0$. Hence by exactness there exists a cocycle $e \in B'$ such that

$$k(e) = \delta[f(a') - b\cdot f(v')],$$

and e is a representative of $\Delta_2 L_h(j^*u, j^*v)$. Next, note that

$$j(u'v' - \delta a') = 0,$$

hence by exactness there exists a cocycle $c \in A'$ such that $i(c) = u'v' - \delta a'$. Then c is a representative of $i^{*-1}(uv)$. The proof is now completed by showing that $g(c) = -e$; to prove this, it suffices to show that $kg(c) = -k(e)$, since k is an isomorphism. The verification of the fact that $kg(c) = -k(e)$ is straightforward, and is left to the reader.

We will now apply this lemma to the problem of determining the cup products in the cohomology ring of a cell complex. Let K be a finite cell complex,† and let K^n denote the n-skeleton of K. Let $e_1, \ldots, e_m$ denote the open n-cells of K. Then K^n is obtained from K^{n-1} by adjoining the cells e_i. For $i = 1, \ldots, m$ let E_i denote a topological space which is homeomorphic to the unit n-cell in Euclidean n-space, and let S_i denote its boundary $(n-1)$-sphere. Then there exist the so-called 'characteristic maps' ϕ_i: $(E_i, S_i) \to (K^n, K^{n-1})$ such that ϕ_i maps $E_i - S_i$ homeomorphically onto e_i. We will suppose that the cells E_i are mutually disjoint, and let

$$E = \bigcup_i E_i, \quad S = \bigcup_i S_i \quad \text{and} \quad \phi\colon (E, S) \to (K^n, K^{n-1})$$

be the map defined by $\phi \mid E_i = \phi_i$. Let $\psi\colon S \to K^{n-1}$ denote the map defined by ϕ.

We are now in a position to state our next result, which gives the precise relationship between cup products in K and the functional cup product of the maps such as ψ by which the cells are attached.

† For the definition of a cell complex, see J. H. C. Whitehead[8]. Note that a finite cell complex is automatically a CW-complex.

Consider the following commutative diagram involving the cohomology groups of K and E (with an arbitrary ring as coefficients):

$$\begin{array}{ccccc} H^{n-1}(K^{n-1}) & \xrightarrow{\Delta_1} & H^n(K^n, K^{n-1}) & \xrightarrow{i^*} & H^n(K^n) \\ \downarrow \psi^* & & \downarrow \phi^* & & \\ H^{n-1}(S) & \xrightarrow{\Delta_2} & H^n(E, S) & & \end{array}$$

Here Δ_2 and ϕ^* are isomorphisms onto. Let

$$\Phi\colon\ H^{n-1}(S)/\psi^* H^{n-1}(K^{n-1}) \to H^n(K^n)$$

be the homomorphism defined by $i^* \cdot (\phi^*)^{-1} \cdot \Delta_2$. Let p and q be positive integers such that $p+q=n$, and let $u \in H^p(K^{n-1})$, $v \in H^q(K^{n-1})$ be cohomology classes such that $\psi^*(u) = \psi^*(v) = 0$ (this condition is automatically satisfied in case p and q are greater than 1). The condition $u \cdot v = 0$ is automatically satisfied because $H^n(K^{n-1}) = 0$, and hence the functional cup product $L_\psi(u, v)$ is defined. Let $u' \in H^p(K^n)$ and $v' \in H^q(K^n)$ be the cohomology classes such that $j^*(u') = u$, $j^*(v') = v$, where $j^*\colon H^*(K^n) \to H^*(K^{n-1})$ is the injection.

THEOREM III. $\Phi L_\psi(u, v) = -u' \cdot v'$.

This theorem is a direct application of Lemma 6. One chooses A = ring of cochains on K^n, A' = ring of cochains on K^n modulo K^{n-1}, A'' = ring of cochains on K^{n-1}, B = ring of cochains on E, B' = ring of cochains on E modulo S, and B'' = ring of cochains on S. In this proof for 'ring of cochains' one may use singular cochains, Alexander-Spanier cochains, or assume that E, S, K^n and K^{n-1} are simplicially subdivided and ϕ is a simplicial map and then use simplicial cochains. Note that E is contractible, and hence has trivial cohomology.

REMARK. This theorem can be looked on as the natural generalization of a result of J. H. C. Whitehead ([9], Theorem 5 (a)).

We now apply this theorem to the problem of determining the functional cup product of a map representing a Whitehead product. Let $X = S_1 \vee S_2$ be a space consisting of the union of two oriented spheres of dimensions p and q respectively with a single point in common, and let $\alpha \in \pi_p(X)$, $\beta \in \pi_q(X)$ be the elements represented by inclusion maps $S_1 \to X$ and $S_2 \to X$. Let $\phi\colon S^r \to X$ be a map of an oriented r-sphere $(r = p+q-1)$ into X which represents the Whitehead product $[\alpha, \beta]$. If $u \in H^p(X)$ and $v \in H^q(X)$, then the functional cup product $L_\phi(u, v)$ is an element of $H^r(S^r)$. Our problem is to determine $L_\phi(u, v)$ precisely.

For the sake of convenience, we will regard the homology and cohomology groups of S_1 and S_2 as subgroups of those of X. Then in any dimension greater than zero, the cohomology and homology groups of X are the direct sum of these subgroups. In case $p=q$ we assume that $u \in H^p(S_1)$ and $v \in H^q(S_2)$ (in case $p \neq q$, this requirement is automatically satisfied). Let $w \in H_r(S^r)$, $w_1 \in H_p(S_1)$, $w_2 \in H_q(S_2)$ be integral homology classes which represent the orientations of the respective spheres.

Theorem IV. $L_\phi(u,v) \cap w = -[u \cap w_1] \cdot [v \cap w_2]$ (here the symbol '$\cap$' is used to denote the inner product of a cohomology class and a homology class).

Proof. As usual, we will consider $S_1 \vee S_2 = (S_1 \times b) \cup (a \times S_2)$ as a subset of the product $S_1 \times S_2$, where $a \in S_1$ and $b \in S_2$. Let E_1 and E_2 be closed oriented cells of dimensions p and q respectively, and let

$$f\colon (E_1, \dot{E}_1) \to (S_1, a),$$

$$g\colon (E_2, \dot{E}_2) \to (S_2, b),$$

be functions which map $E_1 - \dot{E}_1$ homeomorphically onto $S_1 - a$ and $E_2 - \dot{E}_2$ homeomorphically onto $S_2 - b$, and preserve orientations (here $\dot{E}_1$ and $\dot{E}_2$ are the bounding spheres of the cells E_1 and E_2). This means that if $\tau_1 \in H_p(E_1, \dot{E}_1)$ and $\tau_2 \in H_q(E_2, \dot{E}_2)$ denote the respective orientations, then

$$f_*(\tau_1) = w_1, \quad g_*(\tau_2) = w_2.$$

Let $S^r = (E_1 \times \dot{E}_2) \cup (\dot{E}_1 \times E_2)$ be the bounding sphere of the cell $E_1 \times E_2$, and $f \times g\colon (E_1 \times E_2, S^r) \to (S_1 \times S_2, S_1 \vee S_2)$ be the product map. If we choose $\partial(\tau_1 \times \tau_2)$ as orientation for S^r, then the map $\phi\colon S^r \to S_1 \vee S_2$ represents the Whitehead product $[\alpha, \beta]$.

In order to complete the proof, we consider $S_1 \times S_2$ as a cell complex obtained by adjoining a $(p+q)$-cell to $S_1 \vee S_2$. This cell is adjoined by the map $f \times g\colon (E_1 \times E_2, S^r) \to (S_1 \times S_2, S_1 \vee S_2)$. One now applies Theorem III to this situation. In applying this theorem, one can make use of the fact that the cohomology ring of $S_1 \times S_2$ is the tensor product of the cohomology rings of S_1 and S_2, and similarly for the cohomology rings of the various other product spaces occurring. The details are left to the reader. An important step in the proof is the following observation: the injection $H^*(S_1 \times S_2) \to H^*(S_1 \vee S_2)$ maps $u \otimes 1$ and $1 \otimes v$ onto u and v respectively.

Remark. An immediate corollary of Theorem IV is the following statement. Let X be an arbitrary topological space, and let $h\colon \pi_n(X) \to H_n(X)$ be the natural homomorphism of the homotopy groups into the integral homotopy groups. Let p and q be positive integers such that $p \neq q$, $u \in H^p(X)$, $v \in H^q(X)$, $u \cdot v = 0$, $\alpha \in \pi_p(X)$, $\beta \in \pi_q(X)$, and let $\phi: S^r \to X$ be a map representing $[\alpha, \beta]$. Then

$$L_\phi(u, v) \cap w = -[u \cap h(\alpha)] \cdot [v \cap h(\beta)].$$

The statement of the analogous result for the case $p = q$ is more complicated and is left to the reader.

4. Proof of Theorem II

Before we can prove Theorem II, we must prove some preliminary lemmas. Throughout this section, we let $X = S_1 \vee S_2 \vee S_3$ be the union of three oriented spheres, S_1, S_2 and S_3 respectively, having a single point in common. The dimensions of these spheres are denoted by p, q and r respectively.† When convenient, we will regard X as a cell complex having a single vertex and three cells of dimension > 0. Let $\alpha \in \pi_p(X)$, $\beta \in \pi_q(X)$ and $\gamma \in \pi_r(X)$ denote the elements represented by the inclusion maps $S_1 \to X$, $S_2 \to X$ and $S_3 \to X$ respectively.

For the statement of the first lemma, let K be a cell complex obtained by adjoining to X an oriented cell e of dimension

$$\rho - 1 = p + q + r - 1$$

by a map representing $[\alpha, [\beta, \gamma]]$. Let $u \in H^p(K)$, $v \in H^q(K)$, $w \in H^r(K)$ and $z \in H^{\rho-1}(K)$ denote the following integral cohomology classes: u is represented by a cocycle which assigns the value $+1$ to the cell S_1, and 0 to all other cells; v is represented by a cocycle which assigns the value $+1$ to the cell S_2 and 0 to all other cells; w is represented by the cocycle which assigns the value $+1$ to the cell S_3, and 0 to all other cells; z is represented by the cocycle which assigns the value $+1$ to e and 0 to all other cells.

Lemma 7. *The triple products in K are as follows:*

$$\langle u, v, w \rangle = (-1)^p z,$$

$$\langle v, w, u \rangle = -(-1)^{pq+pr+p} z,$$

$$\langle w, u, v \rangle = 0.$$

Proof. Let S be an oriented sphere of dimension $q + r - 1$ and let $g: S \to S_2 \vee S_3$ be a map which represents the Whitehead product $[\beta, \gamma]$. We will consider $S_1 \times S$ as a cell complex consisting of a single vertex, oriented cells of dimensions p and $q + r - 1$ respectively whose closures are S_1 and S, and an oriented cell e_1 of dimension $\rho - 1$. We

† It is not necessary to make any assumptions on the integers p, q and r, other than their being positive.

may consider that the orientations are chosen so that e_1 is attached to $S_1 \vee S$ by a map representing $[\xi, \eta]$, where ξ and η are the homotopy classes of the inclusion maps $S_1 \to S_1 \vee S$ and $S \to S_1 \vee S$ respectively. Let $u' \in H^p(S_1 \times S)$, $y \in H^{q+r-1}(S_1 \times S)$ and $z' \in H^{\rho-1}(S_1 \times S)$ be the integral cohomology classes which are represented by cocycles which take the values $+1$ on the cells S_1, S and e_1 respectively, and 0 on all other cells. Then in the cohomology ring of $S_1 \times S$,

$$u' \cdot y = z', \quad y \cdot u' = (-1)^{p(q+r-1)} z'.$$

Consider the map f': $S_1 \vee S \to K$ defined by $f' \mid S_1 =$ the identity map $S_1 \to S_1$, and $f' \mid S = g$. By a theorem of J. H. C. Whitehead ([9], Lemma 7) f' can be extended to a cellular map f: $S_1 \times S \to K$ such that e_1 is mapped onto e with degree $+1$. Consider the induced homomorphism f^*: $H^*(K) \to H^*(S_1 \times S)$; we have

$$f^*u = u',$$

$$f^*v = f^*w = 0,$$

$$f^*z = z'.$$

Also, by use of Theorem IV one readily proves that

$$R_f(v, w) = L_f(v, w) = -y.$$

Now apply Lemma 4. The result is that

$$\begin{aligned} f^*\langle u, v, w\rangle &= R_f(u, v) \cdot (f^*w) - (-1)^p (f^*u) L_f(v, w) \\ &= (-1)^p u' \cdot y = (-1)^p z', \\ f^*\langle v, w, u\rangle &= R_f(v, w) \cdot (f^*u) - (-1)^q (f^*v) \cdot L_f(w, u) \\ &= -y \cdot u' = -(-1)^{p(q+r-1)} z'. \end{aligned}$$

Also, by Lemma 5, $\quad f^* \langle w, u, v\rangle = 0.$

The lemma now follows from the fact that f^*: $H^{\rho-1}(K) \to H^{\rho-1}(S_1 \times S)$ is an isomorphism onto, and $f^*(z) = z'$.

LEMMA 8. *Let L be a cell complex obtained by adjoining an oriented cell e_2 of dimension $\rho - 1$ to X by a map representing $m \cdot [\alpha, [\beta, \gamma]]$, where m is an integer. Then*

$$\langle u, v, w\rangle = (-1)^p mz,$$

$$\langle v, w, u\rangle = -(-1)^{pq+pr+p} mz,$$

$$\langle w, u, v\rangle = 0,$$

where z is the cohomology class represented by the cocycle which takes the value $+1$ on e_2 and 0 on all other cells.

PROOF. By the lemma of J. H. C. Whitehead used in the proof of the preceding lemma, there exists a map $f\colon L \to K$ (where K is the complex described in the preceding lemma) such that $f|X$ is the identity map $X \to X$, and f maps e_2 onto e with degree m. One now applies the preceding lemma, and the fact that the triple product is a natural operation.

REMARK. By means of this lemma, one can easily prove that $[\alpha,[\beta,\gamma]]$ is an element of $\pi_{\rho-1}(X)$ of infinite order.

We are now ready to prove Theorem II. Assume that there exists a linear dependence relation

$$l[\alpha,[\beta,\gamma]]+m[\beta,[\gamma,\alpha]]+n[\gamma,[\alpha,\beta]]=0.$$

Let M be the cell complex obtained by attaching oriented cells e_1, e_2 and e_3 of dimension $\rho-1$ to X by maps representing $l[\alpha,[\beta,\gamma]]$, $m[\beta,[\gamma,\alpha]]$ and $n[\gamma,[\alpha,\beta]]$ respectively. Let $M_1=X \cup e_1$, $M_2=X \cup e_2$ and $M_3=X \cup e_3$ denote subcomplexes of M. By using Lemma 7 we can compute triple products in M_1, M_2 and M_3; and then by considering the injections $H^*(M) \to H^*(M_1)$, $H^*(M) \to H^*(M_2)$ and $H^*(M) \to H^*(M_3)$ and the fact that the triple product is a natural operation, we can determine the triple products in M. The result is as follows. Let $z_i \in H^{\rho-1}(M)$ be the cohomology class represented by the cocycle which takes the value $+1$ on the cell e_i and 0 on all other cells ($i=1,2,3$). Then

$$\langle u,v,w\rangle=(-1)^p\, lz_1-(-1)^{rp+rq+r}\, nz_3,$$

$$\langle v,w,u\rangle=(-1)^q\, mz_2-(-1)^{pq+pr+p}\, lz_1,$$

$$\langle w,u,v\rangle=(-1)^r\, nz_3-(-1)^{qr+qp+q}\, mz_2.$$

Now let $N=X \vee S_4$ be the union of X and an oriented sphere S_4 of dimension $\rho-1$ with a single point in common. By the lemma of J. H. C. Whitehead used twice previously, the identity map $X \to X$ can be extended to a map $h\colon N \to M$ such that the induced homomorphism of chains, $h_\#\colon C_{\rho-1}(N) \to C_{\rho-1}(M)$ is defined by

$$h_\#(S_4)=e_1+e_2+e_3.$$

Let $z_4 \in H^{\rho-1}(N)$ be the cohomology class of the cocycle which takes the value $+1$ on S_4. Then the homomorphism $h^*\colon H^{\rho-1}(M) \to H^{\rho-1}(N)$ is defined by

$$h^*(z_1)=h^*(z_2)=h^*(z_3)=z_4.$$

Now the triple product operation is obviously 0 in the cell complex N; therefore

$$h^*\langle u,v,w\rangle=h^*\langle v,w,u\rangle=h^*\langle w,u,v\rangle=0.$$

Hence
$$(-1)^p l-(-1)^{rp+rq+r} n=0,$$
$$(-1)^q m-(-1)^{pq+pr+p} l=0,$$
$$(-1)^r n-(-1)^{qr+qp+q} m=0.$$

From these equations Theorem II follows at once. This completes the proof.

BROWN UNIVERSITY

REFERENCES

[1] A. L. BLAKERS and W. S. MASSEY, *The homotopy groups of a triad. I*, Ann. of Math. 53 (1951), pp. 161–205.

[2] —— ——, *The homotopy groups of triad. II*, Ann. of Math., 55 (1952), pp. 192–201.

[3] —— ——, *The homotopy groups of a triad. III*, Ann. of Math., 58 (1953), pp. 409–417.

[4] —— ——, *Products in homotopy theory*, Ann. of Math., 58 (1953), pp. 295–324.

[5] J. C. MOORE, *Some applications of homology theory to homotopy problems*, Ann. of Math., 58 (1953), pp. 325–350.

[6] N. E. STEENROD, *Cohomology invariants of mappings*, Ann. of Math., 50 (1949), pp. 954–988.

[7] G. W. WHITEHEAD, *A generalization of the Hopf invariant*, Ann. of Math., 51 (1950), pp. 192–237.

[8] J. H. C. WHITEHEAD, *Combinatorial homotopy. I*, Bull. Amer. Math. Soc., 55 (1949), pp. 213–245.

[9] ——, *On simply-connected, 4-dimensional polyhedra*, Comment. Math. Helv., 22 (1949), pp. 48–92.

Some Mapping Theorems with Applications to Non-Locally Connected Spaces

R. L. Wilder

Introduction

ONE of Professor Lefschetz's main contributions to topology centered about the theory of local connectedness. In a series of papers and a monograph† he developed the related global concepts, especially the concept of chain-realization of a complex, and applied them to the study of retracts, homology characters of a space, generalized manifolds, and extensions of his fixed-point theorem. Throughout this work, there clearly existed the feeling that through the device of local connectedness, many of the properties of finite complexes, in the classical sense, can be carried over to topological spaces.

It is my purpose in this paper to indicate how, through mapping devices, one can go even further, extending such theorems as that of the complex-like character of a space which is locally connected in various or all dimensions, and the fixed point theorem, to spaces which are not locally connected, but which possess properties which allow of application of the known theorems on locally connected spaces.

It will be of assistance to consider some examples:

EXAMPLE 1. Let M denote the well-known sine curve configuration, consisting of the following two subsets of the cartesian plane:

$$K=\{(0,y)\mid -1\leqq y\leqq 1\},\quad L=\{(x,y)\mid y=\sin 1/x,\, 0<x\leqq 1/\pi\}.$$

Let N be a simple arc in the fourth quadrant of the plane, joining $(0,-1)$ to $(1/\pi,0)$, which, except for these two points, does not meet M. Let $S=M\cup N$. In terms of either Vietoris or Čech homology, S has the same homology character, globally speaking, as the circle S^1; it is therefore 'complex-like'. An easy way to check the assertion

† See [7] and papers referred to therein.

just made is to notice that there exists a continuous mapping $f: S \to S^1$ which carries all of K into one point of S^1 and is elsewhere on S a homeomorphism, and which therefore constitutes a mapping satisfying the well-known Vietoris theorem according to which, if f has counter-images $f^{-1}(y)$, $y \in S^1$, which are all acyclic (in the sense of homology), then S and S^1 have isomorphic homology groups. Thus, although S is not even 0-lc, we are able to relate its homology groups to those of a known locally connected space in such a way as to recognize the complex-like character of S.

EXAMPLE 2. Again, the space S of Example 1, not being 0-lc, is not subject to the fixed-point theorem. However, as we shall show below, there is a sense in which the fixed-point theorem applies. Moreover, in contrast to its homological similarity to S^1, *every* continuous mapping of S into itself has a fixed point.

EXAMPLE 3. On the other hand, consider the following example: Using cylindrical coordinates (r, θ, z) in 3-space, let K denote the circle $r = 1$ in the (r, θ)-plane. On the cylinder $r = 1$, let L denote the portion of the curve $\theta = 1/z$ between the points $p = (1, 1, 1)$ and $q = (1, -1, -1)$. And let A be an arc with endpoints p and q, which, except for these points, lies entirely in the part of 3-space for which $r > 1$. Finally, let $M = K \cup L \cup A$. Then there exist continuous mappings $f: M \to M$ for which the Lefschetz number $\theta(f) \neq 0$ and which have no fixed points (these are mappings which carry K into itself in sense-preserving fashion and slightly displace each point of $L \cup A$).

1. Generalizations of the Vietoris mapping theorem

Example 1 is a special case of the fact that any compact space S which can be mapped onto an lc^n space by an n-monotone mapping must be complex-like, in dimensions 0 to n; for by the Vietoris theorem, over this range of dimensions its homology groups are isomorphic to those of the lc^n space.†

There are, however, ways in which the Vietoris theorem can be extended to non-monotone cases which allow of a much wider degree of application. First let us make the convention that if S is any locally compact space, then $\hat{S}$ denotes its ordinary one-point compactification

† For extension of the Vietoris theorem to the general compact topological space, see [3]. As noticed by Begle (loc. cit.), when the coefficient domain is not a field or a compact topological group, the 'monotone' condition takes a different form; as we use it, *n-monotone* signifies that the counter-images of all points are r-acyclic in the homology sense in all dimensions 0 to n. For this reason, and because of other results of which we make use below, we shall assume the coefficients domain always to be a field.

by the addition of a point $\hat{p}$. Let $\mathfrak{H}^r(S)$ be the r-dimensional homology group of S based on infinite cycles; this is a group that is isomorphic with the group $H^r(S; \hat{p})$.† Then we can state:

THEOREM 1.1. *Let S and T be locally compact spaces, and $f(S)=T$ a continuous mapping such that counter-images of compact subsets of T are compact in S. Then if f is n-monotone, $n>0$, $\mathfrak{H}^n(S) \cong \mathfrak{H}^n(T)$.*

PROOF. Since, if either S or T is compact the other must be, and the theorem then follows from the Vietoris theorem, we assume neither is compact.

We extend f to a mapping $\hat{f}(\hat{S})=\hat{T}$, such that $\hat{f}(\hat{p})=\hat{q}$, where $\hat{q}$ is the point of compactification for T; $\hat{f}$ is continuous at $\hat{p}$ under the compactness condition imposed on f. Also, $\hat{f}$ is n-monotone so that $H^n(\hat{S}) \cong H^n(\hat{T})$. But in the exact sequence

$$(1.1) \qquad H^n(\hat{p}) \xrightarrow{i} H^n(\hat{S}) \xrightarrow{j} H^n(\hat{S}; \hat{p}) \to H^{n-1}(\hat{p}),$$

both $H^n(\hat{p})$ and $H^{n-1}(\hat{p})$ are trivial so that $H^n(\hat{S}) \cong H^n(\hat{S}; \hat{p})$. And since, similarly, $H^n(\hat{T}) \cong H^n(\hat{T}; \hat{q})$, the theorem follows.

THEOREM 1.2. *Let S be a compact space and $f(S)=T$ a continuous mapping. Then if there exists a closed subset D of T such that both $C=f^{-1}(D)$ and D are $(n-1)$- and n-acyclic, and f is n-monotone on $S-C$, then $H^n(S) \cong H^n(T)$.*

PROOF. By Theorem 1.1, $\mathfrak{H}^n(S-C) \cong \mathfrak{H}^n(T-D)$. But

$$\mathfrak{H}^n(S-C) \cong H^n(S; C) \quad \text{and} \quad \mathfrak{H}^n(T-D) \cong H^n(T; D).$$

And by use of an exact sequence similar to that of (1.1), but with C replacing $\hat{p}$, we get $H^n(S) \cong H^n(S; C)$, etc., and the theorem follows by combining isomorphisms.

Since, if a compact space is at most n-dimensional, its homology groups of dimension $>n$ are trivial, an interesting corollary of Theorem 1.2 is:

COROLLARY 1.1. *Let S be compact and $f(S)=T$ a continuous mapping. Then if D is a closed subset of T such that both $C=f^{-1}(D)$ and D are at most $(n-2)$-dimensional, and f is n-monotone on $S-C$, then*

$$H^n(S) \cong H^n(T).$$

EXAMPLE 4. Let C be a 'Cantor set' on a 2-sphere S^2, and f a mapping such that $f(C)$ is a point and on $S-C$, f is a homeomorphism (in other words, f is the mapping which identifies all points of C). Then by the Corollary, $f(S)$ has 2-dimensional Betti number equal to 1.

† See H. Cartan [4]; also Wilder [12]. We shall use the symbol $H^r(A; B)$ to denote the rth homology group of A modulo B.

Example 5. Let S be an S^2 and C a circular disk on S with boundary J. Let T consist of two 2-spheres D and L tangent at a point p. Let $f(S)=T$ be such that $f(J)=p$, and on $S-J$, f is a homeomorphism carrying $C-J$ onto $D-p$ and $S-C$ onto $L-p$. Clearly f is not 1-monotone, but Theorem 1.2 applies to show $H^1(S)$ and $H^1(T)$ must be isomorphic.

Example 4 is also a special case of the following corollary of Theorem 2:

Corollary 1.2. *If C is any closed proper subset of S^2 which does not separate S^2, and all points of C are identified, then the 2-dimensional homology group of the resulting configuration is isomorphic with that of S^2.*

Corollary 1.3. *If C is a closed proper subset of an n-gcm M (orientable or not)† such that $H^{n-1}(C)$ is trivial, then the nth homology group of the space obtained by identifying all points of C is isomorphic with that of M.*

Corollary 1.4. *Let C be an arc on a 2-sphere S. Let $\phi\colon C \to P$ be a continuous mapping onto a Peano space P which is 1- and 2-acyclic. Let two points $p, q \in C$ be identified if $\phi(p)=\phi(q)$. The mapping f of S so induced yields a space T such that $H^2(T)$ is isomorphic with $H^2(S^2)$, $i \leqq 2$.*

For non-compact spaces one can state:

Theorem 1.3. *Let S and T be locally compact spaces and $f(S)=T$ a continuous mapping such that counter-images of compact subsets of T are compact. Then if D is a closed subset of T such that both $C=f^{-1}(D)$ and D are n- and $(n-1)$-acyclic in terms of infinite cycles, $n>1$, and f is n-monotone on $S-C$, then $\mathfrak{H}^n(S) \cong \mathfrak{H}^n(T)$.*

Proof. Evidently

$$H^n(\hat{C}) \cong H^n(\hat{C}; \hat{p}) \cong H^n(C) = 0,$$

and

$$H^{n-1}(\hat{C}) \cong H^{n-1}(\hat{C}; \hat{p}) \cong \mathfrak{H}^{n-1}(C) = 0;$$

and similar relations hold on T and D. Hence $f(\hat{S})=\hat{T}$, defined as in the proof of Theorem 1.1, is n-monotone on $\hat{S}-\hat{C}$, and by Theorem 1.2, $H^n(\hat{S}) \cong H^n(\hat{T})$. As a consequence, $\mathfrak{H}^n(S) \cong \mathfrak{H}^n(T)$.

Remark. If in Theorems 1.1, 1.2, 1.3 and Corollary 1.1 the 'n-monotone' condition on f is replaced by '$(n-1)$-monotone', then (in the case of Theorems 1.1 and 1.3) f induces a homomorphism of $\mathfrak{H}^n(S)$ onto $\mathfrak{H}^n(T)$, and (in the case of Theorem 1.2 and Corollary 1.1) a homomorphism of $H^n(S)$ onto $H^n(T)$.‡

† By 'n-gcm' is meant an n-dimensional compact generalized manifold; see [12], p. 244.

‡ The extension of Vietoris's Theorem (9) (loc. cit.) to general compact spaces is assumed here. This extension has not yet been published, however, and without it conditions of metrizability should presumably be imposed on the spaces considered

Inasmuch as all that is desired in many applications of theorems of the Vietoris type is the finiteness of the Betti number of the image space, it is worthwhile noting that for this purpose the conditions imposed in the preceding theorems can be considerably lightened. For example:

THEOREM 1.4. *Let $f(S)=T$ be a continuous mapping of a compactum S that is $(n-1)$-monotone $(n>0)$, except possibly for points of a closed subset D of T such that $p^n(D)$ and $p^{n-1}(C)$ are finite, where $C=f^{-1}(D)$. If $p^n(S)$ is finite, then $p^n(T)$ is finite.*

PROOF. Consider the following diagram

$$\begin{array}{ccccc} & & H^n(S) \xrightarrow{j} & H^n(S; C) & \xrightarrow{\partial} H^{n-1}(C) \\ & & & \downarrow \Phi & \\ H^n(D) \xrightarrow{i'} & H^n(T) & \xrightarrow{j'} & H^n(T; D) & \end{array}$$

in which the two sequences are portions of the homology sequences of the pairs $(S; C)$ and $(T; D)$. We observe: (1) From the Remark above concerning Theorem 1.1 and the fact that f is $(n-1)$-monotone on $S-C$, it follows that f induces a homomorphism Φ of $\mathfrak{H}^n(S-C)$ onto $\mathfrak{H}^n(T-D)$, and accordingly the homomorphism Φ of the diagram is onto; (2) since both $H^n(S)$ and $H^{n-1}(C)$ are finite dimensional, so must $H^n(S; C)$ be finite dimensional; (3) since (1) together with (2) implies that $H^n(T; D)$ is finite dimensional, and $H^n(D)$ is finite dimensional by hypothesis, it follows that $H^n(T)$ is finite dimensional.

It should be noted that similar considerations yield the following theorem:

THEOREM 1.5. *Let $f(S)=T$ be a continuous mapping of a compact space S, where $p^n(T)$ is finite, and suppose that there exists a closed subset D of T such that both $p^n(C)$ and $p^{n-1}(D)$, $C=f^{-1}(D)$, are finite and f is n-monotone on $S-C$. Then $p^n(S)$ must be finite.*

REMARK. Both Theorems 1.4 and 1.5 continue to hold if S and T are only assumed to be locally compact, f is such that inverses of compact subsets of T are compact, and the Betti numbers involved are all dimensions of homology groups based on infinite cycles; i.e. $p^n(C)$ is the dimension of $\mathfrak{H}^n(C)$, etc. The proofs are based on considerations such as those employed in the proof of Theorem 1.3.

here. For this reason, we shall frequently use the term 'compactum' (=compact metric space) where the above Remark is utilized—or in any case where the Vietoris theorem cited enters the picture.

2. The complex-like character of certain non-locally connected spaces

Given a mapping $f(S) = T$, either S or T may be lc^n. If S is a compact, lc^n space, and f if $(n-1)$-monotone, then T is lc^n—a result which we have previously established [11], and which generalizes the classical theorem on mappings of 0-lc spaces (for $n = 0$, the '(-1)-monotone' condition imposes no monotoneity on the mapping). It follows trivially in this case that T is complex-like in dimensions 0 to n. In view of Theorem 1.2 we can state:

THEOREM 2.1. *Let S be an lc^n compactum, and $f(S) = T$ a continuous mapping. If there exists a closed subset D of T such that both $C = f^{-1}(D)$ and D are n- and $(n-1)$-acyclic, and f is $(n-1)$-monotone on $S - C$, then T is complex-like in dimensions* 0 *to* n. [See Remark, end of § 1.]

THEOREM 2.2. *If a compact space S can be mapped by a continuous mapping f onto an lc^n space T in such a way that the set of points at which f is not n-monotone maps into a subset of a closed set D which, together with $C = f^{-1}(D)$, is n- and $(n-1)$-acyclic, then $p^n(S)$ is finite.*

For certain purposes it is desirable not to have to assume that, as in Theorem 2.1, a space S is lc^n at all points; in such cases, the following theorem and corollary will be found useful:

THEOREM 2.3. *Let S be a compactum such that $p^n(S)$ is finite* $(n > 0)$, *and $f(S) = T$ a continuous mapping; and let D be a closed, totally disconnected subset of T such that if $C = f^{-1}(D)$, S is lc^n at all points of $S - C$, $p^{n-1}(C)$ is finite, and f is $(n-1)$-monotone on $S - C$. Then T is n-*lc.

PROOF. By Theorem 1.4, $p^n(T)$ is finite. Hence T is semi-n-connected ([12], p. 168). Let $x \in T$ and suppose that T is not n-lc at x. Then there exists a neighborhood U of x such that for every neighborhood V of x in U, infinitely many n-cycles in V are lirh in U. We may suppose U sufficiently small so that all n-cycles in U bound on T, since T is semi-n-connected. And since D is totally disconnected, there exists a neighborhood V of x with boundary F such that $\bar{V} \subset U$ and $D \cap F = \phi$. Let R be an open subset of $U - x$ containing F such that $x \notin \bar{R}$ and $R \cap D = \phi$, and let $P = V \cup R$, $Q = V - R$.

Since there exist infinitely many n-cycles in Q that are lirh in U, and since such cycles all bound on S, they are homologous on V to cycles Z_i^n on F;† and such cycles are lirh in $P - Q$. However, since $P - Q$ does not meet D, f is $(n-1)$-monotone at all points of $f^{-1}(P - Q)$.

† See ([12], p. 203), Lemma 1.13. In the statement of this lemma, incidentally, the bar over '$S - K$' should be deleted.

Consequently, by a theorem of Vietoris (loc. cit., [6]), for each Z_i^n there exists a cycle C_i^n of $f^{-1}(F)$ such that $f(C_i^n) \sim Z_i^n$ on F. But the set $f^{-1}(P-Q)$ is an open subset of S containing $f^{-1}(F)$, and since S is lc^n at all points of $f^{-1}(P-Q)$, at most a finite number of the cycles C_i^n are lirh in $f^{-1}(P-Q)$. (See [12], p. 183, Corollary 3.8.) It follows that at most a finite number of the cycles Z_i^n are lirh in $P-Q$, in contradiction to their independence therein.

COROLLARY 2.1. *With S, T, f, C and D as in Theorem 2.3, but with the assumption that $p^r(S)$ and $p^s(C)$ are finite for $r=0,1,\ldots,n$ and $s=0,1,\ldots,n-1$, and $n>-1$, then T is* lc^n.

That T is 0-lc follows from the fact that no continuum can fail to be 0-lc at only a totally disconnected set, and T is easily shown θ-lc at points of T-D. One can state:

LEMMA. *Let S be a compact space such that $p^0(S)$ is finite and $f(S)=T$ a continuous mapping, and suppose there exists a closed, totally disconnected subset D of T such that if $C=f^{-1}(D)$, then S is* 0-lc *at all points of $S-C$. Then T is* 0-lc.

An example or two here would be instructive. For instance, consider Example 3: If all points of K are identified, the resulting mapping of M provides an instance where the Corollary or the Lemma applies. Also consider:

EXAMPLE 6. In the coordinate place, let

$$S_n=\{(x,y)\mid (x=1/n) \text{ and } (-1/n \leqq y \leqq 1/n)\},$$

$$L_n=\{(x.y)\mid 1/(n+1)\leqq x\leqq 1/n) \text{ and } (y=0)\},$$

and S_0 consist of $(0,0)$ alone. Let $S=\bigcup_{n=0}^{\infty} S_n \cup \bigcup_{n=1}^{\infty} L_n$. Let f be a mapping which identifies each pair of points $(1/n,-1/n)$, $(1/n,1/n)$ in S; then $f(S)$ is 0-lc but not 1-lc—it fails to satisfy Theorem 2.3, with $n=1$ and $D=f((0,0))$ only in that $p^0(C)$ is not finite.

If $p_n=(1/n,0)$, then the identity mapping of $M=S_0\cup \bigcup p_n$ onto itself shows the necessity for the condition $p^0(S)$ finite in the Lemma.

In a previous paper [10], I have discussed a canonical procedure for mapping certain types of non-lc spaces onto lc spaces.† For our purposes a brief summary of this will suffice: Let π denote a local topological property. If a space fails to have property π at a point x, then x is called a π-singular point of S. For a certain class Γ of spaces, the existence of a π-singular point may imply that the closure of the set of all π-singular points contains a continuum; in this case we call π

† For the case 0-lc, this procedure was initiated by Hahn [5] and studied extensively by R. L. Moore [8] and G. T. Whyburn [9].

expansive relative to Γ. For example, if a continuum fails to be 0-lc at some point, then it fails to be 0-lc at a subcontinuum; thus the 0-lc property is expansive relative to the class of all continua, or more generally relative to the class of all compact spaces with finite 0-dimensional Betti numbers. And still more generally, if we let C_{-1}^n denote the class of all compact spaces that are complex-like in all dimensions from 0 to n, then the property of being lc^n is expansive relative to C_{-1}^n. And if we let C_m^n $(m < n)$, denote the class of all compact spaces that are lc^m and complex-like in dimensions $m+1$ to n, then the property lc_{m+1}^n of being r-lc for $r = m+1, m+2, \ldots, n$ is expansive relative to C_m^n.

If we call each component of the closure of the set of π-singular points a *π-prime part*,† then it is a general theorem ([10], p. 691) that if a property π is expansive relative to a class Γ of compact spaces, and $S \in \Gamma$, then the space S' obtained by identifying all points in each π-prime part will, if in the class Γ, have property π at every point. In particular, if π is lc_{m+1}^n, $S \in C_m^n$, and each lc_{m+1}^n-prime part is r-acyclic in dimensions 0 to $n-1$, then the space S' of lc_{m+1}^n-prime parts of S will be in the class C_m^n and hence a lc^n space. Evidently in view of the Remark following Theorem 1.3, we may state a very general theorem along these lines:

THEOREM 2.4. *Let $S \in C_m^n$, and suppose that there exists a closed subset C of S which consists of lc_{m+1}^n-prime parts of S and moreover contains all those which are not acyclic in dimensions $m+1$ to $n-1$, but which together with its image in S' is acyclic in dimensions m to n.‡ Then in case the lc_{m+1}^n-prime parts are acyclic in dimensions 1 to m, the space of lc_{m+1}^n-prime parts of S is lc^n.*

The combination of Theorem 1.2 and Corollary 2.1 gives an interesting variant of the preceding theorem:

THEOREM 2.5. *Let $S \in C_m^n$ $(m > 1)$, and suppose there exists a closed subset C of S which consists of lc_{m+1}^n-prime parts of S, contains all those that are not acyclic in dimensions 1 to $n-1$, and such that $p^r(C)$ is finite for $r \leqq m-1$ and $=0$ for $m \leqq r \leqq n$. If the image of C in the space S' of lc_{m+1}^n-prime parts of S is closed and totally disconnected, then S' is lc^n.*

PROOF. It is only necessary to show that $S' \in C_m^n$. We apply Corollary 2.1 to show S' is lc^m; and for this purpose we note that $p^r(S)$ is finite for $r \leqq m$ because of the complex-like character of the lc^m space S, and $p^r(C)$ is finite for $r \leqq m-1$ by hypothesis. We then apply Theorem

† The term 'prime part' ('Primteil') was introduced by Hahn (loc. cit.).

‡ S' as defined above.

1.2 (and the Remark following Theorem 1.3) to show that $p^r(S')$ is finite for $m+1 \leqq r \leqq n$; for this purpose we note that since D is closed and totally disconnected, it is acyclic in dimensions $\geqq 1$, and C is acyclic in dimensions m to n by hypothesis.

Finally, let lc_0^∞ or lc^∞ denote the property of being r-lc for all dimensions r; and in general lc_m^∞ denote the property of being r-lc for all dimensions $r \geqq m$. Then by the same methods as were used in proving Theorem 4 of [10], we may show that the property lc_{m+1}^∞ is expansive relative to the class C_m^∞ of all compact spaces that are lc^m and have finite r-dimensional Betti numbers for all $r > m$ (we make the convention that lc^{-1} imposes no local connectedness conditions). And we can now state that *Theorem* 2 *continues to hold for* $n = \infty$, *and* m *any integer* $\geqq -1$.

3. Fixed point theorems for non-locally connected continua

Utilizing the notions described in the latter part of § 2, we may obtain fixed point theorems for certain types of non-locally connected spaces. First we make the following definition:

Definition 3.1. If $f: S \to T$ is a mapping and π a local topological property such that every π-prime part of S is mapped by f into a subset of a π-prime part of T, then f will be called *π-prime part preserving*. We shall say f has a fixed π-prime part if it maps some π-prime part into itself.

Theorem 3.1. *Let S be a compact space of finite dimension all of whose Betti numbers are finite and whose* lc^∞*-prime parts are acyclic. If f is a continuous* lc^∞*-prime part preserving mapping of S into itself whose Lefschetz number* $\theta(f) \neq 0$, *then f has a fixed* lc^∞*-prime part; and in particular if the acyclic character of the prime parts is such as to make them have the fixed point property, f has a fixed point.*

Proof. Let S' be the space of lc^∞-prime parts of S. As noted above (§ 2), S' is lc^∞. Let $\phi: S \to S'$ be the mapping that maps each point x of S into the prime part, x', that contains x. Since f carries all points of a prime part into a subset of some prime part x', the mapping $g = \phi f \phi^{-1}(x')$ is a well-defined mapping of S' into itself. It is elementary to show that this mapping is continuous. We shall show that $\theta(g) = \theta(f)$, from which it will follow ([2], p. 556) that g has a fixed point, and hence that f maps some prime part into itself.

We recall that since the prime parts of S are acyclic, ϕ induces an isomorphism $H^r(S) \cong H^r(S')$ which we denote by Φ. Let the homomorphism of $H^r(S)$ into $H^r(S)$ induced by f be denoted by F, and that of $H^r(S')$ into $H^r(S')$ induced by g be denoted by G. Then in the diagram

$$\begin{array}{ccc} H^r(S) & \xrightarrow{F} & H^r(S) \\ \downarrow \Phi & & \downarrow \Phi \\ H^r(S') & \xrightarrow{G} & H^r(S') \end{array}$$

the mappings commute: $G\Phi = \Phi F$.

Let $\{Z_i^r\}$ and $\{\Gamma_i^r\}$ denotes bases for $H^r(S)$ and $H^r(S')$ respectively, such that $\Gamma_i^r = \Phi(Z_i^r)$. And let the relations between basic elements of $H^r(S)$ induced by F be denoted by $F(Z_i^r) = \sum a_{ij}^r Z_j^r$, so that

$$\theta(f) = \sum (-1)^r \operatorname{trace} \| a_{ij}^r \|.$$

Then $\Phi F(Z_i^r) \sim \sum a_{ij}^r \Phi(Z_j^r)$, so that $G\Phi(Z_i^r) \sim \sum a_{ij}^r \Phi(Z_j^r)$; that is, $G(\Gamma_i^r) \sim \sum a_{ij}^r \Gamma_j^r$, so that $\theta(g) = \sum (-1)^r \operatorname{trace} \| a_j^r \|$. Hence $\theta(f) = \theta(g)$.

In a similar manner it may be shown:

THEOREM 3.2. *Under the same hypotheses regarding S as in Theorem 3.1, if f is an upper semi-continuous transformation that assigns to each point x of S the union of a collection $P(x)$ of prime parts of S whose union is acyclic, and such that if x and y lie in the same prime part of S then $P(x) = P(y)$, then $\theta(f) \neq 0$ implies that $f(x)$ contains x for some x.*†

[That the mapping g, defined as g was defined in the proof of Theorem 3.1, is in the present case upper semi-continuous is easily seen as follows: Consider the mapping ψ: $S \times S \rightarrow S' \times S'$ which is defined by $\psi(x, y) = (x', y')$, where x' denotes the prime part containing x. Then ψ is a closed mapping since $S \times S$ is compact. The graph of g in $S' \times S'$ is the image under ψ of the graph of f in $S \times S$. The latter is closed by the upper semi-continuity of f and hence the former is closed, and thus g is upper semi-continuous.]

REMARK. Consider the space S of Example 1 again. We noted that its homology groups are the same as those of S^1, observing that this follows immediately from a mapping onto S^1 which is n-monotone for all n. This mapping was equivalent to the mapping of S into its space S' of lc^0-prime parts, the latter forming an S^1. Let us consider the general prime part preserving mappings of S into itself. Since the space S' of prime parts is an S^1, it might be expected that these mappings would satisfy the same fixed point properties as S^1, viz. that in general only the 'sense-reversing' mappings have fixed points. However, it is the case that *every* prime part preserving mapping f of S into itself has a fixed point.

† See [1] for the theorem on which the proof is based and for the definition of $\theta(f)$ in this case.

To see this, note first that if the prime part K maps into itself, it is trivial that there exists a fixed point. Suppose K maps into a point p not in K. Let $L_a = \{(x, y) \mid y = \sin 1/x, 0 < x < a\}$. Then for every neighborhood $U(p)$, the set L_a, for suitable a, maps into $U(p)$. It follows that for a sufficiently small, $f(S)$ cannot intersect L_a, since the complement of L_a in S is an arc A and $f(A)$ is peanian. Hence $\theta(f) = 1$ and by Theorem 3.1, f has a fixed point. Similar considerations show that every continuous map of S into itself has a fixed point.

In both Theorems 3.1 and 3.2, the prime parts are assumed to be acyclic. An examination of examples (such as Example 3) of non-locally connected spaces which fail to satisfy the Lefschetz fixed point theorem reveals that in all cases the lc^r-prime parts, for some r, are not acyclic. The same holds in recent examples (see [6] for instance) of contractible continua not satisfying the fixed point theorem. It seems likely that if the lc^∞-prime parts of such spaces are acyclic, then the fixed point theorems apply for unrestricted continuous mappings. I hope to return to this question in a later paper.

UNIVERSITY OF MICHIGAN

REFERENCES

[1] E. G. BEGLE, *A fixed point theorem*, Ann. of Math., 51 (1950), pp. 544–550.

[2] ——, *Locally connected spaces and generalized manifolds*, Amer. J. Math., 64 (1942), pp. 553–573.

[3] ——, *The Vietoris mapping theorem for bicompact spaces*, Ann. of Math., 51 (1950), pp. 534–543.

[4] H. CARTAN, *Méthodes modernes en topologie algébrique*, Comment. Math. Helv., 18 (1945–46), pp. 1–15.

[5] H. HAHN, *Über irreduzible Kontinua*, Preuss. Akad. Wiss. Sitzungsber., Math.-Naturw. Kl., Abt. IIa, 130 (1921), pp. 217–250.

[6] S. KINOSHITA, *On some contractible continua without fixed point property*, Fund. Math., 40 (1953), pp. 96–98.

[7] S. LEFSCHETZ, Topics in Topology, Princeton University Press, 1942.

[8] R. L. MOORE, *Concerning the prime parts of a continuum*, Math. Zeit., 22 (1925), pp. 307–315.

[9] G. T. WHYBURN, *A decomposition theorem for closed sets*, Bull. Amer. Math. Soc., 41 (1935), pp. 95–96.

[10] R. L. WILDER, *Decompositions of compact metric spaces*, Amer. J. Math., 63 (1941), pp. 691–697.

[11] ——, *Monotone mappings of manifolds*, Bull. Amer. Math. Soc., 54 (1948), p. 1086 (abstract).

[12] ——, Topology of Manifolds, Amer. Math. Soc. Colloquium Publications, 32 (1949).

Intercept-Finite Cell Complexes

S. Wylie

Introduction

In this paper a class of abstract cell complexes is defined which includes both the star-finite and the closure-finite complexes. Two kinds of homology theory are defined for any such complex; in the case, for instance, of a closure-finite complex, one of these reduces to the familiar theory based on finite chains; the other appears to be unfamiliar.

Infinite complexes are usually stubborn against duality theorems. Here an attempt has been made to impose a duality theorem by using for coefficients not a topological group but a modification called a congregation: a congregation is an Abelian group fortified by a non-finite additive structure. The result is that, for suitable coefficient congregations, the chains and cochains of each kind form congregations that enjoy a mutual duality relation of the familiar kind. Nothing, however, is proved about similar relations for homology and cohomology congregations.

The definitions of a cell complex and much of the terminology has been taken from Lefschetz ([2], Ch. III). Any value there may be in the paper is due, directly or indirectly, also to the same author.

1. Intercept-finite cell complexes

The cell complexes to be discussed satisfy axioms equivalent to those of Lefschetz [2], strengthened by an additional axiom, one of *intercept-finiteness*. This term is derived from the use of the word 'intercept' by Tucker [3].

Definition. An intercept-finite cell complex K is a set of cells x, partially ordered by a relation $x \prec x'$, called the incidence relation. To each cell x is assigned an integer p, its dimension, written as a subscript, as in x_p. To each pair x_p, x'_{p+1} is assigned an integer $[x_p : x'_{p+1}]$, the incidence number of the pair. These are subject to the following axioms:

(1) For each pair x, x', there is but a finite number of cells y of K satisfying $x \prec y \prec x'$.

(2) If $x_p \prec x'_q$, $p \leqq q$.

(3) If $[x_p : x'_{p+1}] \neq 0$, $x_p \prec x'_{p+1}$.

(4) For each pair x_{p-1}, x''_{p+1}, $\sum_{x'_p \in K} [x_{p-1} : x'_p][x'_p : x''_{p+1}] = 0$.

The only new feature of these axioms is the first; the set of cells y satisfying the given relation is called by Tucker the intercept of x and x'; this axiom asserts that all intercepts are finite. Clearly all star-finite and all closure-finite complexes are intercept-finite.

The symbol $\prec$ is taken to be reflexive; for each $x \in K$, $x \prec x$.

The term 'finite' will always mean 'finite or vacuous', or 'finite or zero'. The term 'complex' will mean an intercept-finite cell complex.

A *closed subcomplex* of K is defined to be *of the first kind* if it intersects each closure-finite open subcomplex in a finite number of cells. This class includes cl (x) for each $x \in K$ and any closed subcomplex of cl (x). Closed subcomplexes of the first kind form an ideal in the lattice of closed subcomplexes; for the union of two is also of the first kind and any closed subcomplex of a closed subcomplex of the first kind is itself of the first kind.

A *closed subcomplex* of K is defined to be *of the second kind*, if it is itself star-finite: this is equivalent to saying that it intersects st (x) in a finite set of cells for each $x \in K$. The closed subcomplexes of the second kind also form an ideal in the lattice of all closed subcomplexes.

Since, for any $x \in K$, st (x) is closure-finite, in view of the intercept-finiteness of K, any closed subcomplex of the first kind is also of the second kind.

The definition of the open subcomplexes of the two kinds places them, for reasons related to homology and cohomology, in the opposite order. An *open subcomplex* is *of the first kind* if it intersects cl (x) finitely for each $x \in K$, or, equivalently, if it is closure-finite. An *open subcomplex* is *of the second kind* if it intersects finitely each closed subcomplex of the second kind. Each of these two kinds of open subcomplex forms an ideal in the lattice of open subcomplexes and each of the second kind is also of the first kind.

The purpose of these definitions is to provide suitable carriers for chains and cochains; closed subcomplexes of the first kind carry a class of chains that generalize rather naturally the finite chains; those of the second kind carry a class that restricts suitably the arbitrary chains.

2. Homology and cohomology of two kinds

Integral chains on K are defined in the usual way. The group of p-chains with arbitrary assignments of coefficients will be called $C_p^{(a)}$;

the subgroup of p-chains with but a finite number of non-zero coefficients will be called $C_p^{(f)}$. For any p-chain c_p, $\phi(c_p)$, the *domain of* c_p is defined to be the union of cells x_p for which c_p has non-zero coefficient; $|c_p|$, the carrier of c_p is defined to be $\mathrm{cl}\,\phi(c_p)$. A p-chain whose carrier is a closed subcomplex of the first kind is called a p-chain of the first kind and these form a subgroup $C_p^{(1)}$ of $C_p^{(a)}$; the subgroup $C_p^{(2)}$ consists of p-chains with carriers of the second kind. The chains of each kind form a group, from the property that closed subcomplexes of each kind form an ideal. The groups satisfy

$$C_p^{(f)} \subset C_p^{(1)} \subset C_p^{(2)} \subset C_p^{(a)}.$$

It is possible to define the boundary $c_p \mathfrak{d}$ of a p-chain c_p in the usual way if, and only if, for each $x_{p-1} \in K$ there is a finite number of $x'_p \in K$ such that $[x_{p-1} : x'_p]$ and the coefficient of x'_p in c_p both differ from zero. For an arbitrary chain this condition may not be satisfied; but if $c_p \in C_p^{(2)}$, $|c_p| \cap \mathrm{st}\,(x_{p-1})$ is finite and $c_p \mathfrak{d}$ is therefore defined. Clearly $|c_p \mathfrak{d}| \subset |c_p|$. Hence it follows that $\mathfrak{d}$ determines a homomorphism,

$$\mathfrak{d}\colon\ C_p^{(2)} \rightarrow C_{p-1}^{(2)}. \tag{2.1}$$

The same reasoning shows that it also determines

$$\mathfrak{d}\colon\ C_p^{(1)} \rightarrow C_{p-1}^{(1)}. \tag{2.2}$$

It is noteworthy that the image under $\mathfrak{d}$ of $C_p^{(f)}$ does not, except in closure-finite complexes, in general lie on $C_{p-1}^{(f)}$. This is the reason for transferring interest from the finite chains to chains of the first kind.

The homomorphisms (2.1), (2.2), ensure that homology groups in each dimension of the first and of the second kind can be defined, $H_p^{(1)}(K)$, $H_p^{(2)}(K)$ based respectively on chains of the first and second kind. There is a natural homomorphism of $H_p^{(1)}(K)$ into $H_p^{(2)}(K)$ induced by the injection of $C_p^{(1)}$ in $C_p^{(2)}$. Evidently homology groups of each kind can be similarly defined using an arbitrary Abelian coefficient group G; these will be written as $H_p^{(1)}(K; G)$, $H_p^{(2)}(K; G)$.

These groups can be defined in a different way.† The closed subcomplexes of the second kind in K form a directed system under inclusion. If L_1, L_2 are two such subcomplexes and $L_1 \subset L_2$, the injection of L_1 in L_2 determines a homomorphism

$$h\colon\ H_p^{(2)}(L_1; G) \rightarrow H_p^{(2)}(L_2; G).$$

† I am glad to acknowledge the kindness of C. H. Dowker, to whom this observation is due.

The groups $H_p^{(2)}(L; G)$ and the homomorphisms h determine a direct system of groups whose limit group is $H_p^{(2)}(K; G)$ as already defined. One advantage of this definition is that, since L is in each case star-finite, the group $H_p^{(2)}(L; G)$ is a group of a familiar kind.

Cohomology groups of each kind are defined in the analogous way. A p-cochain is defined as a function from the p-cells to the integers and their group is written $C^p_{(a)}$. For a p-cochain c^p, $\phi(c^p)$, its domain, is the set of p-cells whose value in c^p is non-zero and $|c^p|$, the carrier of c^p, is defined to be st $\phi(c^p)$. Thus $C^p_{(i)}$ $(i=1, 2)$, is the group of p-cochains whose carriers are open subcomplexes of the ith kind and

$$C^p_{(a)} \supset C^p_{(1)} \supset C^p_{(2)} \supset C^p_{(f)},$$

where $C^p_{(f)}$ is defined in the natural way.

The coboundary operator δ is defined on each cochain of $C^p_{(1)}$ and $|\delta c^p| \subset |c^p|$. We have therefore homomorphisms analogous to (2.1), (2.2) and can define integral cohomology groups of each kind, $H^p_{(1)}(K)$, $H^p_{(2)}(K)$. Again an arbitrary Abelian group G can be taken as the value group and the resulting cohomology groups will be written as

$$H^p_{(1)}(K; G), \quad H^p_{(2)}(K; G).$$

In case K is closure-finite, every open subcomplex is closure-finite and the closed subcomplexes of the first kind are just the finite closed subcomplexes. $H_p^{(1)}(K)$ is then the homology group based on finite chains, $H^p_{(1)}(K)$ the cohomology group based on arbitrary cochains. If K is, for instance, a simplicial complex, these are the usual groups. In this case $H_p^{(2)}(K)$ is a homology group based on locally finite chains, that is to say, on chains whose carriers are locally compact in the weak topology for the polyhedron of K.

3. Kronecker indices

The family of subcollections of p-cells formed by domains of chains of the ith kind will be called $\Phi_p^{(i)}$. A collection of p-cells belongs to $\Phi_p^{(i)}$ if it is the set of p-cells of some closed subcomplex of the ith kind: this follows from the definitions without difficulty. Similarly $\Phi^p_{(i)}$ stands for the family of domains of p-cochains of the ith kind and a similar observation holds also for these families.

If Θ is a family of subcollections of a set of objects, Θ^* denotes the family of all subcollections intersecting finitely each element of Θ.

PROPOSITION 1. $(\Phi_p^{(i)})^* = \Phi^p_{(i)}$ *and* $(\Phi^p_{(i)})^* = \Phi_p^{(i)}$.

From the finite intersection property of the definitions, $\Phi_p^{(i)} \subset (\Phi^p_{(i)})^*$;

it will be enough then to prove that $\Phi_p^{(1)} \supset (\Phi_{(1)}^p)^*$ and $\Phi_p^{(2)} \supset (\Phi_{(2)}^p)^*$. The full result then follows by symmetry.

If a subcollection θ of p-cells $\in (\Phi_{(1)}^p)^*$, $\mathrm{cl}(\theta)$ intersects each open subcomplex of the first kind finitely; for θ intersects the p-section of each finitely and each is closure-finite. Hence $\theta \in \Phi_p^{(1)}$ and therefore $\Phi_p^{(1)} \supset (\Phi_{(1)}^p)^*$. Now if $\theta \in (\Phi_{(2)}^p)^*$, θ intersects $\mathrm{st}(x)$ finitely for each x, since $\mathrm{st}(x)$ is open of the second kind; it follows from intercept-finiteness that $\mathrm{cl}(\theta)$ intersects each $\mathrm{st}(x)$ finitely, so that $\mathrm{cl}(\theta)$ is closed of the second kind and $\theta \in \Phi_p^{(2)}$. Hence $\Phi_p^{(2)} \supset (\Phi_{(2)}^p)^*$.

Let Q, R be abstract Abelian groups with a bilinear multiplication pairing them to the abstract Abelian group P. Then, in view of the finiteness properties, a Kronecker index can be defined between an element of $C_p^{(i)}(K; Q)$ and an element of $C_{(i)}^p(K; R)$ which provides a pairing of these two groups to P. This induces in the usual way a pairing of $H_p^{(i)}(K; Q)$ and $H_{(i)}^p(K; R)$ to P.

If, on the other hand, P, Q, R are topological groups and the pairing of Q, R to P is continuous, it is not in general true that the pairing of $C_p^{(i)}(K, Q)$, $C_{(i)}^p(K; R)$ is continuous in any natural topologies for the two groups. For example, let K be closure-finite and have an uncountable set of p-cells. Then $C_p^{(1)} = C_p^{(f)}$, $C_{(1)}^p = C_{(a)}^p$. If P is the group of reals mod 1, $Q \cong P$, and R the group of integers, with the usual pairing of Q, R to P, then the element of $C_{(1)}^p(K; R)$ which has value 1 on each x_p provides a non-continuous homomorphism of $C_p^{(1)}(K; Q)$ to P, even if $C_p^{(1)}$ is given the finest product topology.

4. Congregations

The difficulty mentioned in the previous paragraph provides the motive for defining a *congregation*, which is an Abelian group enriched with a structure which falls short of forming it into a topological group.

A *family* $\{g^{(\mu)}\}^{\mu \in M}$ of elements of a group is given by any transformation of any unordered indexing set M with elements μ into the group; different elements of M can be transformed into the same element of the group.

DEFINITION. An Abelian group is said to form a *congregation* Γ when certain families called its *nuclear families*, have been selected, subject to:

(1) If from a nuclear family is removed a subfamily each element of which is 0, the remaining family is nuclear.

(2) A finite family is nuclear if and only if the sum of its elements is 0.

(3) If $\{g^{(\mu)}\}$ is a nuclear family, so is $\{-g^{(\mu)}\}$.

(4) The union of two nuclear families (not necessarily with the same indexing sets) is a nuclear family.

(5) If a nuclear family is partitioned (or bracketed) into finite subsets and each subset is summed, the family of these sums is nuclear.

A nuclear family may be thought of as a family of elements summing to zero: such a family generalizes the set of terms of a series of numbers which is absolutely convergent with sum zero.

A family of elements of Γ is said to be *summable* to sum g, if, by the adjunction to the family of the element $(-g)$, a nuclear family is formed.

Any topological Abelian group determines a congregation on its elements if its nuclear families are taken to be those of its summable families whose sum is zero. Here the term 'summable family' is defined as in Bourbaki [1].

A homomorphism of a congregation Γ into a congregation Δ is an algebraic homomorphism under which each nuclear family of Γ transforms into a nuclear family of Δ. A homomorphism of Γ into a topological group G means a homomorphism into the congregation of G.

If P is the group of reals mod 1, the set of all homomorphisms of Γ into P can be formed into a congregation $\Gamma^{\dagger}$, the *character-congregation* of Γ, by defining as nuclear just those families of homomorphisms under which for each $g \in \Gamma$ the family of images of g is nuclear in P.

THEOREM 2. *If G is a topological Abelian group with a countable base of nuclei and H is any topological Abelian group, any homomorphism of the congregation of G into H is a continuous homomorphism.*

The proof of this theorem will be found in § 6.

The symbol P will in future always stand for the group of reals mod 1 and the symbol Π for its congregation. By Theorem 2, the elements of $\Pi^{\dagger}$ are those of P^*, namely, say, the elements of Z, the group of integers: it can be proved that no family of elements of $\Pi^{\dagger}$ containing an infinite number of non-zero homomorphisms can form a nuclear family, so that $\Pi^{\dagger}$ is a 'discrete' congregation isomorphic to Z. Also $Z^{\dagger}$ has as elements the elements of Z^* that is of P itself; the nuclear families of $Z^{\dagger}$ are those of P, so that $Z^{\dagger}$ can be identified with Π. A congregation Γ such that the natural injection of Γ in $\Gamma^{\dagger\dagger}$ is an isomorphism on $\Gamma^{\dagger\dagger}$ is said to be *reflexive*. What has been argued is that Z and Π are reflexive, each being isomorphic to the character-congregation of the other.

5. Chain and cochain congregations

If Γ is a congregation and a Φ a family of p-cells of K that includes all finite subcollections and forms an ideal, we define a congregation $C_p^\Phi(K; \Gamma)$. Its elements are those p-chains with coefficients in Γ for which $\phi(c_p) \in \Phi$; these form an Abelian group. Its nuclear families are defined to be those families $\{c_p^{(\mu)}\}$ for which

(i) the family of coefficients of each $x_p \in K$ is a nuclear family of Γ, and

(ii) $\bigcup^{\mu \in M} \phi(c_p^{(\mu)}) \in \Phi$.

A similar definition establishes a cochain congregation $C_\Phi^p(K; \Gamma)$. The chain and cochain congregations of the two kinds are defined in the obvious way; for instance,

$$C_p^{(i)}(K; \Gamma) = C_p^{\Phi_p^{(i)}}(K; \Gamma).$$

Let the p-cells of K be indexed by $\lambda \in L$ and be written as $\{x^\lambda\}^{\lambda \in L}$, the dimension being left out of the symbol. The chain whose coefficient of x^λ is g^λ will be written as $\mathbf{g} = \{g^\lambda\}$ and similarly for a cochain.

There is a natural pairing of $C^\Phi(K; \Gamma)$ and $C_{\Phi^*}(K; \Gamma^\dagger)$ to P. If $g \in \Gamma$ and $g^\dagger \in \Gamma^\dagger$, their product is written as $g \cdot g^\dagger$, the value of g under the homomorphism $g^\dagger$. Then $\mathbf{g} \cdot \mathbf{g}^\dagger$ is defined by

$$\mathbf{g} \cdot \mathbf{g}^\dagger = \Sigma_{\lambda \in L}\, g^\lambda \cdot g_\lambda^\dagger,$$

which is a finite sum of elements of P, when zero terms are ignored, since the domain of $\mathbf{g}^\dagger \in \Phi^*$.

PROPOSITION 3. *The product* $\mathbf{g} \cdot \mathbf{g}^\dagger$ *determines for fixed* $\mathbf{g}$ *a homomorphism of* $C_{\Phi^*}(K; \Gamma^\dagger)$ *into* P, *and for fixed* $\mathbf{g}^\dagger$ *a homomorphism of* $C^\Phi(K; \Gamma)$ *into* P.

The proof is to be found in § 6.

This is a preliminary result which is included in a stronger assertion:

THEOREM 4. *Under a natural correspondence*

$$(C^\Phi(K; \Gamma))^\dagger \cong C_{\Phi^*}(K; \Gamma^\dagger).$$

Again the proof is deferred to § 6.

THEOREM 5. *If* Γ *is a reflexive congregation, so is* $C_p^{(i)}(K; \Gamma)$ *and its character-congregation is naturally isomorphic to* $C_{(i)}^p(K; \Gamma^\dagger)$.

This follows easily using Proposition 1 and Theorem 4, both as stated and in an obvious dual form.

Theorem 5 is like the theorem that, if G is a reflexive topological group and K is a finite complex, $C_p(K; G)$ is reflexive and its character group is naturally isomorphic to $C^p(K; G^*)$. In that case the usual deduction about $H_p(K; G)$ and $H^p(K; G^*)$ can be carried out for G locally compact by making use of the powerful theorems established for such groups. I have not been able to find a proof of a corresponding result in the present case.

6. Proofs

PROOF OF THEOREM 2.† If $\{g^{(\mu)}\}^{\mu \in M}$ is a nuclear family of G, then the directed set of finite partial sums of the $g^{(\mu)}$'s is residual in each nucleus of G. (See Bourbaki [1].) Let the congregation of G be called Γ and consider a homomorphism $\alpha: \Gamma \to H$. If V is a nucleus of H and $\{g^{(\mu)}\}$ a nuclear family of Γ, then $\{(g^{(\mu)})\alpha\}$ provides a directed set of partial sums which is residual in V. The counter-image $\alpha^{-1}V$ in G is therefore a subset of G in which the directed set of partial sums of each nuclear family is residual. It will be proved that, if A is any subset of G with this property, then A is a nucleus (not necessarily open) of G. This will prove that α is continuous.

Suppose the contrary. Take a base of symmetrical nuclei, $U_1, U_2, \ldots$, of G such that $U_r + U_r \subset U_{r-1}$. Since G has a countable base of nuclei, such a base can be chosen. Since A is not a nucleus, it contains no U_r and we can find $a_r \in U_r$ and $\notin A$. Consider the family $\{g^{(r)}\}$, $r = 1, 2, \ldots$ defined by

$$g^{(r)} = a_r - a_{r-1} \quad \text{for} \quad r > 1.$$

$$g^{(1)} = a_1.$$

It is asserted that $\{g^{(r)}\}$ is a nuclear family whose partial sums are not residual in A. The latter half of the assertion is easy to see: let F be any finite set of integers with largest element f; then $\sum_{r=1}^{f} g^{(r)} = a_f \notin A$ and this is the sum of a finite set of elements which contains F. Hence $\{g^{(r)}\}$ has partial sums not residual in A.

To prove $\{g^{(r)}\}$ nuclear it is sufficient to define for each r a finite set of positive integers F_r and to establish that for each finite set $F \supset F_r$,

$$\textstyle\sum_{s \in F} g^{(s)} \in U_r.$$

The set F_r is the set $1 \leq s \leq r+3$. Let $F = F_r \cup E$. Then

$$\textstyle\sum_{s \in F} g^{(s)} = \sum_{s \in F_r} g^{(s)} + \sum_{s \in E} g^{(s)} = a_{r+3} + \sum_{s \in E} g^{(s)}.$$

Now $g^{(s)} = a_s - a_{s-1} \in U_{s-2}$. It follows from an easy induction argument that

$$\textstyle\sum_{s \in E} g^{(s)} \in U_{r+1}.$$

Since $a_{r+3} \in U_{r+3}$, then $\sum_{s \in F} g^{(s)} \in U_r$, as required. This contradiction proves that A is a nucleus.

† I am greatly indebted to J. F. Adams for this proof of the theorem.

PROOF OF PROPOSITION 3. Consider a fixed $\mathfrak{g}$ and a nuclear family $\{^{(\mu)}\mathfrak{g}^{\dagger}\}$; it is to be proved that $\{\mathfrak{g}\cdot{}^{(\mu)}\mathfrak{g}^{\dagger}\}$ is nuclear in P. Since $\{^{(\mu)}\mathfrak{g}^{\dagger}\}$ is nuclear, $\psi = \bigcup^{\mu} \phi(^{(\mu)}\mathfrak{g}^{\dagger}) \in \Phi^*$; let $\phi(\mathfrak{g}) \cap \psi = F$, a finite subset of L. For each $\lambda \in L$, $\{^{(\mu)}g_{\lambda}^{\dagger}\}^{\mu\in M}$ is a nuclear family of $\Gamma^{\dagger}$ and so $\{g^{\lambda}\cdot{}^{(\mu)}g_{\lambda}^{\dagger}\}^{\mu\in M}$ is nuclear in P. Using Axiom 4 for congregations repeatedly it follows that the product family $\{g^{\lambda}\cdot{}^{(\mu)}g_{\lambda}^{\dagger}\}_{\lambda\in F}^{\mu\in M}$ is nuclear in P. By summing on λ and appealing to Axiom 5 for congregations, it can be deduced that $\{\sum_{\lambda\in F} g^{\lambda}\cdot{}^{(\mu)}g_{\lambda}\}^{\mu\in M}$ is a nuclear family of P. But

$$\mathfrak{g}\cdot{}^{(\mu)}\mathfrak{g}^{\dagger} = \textstyle\sum_{\lambda\in L} g^{\lambda}\cdot{}^{(\mu)}g_{\lambda}^{\dagger} = \sum_{\lambda\in F} g^{\lambda}\cdot{}^{(\mu)}g_{\lambda}^{\dagger}$$

and the proof is finished. Next consider a fixed $\mathfrak{g}^{\dagger}$ and a nuclear family $\{\mathfrak{g}_{(\mu)}\}_{\mu\in M}$. The same proof with obvious modifications is valid.

PROOF OF THEOREM 4. By Proposition 3 each element of $C_{\Phi^*}(K; \Gamma^{\dagger})$ determines an element of $(C^{\Phi}(K; \Gamma))^{\dagger}$. Distinct elements of C_{Φ^*} differ in respect to some coordinate λ. The distinct homomorphisms for this coordinate provide distinct images for some element of Γ, and so the distinct elements provide distinct images for some element of C^{Φ} whose domain is x^{λ}. There is then a 1-1 transformation α from the elements of C_{Φ^*} into a subset of the elements of $(C^{\Phi})^{\dagger}$. It will be shown

(a) that α is onto,

(b) that α is a homomorphism of the congregation C_{Φ^*},

(c) that α^{-1} is a homomorphism of the congregation $(C^{\Phi})^{\dagger}$.

(a) Let $\boldsymbol{\beta}$: $C^{\Phi} \to P$ be any homomorphism; it is required only to prove that there is an element $\mathfrak{g}^{\dagger}(\beta) \in C_{\Phi^*}$ such that, for each $\mathfrak{g} \in C^{\Phi}$, $(\mathfrak{g})\boldsymbol{\beta} = \mathfrak{g}\cdot\mathfrak{g}^{\dagger}(\beta)$. By restricting $\boldsymbol{\beta}$ to elements whose domain is x^{λ}, a homomorphism β_{λ}: $\Gamma \to P$ is determined; hence $\beta_{\lambda} \in \Gamma^{\dagger}$. Define $\mathfrak{g}^{\dagger}(\beta)$ by $g_{\lambda}^{\dagger} = \beta_{\lambda}$. First it must be shown that $\mathfrak{g}^{\dagger}(\beta) \in C_{\Phi^*}$ by showing that $\phi(\mathfrak{g}^{\dagger}(\beta)) \in \Phi^*$. If not, there is an infinite set $\phi \in \Phi$ such that $\phi \subset \phi(\mathfrak{g}^{\dagger}(\beta))$. For each $\lambda \in \phi$, g^{λ} may be chosen so that $|(g^{\lambda})\beta_{\lambda}| > \frac{1}{4}$, since $(\Gamma)\beta_{\lambda}$ is a non-zero subgroup of P. Let $\mathfrak{g}^{(\lambda)}$ be the element of C^{Φ} with domain x^{λ} and λ-coordinate g^{λ}; let $\mathfrak{g}$ be the element with λ-coordinate 0 for $\lambda \notin \phi$ and equal to g^{λ} for $\lambda \in \phi$. Then $\{\mathfrak{g}^{(\lambda)}\}^{\lambda\in\phi} \cup (-\mathfrak{g})$ is a nuclear family of C^{Φ}. But $\{(\mathfrak{g}^{(\lambda)})\boldsymbol{\beta}\}^{\lambda\in\phi} = \{g^{\lambda}\beta_{\lambda}\}^{\lambda\in\phi}$ is not summable in P since it fails to satisfy the Cauchy criterion (see [1]), having an infinite set of elements outside $-\frac{1}{4} < p < \frac{1}{4}$. This contradiction to the statement that $\boldsymbol{\beta}$ is a homomorphism establishes that $\phi(\mathfrak{g}^{\dagger}(\beta)) \in \Phi^*$. Next it must be shown that, for each $\mathfrak{g} \in C^{\Phi}$, $(\mathfrak{g})\boldsymbol{\beta} = \mathfrak{g}\cdot\mathfrak{g}^{\dagger}(\beta)$. Given $\mathfrak{g}$ let $\mathfrak{g}^{(\lambda)}$ be, as above, the element of domain x^{λ} with λ-coordinate that of $\mathfrak{g}$: then $\{\mathfrak{g}^{(\lambda)}\}^{\lambda\in L} \cup (-\mathfrak{g})$ is nuclear in C^{Φ}. Since $\boldsymbol{\beta}$ is a homomorphism $\{(\mathfrak{g}^{(\lambda)})\boldsymbol{\beta}\}^{\lambda\in L} \cup (-\mathfrak{g})\boldsymbol{\beta}$ is nuclear in P. But $\phi(\mathfrak{g}) \cap \phi(\boldsymbol{\beta}) = F$

has been proved finite, so that $\{(\mathbf{g}^{(\lambda)})\,\boldsymbol{\beta}\}^{\lambda\in L}=\{(g^{\lambda})\,\beta_{\lambda}\}^{\lambda\in L}$ has a finite number of non-zero elements. Using Axioms 1, 2 for a congregation, known to be valid for the congregation of a group, it can be seen that $\{(g^{\lambda})\,\beta_{\lambda}\}^{\lambda\in F}\cup(-\mathbf{g})\,\boldsymbol{\beta}$ is a finite family whose sum is zero. Therefore $(\mathbf{g})\,\boldsymbol{\beta}=\sum_{\lambda\in F}(g^{\lambda})\,\beta_{\lambda}=\mathbf{g}\cdot\mathbf{g}^{\dagger}(\beta)$.

(b) No proof is needed that α is a homomorphism of the algebraic structure of C_{Φ^*}. It is to be proved that, if $\{^{(\mu)}\mathbf{g}^{\dagger}\}^{\mu\in M}$ is a nuclear family of $C_{\Phi^*}(K;\Gamma^{\dagger})$, then $\{(^{(\mu)}\mathbf{g}^{\dagger})\,\alpha\}$ is a nuclear family of $(C^{\Phi})^{\dagger}$ or that, for each $\mathbf{g}\in C^{\Phi}$, $\{\mathbf{g}\cdot{}^{(\mu)}\mathbf{g}^{\dagger}\}^{\mu\in M}$ is nuclear in P. This is part of the content of Proposition 3.

(c) It is to be proved that, if $\{\boldsymbol{\beta}^{(\mu)}\}^{\mu\in M}$ is a nuclear family of elements of $(C^{\Phi})^{\dagger}$, then $\{\alpha^{-1}(\boldsymbol{\beta}^{(\mu)})\}=\{\mathbf{g}^{\dagger}(\beta^{(\mu)})\}$ is nuclear in C_{Φ^*}. Two conditions must be verified: (i) that for each $\lambda\in L$, $\{\mathbf{g}_{\lambda}^{\dagger}(\beta^{(\mu)})\}^{\mu\in M}=\{\beta_{\lambda}^{(\mu)}\}^{\mu\in M}$ is nuclear in $\Gamma^{\dagger}$, and (ii) that

$$\bigcup^{\mu}\phi(\mathbf{g}^{\dagger}(\beta^{(\mu)}))=\bigcup^{\mu}\phi(\boldsymbol{\beta}^{(\mu)})\in\Phi^*.$$

To verify (i) it is only necessary to observe that, since $\{\beta^{(\mu)}\}$ is nuclear in $(C^{\Phi})^{\dagger}$, for any $\mathbf{g}^{(\lambda)}$ with domain x^{λ} $\{(\mathbf{g}^{(\lambda)})\,\boldsymbol{\beta}^{(\mu)}\}^{\mu\in M}=\{(g^{\lambda})\,\beta_{\lambda}^{(\mu)}\}^{\mu\in M}$ is nuclear in P. This holds for any $g^{\lambda}\in\Gamma$, so that $\{\beta_{\lambda}^{(\mu)}\}^{\mu\in M}$ is nuclear in $\Gamma^{\dagger}$.

It is rather more troublesome to establish (ii). Suppose otherwise; then there is an infinite subset $\theta\in\Phi$ such that $\theta\subset\bigcup^{\mu}\phi(\boldsymbol{\beta}^{(\mu)})$. For each μ, $\theta^{(\mu)}=\theta\cap\phi(\boldsymbol{\beta}^{(\mu)})$ is finite since $\phi(\boldsymbol{\beta}^{(\mu)})\in\Phi^*$. Then $\theta=\bigcup^{\mu}\theta^{(\mu)}$, each component being finite and the union infinite. A countable subset $\mu_1,\mu_2,\ldots$ of M can be chosen so that, for each r, $\theta^{(\mu_r)}$ is not contained in $\bigcup_{i=1}^{r-1}\theta^{(\mu_i)}$; for each r, $\lambda_r\in L$ can then be chosen so that $\lambda_r\in\theta^{(\mu_r)}$ and $\notin\bigcup_{i=1}^{r-1}\theta^{(\mu_i)}$. Elements of Γ, $g^{\lambda_1}, g^{\lambda_2},\ldots$ may be chosen in turn such that, for each r,

$$\left|\sum_{i=1}^{r}(g^{\lambda_i})\,\beta_{\lambda_i}^{(\mu_i)}\right|\geqq\tfrac{1}{4}.$$

The element g^{λ_r} has only to be chosen so that its image under the non-zero homomorphism $\beta_{\lambda_r}^{(\mu_r)}$ is at a distance $\geqq\frac{1}{4}$ from an element of P already determined. The element $\mathbf{g}$ is defined to have λ-coordinate 0 for $\lambda\neq\lambda_1,\lambda_2,\ldots$ and to have λ_r-coordinate g^{λ_r}. Then $\phi(\mathbf{g})\subset\theta\in\Phi$, so that $\mathbf{g}\in C^{\Phi}$. But $(\mathbf{g})\,\boldsymbol{\beta}^{(\mu_r)}=\sum_{\lambda\in L}(g^{\lambda})\,\beta_{\lambda}^{(\mu_r)}=\sum_{i=1}^{r}(g^{\lambda_i})\,\beta_{\lambda_i}^{(\mu_r)}$, since, for $i>r$, $\lambda_i\notin\theta^{(\mu_r)}$. Hence $|(\mathbf{g})\,\boldsymbol{\beta}^{(\mu_r)}|\geqq\frac{1}{4}$ and the infinite subfamily $\{(\mathbf{g})\,\boldsymbol{\beta}^{(\mu_r)}\}$ of $\{(\mathbf{g})\,\boldsymbol{\beta}^{(\mu)}\}$ fails to satisfy the Cauchy criterion and $\{(\mathbf{g})\,\boldsymbol{\beta}^{(\mu)}\}$ is therefore not nuclear. This contradicts the assumption that $\{\boldsymbol{\beta}^{(\mu)}\}$ is nuclear in $(C^{\Phi})^{\dagger}$ and hence verifies (ii). The proof of the theorem is thus complete.

Trinity Hall
Cambridge University

References

[1] N. Bourbaki, Topologie Générale (Actualités Series, 916), Hermann, Paris, 1953, chap. III, section 4.

[2] S. Lefschetz, Algebraic Topology, Amer. Math. Soc. Colloquium Publications, vol. XXVII, 1942.

[3] A. W. Tucker, *An abstract approach to manifolds*, Ann. of Math., 34 (1933), pp. 191–243.